Hydrometry: Principles and Practice

Hydrometry: Principles and Practice

Editor: Danny Fuller

www.callistoreference.com

Callisto Reference,
118-35 Queens Blvd., Suite 400,
Forest Hills, NY 11375, USA

Visit us on the World Wide Web at:
www.callistoreference.com

ISBN: 978-1-64116-299-9 (Hardback)

Cataloging-in-Publication Data

Hydrometry : principles and practice / edited by Danny Fuller.
p. cm.
Includes bibliographical references and index.
ISBN 978-1-64116-299-9
1. Hydraulic measurements. 2. Hydraulic engineering--Instruments. 3. Physical measurements. I. Fuller, Danny.
TC177 .H93 2020
681.28--dc23

Table of Contents

Preface

The purpose of the book is to provide a glimpse into the dynamics and to present opinions and studies of some of the scientists engaged in the development of new ideas in the field from very different standpoints. This book will prove useful to students and researchers owing to its high content quality.

The monitoring of the components of the hydrological cycle is known as hydrometry. It includes rainfall, groundwater characteristics, water quality and flow characteristics of surface waters. Hydrometrics is a field of applied science and engineering that deals with hydrometry. It primarily measures the components of the hydrological cycle such as the bulk quantification of water resources. Hydrology, structures, control systems, computer sciences, data management and communications are various traditional engineering practices that fall under this domain. This book is compiled in such a manner, that it will provide in-depth knowledge about the theory and practice of hydrometry. It includes some of the vital pieces of work being conducted across the world, on various topics related to hydrometry. This book, with its detailed analyses and data, will prove immensely beneficial to professionals and students involved in this area at various levels.

At the end, I would like to appreciate all the efforts made by the authors in completing their chapters professionally. I express my deepest gratitude to all of them for contributing to this book by sharing their valuable works. A special thanks to my family and friends for their constant support in this journey.

Editor

1

A comparison of methods for determining field evapotranspiration: photosynthesis system, sap flow, and eddy covariance

Z. Zhang[1], **F. Tian**[1,*], **H. Hu**[1], **and P. Yang**[1]

[1]State Key Laboratory of Hydroscience and Engineering, Department of Hydraulic Engineering, Tsinghua University, Beijing, 100084, China
[*]now at: State Key Laboratory of Hydroscience and Engineering, Department of Hydraulic Engineering, Tsinghua University, Beijing, China

Correspondence to: F. Tian (tianfq@tsinghua.edu.cn)

Abstract. A multi-scale, multi-technique study was conducted to measure evapotranspiration and its components in a cotton field under mulched drip irrigation conditions in northwestern China. Three measurement techniques at different scales were used: a photosynthesis system (leaf scale), sap flow (plant scale), and eddy covariance (field scale). The experiment was conducted from July to September 2012. To upscale the evapotranspiration from the leaf to plant scale, an approach that incorporated the canopy structure and the relationships between sunlit and shaded leaves was proposed. To upscale the evapotranspiration from the plant to field scale, an approach based on the transpiration per unit leaf area was adopted and modified to incorporate the temporal variability in the relationship between leaf areas and stem diameter. At the plant scale, the estimate of the transpiration based on the photosynthesis system with upscaling was slightly higher (18 %) than that obtained by sap flow. At the field scale, the estimates of transpiration derived from sap flow with upscaling and eddy covariance showed reasonable consistency during the cotton's open-boll growth stage, during which soil evaporation can be neglected. The results indicate that the proposed upscaling approaches are reasonable and valid. Based on the measurements and upscaling approaches, evapotranspiration components were analyzed for a cotton field under mulched drip irrigation. During the two analyzed sub-periods in July and August, evapotranspiration rates were 3.94 and 4.53 m day^{-1}, respectively. The fraction of transpiration to evapotranspiration reached 87.1 % before drip irrigation and 82.3 % after irrigation. The high fraction of transpiration over evapotranspiration was principally due to the mulched film above the drip pipe, low soil water content in the inter-film zone, well-closed canopy, and high water requirement of the crop.

1 Introduction

Evapotranspiration (ET) is a major component in energy balance and water cycling (Katul et al., 2012). Much effort has been devoted to the measurement of ET because it is a critically important process in many fields, including hydrology, ecology, agriculture, forestry, and horticulture. Over the past few decades, several different techniques, including the use of eddy covariance, lysimeter, Bowen ratio, soil water budget, large-aperture scintillometer, sap flow, and photosynthesis system (also known as leaf gas exchange instrument), have been developed (Evett et al., 2012; Lei and Yang, 2010; MacKay et al., 2002). In general, transpiration at the leaf scale can be reliably measured through a photosynthesis system using the high-quality humidity sensors in the leaf chamber. At the plant scale, sap flow based on stem energy balance theory is widely applied to measure transpiration, particularly in herbaceous plants. Lysimeter and soil water budget methods can directly estimate ET based on the mass balance principle, but representativeness of the control volume is still dubious, especially under conditions of inhomogeneous soil moisture distribution caused by drip irrigation. Although ET can be obtained by Bowen ratio and the large-aperture scintillometer, eddy covariance is generally considered the most reliable and state-of-the-art technique for the accurate measurement of ET at the field scale.

The abovementioned measurement techniques are essentially different in terms of instrumentation, applicable spatial scale, and theoretical background (Alfieri et al., 2012). Due to the different spatial scales to which ET measurement methods are applied, scale transformation approaches should be used to make ET values measured by different methods comparable (Evett et al., 2012). Additionally, through comparisons, scale transformation approaches can be validated and improved.

Using valid scale transformation approaches, ET values can be inferred outside of their observed scales and compared at the same scale (Evett et al., 2012). For instance, field evapotranspiration can be obtained after upscaling the measurements obtained using the photosynthesis system and sap flow. A comparison of ET measured at different scales can not only allow for the determination of the accuracy and uncertainty of these independent measurements but also provide solid and reliable ET estimates (Allen et al., 2011b). Additionally, different techniques are often combined with appropriate scale transformation approaches in water research, such as the partitioning of evaporation and transpiration in an ecosystem, and the development of ET models from ground-based data or remote sensing images (Alfieri et al., 2012; Williams et al., 2004). In addition, the extrapolations of water use from the level of individual leaf to the whole plant, as well as the extrapolations from individual plant to a stand of plants by using upscaling approaches, represent a critical step in the linking of plant physiology and hydrology (Hatton and Wu, 1995).

Several studies have compared sap flow, soil water budget, Bowen ratio, and eddy covariance measurements in a forest ecosystem (Granier et al., 2000; Silberstein et al., 2001; Williams et al., 2004; Wilson et al., 2001). These studies have primarily focused on the applicability of these techniques, evapotranspiration components, and the energy balance in the forest ecosystem. The approaches used in these studies to upscale from the plant (sap flow) to field scale (eddy covariance) were mainly based on plant population and the size of plant stems.

Cotton is one of the most important fiber economic crops (Ashraf, 2002). A number of ET measurements in cotton fields have been performed using one of these different techniques, such as eddy covariance (Zhou et al., 2011), lysimeter (Howell et al., 2004; Ko et al., 2009; Tolk et al., 2006) and sap flow (Dugas et al., 1994; Tang et al., 2010). Several comparisons of ET measurements in cotton field have also been carried out. Comparisons of ET measurements using the sap flow and lysimeter methods (Dugas, 1990) or the sap flow and Bowen ratio methods (Ham et al., 1990) have been implemented under flood irrigation conditions. The approaches used to upscale ET from the plant to field scale were based on plant population and stem size (similarly to studies conducted in the forest ecosystem) (Dugas, 1990) or on plant population and sampled plant leaf area (Ham et al., 1990). Both of these approaches demonstrated that the cotton transpiration measured by sap flow was higher than those measured by the lysimeter and the Bowen ratio. Additionally, these studies suggested that sap flow should be expressed per unit leaf area to improve field ET estimates. It was hypothesized that the upscaling approaches based on an accurate estimate of field leaf area would provide reliable results. Alfieri et al. (2012) and Chavez et al. (2009) compared ET values obtained by eddy covariance with that measured by lysimeter, and discussed the causes of discrepancy between them. However, comparison of ET measurements in agricultural crop fields under water-saving irrigation conditions is limited. In addition, a comparison of the photosynthesis-system-based method with other techniques has rarely been performed in previous studies. The partitioning of ET under mulched drip irrigation using these methods is seldom reported.

Mulched drip irrigation, which is a new micro-irrigation approach that incorporates the surface drip irrigation method and the film mulching technique, has been widely applied in northwestern China (Wang et al., 2011). Using this irrigation method, the fraction of transpiration over ET can be markedly increased by delivering water precisely to the root zone and by eliminating of the majority of useless soil evaporation via mulching (Zhang et al., 2014). Soil thermal conditions are also improved by mulching to ensure crop germination and seedling growth (Bonachela et al., 2001; Hou et al., 2010; Li et al., 2008). In 2009, mulched drip irrigation was adopted in fields with an area amounting to more than 1.2 million hectares in the Xinjiang Province of China. Mulched drip irrigation is also potentially applicable to other arid and semi-arid regions with similar climatic and farming conditions based on the abovementioned noteworthy advantages (Zhang et al., 2014). Because matter/energy exchanges on the land surface, including those of water and heat, are significantly altered by mulched drip irrigation (Zhou et al., 2011), a comprehensive study of ET using integrated measurements should be conducted in order to obtain a more thorough understanding of this process. In this study, three different ET measurement methods (i.e., photosynthesis system at the leaf scale, sap flow at the plant scale, and eddy covariance at the field scale) were compared in a crop field under mulched drip irrigation condition. The approaches for upscaling ET from the leaf scale to the plant scale and from the plant scale to the field scale were discussed and improved, and evapotranspiration and its components were determined for the analyzed periods.

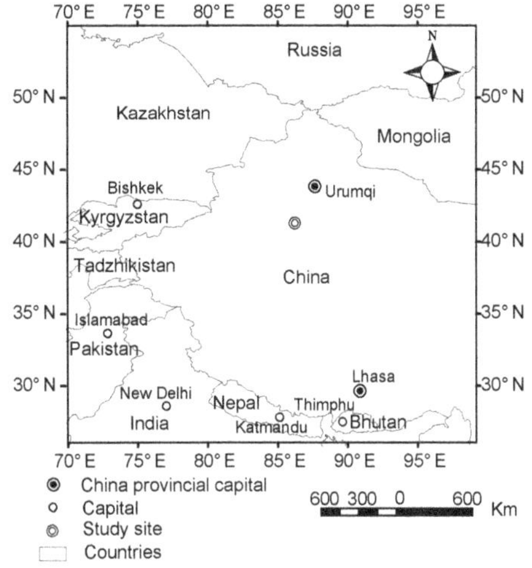

(a) Geographic location of the study site

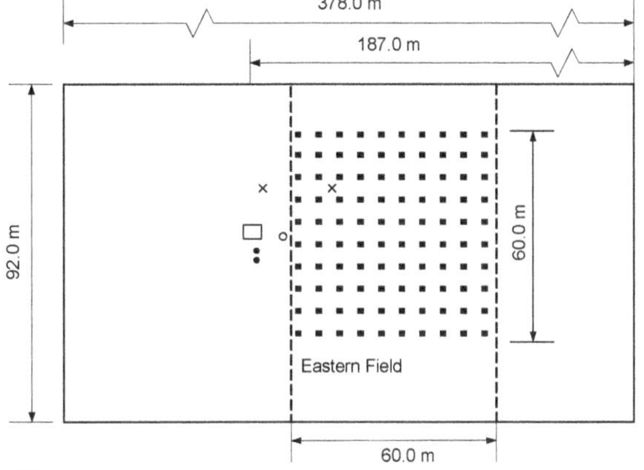

(b) Experimental field layout

Fig. 1. Geographic location of the study site and experimental field layout.

2 Methods and materials

2.1 Experimental site and cotton planting

The experimental site (86°12′ E, 41°36′ N; 886 m a.s.l.; see Fig. 1) is located on the northeast edge of Taklimakan Desert, which belongs to the Bayangol Prefecture of Xinjiang Province in northwestern China. The study area is characterized by a typical inland arid climate with strong diurnal temperature fluctuation and scarce precipitation. The mean annual precipitation is approximately 60 mm. The annual mean temperature is 11.48 °C, and the annual total sunshine duration is 3036 h, which is favorable for cotton growth. The mean annual potential evaporation measured with a Φ20 evaporation pan is 2788 mm (Zhang et al., 2014). The major soil type in the experimental region is silt loam, and saturated volumetric water content is 0.42. The planted

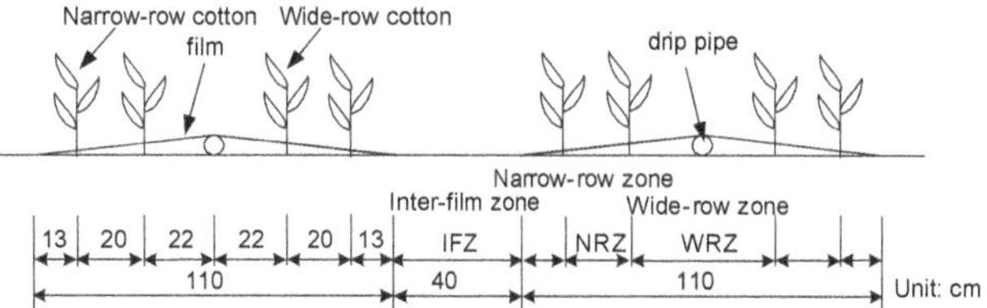

Fig. 2. One pipe, one film, and four rows of cotton arrangement.

crop is cotton (*Gossypium hirsutum* L.). It is the predominant economic crop in Xinjiang Province, which contributes nearly 50 % of the total lint yield of China, with approximately 3.2 million tons produced in 2012 (http://english.gov.cn/2012-10/14/content2242953.htm).

The experimental cotton field had an area of 3.48 ha. A 10 m stationary tower was erected in the middle of the field to mount flux and meteorological instruments. Because the prevailing wind blows from the northeast, sap flow and photosynthesis system measurements were both conducted on the north side of the tower, where the potential source of the water flux was measured through eddy covariance. The surrounding field had the same cotton planting and irrigation conditions as the experimental field, which provided adequate fetch for the meteorological measurements. The profiles for soil water content measurements were located on the south side of the tower. The east part of the field, which was denoted the Eastern Field, was divided into 100 sub-plots with an area of $6 \times 6\,m^2$ to measure the spatial distribution of cotton (Fig. 1).

The style of cotton planting and drip pipe arrangement is referred to as the "one pipe, one film, and four rows of cotton arrangement" (Hu et al., 2011), which indicates that one drip pipe beneath the mulched film is in the middle of four rows of cotton. The width of the film is 110 cm, and the inter-film zone is 40 cm. The three soil profile terms, i.e., wide-row zone, narrow-row zone, and inter-film zone, are defined as shown in Fig. 2.

In the experimental field, cotton was planted on 23 April 2012 and harvested from 20 September 2012 to 20 November 2012. The seeds were sown at 0.1 m intervals in each row to yield an anticipated population of 260 000 plants ha^{-1}. However, the emergence rate in 2012 was 46.3 % due to sandstorm and freezing damage, and actual plant density was 120 000 plants ha^{-1}. Groundwater table depth varied from 2.09 to 3.27 m during the cotton growth period. The amount of irrigated water was 540.23 mm in total throughout the growth period, and the irrigation schedules adopted in 2012 are summarized in Table 1. To meet the plant requirements for nutrients, 173 kg ha^{-1} compound fertilizers (14 % N, 16 % P_2O_5, and 15 % K_2O), 518 kg ha^{-1} calcium superphosphate (18 % N and 40 % P_2O_5), and 288 kg ha^{-1} diammonium phosphate (P_2O_5 > 16 %) were applied as the basic fertilizer before plowing. As supplemental fertilizers during the growth period, approximately 293 kg ha^{-1} urea

Table 1. Irrigation schedule adopted for experiments in 2012.

Cotton growth stage	Squaring stage				Flower stage			Bolls stage					Total
Irrigation date	6-10/11 6-14/15	6-21	6-28	7-6/7	7-15/16	7-26	8-4/5	8-8	8-12/13	8-17	8-22/23	8-27/28	
Amount (mm)	65.17	34.35	35.32	36.77	33.26	44.10	40.00	59.28	46.73	42.19	50.84	52.22	540.23

(46 % N) and 586 kg ha^{-1} drip compound fertilizer (13 % N, 18 % P_2O_5, and 16 % K_2O) were applied through the fertigation method, and 27 kg ha^{-1} foliar fertilizer ($K_2O > 34$ % and $P_2O_5 > 52$ %) was applied through the sprinkle method.

2.2 Instruments

2.2.1 Photosynthesis system

Leaf transpiration occurs simultaneously with photosynthesis, and photosynthesis system can be used as a reliable and accurate tool for the measurement of transpiration (Mahouachi et al., 2006; Mengistu et al., 2011). In this study, an LCpro+ photosynthesis system (model LCpro+, ADC BioScientific Ltd., Hertfordshire, England) was used to measure transpiration at the leaf scale.

The basic components of LCpro+ are a broad leaf chamber, an infrared gas analyzer, two high-quality humidity sensors, an air probe, and a console with a keyboard, display, and memory. The selected leaf was placed in the leaf chamber with a known area of the leaf (6.25 cm^2) enclosed in the broad leaf chamber. The measurements were conducted in an open system configuration in which fresh gas was continually passed through the plant leaf chamber. The transpiration rates were calculated from the differences in the H_2O concentration between the incoming gas (the reference levels) and the gas after passing the leaf specimen (the analysis levels). H_2O concentration was measured using two high-quality humidity sensors contained inside the plant leaf chamber. The increasing concentration of water vapor can be converted to transpiration rate by the following equation (ADC BioScientific Ltd., 2004):

$$M = \frac{(e_{\mathrm{an}} - e_{\mathrm{ref}}) \cdot u_{\mathrm{s}}}{P_{\mathrm{a}}}, \tag{1}$$

where M represents the transpiration rate of the measured leaf (mmol m^{-2} s^{-1}), e_{an} is the water vapor pressure leaving the leaf chamber after dilution correction (mbar), e_{ref} is the water vapor pressure entering the leaf chamber (mbar), u_{s} is the mass flow of air entering the leaf chamber per square meter of leaf area (mmol m^{-2} s^{-1}), and P_{a} is the atmospheric pressure (mbar). For a typical leaf, the H_2O flux M lies between 0 and 15 mmol m^{-2} s^{-1}.

2.2.2 Sap flow

To measure the water use of individual plants and estimate the transpiration of the crop, sap flow gauges were used for stems that were 8–16 mm in diameter (model SGA9, SGA13, Dynamax, Inc., Houston, TX, USA); this measurement approach is based on the stem energy balance theory. This model of sap flow gauges was chosen because it is well adapted to small, non-ligneous stems and has been shown to be accurate in several important economic crops, including cotton (Baker and Vanbanel, 1987; Ham et al., 1990; Tang et al., 2010). The stem water flow rate is calculated using the following equation (Sakuratani, 1981, 1984):

$$F_{\mathrm{p}} = 3.6 \times 10^6 \times \left[P_{\mathrm{in}} - \frac{K_{\mathrm{ST}} \cdot A_{\mathrm{stem}} \cdot (\mathrm{d}T_{\mathrm{u}} + \mathrm{d}T_{\mathrm{d}})}{\mathrm{d}x} - K_{\mathrm{SH}} \cdot \mathrm{CH} \right] / (C_{\mathrm{w}} \cdot \mathrm{d}T), \tag{2}$$

where F_{p} is the stem water flow rate (g h^{-1}), P_{in} is a fixed amount of heat powered by a DC supply (W), K_{ST} is the thermal conductivity of the stem (W m^{-1} K^{-1}), A_{stem} is the stem cross-sectional area (m^2), $\frac{\mathrm{d}T_{\mathrm{u}}}{\mathrm{d}x}$ (K m^{-1}) and $\frac{\mathrm{d}T_{\mathrm{d}}}{\mathrm{d}x}$ (K m^{-1}) are the temperature gradients in the up and down directions, respectively, $\mathrm{d}x$ is the spacing between the thermocouple junctions (m), K_{SH} is the sheath conductivity (W mV^{-1}), CH is the radial-heat thermopile voltage (mV), C_{w} is the specific heat of water (J kg^{-1} K^{-1}), $\mathrm{d}T$ is the temperature increase of the sap (K), and 3.6×10^6 is a unit conversion factor. The second part of the equation, shown in square brackets, represents the axial heat conduction through the stem, and the third part represents the radial heat conducted through the gauge to the ambient air. Hence, the value enclosed in square brackets is heat convection carried by the sap. After dividing by the specific heat of water and the temperature increase of sap, the heat flux is directly converted to water flow rate. In particular, heat storage of the stem is assumed to be zero (Dugas, 1990).

2.2.3 Eddy covariance

The eddy covariance (EC) is known to be a reliable method for obtaining direct field ET measurements (Baldocchi et al., 2001). In this study, the EC system consists of a fast-response 3-D sonic anemometer (model CSAT3, Campbell Scientific

Inc., Logan, UT, USA), a fast-response open-path infrared gas (H_2O and CO_2) analyzer (model EC150, Campbell Scientific Inc., Logan, UT, USA), an air temperature/humidity sensor (model HMP155A, Vaisala Inc., Woburn, MA, USA), and a micrologger (model CR3000, Campbell Scientific Inc., Logan, UT, USA). The CSAT3 sensor was oriented toward the predominant wind direction with an azimuth angle of 50° from true north. The net radiation at a height of 2.25 m (model LITE2, Kipp & Zonen, Delft, the Netherlands) and soil heat flux (model HFP01SC, Hukseflux, the Netherlands; two plates were placed 0.05 m below the ground surface in the wide-row zone and inter-film zone, respectively) were measured to test the data quality based on energy balance closure.

Multiplying the vertical velocity fluctuations by a scalar (e.g., water vapor, carbon dioxide, and air temperature) concentration fluctuation can provide a direct estimate of the latent heat (LE), CO_2, and sensible heat (H) fluxes (see Eqs. 3 through 5) (van Dijk et al., 2004). The EC data were corrected in the post-processing calculations through the following methods: linear de-trend, tilt correction through the yaw and pitch rotation, density fluctuation correction, and correction of the sonic temperature for humidity (van Dijk et al., 2004; Webb et al., 1980). The missing data due to system failures or data rejection were filled using two strategies. Short gaps (less than 2 h) were filled through a linear interpolation, and larger data gaps (more than 2 h and less than 1 day) were filled using the mean diurnal average method (Falge et al., 2001).

$$\mathrm{FL} = \overline{\rho_\mathrm{a}}\,\overline{w's'}, \tag{3}$$

$$\lambda \mathrm{ET} = \lambda \overline{\rho_\mathrm{a}}\,\overline{w'q'}, \tag{4}$$

$$H = C_\mathrm{P}\overline{\rho_\mathrm{a}}\,\overline{w'T'} \tag{5}$$

In the general equation presented as Eq. (3), FL is the flux of specific mass, ρ_a is the air density ($\mathrm{kg\,m^{-3}}$) at a given air temperature, and $\overline{w's'}$ is the covariance between the fluctuations in the vertical wind speed w' ($\mathrm{m\,s^{-1}}$) and the fluctuations in a scalar concentration s'. In particular, when the instantaneous deviation of the specific humidity from mean specific humidity (q), which is denoted q' ($\mathrm{kg\,kg^{-1}}$), is used in the general equation, ET can be derived from Eq. (4). λET is the latent heat flux ($\mathrm{W\,m^{-2}}$), and λ is the latent heat of water vaporization ($\mathrm{J\,kg^{-1}}$). The sensible heat fluxes H ($\mathrm{W\,m^{-2}}$) can also be calculated using the instantaneous deviation of the air temperature T' (K). C_P is the specific heat of dry air at constant pressure ($\mathrm{J\,kg^{-1}\,K^{-1}}$).

2.3 Evapotranspiration measurements and upscaling approaches

2.3.1 Evapotranspiration measurements

The experiment was conducted during summer 2012 in this cotton field. Three sub-periods representing the typical cotton growth stages were selected for comparison analysis of sap flow and eddy covariance analysis, that is, sub-period 1 (SP1) from 23 to 25 July in the flower stage, sub-period 2 (SP2) from 9 to 11 August in the bolling stage, and sub-period 3 (SP3) from 16 to 18 September in the open boll stage. In addition, photosynthesis system measurements were performed on 3 days (i.e., 23 July, 27 July, and 10 August) to compare with sap flow results. There was no irrigation during these sub-periods and days.

Four sap flow gauges were installed on two wide-row cottons and two narrow-row cottons on the north side of the tower (see Fig. 1). All of the gauges were sampled every 10 min, and data were stored in a CR1000 data logger (model CR1000, Campbell Scientific, Inc., Logan, UT, USA). Representative plants that had the averaged plant height and leaf area index (LAI) of the field were selected for measurements, and the averaged value of four gauges was used to represent the individual plant transpiration rate. The stem diameter of each gauged plant at 5 cm above the soil surface was measured every 2 days, and the leaf area of each gauged plant was measured at the time of gauge removal. P_in varied from 80 to 150 mW due to gauge size, and K_ST was assumed to be $0.54\,\mathrm{W\,m^{-1}\,K^{-1}}$ (Sakuratani, 1984). The value of K_SH was unique to each configuration, with a different gauge and a different stem diameter, and was determined by solving Eq. (2) under the zero flow condition ($F_\mathrm{p} = 0$) using the data obtained each day. Previous studies have assumed that the transpiration should be zero before dawn (Chabot et al., 2005; Dugas et al., 1994; Kigalu, 2007). Such a condition was assumed to be achieved from 03:00 to 05:00 (UTC + 6) in this study, given that sunrise occurred between 05:00 and 06:00 LT during study periods. The stem energy balance method required a steady state and a constant energy input from the heater strip inside the gauge. Therefore, in practice, we installed aluminum bubble foil shields around the gauges to insulate the stem section from changes in the environment.

The EC system was installed 2.25 m a.g.l. (above ground level) on the stationary tower and maintained at the same height throughout the experiment (cotton canopy height reached 60 cm on 1 July 2012 and 67 cm on 30 September 2012). The measurements were conducted at a frequency of 10 Hz, and 30 min-averaged fluxes were computed. Eddy covariance provided continuous ET data for the whole study period.

The LCpro+ photosynthesis system measurements were conducted at 08:00, 10:00, 13:00, 16:00, and 18:00 (UTC + 6) on the 3 days (23 July, 27 July, and 10 August). On these days, LCpro+ was applied to four plants on which sap flow gauges were installed. For each plant, six sunlit leaves located at the top, middle, and bottom layers of the canopy (i.e., two sunlit leaves in each layer) were selected for LCpro+ measurements. Five samples for each leaf were measured and the averaged value was the representative transpiration of this leaf.

Table 2. The ratio of the sunlit (α) or shaded ($1-\alpha$) leaf area to the total leaf area and the ratio of transpiration rate of a shaded leaf to that of a sunlit leaf (β) at a specific time and canopy layer.

Time	Top layer (occupied 10.1 % of the leaf area[a])			Middle layer (occupied 60.5 % of the leaf area[a])			Bottom layer (occupied 29.4 % of the leaf area[a])		
	α[b]	$1-\alpha$	β[b]	α[b]	$1-\alpha$	β[b]	α[b]	$1-\alpha$	β[b]
08:00	0.29	0.71	0.55	0.21	0.79	0.44	0.17	0.83	0.26
10:00	0.33	0.67	0.64	0.23	0.77	0.54	0.20	0.80	0.46
13:00	0.34	0.66	0.58	0.24	0.76	0.45	0.21	0.79	0.65
16:00	0.29	0.71	0.39	0.21	0.79	0.46	0.17	0.83	0.34
18:00	0.14	0.86	0.47	0.17	0.83	0.40	0.12	0.88	0.40

[a] Zhang et al. (2007); [b] Tao (2007).

To understand the variation and uncertainty introduced through LCpro+ measurements and upscaling approaches, a variability analysis of the transpiration at leaf scale at three different levels (i.e., leaf level, layer level and plant level) was conducted in the morning, noon, and afternoon on 23 July. All of the tested leaves were sunlit leaves. At the leaf level, five samples were measured on one typical leaf, and the mean, standard deviation (SD), and coefficient of variation (CV) were calculated based on the five samples. At the layer level, five different leaves in the same canopy layer were selected. The transpiration for each leaf was obtained by averaging five samples. The mean, SD, and CV associated with the layer level were calculated based on the transpiration of the five tested leaves. At the plant level, five different leaves were randomly selected from the whole plant. Additionally, the transpiration for each leaf was obtained by averaging five samples, and the mean, SD, and CV were calculated based on the transpiration of these five leaves.

2.3.2 Upscaling approaches

The inter-comparison of multi-scale ET can validate ET estimates and provide ET components. However, due to the particular spatial scales at which the different ET are measured, as well as the variation in the samples (e.g., leaves and plants), it is necessary to utilize appropriate upscaling approaches before performing the inter-comparison (Evett et al., 2012; Hatton and Wu, 1995).

To obtain ET at the plant scale, transpiration can be simply upscaled from ET at the leaf scale by multiplying the average transpiration rate of a unit leaf area by the total plant leaf area (Approach 1). Due to the enormous variability in leaf transpiration at the plant level, as well as the marked differences in transpiration between shaded and sunlit leaves, this approach is hypothesized to induce significant errors (Petersen et al., 1992).

The ratio of the shaded or sunlit leaves to the total leaves is associated with the canopy structure, and the diurnal trend varies due to sun position in the different canopy layers (Sarlikioti et al., 2011; Thanisawanyangkura et al., 1997). Therefore, a new upscaling approach (Approach 2) is proposed. This approach incorporates the canopy structure and the relationships between sunlit and shaded leaves, and plant transpiration rate can be calculated based on the following equation:

$$M_P = 6.48 \times 10^{-3} \sum_{1}^{m} \{M_k \cdot (\alpha_k \cdot A_k) + [(M_k \cdot \beta_k) \cdot (1-\alpha_k) \cdot A_k]\}, \quad (6)$$

where M_P is the representative plant transpiration rate ($\mathrm{g\,h^{-1}}$), m is the number of canopy layers (denoted k, 1 to m), M_k is the LCpro+ measurement value for the sunlit leaf in layer k (see Eq. 1), $\mathrm{mmol\,m^{-2}\,s^{-1}}$), α_k and β_k are the ratio of sunlit leaf area to total leaf area and the ratio of transpiration rate of a shaded leaf to that of a sunlit leaf in layer k, respectively, A_k is the leaf area in layer k ($\mathrm{cm^2}$), and 6.48×10^{-3} is a unit conversion factor.

In this study, when using Approach 2, the cotton canopy was divided into three layers ($m = 3$), and two sunlit leaves at each layer were selected to be measured. The averaged value was the representative transpiration rate for a sunlit leaf at the indicated layer, whereas the representative transpiration rate for a shaded leaf at this layer was calculated based on the ratio of transpiration rate of a shaded leaf to that of a sunlit leaf (Tao, 2007).

Because we did not measure the parameters of the cotton canopy, we used the typical parameters reported in the literature. A stable canopy structure was formed prior to the measurement days of 23 July, 27 July, and 10 August, thus the canopy structure was assumed to be identical for the analysis (Zhang et al., 2007). Based on the study conducted by Tao (2007) on the physiological properties of shaded and sunlit leaves of cotton, the ratio of the shaded to the sunlit leaves and the ratio of transpiration rate of a shaded leaf to that of a sunlit leaf can be obtained at a specific time and layer (Table 2).

Traditionally, to obtain field ET from plant scale, we can multiply the average sap flow per plant by the population of plants in the experimental field (Approach 3; for more details, see Dugas et al., 1994). Although we selected the gauged plants as typical representative plants, the limited samples and large variability between the plants results in a large error in the estimation of field ET using this approach (Ham et al., 1990). Reliable field transpiration estimates require additional plant attributes, such as stem diameter and leaf area, to construct a relationship between individual (sap flow) and population (field) transpiration.

Some studies have reported that sap flow is proportional to the stem diameter of a plant (Wilson et al., 2001; Granier et al., 2000). Because the measurement of a stem diameter is a simple and rapid process, we can easily obtain the representative stem diameter for a field and then calculate the representative plant transpiration. The field transpiration can then be directly estimated by multiplying the representative plant transpiration by the plant density (Approach 4; for more details, see Dugas, 1990).

Since transpiration represents the water vapor lost from leaf surfaces, the upscaling approach would be improved if the adjustment of the sap-flow-based ET estimate is based on the leaf area (Heilman and Ham, 1990). However, measurement of the leaf area may require additional work compared with measurement of the stem diameters, and is time-consuming and impractical if the number of samples is too large. With comprehensive respect to feasibility and accuracy, an integrated upscaling approach of ET from plant- to field scale was developed by Chabot et al. (2005). A relationship (function $A = (D)$) between leaf area and stem diameter of sugarcane was developed based on 100 plant samples. Based on the investigations of the stem diameters and the plant densities in 12 1 m-long sub-plots distributed throughout the field, the total leaf area can be calculated using the abovementioned relationship. The sap flow was expressed per unit leaf area, and the transpiration can then be obtained by multiplying the sap flow per unit leaf area by the total leaf area in the field (Approach 5; for more details, see Chabot et al., 2005).

In consideration of annual crops growing quickly and the relationship between leaf area and stem diameter changing rapidly, Approach 5 is modified to incorporate the temporal variability in the relationship between leaf area and stem diameter to obtain Approach 6. The different relationships between leaf area and stem diameter $A_j = f_j(D_j)$ are used for different cotton growth stages j in Approach 6. The total leaf area in the field can be estimated using the following equation:

$$A_{\mathrm{total},j} = f_j\left(\overline{D_j}\right) \cdot n, \tag{7}$$

where A_{total} is total leaf area (cm^2) in the experimental field and n represents the number of plants.

The sap flow is assumed to be proportional to the leaf area; hence, field transpiration rate E_{SF} can be calculated by the following equation:

$$E_{\mathrm{SF}} = \frac{F_{\mathrm{P}}}{A_{\mathrm{g}}} \cdot \frac{A_{\mathrm{total}}}{1000\,Q}, \tag{8}$$

where E_{SF} is the field transpiration rate derived from the sap flow measurements (mm h^{-1}), F_{p} is the plant sap flow rate (g h^{-1}), A_{g} is leaf area of the plant on which sap flow measurements are performed (cm^2), Q is the field area (m^2), and 1000 is a unit conversion factor.

Using the similar approach, the field transpiration E_{PS} (mm h^{-1}) can also be obtained from the LCpro+ measurements through the following equation, the results of which are presented in Sect. 3.3.5 for comparison:

$$E_{\mathrm{PS}} = \frac{M_{\mathrm{P}}}{A_{\mathrm{g}}} \cdot \frac{A_{\mathrm{total}}}{1000\,Q}. \tag{9}$$

Through the use of upscaling approaches, ET results measured using different methods can be compared at plant or field scale.

2.4 Other measurements

In addition to the ET measurements described above, soil moisture and crop attributes (e.g., leaf area and stem diameter) were also measured in this study. Thirty soil sensors (three models, i.e., Hydra Probe, Stevens Water Monitoring System, Inc., Beaverton, OR, USA; Digital TDT, Acclima Inc., Meridian, ID, USA; CS616, Campbell Scientific Inc., Logan, UT, USA) were placed in the wide-row, narrow-row, and inter-film zones at 0.05, 0.10, 0.15, 0.20, 0.30, 0.40, 0.50, 0.60, 0.70, and 0.80 m below the ground to obtain a general view of field soil water condition. The data were stored every 5 min in a CR1000 data logger.

In order to obtain the relationships between leaf area and stem diameter $A_j = f_j(D_j)$ used in Approach 6, 10 typical cotton plants of an averaged size (compared with the plants throughout the field) were randomly selected for the leaf area measurements every 2 weeks and stem diameters of selected plants were recorded at the same time. All of the leaves were stripped from each plant, and the leaf area was then obtained by directly scanning all of the leaves using a leaf area meter (model Yaxin-1241, Beijing Yaxinliyi Science and Technology Co., Ltd., China). The LAI was calculated by dividing the leaf area by the area that each plant occupied.

The plant density and cotton stem diameters were investigated inside 100 sub-plots of the Eastern Field on 1 July and 12 September. We selected six 0.6 m^2 quadrats distributed throughout each sub-plot to count the number of plants, and we measured the stem diameters of 20 plants in each sub-plot. The dynamic change of stem diameter was measured with 10 fixed plants (typical ones with the averaged plant height and LAI of the whole field) located in the Eastern

Table 3. Variability in transpiration at the leaf scale on 23 July 2012.

Time	Morning (07:30–08:30)			Noon (11:30–12:30)			Afternoon (16:30–17:30)		
Level of analysis	Leaf	Layer	Plant	Leaf	Layer	Plant	Leaf	Layer	Plant
Mean (mm h^{-1})	1.09	1.07	0.75	1.68	1.87	1.39	1.14	1.15	1.09
Standard deviation (mm h^{-1})	0.04	0.14	0.22	0.06	0.34	0.38	0.05	0.19	0.40
Coefficient of variation (%)	3.63	12.85	29.09	3.28	17.96	27.58	4.74	16.94	36.30

Field every 2 weeks during cotton growth period. In addition, all stem diameters were measured at 5 cm above the soil surface.

3 Results

3.1 Meteorological conditions

The meteorological conditions during the study period, including air temperature, net radiation, vapor pressure deficit (VPD), and wind speed, are shown in Fig. 3. The air temperature and net radiation were considerably higher during SP1 and SP2 than during SP3, whereas the VPD and wind speed showed no significant difference among the three sub-periods. In addition, no precipitation occurred on these days.

Frequent drip irrigation at 5- to 10-day intervals was implemented during July and August (see Table 1) and resulted in high air relative humidity in SP1 and SP2 with a value of approximately 50 %. Because irrigation was terminated in September, the soil surface became dry, and the air relative humidity dropped to 34 % in SP3. However, the VPD during these three sub-periods did not change significantly due to the change of saturation vapor pressure, which was lowest in SP3.

The 3 days (23 July, 27 July, and 10 August) that were chosen for the LCpro+ measurements were sunny days with the highest net radiation. In contrast, on cloudy days, such as 24 July, 25 July, and 9 August, the net radiation was relatively low.

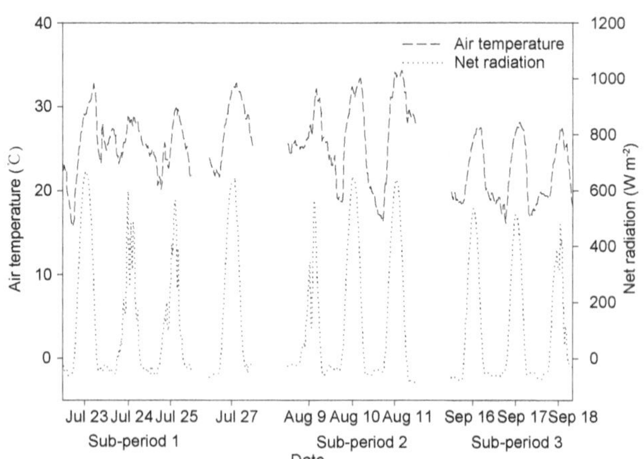

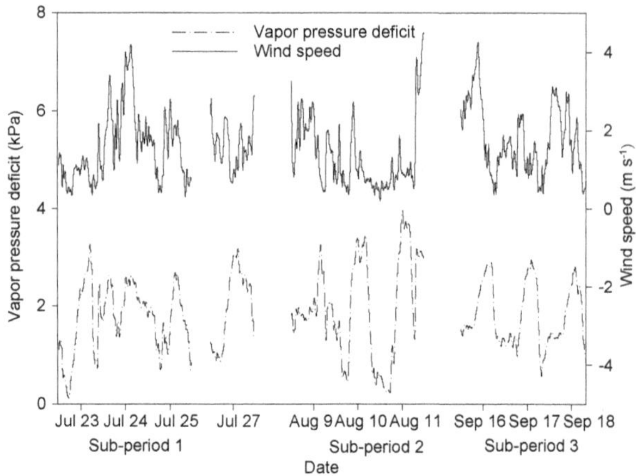

Fig. 3. Diurnal trends of air temperature, net radiation, vapor pressure deficit, and wind speed measured 2.25 m above the ground.

3.2 Comparison at the plant scale

3.2.1 Variability in transpiration at the leaf scale

The variability analysis results at the leaf scale are shown in Table 3. The CVs at different times (morning, noon, and afternoon) reveal the consistent trend obtained for each level of analysis. The averaged CV at the leaf level, which had a magnitude of 3.89 %, is hypothesized to reflect the random error. The CV at the layer level, which had an averaged value of 15.91 %, was greater than that at the leaf level and less than that at the plant level. Regardless of the different transpiration rates between sunlit and shaded leaves, the CV of the whole plant was 30.99 %, which suggests a large variability in the leaf transpiration rate throughout the plant. In addition, the difference in the transpiration rate between a sunlit leaf and a shaded leaf can be significant, e.g., as high as fourfold at 8 a.m. (Tao, 2007).

3.2.2 Upscaling from the leaf to the plant scale

Based on the upscaling approaches from leaf to plant scale and the data obtained from the measurements and literature, the scaled plant transpiration can be determined using Approaches 1 (M_s) and 2 (M_p) (Table 4). In general, the value of M_s was as high as 1.69-fold of M_p. In consideration of the

Table 4. Plant transpiration derived using Approach 1 (M_s) and Approach 2 (M_p).

Date	23 Jul		27 Jul		10 Aug	
Time	M_s ($g\,h^{-1}$)	M_p ($g\,h^{-1}$)	M_s ($g\,h^{-1}$)	M_p ($g\,h^{-1}$)	M_s ($g\,h^{-1}$)	M_p ($g\,h^{-1}$)
08:00	95.17	47.32	Data missing		82.91	44.53
10:00	164.83	110.06	156.82	100.82	107.64	67.33
13:00	168.27	106.28	121.19	75.38	174.29	103.65
16:00	101.35	59.76	135.15	75.54	148.53	84.44
18:00	53.39	26.42	30.55	13.79	54.11	26.77

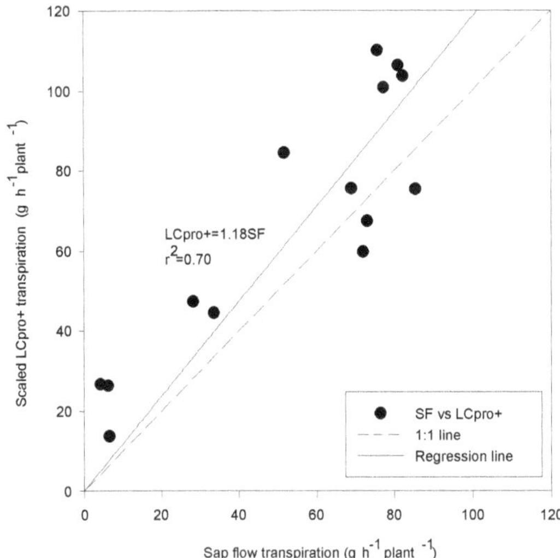

Fig. 4. Correlation between the transpiration measured by sap flow and the scaled transpiration of LCpro+ measurements on 23 July, 27 July, and 10 August.

difference between the sunlit and shaded leaves, Approach 1 takes all the leaves as sunlit ones, and likely overestimates plant transpiration.

3.2.3 Comparison of sap flow and the scaled LCpro+ measurements

The scaled transpiration obtained using Approach 2 (M_P) was compared with the results measured through sap flow (F_P) for the same cotton plants. The results are shown in Fig. 4. In general, the value of M_P was slightly higher than that of F_P, and the slope of the regression line was 1.18 ($r^2 = 0.70$). Biases clearly existed when the transpiration rate was too low or too high, which indicates that the LCpro+ measurement may most likely disturb the normal status of leaf transpiration due to contact. In contrast, the comparison of Approach 1 results and sap flow measurements resulted in slope and r^2 values of 1.94 and 0.52, respectively (data not shown in Fig. 4, see Table 5). Thus, Approach 2 exhibits significantly improved upscaling results.

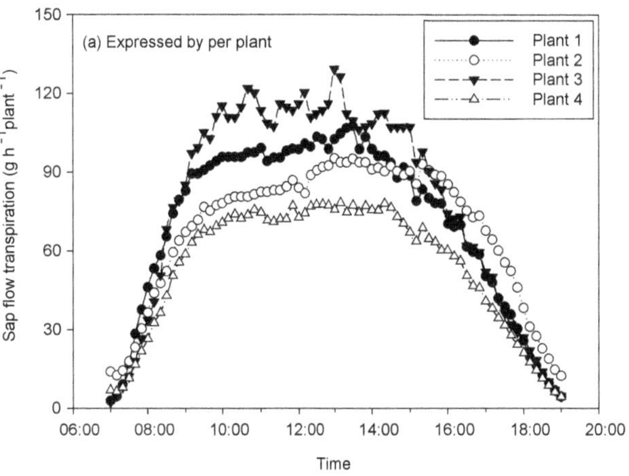

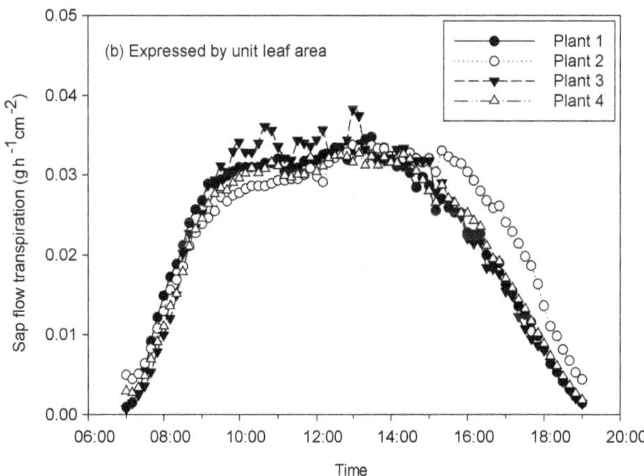

Fig. 5. Transpiration estimates based on sap flow and expressed by unit leaf area and per plant.

3.3 Comparison at the field scale

3.3.1 Variability in transpiration at the plant scale

Large differences in the plant transpiration were observed among four plants that were gauged based on sap flow (Fig. 5a). On 9 July, the cumulative sap flows obtained for plants 1 through 4 were 866, 840, 959, and 659 $g\,day^{-1}$,

Table 5. The slope and coefficient of determination (r^2) for the different upscaling approaches.

Upscaling approach		Equation	Slope	r^2	Brief description	Error analysis ($\frac{\sigma_{M_P}}{M_P}$ or $\frac{\sigma_{E_{SF}}}{E_{SF}}$)*
From leaf- to plant scale	Approach 1	LCpro+ = 1.94 SF	1.94	0.52	Total leaf area and uniform transpiration (T) rate	0.311
	Approach 2	LCpro+ = 1.18 SF	1.18	0.70	Canopy structure, sunlit and shaded leaves	0.161
From plant- to field scale	Approach 3	SF = 1.61 EC	1.61	0.88	Plant population (PP)	0.156
	Approach 4	SF = 1.31 EC	1.31	0.88	PP, T is proportional to the stem diameter (SD)	0.123
	Approach 5	SF = 1.33 EC	1.33	0.87	PP, T is proportional to the leaf area (LA), Fixed relationship between LA and SD	0.113
	Approach 6	SF = 1.10 EC	1.10	0.87	PP, T is proportional to LA, Dynamic relationship between LA and SD	0.073

* M_P is upscaled plant transpiration, E_{SF} is upscaled field transpiration, σ_{M_P} and $\sigma_{E_{SF}}$ are the standard error for M_P and E_{SF}, respectively.

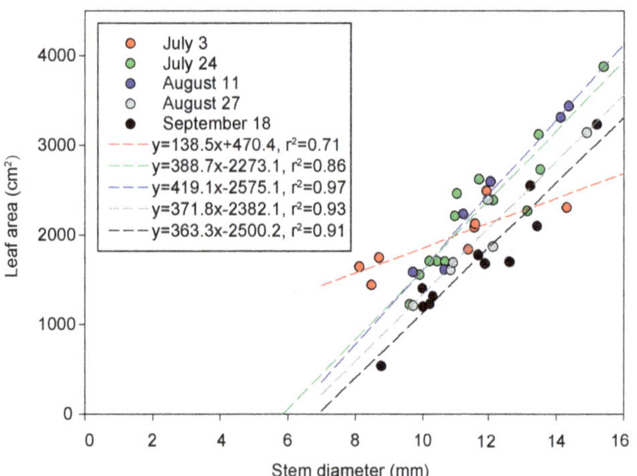

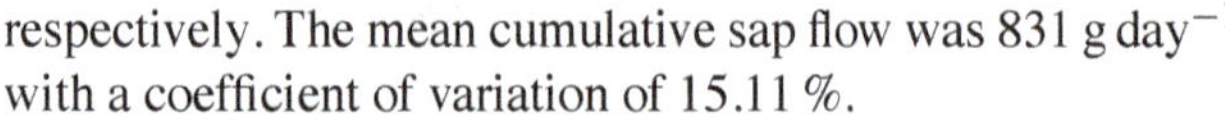

Fig. 6. Relationships between leaf area and stem diameter.

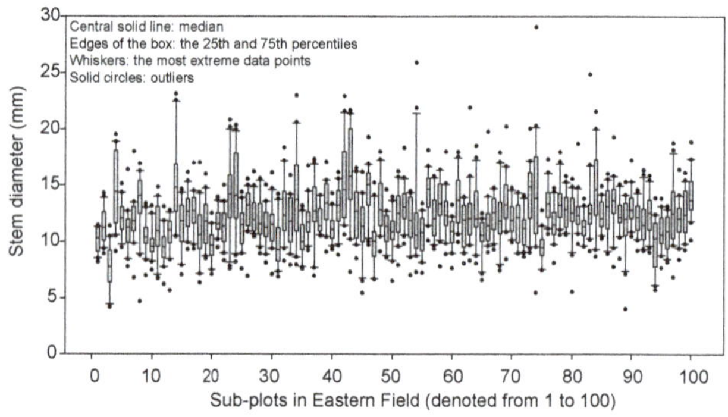

Fig. 7. Stem diameter variability in 100 sub-plots located in the Eastern Field on 12 September 2012.

respectively. The mean cumulative sap flow was 831 g day^{-1} with a coefficient of variation of 15.11 %.

Because the sap flow was expressed per unit leaf area (Fig. 5b), the errors were markedly reduced (Heilman and Ham, 1990). The cumulative sap flows per unit leaf area on 9 July for plants 1 through 4 were 0.280, 0.299, 0.284, and 0.275 g cm^{-2} day^{-1}, respectively. The mean cumulative sap flow was 0.285 g cm^{-2} day^{-1} with a coefficient of variation of 3.53 %. The results are consistent with the findings reported by Ham et al. (1990), who observed sap flow CV values expressed per plant and unit leaf area of 13 and 7.7 %, respectively. The results indicate that, although the measurements of the leaf area may require additional work, they may reduce the number of devices required to represent the field condition and are thus worth the effort (Dugas, 1990). Therefore, it is necessary to account for plant variability in sap flow measurements, even in homogenous cotton farmland.

3.3.2 Upscaling from the plant to the field scale

When using Approach 6, a series of relationships $A_j = {}_j(D_j)$ with correlation coefficients (r^2) ranging from 0.71 to 0.97 were developed based on the experiments to represent different cotton growth stages j, including the three sub-periods (Fig. 6). The slope increased rapidly from 3 to 24 July, which suggests that the leaf area changed rapidly in July. The slope was then fairly stable throughout the remaining growth period, whereas the intercept gradually became small over time, which demonstrates that the rate of defoliation gradually exceeded the rate of leaf area growth.

As described in Sect. 2.4, 2000 plants were randomly selected from 100 sub-plots (denoted i from 1 to 100) in the Eastern Field to determine the plant stem diameter at the end of the cotton growth period on 12 September 2012 (Fig. 7). The average value of the stem diameter was 12.18 mm, and the standard deviation was 2.64 mm, which suggests a notable variability (CV = 21.7 %) among the plants under growth conditions. In addition, the dynamic changes in stem diameters measured by 10 fixed plants every 2 weeks are illustrated in Fig. 8. The cotton stem grew rapidly in the vegetative stage after seed germination, and became stable in the

Table 6. Upscaled field transpiration derived through Approach 6 (E_{SF}, mm day^{-1}) and Approach 3 (E_s, mm day^{-1}).

Sub-period 1	E_{SF}	E_s	Sub-period 2	E_{SF}	E_s	Sub-period 3	E_{SF}	E_s
23 Jul	5.88	8.97	9 Aug	3.42	5.43	16 Sep	3.36	4.78
24 Jul	4.72	7.21	10 Aug	5.54	8.76	17 Sep	3.35	4.81
25 Jul	4.62	7.10	11 Aug	5.31	8.41	18 Sep	2.67	3.86

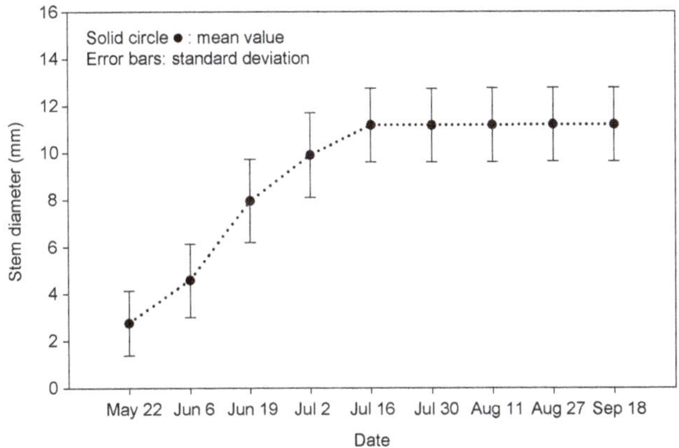

Fig. 8. Dynamic changes in the stem diameter. The stem diameters of 10 fixed plants were measured every 2 weeks.

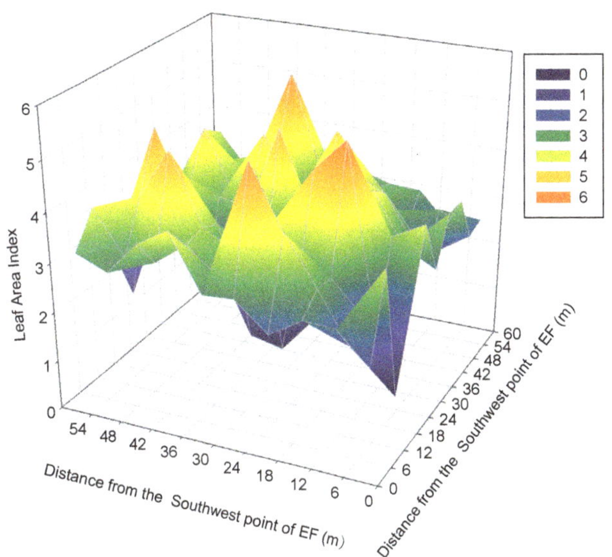

Fig. 9. Example of leaf area index (LAI) distribution in the Eastern Field (EF) on 11 August 2012.

reproductive stage after 16 July. Therefore, we can predict the stem diameter (D_i, mm) for any of the cotton growth stages based on the data shown in Figs. 7 and 8.

The number of plants n_i in each sub-plot was counted on six random 0.6 m^2 quadrats. Based on the dynamic relationships between the stem diameter and the leaf area, as well as the stem diameter distribution, we can obtain the leaf area distribution in the field for a specific time during the cotton growth period using Eq. (7). For instance, the results of the leaf area index distribution in the Eastern Field on 11 August 2012 are shown in Fig. 9.

Based on sap flow measurements and total leaf area of the field, we can obtain the scaled field transpiration (E_{SF}) using Approach 6. The results are shown in Table 6. The transpiration derived from Approach 3 (E_s) was higher than E_{SF} by a factor of 1.52, which indicates that the gauged plants were probably stronger than the representative plant size. The results agree with those reported by Ham et al. (1990), who observed that E_s was as high as 1.63-fold of E_{SF}. The results derived from the other approaches are not shown in Table 6.

3.3.3 Energy balance closure of eddy covariance

Energy balance closure is one approach that can be used to evaluate the reliability of eddy covariance (EC) measurements (Wilson et al., 2002). Using all valid half-hourly data in the three sub-periods (data points, $n = 399$), the slope between the available energy flux ($R_n - G$) and the sum of sensible and latent heat fluxes (LE + H) for this site was 0.70, the intercept was 8.01 W m^{-2}, and the coefficient of determination (r^2) was 0.90, as shown in Fig. 10. The reasons underlying the energy imbalance has been investigated by numerous researchers over the past few decades (Franssen et al., 2010; Leuning et al., 2012; Stoy et al., 2013); however, these are complicated and not yet fully understood.

Under mulched drip irrigation, general factors accounting for the lack of energy balance closure, including the mismatch in the source area for different measurements, sampling errors, systematic bias, neglected energy sinks (e.g., energy storage in cotton biomass), the loss of low/high-frequency contributions to the turbulent flux, and neglected advection of scalars, still make sense. However, the plastic mulching film likely increases the probability and magnitude of the imbalance (Ding et al., 2010; Zhou et al., 2011). The study conducted by Zhou et al. (2011), who analyzed mulched drip irrigation in a cotton field, suggested that the turbulent fluxes (LE + H) could be blocked by more than 11% relative to the available energy ($R_n - G$) due to the impact of mulch. If this is true, the slope between ($R_n - G$) and (LE + H) will increase to 0.81 (present closure of 0.70 plus 0.11) in this study, which is promising based on the previously obtained values of 0.53–0.99 for the energy closure (Wilson et al., 2002). Thus, we are confident that the eddy

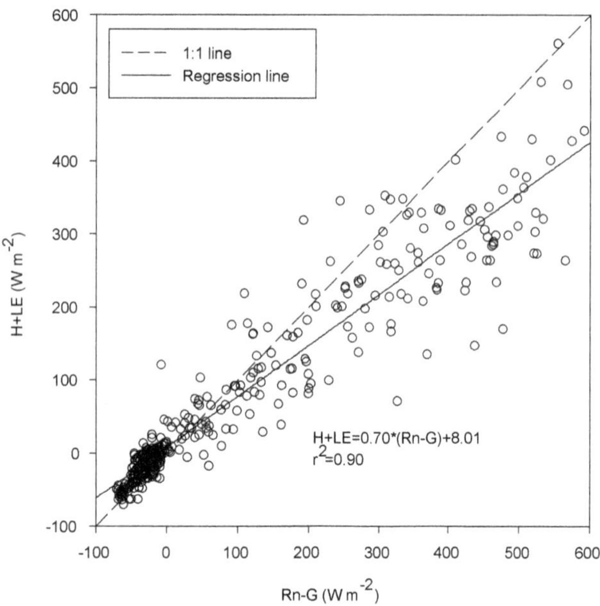

Fig. 10. Energy balance closure of eddy covariance. The data are paired 30 min averages collected during the three sub-periods.

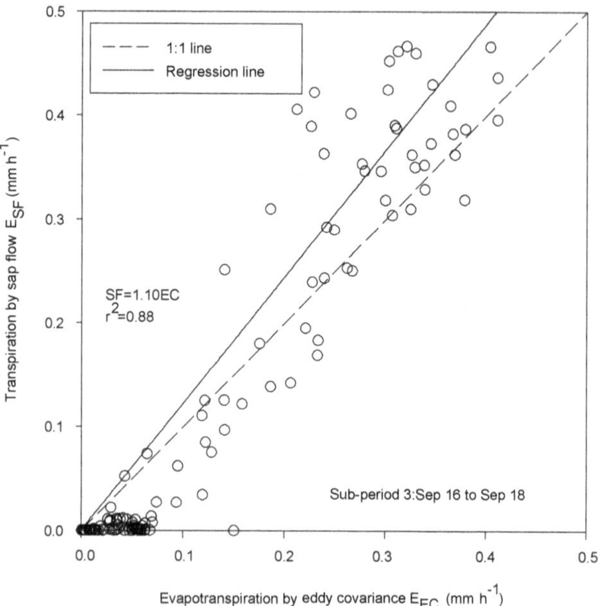

Fig. 11. Correlation between transpiration obtained from the upscaling of the sap-flow-based measurement (Approach 6; E_{SF}) and ET obtained through eddy covariance (E_{EC}) for sub-period 3.

covariance measurements provide an accurate ET estimate at this site.

3.3.4 Comparison of eddy covariance and the scaled sap flow and LCpro+ measurements

In general, drip irrigation systems deliver the limited amount of water directly to the plant root zone; consequently, the soil water content (SWC) in the inter-film zone is very low (Bonachela et al., 2001). In addition, the mulched film eliminates soil evaporation in the wide-row and narrow-row zones (Wang et al., 2001). Therefore, the soil evaporation is expected to be a small portion of ET under mulched drip irrigation, especially when irrigation is stopped for a long time. In this study, LCpro+ measurements were used to measure the bare soil evaporation in the inter-film zone when the soil pot was substituted for the leaf chamber on 20 September (2 days after SP3, no irrigation for 23 days, SWC = 15.5 % within a depth of 20 cm). The measured value was only 0.04 mm day^{-1}. Therefore, we assume that soil evaporation was sufficiently small in SP3 so that it can be neglected. In other words, evapotranspiration measured by eddy covariance in SP3 contained the transpiration component only. Thus, in this study, SP3 was chosen as the period for transpiration comparison at the field scale.

Based on the four upscaling approaches described in Sect. 2.3.2, the correlations between E_{SF} and E_{EC} values were analyzed for SP3. At times, the wind blew from the back of the 3-D sonic anemometer, and the flow distortion caused by the anemometer's arms and other supporting structures was considerable and may have resulted in an underestimation of ET (van Dijk et al., 2004). Therefore, the data obtained when the wind blew from the back of the 3-D sonic anemometer were excluded in our correlation analysis.

The slopes of the regression line were 1.61, 1.31, 1.33, and 1.10 for Approaches 3 through 6, respectively (Table 5). Approach 6 improves the upscaling results significantly. Figure 11 shows a pronounced qualitative similarity for the transpiration obtained through sap flow (Approach 6) (E_{SF}) and through eddy covariance (E_{EC}), which confirms that Approach 6 is a reasonable upscaling approach.

The diurnal trends of the transpiration estimates obtained by sap flow (Approach 6; E_{SF}) and by eddy covariance (E_{EC}) are shown in Fig. 12 for SP3. For convenience, the potential evapotranspiration calculated using the FAO Penman–Monteith equation (E_P; Allen et al., 1998) is also shown in this figure. The E_{SF} and E_{EC} matched E_P well, which suggests that the instruments can respond well to changes in the meteorological conditions of the surrounding environment. Also, the coincidence between potential evaporation (E_P) and measured transpiration shows that the two independent methods can get the similar values for the well-watered crop, which further implies the rationality of our measurements. On 17 September, due to distortion by the anemometer's arms and other supporting structures, E_{EC} was obviously less than E_{SF}.

The results prove that Approach 6, which takes dynamic relationships between leaf area and stem diameter into account, is advanced and reasonable. Using this upscaling approach to obtain field transpiration, the evapotranspiration components are analyzed in the following section.

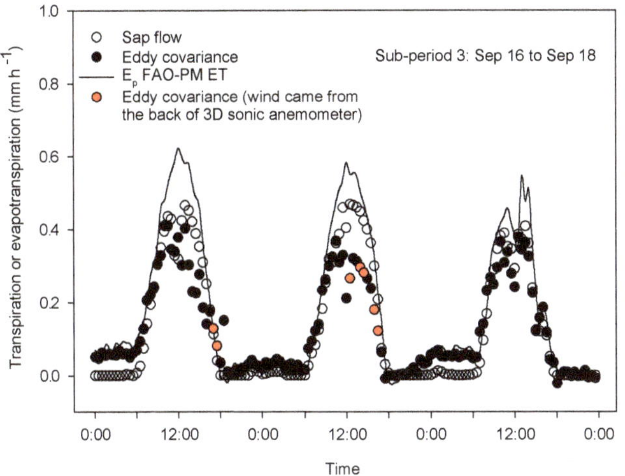

Fig. 12. Diurnal trends of transpiration derived from the upscaling of the sap flow measurements (Approach 6) and measured by eddy covariance during sub-period 3.

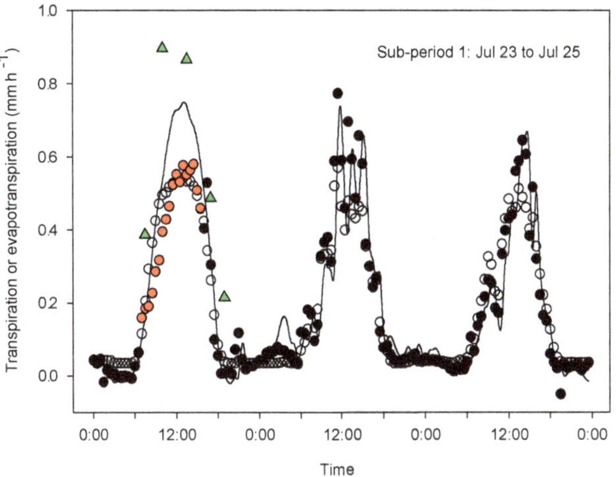

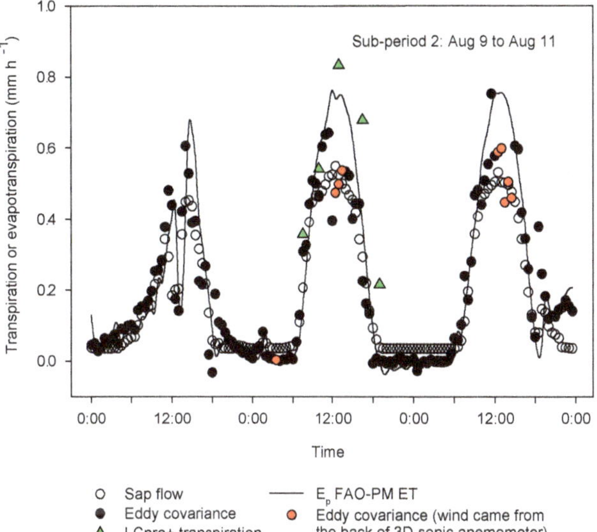

Fig. 13. Diurnal trends of transpiration determined by sap flow measurements (upscaled using Approach 6 and calibrated), the LCpro+ photosynthesis system (upscaled using Approach 2), and evapotranspiration determined by eddy covariance during sub-periods 1 and 2.

3.3.5 Evapotranspiration components under mulched drip irrigation conditions

The partitioning of the evapotranspiration flux is important for understanding the water exchange and optimizing water management in arid and semi-arid areas. In previous studies, the stable isotope technique has been widely applied to the evapotranspiration partitioning (Wang et al., 2010, 2013). Meanwhile, the difference between E_{EC} and E_{SF} also provides one useful approach for partitioning these fluxes and reflects the contribution of soil evaporation to the total ET within the flux footprint of eddy covariance (Williams et al., 2004; Wilson et al., 2001).

As described in Sect. 3.3.4, soil evaporation can be neglected in SP3 due to the dry soil surface in the inter-film zone (SWC = 15.6 % within a depth of 20 cm), the relatively low evaporative demand ($E_P = 4.396$ mm day^{-1}), and the fully closed canopy. Therefore, the difference between E_{SF} and E_{EC} in SP3 was regarded as a systematic error induced by both inherent error of methods (i.e., underestimate/overestimate of T and ET) and upscaling approaches. Since the measurements and upscaling approaches were completely identical in the three sub-periods, the systematic error of SP3 could be consistent with that observed for the other two sub-periods. The systematic error in SP1 and SP2 was then overcome by using SP3 to calibrate the sap flow. We recalculated all of the upscaled sap flow data in SP1 and SP2 using the regression model between E_{EC} and E_{SF} derived from SP3 (Transpiration = 0.737 × [upscaled sap flow] + 0.035). After the recalculation, the soil evaporation under mulched drip irrigation in this region can be evaluated by the difference between E_{EC} and the recalculated E_{SF}. This method was adopted and proved to be valid in the study conducted by Williams et al. (2004) in an olive orchard (Williams et al., 2004).

The diurnal trends of evapotranspiration after upscaling and calibration are shown in Fig. 13. As shown in this figure, E_{EC} was fairly high in SP1 and SP2, reaching up to 0.7 mm h^{-1} at noon due to the favorable soil moisture conditions, high LAI and evaporation potential. In contrast, the ET value was only 0.4 mm h^{-1} at noon in SP3. The gap between E_{EC} and E_{SF} was the component of soil evaporation. At noon, the soil evaporation was appreciable, whereas it was quite small at night. We also plot the data obtained by applying LCpro+ on 23 July and 10 August in this figure. The results show that E_{PS} was higher than E_{EC} most of the time, which is consistent with the conclusion obtained in Sect. 3.2.3.

Figure 14 shows the correlation between transpiration obtained from sap flow measurement (after upscaling and calibration) and ET obtained through eddy covariance. Evapotranspiration by EC and transpiration by sap flow agree well for low- and mid-rates, but disagree for higher flux rates. There may be two potential reasons to explain this phenomenon: the soil evaporation was probably more intense at noon due to the higher temperature and radiation, or there was a saturation level for plant transpiration above which transpiration stayed constant and more evaporation occurred. However, it is still not clear based on this study.

In general, the slopes were 0.871 and 0.823 for SP1 and SP2 in Fig. 14, that is, T/ET ($E_{\mathrm{SF}}/E_{\mathrm{EC}}$) was 87.1 and 82.3 % for these two sub-periods, respectively. The results suggested that the fraction of soil evaporation to evapotranspiration was greater in SP2 than in SP1. This difference might be due to the fact that soil water content (SWC), which significantly affected soil evaporation in the cotton growth period, was higher in SP2 than in SP1 due to drip irrigation (Table 7). In fact, irrigation occurred more than one week before 23 July (33.26 mm irrigation on 15/16 July). In contrast, 59.28 mm drip irrigation was implemented on 8 August, which was only one day before SP2. The magnitudes of the soil evaporation were 0.508 mm day^{-1} in SP1 and 0.801 mm day^{-1} in SP2. The results confirm that transpiration constitutes the largest portion of ET under mulched drip irrigation when the canopy is closed and provide quantitative estimates of the soil evaporation before (SP1) and after (SP2) irrigation at this site during the cotton flower and bolling stages. However, the results of ET components are only based on the short period observation. More data is needed if the fraction of transpiration over ET for the whole cotton growth period is to be determined.

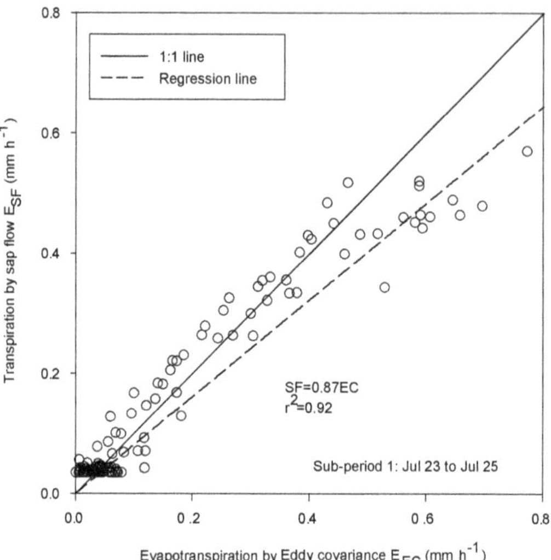

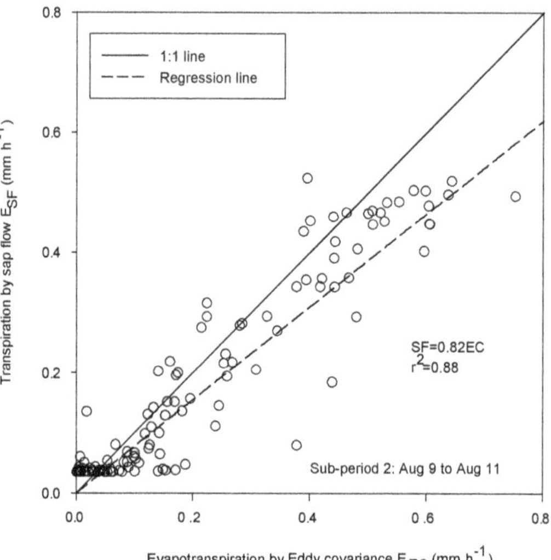

Fig. 14. Correlation between the transpiration obtained through sap flow measurements (upscaled using Approach 6 and calibrated; E_{SF}) and the evapotranspiration obtained through eddy covariance (E_{EC}) for sub-periods 1 and 2.

3.4 Error analysis

In order to assess the flux uncertainties, and clarify how instrument precision, vegetative measurements and calculation theory affect the uncertainties (Good et al., 2012), the error analysis of upscaling approaches (Approach 1 to 6) are implemented in this section. Using the consistent manner, the error of soil evaporation estimate is also explored. It is worth noting that since the true values of evapotranspiration are inaccessible, the error analysis below is only based on the standard error, representing the variation relative to the mean, and is not an indication of measurement accuracy.

3.4.1 Plant scale

The plant transpiration is calculated by the following equation in Approach 1:

$$M_{\mathrm{P}} = M \cdot A_{\mathrm{plant}}, \tag{10}$$

where M is the transpiration rate for the sunlit leaf by the LCpro+ measurement, and A_{plant} is leaf area of the plant.

Barry (1978) indicated that when a final result is calculated from direct measurements, its precision is a function of the variability in the direct measurements. The plant transpiration is computed from direct measurements, including leaf transpiration and leaf area. Therefore, the standard error (SE) of M_P can be expressed by SE of the direct measurements:

$$\sigma_{M_{\mathrm{P}}}^2 = \left(\sigma_{\mathrm{M}} \times A_{\mathrm{plant}}\right)^2 + \left(\sigma_{A_{\mathrm{plant}}} \times M\right)^2, \tag{11}$$

where $\sigma_{M_{\mathrm{P}}}$, σ_M, and $\sigma_{A_{\mathrm{plant}}}$ are the standard errors for M_{P}, M, and A_{plant}, respectively. The variability of M and A_{plant} is assumed to be normally distributed and independent since the M and A_{plant} are separately measured (Ham et al., 1990). Then we can rewrite Eq. (11) and express the variability of all parameters relative to their respective mean:

Table 7. Evapotranspiration components (mm day^{-1}) under mulched drip irrigation for the three sub-periods.

Sub-periods	E_P^1	E_{EC}^2	Fraction of transpiration to ET (%)	E_{soil}^3	Whole profile SWC[4] (within 20 cm)	IFZ[5] SWC[4] (within 20 cm)	LAI[6]
23–25 Jul (SP1)	5.004	3.941	87.1 %	0.508	24.2 %	20.0 %	3.080
9–11 Aug (SP2)	5.348	4.527	82.3 %	0.801	31.5 %	26.7 %	3.163
16–18 Sep (SP3)	4.396	3.014	100.0 %	0	17.9 %	15.6 %	2.402

[1] E_P: ET calculated by the FAO Penman–Monteith equation; [2] E_{EC}: ET measured by eddy covariance; [3] E_{soil}: soil evaporation calculated by the multiplication of E_{EC} by the fraction of transpiration to ET; [4] SWC: soil water content; [5] IFZ: inter-film zone; [6] LAI: leaf area index.

$$\frac{\sigma_{M_P}}{M_P} = \left[\left(\frac{\sigma_M}{M}\right)^2 + \left(\frac{\sigma_{A_{plant}}}{A_{plant}}\right)^2\right]^{\frac{1}{2}}. \qquad (12)$$

This analysis shows that both the variability of M and A_{plant} affect the variability of plant transpiration estimate. In this study, $\frac{\sigma_{A_{plant}}}{A_{plant}}$ has been determined based on the data, whose value is 0.018. As shown in Sect. 3.2.1, in Approach 1, $\frac{\sigma_M}{M}$ is the variability in leaf transpiration at the plant level, whose value is 0.310. Therefore, $\frac{\sigma_{M_P}}{M_P}$ is equal to 0.311. It is worth noting that if we take the transpiration difference of sunlit leaf and shaded leaf into account, the $\frac{\sigma_M}{M}$ should be larger and the $\frac{\sigma_{M_P}}{M_P}$ will increase, accordingly.

Similarly, we can assess the flux uncertainties of Approach 2. When we assume that the $\frac{\sigma_M}{M}$ and $\frac{\sigma_{A_{plant}}}{A_{plant}}$ are constant in different canopy layers, the variability of plant transpiration can be obtained by the following equation:

$$\frac{\sigma_{M_P}}{M_P} = \left\{\left[\frac{M_{P1}^2 + M_{P2}^2 + M_{P3}^2}{M_P^2}\right] \cdot \left[\left(\frac{\sigma_M}{M}\right)^2 + \left(\frac{\sigma_{A_{plant}}}{A_{plant}}\right)^2\right]\right\}^{\frac{1}{2}}, \qquad (13)$$

where M_{Pi} is the plant transpiration of the canopy layer i. In this study, $\frac{\sigma_{A_{plant}}}{A_{plant}}$ is 0.018, and $\frac{\sigma_M}{M}$ is the variability in leaf transpiration at the layer level, whose value is 0.160. $\frac{\sigma_{M_P}}{M_P}$ is equal to 0.161 when $[\frac{M_{P1}^2+M_{P2}^2+M_{P3}^2}{M_P^2}]$ is 1. However, since $[\frac{M_{P1}^2+M_{P2}^2+M_{P3}^2}{M_P^2}]$ is always less than 1, $\frac{\sigma_{M_P}}{M_P}$ would be less than 0.161 in Approach 2. The results suggest that the variability of plant transpiration will sharply decrease when the canopy structure has been considered. Compared with Approach 1, Approach 2 provides us more reliable upscaled transpiration at the plant scale.

3.4.2 Field scale

In Approach 3, the field transpiration is calculated by the following equation:

$$E_{SF} = F_P \times n, \qquad (14)$$

where F_P is sap flow value per plant, and n is the plant density. Similarly, we can also rewrite Eq. (14) to express the variability of all parameters relative to their respective mean:

$$\frac{\sigma_{E_{SF}}}{E_{SF}} = \left[\left(\frac{\sigma_{F_P}}{F_P}\right)^2 + \left(\frac{\sigma_n}{n}\right)^2\right]^{\frac{1}{2}}, \qquad (15)$$

where $\sigma_{E_{SF}}$, σ_{F_P} and σ_n are the standard errors for E_{SF}, F_P and n, respectively. Based on the measurements, $\frac{\sigma_{F_P}}{F_P}$ is determined as 0.151 in Sect. 3.3.1 and $\frac{\sigma_n}{n}$ is 0.040. Therefore, $\frac{\sigma_{E_{SF}}}{E_{SF}}$ in Approach 3 has the value of 0.156 in this study.

Similarly, the $\frac{\sigma_{E_{SF}}}{E_{SF}}$ can be calculated using Eq. (16) in Approach 4, and Eq. (17) in Approach 5 and 6, respectively. F_{Pstem} is sap flow value per unit stem diameter, S_{rep} is the representative stem diameter for typical plant, F_{PA} is sap flow value per unit leaf area, and A_{rep} is the representative leaf area for typical plant. $\frac{\sigma_{F_{Pstem}}}{F_{Pstem}}$ and $\frac{\sigma_F}{F_{PA}}$ are determined based on the measurements, whose values are 0.105 and 0.035, respectively. $\frac{\sigma_n}{n}$ is 0.040 as mentioned before. Since we have measured 2000 plants to obtain the representative stem diameter, it is reasonable to assume $\frac{\sigma_{S_{rep}}}{S_{rep}}$ is small. In this study, $\frac{\sigma_{S_{rep}}}{S_{rep}}$ is assigned to 0.05. $\frac{\sigma_A}{A_{rep}}$ is influenced by both variability of the relationship between leaf area and stem diameter, and the stem diameter variability $\frac{\sigma_{S_{rep}}}{S_{rep}}$. In Approach 5, $\frac{\sigma_A}{A_{rep}}$ is assumed to be 0.1. Given that we have adopted dynamic relationships for different cotton growth stages in Approach 6, $\frac{\sigma_A}{A_{rep}}$ is assigned to 0.05.

$$\frac{\sigma_{E_{SF}}}{E_{SF}} = \left[\left(\frac{\sigma_{F_{Pstem}}}{F_{Pstem}}\right)^2 + \left(\frac{\sigma_{S_{rep}}}{S_{rep}}\right)^2 + \left(\frac{\sigma_n}{n}\right)^2\right]^{\frac{1}{2}}, \qquad (16)$$

$$\frac{\sigma_{E_{SF}}}{E_{SF}} = \left[\left(\frac{\sigma_F}{F_{PA}}\right)^2 + \left(\frac{\sigma_A}{A_{rep}}\right)^2 + \left(\frac{\sigma_n}{n}\right)^2\right]^{\frac{1}{2}} \qquad (17)$$

The final results of error analysis are shown in Table 5. $\frac{\sigma_{E_{SF}}}{E_{SF}}$ is 0.123 in Approach 4, 0.113 in Approach 5, and 0.073 in Approach 6, respectively. The results suggest that although Approach 6 introduces more parameters into the field transpiration estimate, the flux uncertainty has been reduced in this approach. That is because the variability of sap flow rates are reduced when the rates are expressed on unit leaf area. Meanwhile, the variability of leaf area estimate has been reduced by the application of a dynamic relationship between leaf area and stem diameter. That is to say, from the statistical perspective, Approach 6 provides us the most reliable upscaled transpiration at the field scale in this study.

3.4.3 Soil evaporation

Soil evaporation is calculated in Sect. 3.3.5 by the following equation:

$$E_{\text{soil}} = E_{\text{EC}} - F_{\text{PA}} \times A_{\text{rep}} \times n. \tag{18}$$

The soil evaporation is computed from direct measurements including eddy covariance, sap flow, leaf area and plant density. Therefore, the standard error (SE) of E_{soil} can be expressed by SE of the direct measurements:

$$\sigma_{\text{soil}}^2 = \sigma_{\text{EC}}^2 + \left(\sigma_{\text{F}} \times A_{\text{rep}} \times n\right)^2 + \left(\sigma_{\text{A}} \times F_{\text{PA}} \times n\right)^2 + \left(F_{\text{PA}} \times A_{\text{rep}} \times \sigma_n\right)^2. \tag{19}$$

We can also rewrite Eq. (19) as follows:

$$\frac{\sigma_{\text{soil}}}{E_{\text{soil}}} = \left\{\left(\frac{E_{\text{soil}}}{E_{\text{EC}}}\right)^{-2}\left(\frac{\sigma_{\text{EC}}}{E_{\text{EC}}}\right)^2 + \left[\left(\frac{E_{\text{soil}}}{E_{\text{EC}}}\right)^{-1} - 1\right]^2 \left[\left(\frac{\sigma_{\text{F}}}{F_{\text{PA}}}\right)^2 + \left(\frac{\sigma_{\text{A}}}{A_{\text{rep}}}\right)^2 + \left(\frac{\sigma_n}{n}\right)^2\right]\right\}^{\frac{1}{2}}. \tag{20}$$

This analysis shows that the variability of E_{EC} plays an important role when $\frac{E_{\text{soil}}}{E_{\text{EC}}}$ is large. When $\frac{E_{\text{soil}}}{E_{\text{EC}}}$ becomes small, the variabilities of F_{PA}, A_{rep} and n are more significant in the estimate of soil evaporation.

As mentioned above, $\frac{\sigma_{\text{F}}}{F_{\text{PA}}}$ and $\frac{\sigma_n}{n}$ are 0.035 and 0.040, respectively (Approach 6). Since ET measured by eddy covariance is relatively stable, we can suppose that $\frac{\sigma_{\text{EC}}}{E_{\text{EC}}}$ is quite small with the value of 0.001. $\frac{\sigma_{\text{A}}}{A_{\text{rep}}}$ is assigned to 0.1 and 0.05 for comparison.

The behavior of Eq. (20) is demonstrated in Fig. 15 when using these typical variance levels mentioned above. When $\frac{E_{\text{soil}}}{E_{\text{EC}}}$ becomes smaller, the expected $\frac{\sigma_{\text{soil}}}{E_{\text{soil}}}$ increases sharply, and the measurements of sap flow, leaf area and plant density are more significant. In this study, the $\frac{E_{\text{soil}}}{E_{\text{EC}}}$ is approximately 15 %, and then the $\frac{\sigma_{\text{soil}}}{E_{\text{soil}}}$ is about 0.64 ($\frac{\sigma_{\text{A}}}{A_{\text{rep}}} = 0.1$) and 0.41 ($\frac{\sigma_{\text{A}}}{A_{\text{rep}}} = 0.05$). The results indicate that the soil evaporation is difficult to evaluate under mulched drip irrigation conditions when E_{soil} is the small component of ET. The comparison of two curves in Fig. 15 shows that the variability of E_{soil} has not been markedly reduced when only $\frac{\sigma_{\text{A}}}{A_{\text{rep}}}$ decreases. That is to say, the variability of E_{soil} will not be reduced until the measurements of sap flow, LAI and plant density are improved simultaneously.

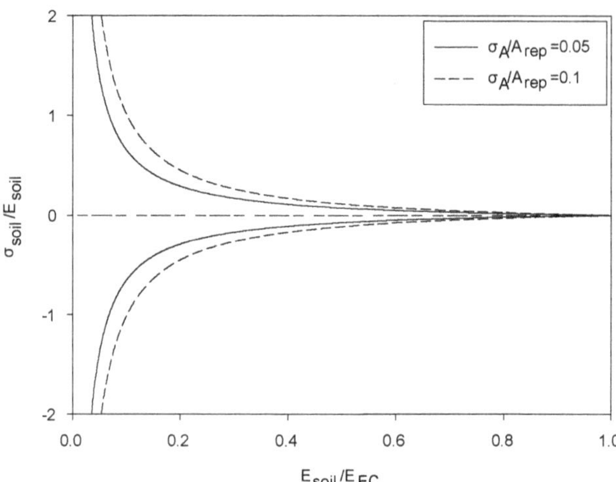

Fig. 15. Expected variability of soil evaporation estimate ($\frac{\sigma_{\text{soil}}}{E_{\text{soil}}}$) in response to the fraction of evaporation over ET (Eq. 20). The curves show the variability for the different $\frac{\sigma_{\text{A}}}{A_{\text{rep}}}$ levels. $\frac{\sigma_{\text{EC}}}{E_{\text{EC}}}$, $\frac{\sigma_{\text{F}}}{F_{\text{PA}}}$ and $\frac{\sigma_n}{n}$ are held constant at 0.001, 0.035 and 0.040, respectively.

4 Discussion

Three different measurement methods, namely the photosynthesis system, sap flow, and eddy covariance, were used in this study to estimate evapotranspiration in a cotton field under mulched drip irrigation. Although these three methods differ significantly in the physical theories on which the measurements are based and the particular spatial and temporal scales to which they pertain, the results derived from each of the measurements after scale transformation show satisfactory consistency when employed during the cotton growth season. The reasonably good agreement between the results obtained using LCpro+, sap flow, and eddy covariance provides some confidence in their reliability for the estimation of the evapotranspiration in an agricultural ecosystem using these three methods and the described upscaling approaches.

In farmland, the partitioning of evapotranspiration components is essential for guiding the irrigation schedule to achieve the dual goals of water saving and high yield (Wang et al., 2001). Since this type of investigation needs the data measured at different spatial scales, scale transformation should be implemented. Different species have different transpiration characters. In addition, the transpiration rates of leaves vary substantially depending on the leaf's position, orientation, age, and size (Sassenrath-Cole, 1995; Thanisawanyangkura et al., 1997), and the transpiration of plants also varies markedly with the heterogeneous soil water availability, the diverse plant age and LAI (Dugas, 1990). Therefore, it is more simple to conduct scale transformation in farmland than in forest due to the single crop species planted, the relatively homogeneous vegetation distribution pattern, and the low spatial variability in the water availability (Loranty et al., 2008), which make it straightforward and feasible to extrapolate point observations to representative area values, and lead to highly credible and reasonable scaled results (Allen et al., 2011a). However, it is also a challenge to conduct scale transformation in farmland due to rapid crop growth, rapid changes in leaf area and stem diameter, and large diversity in growth conditions among plants, all of which affect the results and introduce errors (Chabot et al., 2005).

In this study, taking into account the rapid growth of the plants, we establish links between leaf areas and stem diameters during every sub-period for the scale transformation. This approach overcomes the limitation caused by rapid growth and achieves a good result for the derivation of the field leaf area. Plant transpiration derived from the photosynthesis system has seldom been reported before. Because the number of samples measured by instruments is limited compared to the large number of leaves and because there is considerable variability among the leaves, it is quite difficult to extrapolate photosynthesis system measurements to the plant scale (Dugas et al., 1994; Kigalu, 2007). In this study, the different transpiration rates of sunlit and shaded leaves, as well as canopy structure, were taken into account. This upscaling approach was proven to provide a reasonable estimation of transpiration at the plant scale.

However, discrepancy still exists among ET results obtained using the photosynthesis system, sap flow, and eddy covariance. The upscaling approaches used to transform ET from the leaf- to the plant scale or from the plant- to the field scale may lead to errors and result in discrepancies. First, the photosynthesis system and sap flow methods can measure only a subset of leaves or plants in a field. These limited samples sometimes do not completely capture the variance and the mean response of the overall situation in which the target scaling level method operates. In addition, the canopy parameters and the ratio of the transpiration rate of a shaded leaf to that of a sunlit leaf, which were derived from the literature, may vary from the actual values (Petersen et al., 1992; Thanisawanyangkura et al., 1997). The simultaneous observation of canopy structure is expected to improve the results. Another possible source of divergence between the LCpro+, sap flow, and eddy covariance results could be the unmatched observation area. Although LCpro+ and sap flow measurements and the leaf area estimates were conducted within the flux source footprint of eddy covariance, the changing wind direction and footprint might change the measuring area of eddy covariance and frustrate attempts to match the scaled transpiration to the eddy covariance measurements (Williams et al., 2004).

As with any other measurement techniques, photosynthesis system, sap flow and eddy covariance methods have their own inherent limitations which should be mentioned (Good et al., 2012). It is reported in previous studies that sap flow overestimates transpiration by 7–35 % (Chabot et al., 2005; Ham et al., 1990) due to the stem heat storage, heat dissipation to the ambient environment and accuracy of stem temperature measurements. For eddy covariance, it is a known phenomenon that the observation likely underestimates ET at field scale (Foken, 2008; Wilson et al., 2002). Discrepancy might also come from these inherent factors mentioned above.

Due to the severe lack of water resources in arid and semi-arid regions, mulched drip irrigation has been widely applied as a highly efficient water-saving irrigation method (Wang et al., 2011). As shown in the evapotranspiration partition results in this study, a portion of soil evaporation is significantly reduced through mulched drip irrigation, and most of the water is consumed by plant transpiration during the analysis periods. Because transpiration is accompanied by photosynthesis and plant productivity, higher transpiration indicates a better crop yield (Katul et al., 2012), and mulched drip irrigation tends to improve water use efficiency. Compared to the fraction of cotton transpiration to evapotranspiration of 65 % (Tang et al., 2010) and 56 % (Ham et al., 1990) observed under traditional flood irrigation conditions during the same cotton growth stages (flower and bolling stages), the fraction of 87.1 % before irrigation and 82.3 % after irrigation that were obtained in this study are much higher, which confirms that mulched drip irrigation is a more efficient method for achieving water savings. The quantitative estimation of evaporation and transpiration in this study may provide supports for the application of mulched drip irrigation in the future.

5 Conclusions

A comparison of the methods used to determine evapotranspiration and its components in a cotton field under mulched drip irrigation conditions was conducted in this study. The methods used were based on a photosynthesis system, sap flow, and eddy covariance, which provided information at the leaf-, plant-, and field scale, respectively. The variability in the transpiration at the leaf scale and at the plant scale was discussed. Upscaling approaches were explored to obtain comparable ET estimates from the multi-scale measurements. The results show that ET estimates derived from the three methods agree well after scale transformation, which indicates that, taking into account the variability between individuals, the selection of representative samples, and the adoption of a suitable scale transformation approach, any of these three methods can provide good estimates of field evapotranspiration in farmland.

The comparison of the methods and the discussion of the variability associated with the three ET measurement methods will help researchers assess the quality, validity, and representativeness of ET information derived using these techniques. The upscaling approaches can help other researchers estimate field evapotranspiration from point measurements, such as those obtained based on photosynthesis system and sap flow, and will provide data and precedent for further study on the water cycle and ecological processes in farmland.

Based on the transpiration estimates obtained from the upscaling of sap flow measurements and ET obtained through eddy covariance, the evapotranspiration components were analyzed. The evapotranspiration rates were determined to 3.94 and 4.53 mm day^{-1} during the cotton flower (July) and bolling (August) stages, respectively. The results show that a fraction of transpiration over ET is significantly increased under mulched drip irrigation during cotton flower and bolling stages. The fraction of transpiration to evapotranspiration reached 87.1 % before drip irrigation and 82.3 % after irrigation during the analysis periods. The results might support the popularization of mulched drip irrigation in other arid and semi-arid regions in the future to address the challenge of water scarcity.

Acknowledgements. This study was supported by the National Science Foundation of China (NSFC 51190092, 51109110, and 51222901), SRF for ROCS, SEM, and the Foundation of the State Key Laboratory of Hydroscience and Engineering of Tsinghua University (2012-KY-03). Their support is greatly appreciated. We also thank the staff at Tsinghua University-Korla Oasis Eco-hydrology Experimental Research Station for their excellent field and lab assistance. We thank Murugesu Sivapalan from the University of Illinois at Urbana-Champaign and Huimin Lei from Tsinghua University for their helpful advice. Also, we would like to give our sincere thanks to the reviewers (Xianwen Li, S. P. Good, and B. Mitra) for all their comments which improve the quality of manuscript substantially.

Edited by: L. Wang

References

ADC Bioscientific Ltd.: LCi Portable Photosynthesis System: Instruction Manual, ADC BioScientific Ltd., Hoddesdon, UK, 2004.

Alfieri, J. G., Kustas, W. P., Prueger, J. H., Hipps, L. E., Evett, S. R., Basara, J. B., Neale, C. M. U., French, A. N., Colaizzi, P., Agam, N., Cosh, M. H., Chavez, J. L., and Howell, T. A.: On the discrepancy between eddy covariance and lysimetry-based surface flux measurements under strongly advective conditions, Adv. Water Resour., 50, 62–78, 2012.

Allen, R. G., Pereira, L. S., Raes, D., and Smith, M.: FAO Irrigation and drainage paper No. 56, Food and Agriculture Organization of the United Nations, Roma, Italy, 1998.

Allen, R. G., Pereira, L. S., Howell, T. A., and Jensen, M. E.: Evapotranspiration information reporting: I. Factors governing measurement accuracy, Agr. Water Manage., 98, 899–920, 2011a.

Allen, R. G., Pereira, L. S., Howell, T. A., and Jensen, M. E.: Evapotranspiration information reporting: II. Recommended documentation, Agr. Water Manage., 98, 921–929, 2011b.

Ashraf, M.: Salt tolerance of cotton: Some new advances, Crit. Rev. Plant Sci., 21, 1–30, 2002.

Baker, J. M. and Vanbavel, C.: Measurement of mass-flow of water in the stems of herbaceous plants, Plant Cell Environ., 10, 777–782, 1987.

Baldocchi, D., Falge, E., Gu, L. H., Olson, R., Hollinger, D., Running, S., Anthoni, P., Bernhofer, C., Davis, K., Evans, R., Fuentes, J., Goldstein, A., Katul, G., Law, B., Lee, X. H., Malhi, Y., Meyers, T., Munger, W., Oechel, W., Paw U, K. T., Pilegaard, K., Schmid, H. P., Valentini, R., Verma, S., Vesala, T., Wilson, K., and Wofsy, S.: FLUXNET: a new tool to study the temporal and spatial variability of ecosystem-scale carbon dioxide, water vapor, and energy flux densities, B. Am. Meteorol. Soc., 82, 2415–2434, 2001.

Barry, B. A.: Errors in practical measurement in science, engineering and technology, Wiley, New York, 1978.

Bonachela, S., Orgaz, F., Villalobos, F. J., and Fereres, E.: Soil evaporation from drip-irrigated olive orchards, Irrig. Sci., 20, 65–71, 2001.

Chabot, R., Bouarfa, S., Zimmer, D., Chaumont, C., and Moreau, S.: Evaluation of the sap flow determined with a heat balance method to measure the transpiration of a sugarcane canopy, Agr. Water Manage., 75, 10–24, 2005.

Chavez J. L., Howell, T. A., and Copeland, K. S.: Evaluating eddy covariance cotton ET measurements in an advective environment with large weighing lysimeters, Irrig. Sci., 28, 35-=50, 2009.

Ding, R., Kang, S., Li, F., Zhang, Y., Tong, L., and Sun, Q.: Evaluating eddy covariance method by large-scale weighing lysimeter in a maize field of northwest China, Agr. Water Manage., 98, 87–95, 2010.

Dugas, W. A.: Comparative measurement of stem flow and transpiration in cotton, Theor. Appl. Climatol., 42, 215–221, 1990.

Dugas, W. A., Heuer, M. L., Hunsaker, D., Kimball, B. A., Lewin, K. F., Nagy, J., and Johnson, M.: Sap flow measurements of transpiration from cotton grown under ambient and enriched CO_2 concentrations, Agr. Forest. Meteorol., 70, 231–245, 1994.

Evett, S. R., Kustas, W. P., Gowda, P. H., Anderson, M. C., Prueger, J. H., and Howell, T. A.: Overview of the bushland evapotranspiration and agricultural remote sensing experiment 2008 (BEAREX08): a field experiment evaluating methods for quantifying ET at multiple scales, Adv. Water Resour, 50, 4–19, doi:10.1016/j.advwatres.2012.03.010, 2012.

Falge, E., Baldocchi, D., Olson, R., Anthoni, P., Aubinet, M., Bernhofer, C., Burba, G., Ceulemans, R., Clement, R., Dolman, H., Granier, A., Gross, P., Grunwald, T., Hollinger, D., Jensen, N. O., Katul, G., Keronen, P., Kowalski, A., Lai, C. T., Law, B. E., Meyers, T., Moncrieff, H., Moors, E., Munger, J. W., Pilegaard, K., Rannik, U., Rebmann, C., Suyker, A., Tenhunen, J., Tu, K., Verma, S., Vesala, T., Wilson, K., and Wofsy, S.: Gap filling strategies for defensible annual sums of net ecosystem exchange, Agr. Forest. Meteorol., 107, 43–69, 2001.

Foken, T.: The energy balance closure problem: An overview, Ecol. Appl., 18, 1351–1367, 2008.

Franssen, H. J., Stöckli, R., Lehner, I., Rotenberg, E., and Seneviratne, S. I.: Energy balance closure of eddy-covariance data: A multisite analysis for European FLUXNET stations, Agr. Forest. Meteorol., 150, 1553–1567, 2010.

Good, S. P., Soderberg, K., Wang, L., and Caylor, K. K.: Uncertainties in the assessment of the isotopic composition of surface fluxes: A direct comparison of techniques using laser-based water vapor isotope analyzers, J. Geophys. Res., 117, D15301, doi:10.1029/2011JD017168, 2012.

Granier, A., Biron, P., and Lemoine, D.: Water balance, transpiration and canopy conductance in two beech stands, Agr. Forest. Meteorol., 100, 291–308, 2000.

Ham, J. M., Heilman, J. L., and Lascano, R. J.: Determination of soil-water evaporation and transpiration from energy-balance and stem-flow measurements, Agr. Forest. Meteorol., 52, 287–301, 1990.

Hatton, T. J. and Wu, H. I.: Scaling theory to extrapolate individual tree water-use to stand water-use, Hydrol. Process., 9, 527–540, 1995.

Heilman, J. L. and Ham, J. M.: Measurement of mass-flow rate of sap in ligustrum-japonicum, Hortscience, 25, 465–467, 1990.

Hou, X., Wang, F., Han, J., Kang, S., and Feng, S.: Duration of plastic mulch for potato growth under drip irrigation in an arid region of northwest China, Agr. Forest. Meteorol., 150, 115–121, 2010.

Howell, T. A., Evett, S. R., Tolk, J. A., and Schneider, A. D.: Evapotranspiration of full-, deficit-irrigated, and dryland cotton on the northern Texas high plains, J. Irrig. Drain. E.-ASCE., 130, 277–285, 2004.

Hu, H., Tian, F., and Hu, H.: Soil particle size distribution and its relationship with soil water and salt under mulched drip irrigation in Xinjiang Province of China, Sci. China Tech. Sci., 54, 1–7, 2011.

Katul, G. G., Oren, R., Manzoni, S., Higgins, C., and Parlange, M. B.: Evapotranspiration: A process driving mass transport and energy exchange in the soil-plant-atmosphere-climate system, Rev. Geophys., 50, RG3002, doi:10.1029/2011RG000366, 2012.

Kigalu, J. M.: Effects of planting density on the productivity and water use of tea (Camellia sinensis L.) clones I. Measurement of water use in young tea using sap flow meters with a stem heat balance method, Agr. Water Manage., 90, 224–232, 2007.

Ko, J., Piccinni, G., Marek, T., and Howell, T.: Determination of growth-stage-specific crop coefficients (K_C) of cotton and wheat, Agr. Water Manage., 96, 1691–1697, 2009.

Lei, H. and Yang, D.: Interannual and seasonal variability in evapotranspiration and energy partitioning over an irrigated cropland in the North China Plain, Agr. Forest. Meteorol., 150, 581–589, 2010.

Leuning, R., Van Gorsel, E., Massman, W. J., and Isaac, P. R.: Reflections on the surface energy imbalance problem, Agric. Forest. Meteorol., 156, 65–74, 2012.

Li, S., Kang, S., Li, F., and Zhang, L.: Evapotranspiration and crop coefficient of spring maize with plastic mulch using eddy covariance in northwest China, Agr. Water Manage., 95, 1214–1222, 2008.

Loranty, M. M., Mackay, D. S., Ewers, B. E., Adelman, J. D., and Kruger, E. L.: Environmental drivers of spatial variation in whole-tree transpiration in an aspen-dominated upland-to-wetland forest gradient, Water Resour. Res., 44, W02441, doi:10.1029/2007WR006272, 2008.

MacKay, D. S., Ahl, D. E., Ewers, B. E., Gower, S. T., Burrows, S. N., Samanta, S., and Davis, K. J.: Effects of aggregated classifications of forest composition on estimates of evapotranspiration in a northern Wisconsin forest, Global Change Biol., 8, 1253–1265, 2002.

Mahouachi, J., Socorro, A. R., and Talon, M.: Responses of papaya seedlings (Carica papaya L.) to water stress and re-hydration: Growth, photosynthesis and mineral nutrient imbalance, Plant Soil, 281, 137–146, 2006.

Mengistu, T., Sterck, F. J., Fetene, M., Tadesse, W., and Bongers, F.: Leaf gas exchange in the frankincense tree (Boswellia papyrifera) of African dry woodlands, Tree Physiol., 31, 740–750, 2011.

Petersen, K. L., Fuchs, M., Moreshet, S., Cohen, Y., and Sinoquet, H.: Computing transpiration of sunlit and shaded cotton foliage under various water stresss, Agron. J., 84, 91–97, 1992.

Sakuratani, T.: A heat balance method for measuring water flux in the stem of intact plants, J. Agr. Meteorol., 37, 9–17, 1981.

Sakuratani, T.: Improvement of the probe for measuring water flow rate in intact plants with the stem heat balance method, J. Agrometeorol., 40, 273–277, 1984.

Sarlikioti, V., de Visser, P., and Marcelis, L.: Exploring the spatial distribution of light interception and photosynthesis of canopies by means of a functional-structural plant model, Ann. Bot., 107, 875–883, 2011.

Sassenrath-Cole, G. F.: Dependence of canopy light distribution on leaf and canopy structure for two cotton (Gossypium) species, Agr. Forest. Meteorol., 77, 55–72, 1995.

Silberstein, R., Held, A., Hatton, T., Viney, N., and Sivapalan, M.: Energy balance of a natural jarrah (Eucalyptus marginata) forest in Western Australia: measurements during the spring and summer, Agr. Forest. Meteorol., 109, 79–104, 2001.

Stoy, P. C., Mauder, M., Foken, T., Marcolla, B., Boegh, E., Ibrom, A., Arain, M. A., Arneth, A., Aurela, M., Bernhofer, C., Cescatti, A., Dellwik, E., Duce, P., Gianelle, D., van Gorsel, E., Kiely, G., Knohl, A., Margolis, H., McCaughey, H., Merbold, L., Montagnani, L., Papale, D., Reichstein, M., Saunders, M., Serrano-Ortiz, P., Sottocornola, M., Spano, D., Vaccari, F., and Varlagin, A.: A data-driven analysis of energy balance closure across FLUXNET research sites: The role of landscape scale heterogeneity, Agr. Forest. Meteorol., 171–172, 137–152, 2013.

Tang, L., Li, Y., and Zhang, J.: Partial rootzone irrigation increases water use efficiency, maintains yield and enhances economic profit of cotton in arid area, Agr. Water Manage., 97, 1527–1533, 2010.

Tao, Y.: Contrusting physiological properties of shaded and sunlit leaves, and applying a photosynthesis model for cotton, Nanjing University of Information Science & Technology, Nanjing, 2007.

Thanisawanyangkura, S., Sinoquet, H., Rivet, P., Cretenet, M., and Jallas, E.: Leaf orientation and sunlit leaf area distribution in cotton, Agr. Forest. Meteorol., 86, 1–15, 1997.

Tolk, J. A., Howell, T. A., and Evett, S. R.: Nighttime evapotranspiration from alfalfa and cotton in a semiarid climate, Agron. J., 98, 730–736, 2006.

van Dijk, A., Moene, A. F., and de Bruin, H. A. R.: The principles of surface flux physics: theory, practice and description of the ECPACK library, Internal Report 2004/1, Meteorology and Air Quality Group, Wageningen University, Wageningen, the Netherlands, 99 pp., 2004.

Wang, H., Zhang, L., Dawes, W. R., and Liu, C.: Improving water use efficiency of irrigated crops in the North China Plain – measurements and modeling, Agr. Water Manage., 48, 151–167, 2001.

Wang, L., Caylor, K. K., Villegas, J. C., Barron-Gafford, G. A., Breshears, D. D., and Huxman, T. E.: Partitioning evapotranspiration across gradients of woody plant cover: Assessment of a stable isotope technique, Geophys. Res. Lett., 37, L09401, doi:10.1029/2010GL043228, 2010.

Wang, L., Niu, S., Good, S. P., Soderberg, K., McCabe, M. F., Sherry, R. A., Luo, Y., Zhou, X., Xia, J., and Caylor, K. K.: The effect of warming on grassland evapotranspiration partitioning using laser-based isotope monitoring techniques, Geochim. Cosmochim. Acta, 111, 28–38, 2013.

Wang, R., Kang, Y., Wan, S., Hu, W., Liu, S., and Liu, S.: Salt distribution and the growth of cotton under different drip irrigation regimes in a saline area, Agr. Water Manage., 100, 58–69, 2011.

Webb, E. K., Pearman, G. I., and Leuning, R.: Correction of flux measurements for density effects due to heat and water vapour transfer, Q. J. Roy. Meteorol. Soc., 106, 85–100, 1980.

Williams, D. G., Cable, W., Hultine, K., Hoedjes, J. C. B., Yepez, E. A., Simonneaux, V., Er-Raki, S., Boulet, G., de Bruin, H. A. R., Chehbouni, A., Hartogensis, O. K., and Timouk, F.: Evapotranspiration components determined by stable isotope, sap flow and eddy covariance techniques, Agr. Forest. Meteorol., 125, 241–258, 2004.

Wilson, K. B., Hanson, P. J., Mulholland, P. J., Baldocchi, D. D., and Wullschleger, S. D.: A comparison of methods for determining forest evapotranspiration and its components: sap-flow, soil water budget, eddy covariance and catchment water balance, Agr. Forest. Meteorol., 106, 153–168, 2001.

Wilson, K., Goldstein, A., Falge, E., Aubinet, M., Baldocchi, D., Berbigier, P., Bernhofer, C., Ceulemans, R., Dolmanh, H., Field, C., Grelle, A., Ibrom, A., Lawl, B. E., Kowalski, A., Meyers, T., Moncrieffm, J., Monsonn, R., Oechel, W., Tenhunen, J., Valentini, R., and Verma, S.: Energy balance closure at FLUXNET sites, Agr. Forest. Meteorol., 113, 223–243, 2002.

Zhang, J., Wang, Y., Mao, W., Dong, Q., and Zhao, Y.: Dynamic Simulation of leaf area in cotton canopy, Trans. Chin. Soc. Agric. Mach., 38, 117–120, 2007.

Zhang, Z., Hu, H. C., Tian, F., Hu, H. P., Yao, X., and Zhong, R.: Soil salt distribution under mulched drip irrigation in an arid area of northwestern China, J. Arid. Environ., 104, 23–33, doi:10.1016/j.jaridenv.2014.01.012, 2014.

Zhou, S., Wang, J., Liu, J., Yang, J., Xu, Y., and Li, J.: Evapotranspiration of a drip-irrigated, film-mulched cotton field in northern Xinjiang, China, Hydrol. Process., 26, 1169–1178, doi:10.1002/hyp.8208, 2011.

2

An experimental set-up to measure latent and sensible heat fluxes from (artificial) plant leaves

Stanislaus J. Schymanski, Daniel Breitenstein, and Dani Or

Department of Environmental Systems Science, ETH Zurich, 8092 Zurich, Switzerland

Correspondence to: Stan Schymanski (stan.schymanski@env.ethz.ch)

Abstract. Leaf transpiration and energy exchange are coupled processes that operate at small scales yet exert a significant influence on the terrestrial hydrological cycle and climate. Surprisingly, experimental capabilities required to quantify the energy–transpiration coupling at the leaf scale are lacking, challenging our ability to test basic questions of importance for resolving large-scale processes. The present study describes an experimental set-up for the simultaneous observation of transpiration rates and all leaf energy balance components under controlled conditions, using an insulated closed loop miniature wind tunnel and artificial leaves with pre-defined and constant diffusive conductance for water vapour. A range of tests documents the above capabilities of the experimental set-up and points to potential improvements. The tests reveal a conceptual flaw in the assumption that leaf temperature can be characterized by a single value, suggesting that even for thin, planar leaves, a temperature gradient between the irradiated and shaded or transpiring and non-transpiring leaf side can lead to bias when using observed leaf temperatures and fluxes to deduce effective conductances to sensible heat or water vapour transfer. However, comparison of experimental results with an explicit leaf energy balance model revealed only minor effects on simulated leaf energy exchange rates by the neglect of cross-sectional leaf temperature gradients, lending experimental support to our current understanding of leaf gas and energy exchange processes.

1 Introduction

Most of the precipitation falling on land returns to the atmosphere by the process of transpiration, i.e. passing through the plant vascular system, undergoing phase change in leaves, and diffusing through stomata. Plant transpiration rates and CO_2 uptake are controlled by stomata and by the leaf energy balance, i.e. the partitioning of the absorbed solar irradiance into radiative, sensible, and latent heat fluxes. Present understanding of leaf gas and energy exchange is based on controlled experiments with real and artificial leaves, where the individual components of the energy balance and their sensitivities to environmental forcing were assessed separately. The state-of-the-art measurements of leaf transpiration rates employ a mass balance of an open controlled volume at steady state, i.e. by the difference of the products of air flow rate and humidity between the incoming air and the outgoing air from a control volume containing a transpiring leaf (Field et al., 1982). The transfer of heat between a leaf and the surrounding air is less commonly measured. Studies exist where this heat flux was estimated from cooling curves after a sudden reduction in absorbed radiation (Kumar and Barthakur, 1971), but in order to test our understanding of the leaf energy balance, we need a way to monitor leaf heat and gas exchange simultaneously under controlled steady-state conditions.

Leaf gas exchange and hence the leaf energy balance underly strong biological control by stomata. Only a few studies exist where leaf gas exchange and stomatal apertures were simultaneously observed (e.g. Kappen et al., 1987; Kaiser and Kappen, 1997), but these observations were not used to study physical processes, probably due to strong dynamics and uncertainty related to deduction of stomatal conductance from observed apertures. Therefore, many studies employed leaf replica to gain a better understanding of individual processes related to the leaf energy balance and gas exchange separately from biological control. For example, externally or internally heated plates were employed to estimate sensible heat transfer coefficients as a function of plate size/shape, wind speed, and turbulence (Wigley and Clark, 1974; Thom, 1968; Parkhurst et al., 1968; Grace et al., 1980). Others have used wetted leaf replica and weighing or electrochemical methods using leaf-shaped electrodes to obtain mass transfer coefficients (Schuepp, 1972). In the latter method, dimensional analysis was used to transfer results obtained from a liquid medium to real leaves surrounded by air.

A range of studies employed perforated foils or plates to study the effect of pore size and density on transpiration under steady-state conditions (e.g. Brown and Escombe, 1900; Sierp and Seybold, 1929; Ting and Loomis, 1963; Cannon et al., 1979; Zwieniecki et al., 2016). In most of these experiments, the perforated surface was mounted on a water reservoir and transpiration rate was measured by weighing the water reservoir. So far, studies using artificial leaves with dimensions and pore sizes similar to real leaves have not been published. Morrison Jr. and Barfield (1981) presented a thin artificial leaf design of similar shape and size to a tobacco leaf, consisting of teflon membrane disks sandwiching a filter paper and an external water supply consisting of cotton wicks, with a total leaf thickness of 0.42 mm. This could easily be combined with some of the above-mentioned perforated foils in order to obtain a more realistic physical model of a real leaf, but surprisingly, we have not found any such experiments in the literature.

Even more surprisingly, the simultaneous measurements of radiative, latent, and sensible heat exchange of transpiring leaves or leaf replica have not been presented in the literature. This suggests that leaf energy balance closure has never been used to assess uncertainty in the observations in a similar way to what is commonly done for eddy covariance measurements at the canopy scale (e.g. Wohlfahrt and Widmoser, 2013). In contrast to the canopy scale, leaf-scale processes lend themselves to investigation through controlled experiments, theoretically permitting rigorous testing of our understanding of leaf energy partitioning, which is at the basis of canopy-scale processes.

To improve our experimental and observational capabilities at the leaf scale, the goal of the present study was to design an experimental set-up that permits the direct measurement of all the leaf energy components simultaneously (sensible, latent and radiative exchange) while controlling boundary conditions (air temperature, humidity, wind speed, irradiance).

2 Materials and methods

To separate the physical aspects of leaf energy and gas exchange from biological control, we used artificial leaves with laser-perforated surfaces representing fixed stomatal apertures and embedded thermocouples to obtain the best possible measurements of leaf temperature near the evaporating sites (Fig. 1). We further devised a specialized thermally insulated leaf wind tunnel to control atmospheric conditions including air temperature, humidity, irradiance, and wind speed and allowing measurement of all leaf energy balance components independently, including net radiation, as well as latent and sensible heat flux (Fig. 2). The leaf wind tunnel and the artificial leaves are described in detail below. Details of technical equipment used in this study are listed in Table C1. All variables used in this paper and their descriptions, units, and standard values are given in Table D1. All data, equations, and model code necessary to reproduce the results presented here can be accessed online at https://doi.org/10.5281/zenodo.241217.

2.1 Artificial leaves

Different artificial leaves were constructed, all consisting of a capillary filter paper glued onto aluminium tape, with a water supply tube and a thermocouple sandwiched between the filter paper and the aluminium tape. The water supply tube was flattened at the end and tightly glued to the aluminium tape and filter paper using Araldite epoxy resin (Fig. 1), to prevent intrusion of air along the edges. For some leaves, we used Whatman No. 41 filter paper (0.2 mm thick) and embedded 0.25 mm thick copper-constantan thermocouples (TG-T-30-SLE, Table C1), whereas for others, we used a 0.1 mm thick Durapore membrane filter (type 0.45 µm HV[1]) and 0.08 mm thick thermocouple wire (TG-T-40-SLE). The Durapore membrane filters appeared more homogeneous and tear resistant than the Whatman filter papers, whereas the thinner thermocouple wires produced smaller bumps on the leaf surface. The water supply was connected to a liquid flow meter (SLI-0430, Table C1) and a water supply with a free water surface placed 1–3 cm below the position of the leaf. It had to be lower than the leaf to ensure that the liquid flow did not exceed the transpiration rate (e.g. droplets forming due to positive head between reservoir and leaf) and as high as possible to avoid cavitation and air intrusion along the flow path. Stomatal resistance was introduced by covering the surface of the capillary filter paper with a laser-perforated aluminium foil, attached to the leaf using thin strips of double-sided sticky tape lining the outside of the rectangular leaves. The laser-perforated foils were untreated aluminium of 25 µm thickness. Laser perforations of different sizes and densities produced different effective leaf conductances. Laser perforations were performed by Ralph Beglinger (Lasergraph AG, Würenlingen, Switzerland). The geometry of the laser perforations used for each leaf was measured using a confocal laser scanning microscope (CLSM VK-X200, Keyence, Osaka, Japan) and the specific diffusive conductances for

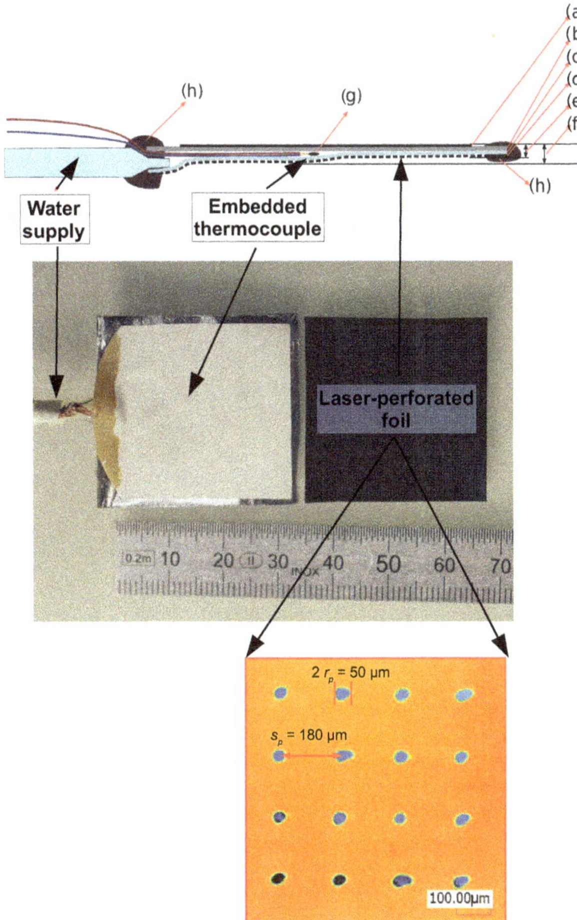

Figure 1. Artificial leaf. Top: cross section of the artificial leaf; centre: leaf image before full assembly; bottom: topography of a laser-perforated foil obtained using confocal laser scanning microscopy (CLSM). **(a)** Black aluminium tape (0.05 mm thick); **(b)** aluminium tape (0.08 mm); **(c)** absorbent filter paper (0.1–0.2 mm); **(d)** laser-perforated foil (0.025 mm); **(e)** min. leaf thickness: 0.3–0.4 mm; **(f)** max. leaf thickness: 0.35–0.65 mm; **(g)** thermocouple; **(h)** glue; **(i)** water supply tube (from the flow meter). Numbers in the CLSM image indicate typical pore diameters ($2r_p$) and pore spacings (s_p) for foils with 7 perforations per mm^2.

the perforated surfaces were estimated based on derivations presented by Lehmann and Or (2015), neglecting any internal resistance (termed "end correction" by Lehmann and Or, 2015), as we assume that the wet filter paper has direct contact with the perforated foil. The relevant equations are described in Appendix A.

2.2 Thermal mapping of artificial leaves

To evaluate the spatial temperature distribution of the artificial leaf surface and to assess how the temperature recorded by the embedded thermocouple may be seen as representative of average leaf temperature, we recorded infrared images of the leaf surface temperature. For this purpose, artificial leaves were placed in a conventional wind tunnel above a heat plate linked to a water bath, thereby providing constant background temperature, and infrared images of the leaf surfaces were obtained using a cryogenically cooled infrared (IR) camera (FLIR SC6000, Table C1) at different wind speeds. For these experiments, we did not use laser-perforated foils, but exposed the wet (or dry) filter paper directly to the IR camera, when the leaf was placed with the evaporating side upwards. We also reversed the leaf to detect any differences in surface temperature between the two leaf sides.

2.3 Leaf wind tunnel

For measurements of the water vapour and energy exchange of artificial leaves under fully controlled conditions, we designed a thermally insulated closed loop wind tunnel with a transparent leaf chamber, allowing control of gas and energy exchange with the surroundings (Fig. 2), as described below.

The main body of the wind tunnel was built of extruded polystyrene foam slabs (Sagex XPS-EN13164-T3-CS, Sager AG, Dürrenäsch, Switzerland[2]) with a low heat capacity ($1400\,\mathrm{J\,kg^{-1}\,K^{-1}}$ at a density of $30\,\mathrm{kg\,m^{-3}}$) and low thermal conductivity ($0.035\,\mathrm{W\,m^{-1}\,K^{-1}}$). The geometry and dimensions of the wind tunnel are given in Fig. B1. It is a closed loop tunnel with a rectangular inner cross section, which varied gradually between 5×3 cm in the leaf chamber and 5×5 cm on the opposite side, where a fan occupies the entire cross section (Fig. 2).

The frame of the transparent leaf chamber was produced by a 3-D printer, using transparent acrylic resin. The walls consist of three layers of 1 mm thickness, separated by 1 mm thick air gaps. On the inner walls of the chamber, a 1 cm thick layer of polystyrol foam was added for improved thermal insulation. At the top and bottom, the chamber was sealed with two layers of transparent PVC coated polypropylene (Propafilm®-C, ICI Americas Inc., Wilmington, DE, USA). The two layers of polypropylene foil were intended to permit the transmission of shortwave and longwave radiation while minimizing conductive heat transfer.

The net radiation of the leaf was measured using Peltier-based heat flux sensors of 1 cm by 1 cm size (gSKIN, Table C1), which were painted black and calibrated against a net radiometer (NR Lite2, Table C1) using a tungsten light source. The sensor response to irradiance varying between 0 and $700\,\mathrm{W\,m^{-2}}$ was linear ($R^2 > 0.99$) and the sensitivity ranged between 0.001 and 0.0013 mV per $\mathrm{W\,m^{-2}}$ net radiation. Three of these heat flux sensors were mounted on retractable wires such that they could be periodically positioned above, beside, and below the artificial leaf for radiation measurement, while being kept out of the chamber during equilibration. Their positions during a measurement were 1 cm above the leaf, 1 cm below the leaf, and one was posi-

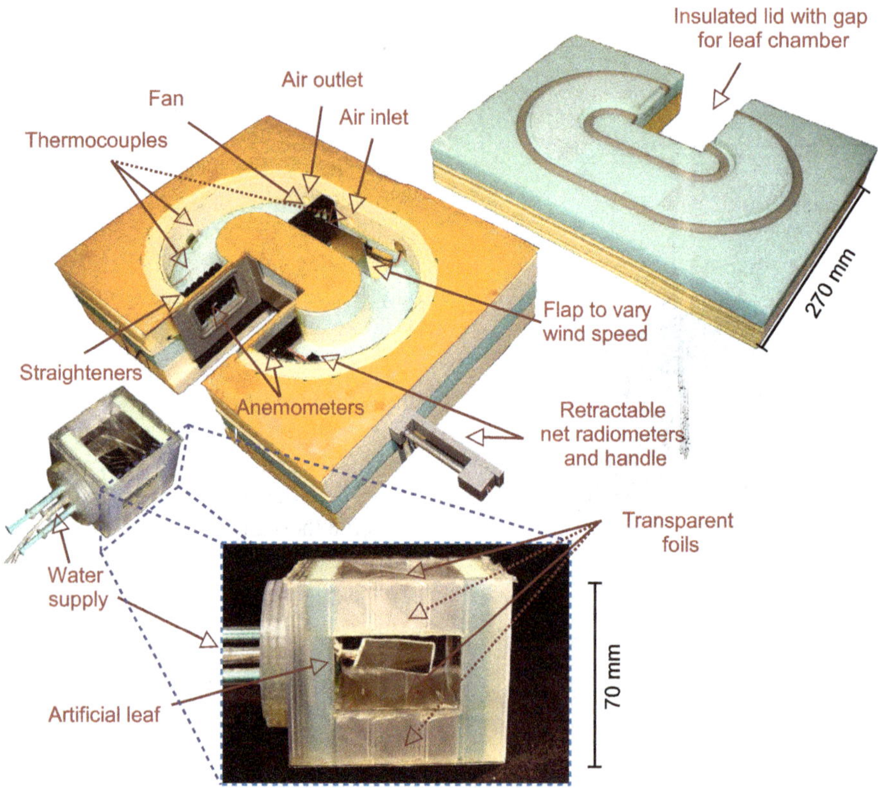

Figure 2. Insulated wind tunnel and leaf chamber. The wind tunnel is photographed with its insulated lid removed (top right). The leaf chamber (inset) fits tightly (as air tight as possible) into the empty slot of the wind tunnel on the left. The perspective through the leaf chamber is along the wind flow path (in the upwind direction), illustrating the smooth flow path of a 5 cm by 3 cm cross section. For detailed dimensions, see Fig. B1. Dashed arrows point to locations of features that cannot be seen in the pictures.

tioned at the same height as the leaf, but 0.5 cm downwind from the leaf (Fig. 3). See Appendix B3 for details on the use of these sensors in the calculations.

Temperature measurements were performed using T-type thermocouples (Table C1), which were placed (a) in the air stream upstream and downstream of the leaf chamber, (b) lightly inserted into the wind tunnel wall on the inside and the outside of the chamber, and (c) in the duct through which air was supplied to the wind tunnel by a humidifier. The air humidifier was a custom assembly by Cellkraft (Table C1) and provided an adjustable flow rate of up to 10 L min^{-1}, with adjustable air temperature and dew point. Air temperature was controlled by an external chiller (MRC300DH2-HT-DV, Laird Technologies, Cleveland OH, USA), supplying the humidifier with cooling liquid (water) between 4 and 40 °C.

Constant wind speed was generated using an axial fan of 5 cm by 5 cm diameter (MULTICOMP – MC35357, Table C1), which produced wind speeds of up to 5.4 m s^{-1} at a power consumption of less than 1.4 W in our wind tunnel, compared to a sensible heat exchange of up to 0.6 W by our 3 cm by 3 cm artificial leaves. A stack of 3 cm long plastic straws (each with a diameter of 7 mm) in the flow path acted as straighteners to reduce spiralling of the air flow caused by the rotating fan. The fan was placed inside the chamber, enabling direct control over the amount of heat injected by the fan into the wind tunnel, deduced from its rate of electrical power consumption. Power consumption by the fan was kept constant using a programmed power controller (NI USB-6008, Table C1), while wind speed was varied by adjusting the position of a flap in the flow path (Fig. 2) and monitored by a miniature thermal wind speed sensor (FS5 Flowmodule, Table C1), which was calibrated in the wind tunnel against a high accuracy air flow sensor (EE75, Table C1). The calibration produced a non-linear relationship between recorded sensor voltage and wind speed, ranging from 1350–1425 mV at 1.2 m s^{-1} wind speed to 1537–1630 mV at 4.4 m s^{-1} wind speed for different sensors. For each sensor, we fitted an exponential relationship between wind speed and voltage with an $R^2 > 0.99$, which had the tendency to over-estimate wind speed at values above 4.2 m s^{-1} and below 1.2 m s^{-1}. The wind speed sensors were only turned on briefly after each recording of chamber steady-state conditions to avoid contamination of the air temperature signal by the sensors' heat production.

All devices were connected to data loggers (CR 1000, Table C1), logged every second, and plotted on computer

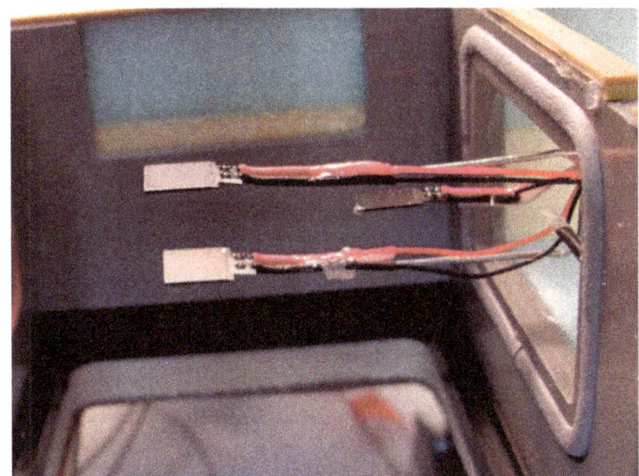

Figure 3. Arrangement of net radiometers in measuring position when the leaf chamber is removed. Two sensors are placed 2 cm apart vertically (1 cm above the leaf and 1 cm below the leaf), and one sensor at the level of the artificial leaf, slightly downwind from the leaf.

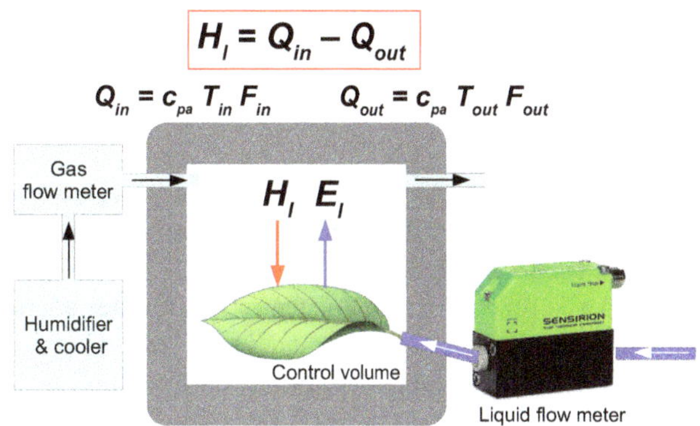

Figure 4. Simplified energy balance of the insulated wind tunnel. Latent heat flux (E_l) is calculated from the liquid flow rate into the leaf, and sensible heat flux (H_l) is calculated from the difference in heat content of incoming and outgoing air (c_{pa}: heat capacity of air; T_{in}, T_{out}: air temperatures of incoming and outgoing air; F_{in}, F_{out}: incoming and outgoing air flow rates). See Appendix B and Eqs. (B1)–(B7) for details.

screens. Steady states were identified visually by examining the graphs and data points were generated by averaging the values of each sensor over 10 s at steady state.

2.4 Calculation of sensible heat flux

The exchange of sensible heat between the artificial leaf and the air was calculated based on the energy balance of the entire wind tunnel, by the difference between the heat contained in the incoming and outgoing air (Fig. 4) and subtracting the heat added by the fan (see Appendix B for details on the thermodynamic calculations). The thermal insulation of the wind tunnel minimized uncontrolled heat exchange with the surroundings. To estimate the rate of conductive heat exchange per temperature difference between the air inside and outside the chamber, we considered the entire air–wall interfacial area at the inner side of the tunnel totalling 868 cm^2 and a minimum wall thickness of 5 cm (Fig. B1). This would result in a conductive heat transfer per Kelvin chamber-lab air temperature difference of roughly $0.035\,W\,m^{-1}\,K^{-1}/0.05\,m \times 0.0868\,m^2 = 0.061\,W\,K^{-1}$. Considering that a 3×3 cm large leaf exchanges 0.09 W heat with the chamber air per 100 $W\,m^{-2}$ sensible heat flux, a 1 K temperature difference between the chamber air and the lab air would roughly add a bias of 70 $W\,m^{-2}$ in our estimation of sensible heat flux. To reduce the impact of this potential bias, we regulated the air temperature within the wind tunnel to track the external air temperature in the lab to within ±0.1 K.

2.5 Leaf gas and energy exchange model

The transpiration rates and energy balance measurements were compared with model simulations based on a steady-state solution of the leaf energy balance, derived from general heat and mass transfer theory (Schymanski et al., 2013; Schymanski and Or, 2016). For comparison, similar simulations were performed using a simplified model of the leaf energy balance, as described in the appendix of Ball et al. (1988).

The models mentioned above assume strong thermal coupling between the surface temperatures on both sides of the leaf and therefore equal leaf temperatures on both sides at steady state, resulting in a single energy balance equation:

$$R_s = R_{ll} + H_l + E_l, \tag{1}$$

where R_s is absorbed shortwave radiation, R_{ll} is the net emitted longwave radiation, i.e. the emitted minus the absorbed, H_l is the sensible heat flux away from the leaf, and E_l is the latent heat flux away from the leaf, all in units of $W\,m^{-2}$.

The net longwave emission is represented by the difference between black-body radiation at leaf temperature (T_l, K) and that at the temperature of the surrounding objects (T_w, in our experiments equal to air temperature, T_a, K) (Monteith and Unsworth, 2007):

$$R_{ll} = 2\epsilon_l \sigma (T_l^4 - T_w^4), \tag{2}$$

where ϵ_l is the leaf's longwave emmissivity (≈ 1) and σ ($5.67 \times 10^{-8}\,W\,K^{-4}\,m^{-2}$) is the Stefan–Boltzmann constant. Sensible heat flux (H_l) is represented as

$$H_l = 2h_c(T_l - T_a), \tag{3}$$

where h_c ($W\,K^{-1}\,m^{-2}$) is the average one-sided convective heat transfer coefficient, determined primarily by leaf size and wind speed (Eqs. B10–B11 in Schymanski and Or, 2017).

The latent heat flux is a function of the transpiration rate ($E_{l,mol}$, mol m^{-2} s^{-1}):

$$E_l = E_{l,mol} M_w \lambda_E, \tag{4}$$

where M_w (kg mol^{-1} is the molar mass of water and λ_E (2.45×10^{-6} J kg^{-1}) is the latent heat of vaporization. In the model used here (Schymanski and Or, 2017), $E_{l,mol}$ is computed as a function of the concentration of water vapour within the leaf (C_{wl}, mol m^{-3}) and in the free air (C_{wa}, mol m^{-3}) (Incropera et al., 2006, Eq. 6.8):

$$E_{l,mol} = g_{tw}(C_{wl} - C_{wa}), \tag{5}$$

where g_{tw} (m s^{-1}) is the total leaf conductance for water vapour, dependent on diffusive (stomatal) (g_{sw}) and aerodynamic (leaf boundary layer) conductance (g_{bw}), expressed as follows:

$$g_{tw} = \frac{1}{\frac{1}{g_{sw}} + \frac{1}{g_{bw}}}. \tag{6}$$

Note that g_{sw} depends on sizes, shapes, and densities of stomata (Eqs. A1–A3), whereas g_{bw} is a function of leaf size and wind speed (Eqs. B2, B10, and B11 in Schymanski and Or, 2017).

As an alternative to Eq. (5), $E_{l,mol}$ is commonly expressed as a function of the vapour pressure difference between the free air (P_{wa}, Pa) and the leaf (P_{wl}, Pa), in which total conductance ($g_{tw,mol}$) is expressed in molar units (mol m^{-2} s^{-1}):

$$E_{l,mol} = g_{tw,mol} \frac{P_{wl} - P_{wa}}{P_a}. \tag{7}$$

Partitioning of $g_{tw,mol}$ into $g_{sw,mol}$ and $g_{bw,mol}$ is done similarly to Eq. (6). Under a few simplifying assumptions, conductance values can be converted between molar units and m s^{-2} in the following way (Schymanski and Or, 2017):

$$g_{sw} = g_{sw,mol} R_{mol} \frac{T_a}{P_a}. \tag{8}$$

R_{ll}, H_l, and E_l (through temperature dependence of C_{wl}) depend on leaf temperature (T_l) in such a way that for any environmental forcing (irradiance, humidity, temperature, and wind speed) and leaf properties (characteristic length scale, leaf diffusive conductance), a steady-state T_l can be found that satisfies the above energy balance equation (Eq. 1). For our artificial leaves, the characteristic length scale is 0.03 m, and the leaf diffusive conductance is deduced from foil thickness, as well as sizes and spacings of pores in the laser-perforated foils, as described in Appendix A.

The model is explained in detail in Schymanski and Or (2017), whereas all data, equations, and model code necessary to reproduce the results presented here can be accessed online at https://doi.org/10.5281/zenodo.241217.

2.5.1 Different temperatures on both sides of the leaf

Many leaf gas and energy exchange models assume a single leaf temperature (T_l) applicable for both sides of a leaf. This assumption is justified for very thin leaves or very high leaf thermal conductivities, whereas our infrared images pointed to significant temperature gradients between the wet and dry sides of an artificial leaf (see Sect. 3.1.2). Hays (1975) measured leaf thermal conductivities (k_l) over a range of leaves and found values in the range between 0.27 and 0.57 W m^{-1} K^{-1}, compared to the thermal conductivity of air at 0.026 and liquid water at 0.59 W m^{-1} K^{-1}. To estimate the potential error introduced by the assumption that both sides of the leaf have the same leaf temperature, we first expressed a conductive heat flux from the upper to lower side of the leaf (Q_l) as

$$Q_l = k_l \frac{T_{l_u} - T_{l_l}}{z_l} \tag{9}$$

where k_l (W m^{-1} K^{-1}) is the leaf thermal conductivity, T_{l_u} and T_{l_l} are the upper and lower sides of the leaf respectively, while z_l is the thickness of the leaf. We then formulated the energy balance equation for the upper and lower leaf sides separately, as

$$R_s = E_{l_u} + H_{l_u} + R_{ll_u} + Q_l \tag{10}$$

and

$$Q_l = E_{l_l} + H_{l_l} + R_{ll_l}, \tag{11}$$

where we assumed that only the upper side of the leaf absorbs shortwave radiation (R_s). Equations (10) and (11) are equivalent to Eq. (1), with the addition of Q_l (Eq. 9) and explicit distinction between the two sides, denoted by the subscript u for the upper and l for the lower side. Since our model is formulated for forced convection, we assume that the heat transfer coefficient (h_c) has the same value on both sides; hence, differences in the sensible heat flux are attributed to different leaf surface temperatures only. However, E_{l_u} and E_{l_l} can differ due to both different surface temperatures and different stomatal conductances. In the extreme case, e.g. for a hypostomatous leaf, $E_{lu} = 0$, while E_{ll} is calculated similarly to one-sided E_l with T_l replaced by T_{ll}. Instead of one equation with one unknown (Eq. 1 with unknown T_l), we now obtain two equations with two unknowns (Eqs. 10 and 11 with T_{l_u} and T_{l_l}), as all variables in these two equations are functions of only measured quantities as well as T_{l_l} and/or T_{l_u}. The equations were solved numerically using the SageMath open source software (SageMath, 2016), and the code is available online at https://doi.org/10.5281/zenodo.241217. In the results section, we compare measured fluxes and leaf temperatures with simulated values using both the uniform temperature model ("bulk") and the model based on different surface temperatures on both leaf sides ("2s").

3 Results

3.1 Characterization of artificial leaves

3.1.1 Pore properties and stomatal resistances

The laser-perforated aluminium foils have a shiny side and a matte side, and confocal laser scanning microscope (CLSM) images of the perforated foils (Fig. 5) were taken on the matte side prior to construction of the artificial leaves. The matte side was facing outwards after construction of the artificial leaves. More detailed analysis of pore geometries was performed on duplicate foils and suggests that the laser perforations were done from the shiny side, resulting in irregular surfaces around the pores and slightly conical pore geometries with smaller diameters on the matte side compared to the shiny side. Therefore, images taken on the matte side may result in under-estimation of the effective pore sizes. When we compared estimations of pore sizes and conductances based on images taken on either side of the aluminium foil, we found higher conductance values by up to 50 % if images were taken on the shiny side compared to the matte side (Fig. A1, Table A1). Detailed analysis of individual pore geometries also revealed that the average cross-sectional pore area over the typical 25 µm pore length could be up to 50 % larger than the areas measured 10 µm below the foil surface (Appendix A). To account for all these uncertainties and potential biases, Table 1 provides ranges of values deduced from at least three images on each side of a foil (Columns 1–4) and a column with stomatal conductance (g_{sw}) values resulting from the assumption that the average cross-sectional pore areas were 50 % larger than deduced from the images (Column 5). The last column in Table 1 represents g_{sw} values deduced from wind tunnel experiments described in Sect. 3.2, which were remarkably consistent with the theoretical values in Column 5.

The perforated foils were glued to the artificial leaf along the edges (Fig. 1, while they only adhered to the wet filter paper by capillary forces if a water film was present between the filter paper and the perforated foil. When carefully saturating an artificial leaf, we found that water could be held within the pores (Fig. A3), which would result in a dramatic shortening of the diffusive path length across the pores, from 25 µm (foil thickness) to less then 1 µm if there were no considerable head loss along the flow path, as the water reservoir was kept only a few centimetres below the leaf to reduce the risk of embolism. To get an appreciation for the maximum effect of capillary rise within the pores on stomatal resistance, we derived stomatal conductance values for the different perforated foils based on both 25 and 0 µm pore length, and found that a reduction of the pore length from 25 to 0 µm could result in a 3-fold increase in estimated stomatal conductance (data not shown). However, as presented in Table 1, the stomatal conductance values deduced from wind tunnel experiments are more consistent with values determined under the assumption that no water was held in the pores.

3.1.2 Leaf thermal mapping

We placed a thick artificial leaf (0.2 mm thick filter paper with a 0.2 mm thick thermocouple fed in from the side) and a thin artificial leaf (0.1 mm thick membrane filter and a 0.13 mm thick thermocouple) under the thermal IR camera and took images of their surface temperatures. Temperatures increased from the leading edge downwind by no more than 1.1 K (Fig. 6). Interestingly, surface temperatures of the wet surfaces were lower wherever it was detached from the underlying aluminium tape, e.g. along the thermocouple wires and air pockets (Figs. 6 and 7). We also found that the surface temperature of the dry side of the leaf was warmer by up to 1.4 K compared to the wet side (Fig. 7). Please refer to the discussion section (Sect. 4.3) for the relevance of these findings.

3.2 Leaf wind tunnel experiments

Experiments were performed using artificial leaves with different perforation densities under varying air humidity or varying wind speed, with and without shortwave radiation. In addition to the artificial leaves with pore densities given in Table 1, we also used artificial leaves without a perforated foil, i.e. with a wet surface on the lower side, producing non-restricted one-sided leaf boundary layer transfer of water vapour. For simulated energy balance components and leaf temperatures, we chose diffusive conductance values (g_{sw}) that best matched the observed transpiration rates and then compared these with conductance values computed from laser perforation analysis (Table 1). Simulations were performed using the original model assuming equal leaf temperatures on both sides of the leaf (bulk) and the two-sided leaf temperature model (2s). We also adjusted leaf thermal conductivities (k_l) within the range between air and water, to best reproduce measured leaf temperatures in the 2s model. Figures 8 and D1–D2 represent experiments in the absence of shortwave irradiance, and their five panels include (from top to bottom) latent and sensible heat flux, sums of latent and sensible heat flux along with net absorbed radiation, leaf–air temperature difference, leaf conductance to water vapour, and the convective heat transfer coefficient. For the latter two, observed values were deduced from observed fluxes and leaf–air temperature and vapour concentration gradients. Figure 9 represents experiments under irradiance. Since sensible heat flux could not be measured accurately under irradiance (see discussion, Sect. 4.1), we left out the bottom panel. As seen in Fig. 9a, the over-estimation of observed sensible heat flux (H_l, empty circles) was likely of the order of 500 W m^{-2}, which also led to a mismatch in the energy balance by a similar amount (second panel from the top).

Table 1. Perforation characteristics and resulting diffusive conductances (g_{sw}, from Eq. A3), either taken the original pore area deduced from CLSM images or 50% increased pore area, taking into account conical shapes of pores. In the last column, we provide effective stomatal conductances deduced from wind tunnel experiments.

Pore density mm^{-2}	Pore area μm^{-2}	Pore radius μm	g_{sw} $m\,s^{-1}$	g_{sw} (1.5 × area) $m\,s^{-1}$	g_{sw} (wind tunnel) $m\,s^{-1}$
52–68.8	859–1240	16–20	0.032–0.052	0.046–0.076	0.05
27.3–38.2	710–1572	15–22	0.015–0.032	0.022–0.046	0.035–0.042
7.1–7.8	890–1886	16–24	0.004–0.009	0.006–0.012	0.007–0.009

g_{sw}: diffusive (stomatal) conductance for water vapour; d_p: pore depth.

In the absence of shortwave radiation, both sensible and latent heat fluxes were very consistent between observations and model simulations (top panels in plots), no matter whether vapour pressure or wind speed was varied. The sums of observed latent and sensible heat flux varied between 20 and 120 W m^{-2} and were largely consistent with simulated radiative exchange of the leaf in half of the cases, while exceeding the simulated exchange of radiative energy in the other half of the experiments. Our net radiation sensors were not able to confirm such high radiative energy exchange rates, and generally under-estimated the net absorbed longwave radiation by more than half, compared to simulations (red dots in Figs. 8b, 9, and D1b). The observed leaf temperatures were also generally under-estimated by the bulk model (absolute leaf–air temperature difference was over-estimated by 0.5–1 K). However, when solving the leaf energy balance for each leaf side separately (considering conductive heat transport across the leaf towards the transpiring side, Sect. 2.5.1), the observed leaf temperatures were consistent with the simulated leaf temperatures of the non-transpiring side of the leaf (Figs. 8, 9a, D1, and D2). Interestingly, solving for the temperature gradient between the two sides of the leaf did not have much effect on the simulated heat fluxes (top panels in the plots). Note that the values of k_l, chosen to reproduce observed leaf temperatures, varied between experiments, between 0.03 and 0.3 W K^{-1} m^{-1}, compared to values of 0.026 for air and 0.59 for water.

Irradiation of the leaf with 370–550 W m^{-2} shortwave radiation resulted in large over-estimation of sensible heat flux in the observations and hence unrealistically high sums of latent and sensible heat fluxes, while simulated and latent heat fluxes reproduced the observed very accurately (top panels in Fig. 9). Observed leaf temperatures were still higher than the simulated ones, while the observed radiative exchange was relatively close to simulated $R_s - R_{ll}$ (second panels in Fig. 9a and b). Note that the radiation sensor placed downwind of the leaf (Fig. 3) did not produce reliable radiation values, as the readings were affected by leaf temperature (data not shown). To compute $R_{nleaf} = R_s - R_{ll}$, we hence subtracted the reading of the sensor placed below the leaf from the reading of the sensor placed above the leaf.

4 Discussion

4.1 Measurement of leaf energy balance components

Our experimental set-up enables independent measurement of leaf-scale exchange of latent and sensible heat in the absence of shortwave irradiance. This was confirmed on a variety of artificial leaves with different diffusive conductances, under varying vapour pressure and wind speed. Energy balance closure in the absence of light was generally within 60 W m^{-2} s^{-1}, as illustrated in the net energy exchange panels in Figs. 8, D1, and D2.

Under shortwave irradiance, however, sensible heat flux deduced from measurements was largely over-estimated, probably due to absorption of radiation by surfaces within the wind tunnel, despite coating with reflective tape and a second transparent window below the leaf. It is also important to note that, despite construction of the wind tunnel using thermally insulating materials, the internal air temperature had to be kept close to lab air temperature, in order to prevent conductive heat exchange across the chamber walls. We found that a temperature difference of only 2 K between the air within the wind tunnel and outside could result in a bias in estimated sensible heat flux by 300 W m^{-2} in our experimental set-up (data not shown).

The exchange of longwave radiation between the leaf and the surroundings was not captured in a consistent way by our experimental set-up, suggesting that the measurements systematically under-estimated longwave radiation away from the leaf by more than 50 %. Consideration of the viewing angle of the net radiation sensor would only correct the estimates by 20 % (Sect. B3). The reason for the bias is most likely that the miniature radiation sensors were calibrated against an industrial net radiometer using shortwave radiation as the main energy source, while their capability to absorb in the longwave range was not tested. In the presence of shortwave radiation, the sensors were adequate to characterize the radiative load on the leaf, as illustrated by the correct simulation of latent heat flux in Fig. 9. Note, however, that the sensor placed at leaf level, but slightly downwind from the leaf, did not return reliable values, as it was affected by leaf temperature (data not shown). This is likely due to ver-

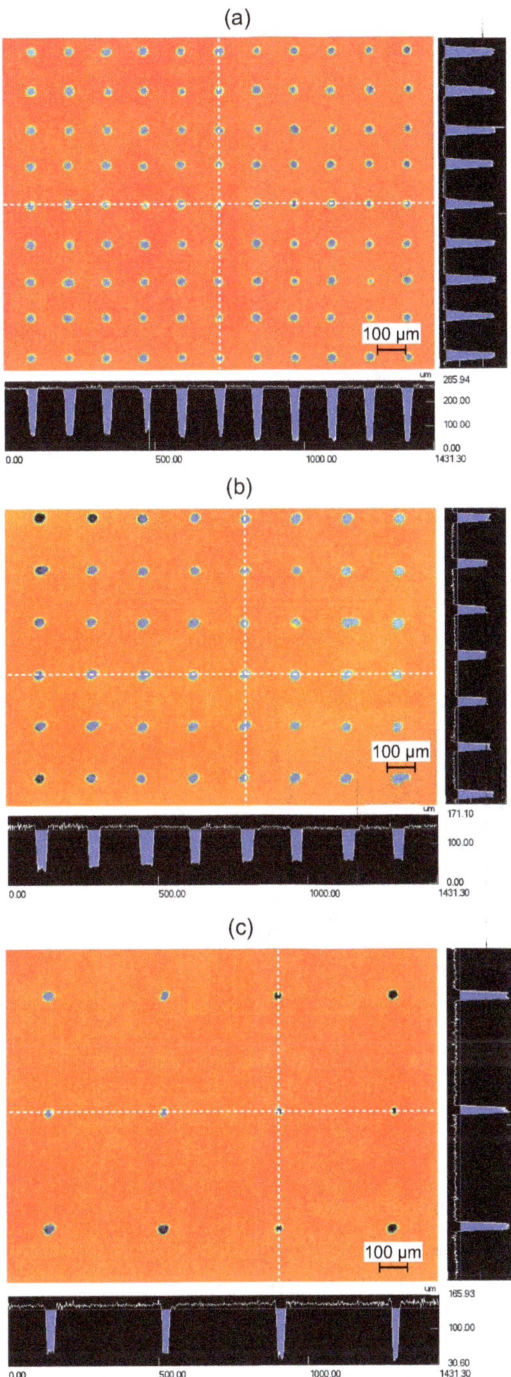

Figure 5. Example confocal laser scanning microscope (CLSM) images of perforated foils summarized in Table 1. **(a)** 64 perforations per mm^2, **(b)** 35 perforations per mm^2, **(c)** 7.8 perforations per mm^2. Black bars at the bottom and on the right of each picture show topographic profiles of transects crossing perforations (white dashed lines in the main images), with the detection thresholds marked as blue-filled areas.

tical temperature gradients in the aerodynamic wake of the leaf, which could result in compensatory heat flux through the sensor in addition to that caused by absorbed radiation. The sensors 1 cm above and below the leaf surface were unaffected by the leaf boundary layer and produced a signal that was weakly affected by the leaf temperature, consistent with the leaf temperature effect on the net emission of longwave radiation.

4.2 Utility of artificial leaves

Despite many inherent limitations in mimicking real leaves, the artificial leaves proved very useful for analysis of steady-state leaf energy balance components under constant leaf properties, in particular leaf diffusive conductance. This is supported by the reproduction of experimental results using a model with constant stomatal conductance (g_{sw}), both for varying vapour pressure and wind speed. Water supply to the evaporating sites via a flow-monitored tube and porous filter paper resulted in relatively homogeneous conditions over the leaf surfaces, as evidenced by our infrared images (Figs. 6 and 7). However, in some cases, water transport to the edges of the artificial leaves ceased over time, which could be seen at the end of the experiment as dry patches on the filter paper. The bottom left corner of the leaf in Fig. 6 shows the onset of such an effect. To detect this effect in artificial leaves where the filter paper was covered by a laser-perforated foil, experiments were run in two directions, first increasing wind or vapour pressure and then decreasing it again. Whenever we found a clear reduction in transpiration at the end of an experiment compared to the start, we discarded the whole data set. We kept the free water surface of the water supply tank only a few centimetres below the position of the leaf in order to facilitate water transport all the way to the edges of the filter paper and to avoid hydraulic failure.

One of our aims was to produce evaporating surfaces with a diffusive conductance (g_{sw}) that is known a priori. However, the uncertainty in the computation of g_{sw} based on CLSM topographical images of our laser-perforated foils was substantial (Tables 1 and A1). Roughly 25 % uncertainty was introduced by irregularities in pore sizes, resulting in different average pore areas in different images of the same foil (see the ranges in the fourth column of Table 1). An additional potential bias by 30 % was caused by the conical shape of the pores. As shown in Fig. A2, the mean diameter of a pore across the whole pore length was likely 25 % larger than the diameter measured at a 10 μm distance from the surface. This would result in 50 % larger mean cross-sectional areas in the pores and 30–50 % larger values for g_{sw}, as shown in the fifth column of Table 1.

We also found that a substantial uncertainty could arise from lack of knowledge about the position of the evaporating sites if they were within the perforations. As illustrated in

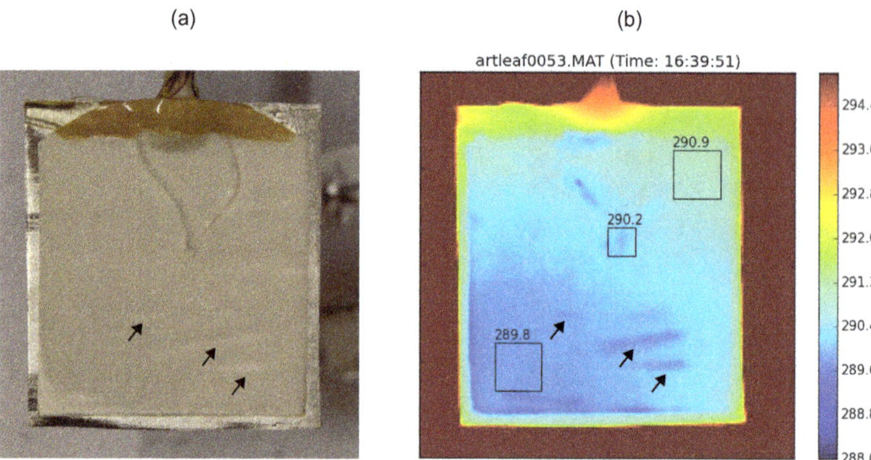

Figure 6. Wet side of the thin artificial leaf at $3\,\mathrm{m\,s^{-1}}$ wind speed. **(a)** Photographic image; **(b)** infrared temperature map, with average temperatures in different sub-areas. Arrows indicate air pockets between the wet filter paper and the underlying aluminium tape. Wind direction is from the bottom to top of the images. The colour bar indicates temperatures in K.

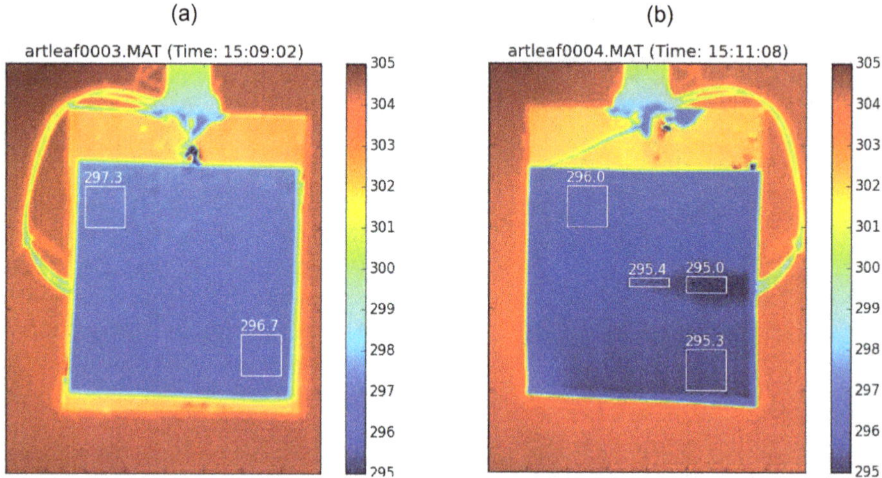

Figure 7. Infrared temperature maps of the thick artificial leaf in the absence of wind. **(a)** dry side, **(b)** wet side. The colour bar indicates temperatures in K, while white labels above the white rectangles indicate average temperature (in K) over the respective rectangles. Note that the wet side is significantly cooler than the dry side of the leaf.

Fig. A3, water could have entered the pores and been drawn up to the surface, reducing the diffusion distance through the pores. In the extreme case, this could have led to a 3-fold increase in g_{sw} compared to assuming that the evaporating sites are below the pores. However, although this was observed after careful saturation of an artificial leaf under the CLSM, the g_{sw} values deduced from wind tunnel measurements (last column in Table 1) were consistent with those calculated under the assumption that the evaporating sites were below the pores. Since emptying of previously water-filled pores is irreversible as long as the hydraulic head in the artificial leaves is negative, we expect most pores to be air-filled in our experiments.

4.3 Leaf temperatures

Despite the thin leaf design, our data suggest that significant temperature gradients can occur between the dry and evaporating sides of the leaf. This was confirmed directly by the infrared images of the wet and dry leaf surfaces (Fig. 7) and by the consistent bias in leaf temperature simulated by the bulk model compared to observed leaf temperatures. Note that the dry leaf surface was black painted aluminium tape, whereas the wet side was paper, which could result in emissivity differences and hence bias in the surface temperature differences deduced from infrared imaging. However, given that the infrared emissivity of a wet surface is close to 1, while that of untreated aluminium is very low, we would expect any bias caused by emissivity differences to result in under-

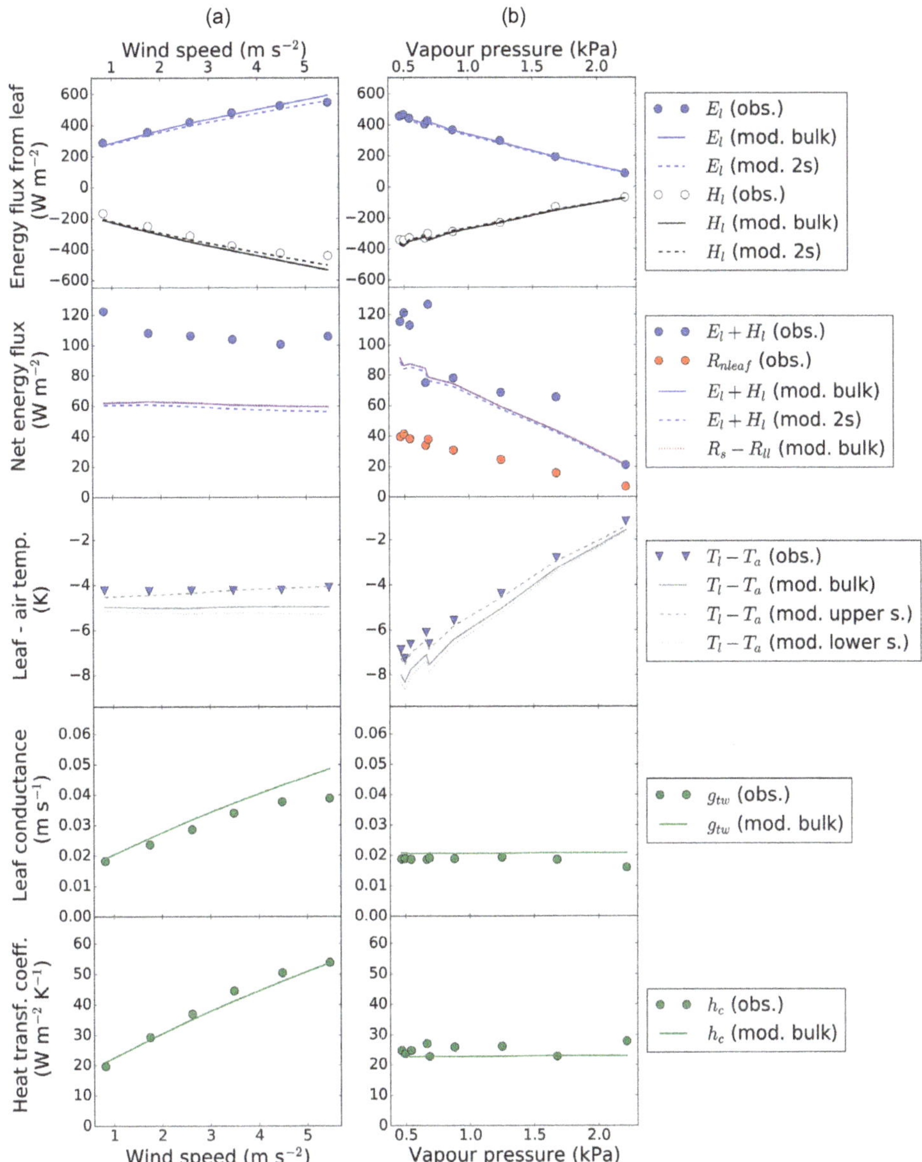

Figure 8. Artificial leaf with wet surface on the lower side (no stomatal resistance), under **(a)** varying wind speed and **(b)** varying vapour pressure. Numerical model results (lines) based on the same boundary conditions as observations (symbols). Red dashed lines indicate conditions in the plots where the forcing was roughly equivalent between Panels **(a)** and **(b)**. E_l: latent heat flux; H_l: sensible heat flux; $R_s - R_{ll}$: absorbed net radiation; $T_l - T_a$: leaf–air temperature difference; g_{tw}: total leaf conductance to water vapour; h_c: convective heat transfer coefficient; "mod. bulk": bulk leaf temperature model; "mod. 2s": model based on different leaf temperatures on both the "upper" and "lower" leaf sides. Boundary conditions: $g_{sw} = 999\,m\,s^{-1}$; $R_s = 0$; $T_a = 295.4$–295.6 K **(a)** and 295.4–296.6 **(b)**; $P_{wa} = 1200$–1342 Pa **(a)**; $v_w = 1.0\,m\,s^{-1}$ **(b)**; $k_l = 0.1\,W\,K^{-1}\,m^{-1}$; $z_l = 0.5$ mm.

estimation of the dry surface temperature, meaning that the temperature difference could even be higher than the 1.4 K we found.

The level of de-coupling between the wet and dry sides of the leaf depends on the leaf thermal conductivity, as illustrated by the cooler surface temperatures wherever little air intrusions between the wet filter paper and dry aluminium tape occurred (Fig. 6). Since the thermocouples within the leaf were in contact with the upper aluminium tape of the leaf, they most likely are strongly coupled to the surface temperature of the dry side of the leaf. This may explain the persistent positive bias of the thermocouple reading compared to model simulations, even if the model simulations reproduced the leaf energy balance components very well. However, the magnitude of the bias is not always explained by decoupling between the two leaf sides, as the corresponding leaf thermal conductance would have to be near that of air in some simulations. More targeted experiments are needed to rule out

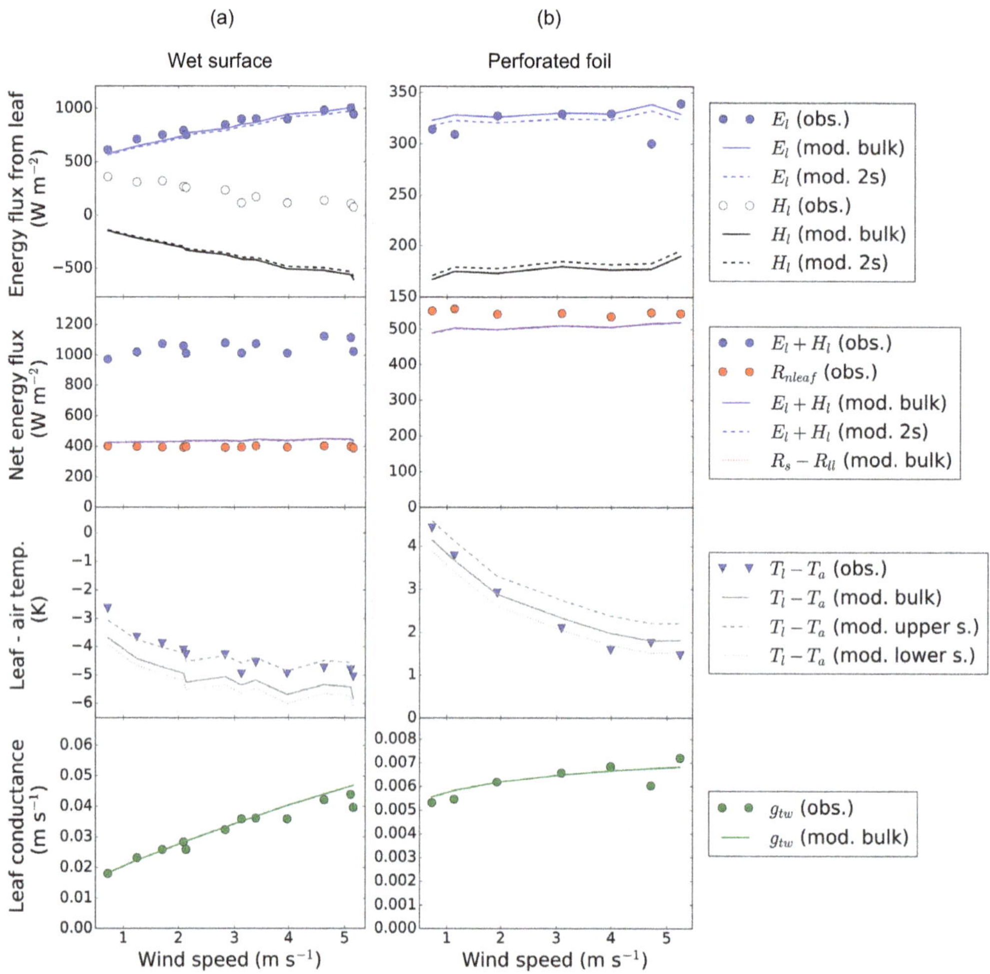

Figure 9. Artificial leaves with **(a)** the wet surface on the lower side and **(b)** 7 perforations mm^{-2}. Numerical model results (lines) based on the same boundary conditions as observations (symbols). E_l: latent heat flux; H_l: sensible heat flux; $R_\text{s} - R_\text{ll}$: absorbed net radiation; $T_\text{l} - T_\text{a}$: leaf–air temperature difference; g_tw: total leaf conductance to water vapour; h_c: convective heat transfer coefficient; "mod. bulk": bulk leaf temperature model; "mod. 2s": model based on different leaf temperatures on both the "upper" and "lower" leaf sides. Boundary conditions: $g_\text{sw} = 999$ **(a)** and 0.008 **(b)** m s^{-1}; $R_\text{s} = 370$–$380\ \text{W m}^{-2}$ **(a)** and 530–550 W m^{-2} **(b)**; $T_\text{a} = 295.8$–296.6 K **(a)** and 295.4–296.6 K **(b)**; $P_\text{wa} = 521$–802 Pa **(a)** and 341–356 Pa **(b)**; $k_\text{l} = 0.27\ \text{W K}^{-1}\ \text{m}^{-1}$ **(a)** and 0.3 **(b)** $\text{W K}^{-1}\ \text{m}^{-1}$; $z_\text{l} = 0.5$ mm.

a systematic error in the representation of the sensitivity of transpiration to leaf temperature.

It is remarkable that consideration of conductive heat flux through the leaf in the 2s model has a strong effect on simulated leaf temperatures, but not so much on the simulated latent and sensible heat fluxes, compared to the bulk leaf temperature model. This is likely because the simulated bulk leaf temperature is between the temperatures of the transpiring and dry side in our hypostomatous leaf replica. Underestimation of sensible heat flux on one side is hence partly compensated for by over-estimation of sensible heat flux on the other side. Consideration of conductive heat flux through the leaf increases leaf temperature on the dry side because of decoupling from evaporative cooling on the wet side. On the wet side, the decoupling reduces heat input from the dry side, but if the leaf is colder than the air, this is partly compensated for by increased uptake of sensible heat as the wet side of the leaf cools. This has the result that the leaf temperature on the wet side is closer to the simulated bulk leaf temperature than the leaf temperature on the dry side, resulting in little difference in latent heat flux between the bulk and the 2s model.

Given that significant temperature differences between the dry and evaporating sides of the leaf were simulated even for leaf thermal conductance values similar to natural leaves (up to 1 K in Fig. 9), we conclude that care must be taken when inferring leaf-internal vapour pressure from leaf temperature measurements. However, for the simulation of latent and sensible heat fluxes, a bulk formulation seems adequate due to the compensating effects of under-estimating leaf temperature on one side and over-estimating on the other.

4.4 Leaf conductances deduced from experiments

By inverting Eqs. (3) and (5), we computed the effective one-sided convective heat transfer coefficients (h_c) and total leaf conductances to water vapour ($g_{tw,mol}$) based on observed latent and sensible heat fluxes, air and leaf temperatures, and air vapour pressure. These are reproduced in Panels (d) and (e) of the results plots and compared with theoretical values based on wind speed and leaf properties. Both theoretical and inferred values were reasonably close to each other and responded to wind speed in a consistent way. However, the theoretical values for g_{tw} were consistently higher than the deduced values, while the theoretical values for h_c were consistently lower than the deduced ones. This can be explained in the context of biased leaf temperature measurements.

Observed and simulated leaf temperatures generally increased with increasing wind speed in the absence of shortwave radiation, except for artificial leaves without perforated foils (i.e. lower wet surface exposed to air), where leaf temperature did not change with wind speed. Consistent with the higher observed leaf temperatures relative to simulations, leaf conductance to water vapour, calculated from observed leaf temperature, vapour pressure, and latent heat flux, was generally lower than the simulated conductances. For the same reason, convective heat transfer coefficients computed from observed sensible heat flux and leaf temperature were generally higher than simulated values. This is because higher leaf temperature implies an increased gradient for latent heat flux and a reduced gradient for sensible heat flux when leaf temperatures are below ambient. Note that this feedback has recently also been shown to result in reduced transpiration and/or increased leaf water use efficiency with increasing wind speed when irradiated leaves are warmer than ambient air (Schymanski and Or, 2016).

4.5 Potential for new insights and limitations

Independent measurement of sensible and latent heat fluxes from artificial leaves with fixed stomatal conductance offers various opportunities to test our understanding of the physics of leaf gas and energy exchange. In addition to the discovery of surprisingly strong temperature gradients between the two sides of a hypostomatous leaf (this study), previous experiments using the same set-up have already led to the discovery of inconsistencies in the widely used Penman–Monteith equation for transpiration, mainly resulting from the neglect of two-sided sensible heat exchange by planar leaves (Schymanski and Or, 2017). The experimental set-up presented here could also be used for detailed studies of the role of stomata sizes, shapes and arrangements for leaf gas and energy exchange, as well as isotope partitioning. Due to the explicit control volume approach with thermal insulation, the set-up could also be used to study physical components of surface–atmosphere feedbacks at the laboratory scale.

However, the estimation of sensible heat flux from the chamber energy balance requires steady-state conditions, where heat exchange between the chamber air and the chamber walls is negligible. In our experience, it takes tens of minutes to hours before a steady state is achieved after an experimental change in the boundary conditions. This is due to the low heat capacity of the chamber air compared to that of the wind tunnel walls and any equipment placed within the wind tunnel, and makes the set-up of limited use for living leaves that vary their stomatal conductance at timescales shorter than the characteristic timescale of the whole chamber. Furthermore, for the estimation of sensible heat flux from an irradiated leaf, it would be necessary to focus a light beam on the leaf surface only and avoid any absorption of stray light by other surfaces inside the wind tunnel. And finally, in order to close the energy balance, it would be desirable to have reliable measurements of the leaf's radiative exchange, including longwave radiation. It may be necessary to develop new sensors that are small enough to fit into the wind tunnel without modifying air flow and the energy balance of the wind tunnel.

5 Conclusions

The experimental set-up presented here allows for the first-time simultaneous and independent measurement of gas and energy exchange by artificial leaves under fully controlled conditions. In the absence of shortwave irradiance, the results from the experimental set-up were remarkably consistent with theoretical predictions for latent and sensible heat fluxes. The experiments presented here also highlight some unexpected difficulties in characterizing leaf temperature due to strong temperature gradients between dry and evaporating leaf sides of planar leaves. More development is needed to achieve reliable measurement of the leaf's radiative energy exchange and sensible heat flux of an irradiated leaf. Preliminary experiments with irradiated leaves suggest that issues related to light absorption by internal wind tunnel surfaces need to be resolved. Additionally, we plan experiments using live leaves to test the energy balance under a range of boundary conditions and effects of leaf shape and surface properties on the partitioning between sensible and latent heat flux. This is not possible using common state-of-the-art leaf gas exchange systems, as they do not permit characterization of the radiative energy load (net radiation) and exchange of sensible heat by the leaf.

Code and data availability. All code and data used to generate the results presented in this paper are available online at https://doi.org/10.5281/zenodo.241217 (Schymanski, 2017).

Appendix A: Calculation of diffusive leaf conductance from pore dimensions

Diffusive conductance (g_{sw}) for the perforated foils was computed based on pore sizes and densities, following the derivations by Bange (1953), as summarized by Lehmann and Or (2015). We assumed that the stomatal conductance results from two resistances in series, the throat resistance (r_{sp}), dependent on the areas of the pores and the thickness of the perforated foil (d_p), and the vapour shell resistance (r_{vs}), dependent on the size and spacing of the stomata, which can be understood as the resistance related to distribution of the point source water vapour over the entire one-sided leaf boundary layer. We hereby neglect any internal resistance (termed "end correction" by Lehmann and Or, 2015), as we assume that the wet filter paper has direct contact with the perforated foil. The throat resistance (r_{sp}, $m^2\,s\,mol^{-1}$) was computed as (Eq. 1 in Lehmann and Or, 2015)

$$r_{sp} = \frac{d_p}{A_p k_{dv} n_p}, \tag{A1}$$

where k_{dv} is the ratio of the vapour diffusion coefficient and the molar volume of air (D_{va}/V_m), A_p is the average pore area, and n_p is the number of pores per surface area. For circular pores the pore area is a geometric function of pore radius (r_p): $A_p = \pi r_p^2$. For the vapour shell resistance (r_{vs}, $m^2\,s\,mol^{-1}$), we use the formulation originally proposed by Bange (1953):

$$r_{vs} = \left(\frac{1}{4r_p} - \frac{1}{\pi s_p}\right)\frac{1}{k_{dv} n_p}, \tag{A2}$$

where s_p (m) is the spacing between stomata, inferred from the images as $s_p = 1/\sqrt{n_p}$. Stomatal conductance ($g_{sw,mol}$, $mol\,m^{-2}\,s^{-1}$) was then calculated as

$$g_{sw,mol} = 1/(r_{sp} + r_{vs}) \tag{A3}$$

and conversion to units of g_{sw} ($m\,s^{-1}$) was done following Schymanski and Or (2017), i.e. $g_{sw} = g_{sw,mol} R_{mol} T_a / P_a$.

At least three confocal laser scanning microscopy (CLSM) images of each perforated foil were examined and perforations were identified based on the 3-D topography, using an elevation threshold of 10 µm below the median to define a pore (Figs. 5, A2). Average pore area (A_p, m^2) and number of pores per surface area (n_p, m^{-2}) were computed for each image using the proprietary Keyence VK Analyzer software, version 3.3.0.0. Pore radius (r_p, m) was deduced from average pore area assuming circular pores, while pore spacing (s_p, m) was computed based on the assumption of regular pore spacing as $s_p = 1/\sqrt{n_p}$. The pore depth (d_p, m) was assumed to be the same as the foil thickness of 25 µm. To assess the effect of water films entering the pores (Fig. A3), calculations were made for both $d_p = 25$ and $d_p = 0$ µm. Note that the foils had a shiny and a matte side and laser cutting of the pores was performed from the shiny side, whereas the foils were mounted on the artificial leaves with the matte side exposed to the air. Due to the procedure of laser cutting, the surface surrounding the pores was not smooth on the shiny side, but featured ridges and a larger indent around the pores. This could result in significantly greater pore sizes when the same image analysis was done on images taken on the shiny side (Fig. A1 and Table A1). To assess how representative the pore cross-sectional areas 10 µm away from the surface were of the average pore cross-sectional areas, we chose a single pore in the same foil as in Fig. A1, scanned it from both sides at a higher magnification and aligned the pore profiles in a way to represent a cross section along the entire pore length (Fig. A2). We found that pore sizes determined at 10 µm depth represented the sizes of the throats of the pores and might under-estimate the mean cross-sectional pore areas when averaged along the pore axis by up to 50 %. Therefore, we computed values of g_{sw} based on the original pore areas determined using the procedure described above (Columns 1–4 in Table 1) and additionally did the same calculations assuming 50 % larger pores (Column 5 in Table 1). The latter were most consistent with conductance values deduced from leaf wind tunnel experiments (last column in Table 1).

Table A1. Perforation characteristics and resulting stomatal conductances for a foil with 7.8 pores mm^{-2}, scanned three times on each side (see example scans in Figs. 5 and A1). For each image, an average pore size and pore density were computed, which were then used to compute stomatal conductance (g_{sw}, assuming that pore length (d_p) equals a foil thickness of 25 µm). Ranges given in the table represent the ranges obtained from three images on each side.

	Pore area μm^{-2}	Pore radius µm	g_{sw} ($d_p = 25 \mu m$) $m\,s^{-1}$
Matte side	890–1231	17–20	0.006–0.008
Shiny side	1376–1886	21–24	0.009–0.011

g_{sw}: diffusive conductance from Eq. (A3); d_p: pore depth.

Appendix B: Details on leaf wind tunnel and computations

The leaf wind tunnel is described in the main text (for detailed dimensions, see Fig. B1), while here we describe the calculations to deduce the quantities of interest from measured quantities.

B1 Inference of sensible heat flux from wind tunnel measurements

Sensible heat exchange between the air and the leaf was inferred from steady-state chamber heat balance, based on the following assumptions.

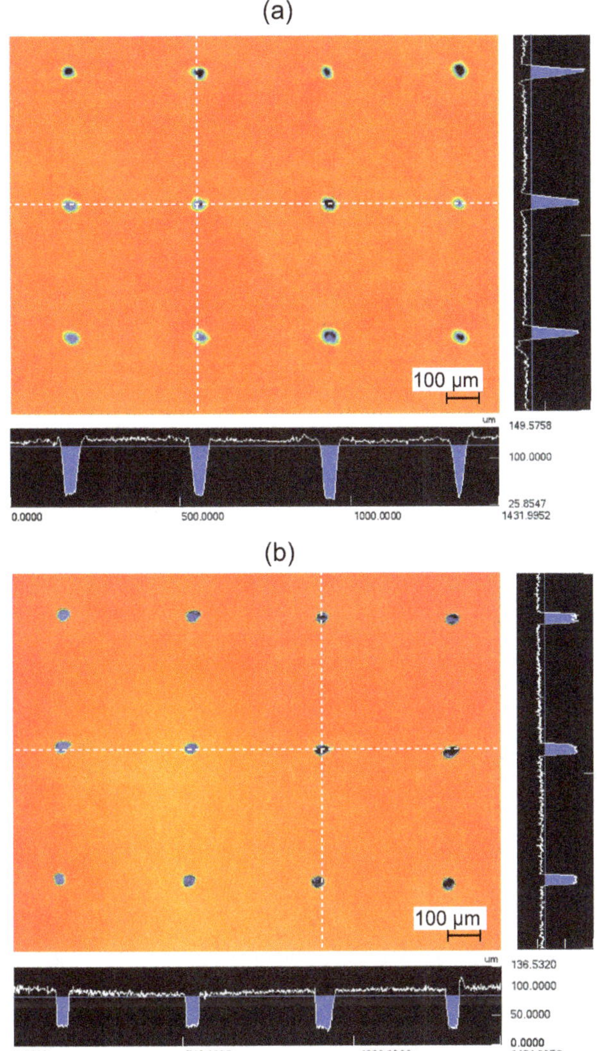

Figure A1. Perforations of the same foil (7 pores mm^2), scanned on the shiny side **(a)** and on the matte side **(b)**. Colours represent surface elevation as shown in the profile below and to the right of each panel. Ranges of pore sizes and densities are given in Table A1.

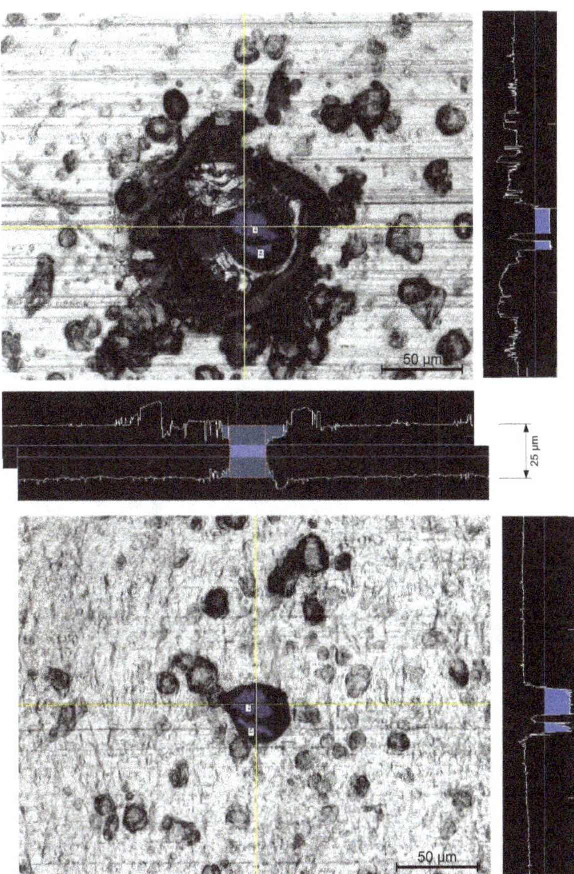

Figure A2. Single pore measured from both directions (top: shiny side; bottom: matte side) for the same foil as in Fig. A1. Black bars illustrate measured height profiles across the images, corresponding to the transects indicated by yellow lines. The horizontal profiles are shown in the middle, with the profile belonging to the bottom image mirrored and aligned to produce an estimated cross section of the foil and the pore, in combination with the height profile of the upper image. The red dashed box represents the hypothetical pore profile deduced from pore diameter 10 µm below the surface, whereas the light blue shaded area represents the detailed profile of the pore. The shaded area is > 20 % larger than the area of the red box, indicating that the average pore diameter is > 20 % larger and the average cross-sectional area 40–50 % larger than that deduced from the lower image alone.

1. Heat conduction through the wind tunnel walls is negligible.

2. Heat input by a fan is equal to its electric power consumption.

3. The molar outflow of dry air equals the molar inflow of dry air, while the molar outflow of water vapour equals the molar inflow plus the evaporation rate.

4. The heat content of the incoming air can be calculated from its flow rate, humidity, and temperature, measured inside the duct through the wind tunnel wall.

5. The heat content of the outgoing air can be calculated from its flow rate, average humidity, and temperature.

Based on these assumptions, the energy balance of the wind tunnel is written as

$$\begin{aligned} 0 = {} & L_A H_l + Q_{in} \\ & + \left(F_{in,mol,a} M_{air} c_{pa} + F_{in,mol,w} M_w c_{pv}\right) T_{in} \\ & - \left(F_{out,mol,a} M_{air} c_{pa} + F_{out,mol,w} M_w c_{pv}\right) T_{out} \end{aligned} \tag{B1}$$

where L_A is the leaf area (m^2), Q_{in} is the heat input by the fan (W), $F_{in,mol}$ and $F_{out,mol}$ ($mol\,s^{-1}$) refer to the incoming and outgoing molar flow rates of dry air ("a" in subscript) or water vapour ("v" in subscript), c_{pv} ($J\,kg^{-1}\,K^{-1}$) is the constant-pressure heat capacity of water vapour, T_{in} refers

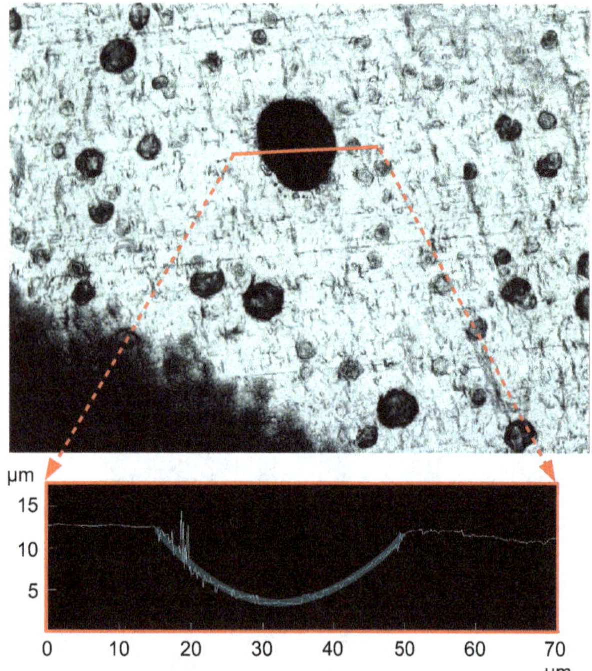

Figure A3. Profile of a water meniscus in an artificial pore at −24.4 cm hydraulic head. A leaf with a pore diameter of 35 µm and very low pore density (1.8 pores mm^{-2}) was used to observe the position of the water meniscus under suction in a single pore as the hydraulic head was varied by progressively lowering the water supply reservoir below the level of the artificial leaf. Top: laser scanning image of the pore with a transect marked across the pore. Note that the smaller black patches are not pores, but surface dents on the surface of the aluminium foil. Bottom: magnified height profile of the section depicted in the top figure, depicting the surface of the water meniscus inside the pore. For clarity, the supposed water surface is enhanced by a thicker semi-transparent blue line.

to the temperature of the incoming air (K), and T_{out} refers to the temperature of the outgoing air (K). At steady state, $F_{out,mol,a} = F_{in,mol,a}$ and

$$F_{out,mol,w} = F_{in,mol,w} + L_A E_{lmol}, \tag{B2}$$

so we can solve the above for H_l as

$$H_l = E_{lmol} M_w T_{out} c_{pv} + \frac{(T_{out} - T_{in}) F_{in,mol,a} M_{air} c_{pa}}{L_A} + \frac{(T_{out} - T_{in}) F_{in,mol,w} M_w c_{pv}}{L_A} - \frac{Q_{in}}{L_A}. \tag{B3}$$

The humidifier producing the air stream into the wind tunnel reported the volumetric flow rate of dry air at $T_r = 273.13$ K temperature and $P_r = 101\,300$ Pa pressure ($F_{in,v,a,n}$, $m^3\,s^{-1}$) and the vapour pressure of the incoming air ($P_{w,in}$, Pa). To convert from volumetric flow rates to molar flow rates, we used the ideal gas law ($P_a V_a = n_a R_{mol} T_a$),

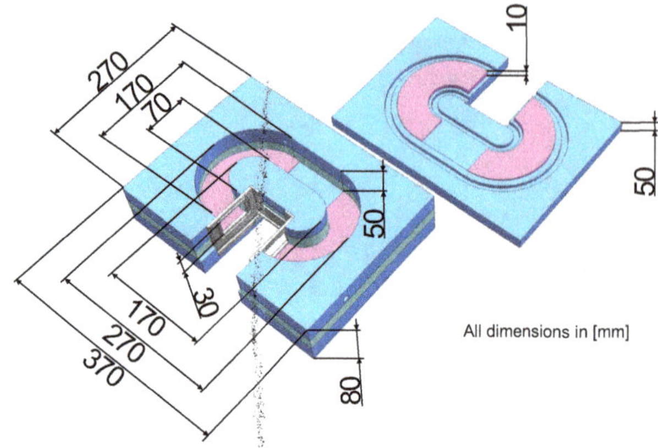

Figure B1. Wind tunnel dimensions. At the back end, the cross section is 5 cm high, to accommodate the fan. The cross section is gradually reduced to 3 cm at the transition to the leaf chamber. See also Fig. 2. The ground area of the circular tunnel is 288 cm^2, which, assuming an average height of 4 cm, results in 1.152 L air volume in the chamber. The circumference of the wind tunnel is 73.4 cm, resulting in a conductive air–wall interfacial area (neglecting the inner wall) of 2 × 288 cm^2 + 73 cm × 4 cm = 868 cm^2.

where the volume (V_a) was replaced by $F_{in,v,a,n}$, the molar amount (n_a) by $F_{in,mol,a}$, and T_a and P_a by their respective reference temperature and pressure (T_r and P_r):

$$F_{in,mol,a} = \frac{P_r F_{in,v,a,n}}{R_{mol} T_r} \tag{B4}$$

The molar flow rate of water vapour is computed along similar lines:

$$F_{in,mol,w} = \frac{F_{in,v} P_{w,in}}{R_{mol} T_{in}}. \tag{B5}$$

Considering Dalton's law of partial pressures, i.e. that the total pressure (P_a) is the sum of the partial pressures of water vapour and dry air, and assuming that both the dry air and the water vapour are well mixed within the same volume (represented by the volumetric flow rate into the chamber, $F_{in,v}$), we write

$$F_{in,v} = \frac{F_{in,mol,a} R_{mol} T_{in}}{P_a - P_{w,in}}. \tag{B6}$$

To obtain the volumetric inflow rate ($F_{in,v}$) from the measured $F_{in,v,a,n}$, $P_{w,in}$, and T_{in}, we can substitute Eq. (B4) into (B6):

$$F_{in,v} = \frac{F_{in,v,a,n} P_r T_{in}}{(P_a - P_{w,in}) T_r}. \tag{B7}$$

Given that L_A, $L_A E_{lmol}$, T_{in}, T_{out}, Q_{in}, $F_{in,v,a,n}$, and $P_{w,in}$ were measured directly, Eqs. (B3)–(B6) could be used to infer the sensible heat flux (H_l) from the chamber energy balance, without any parameter fitting.

B2 Inference of vapour pressure inside the wind tunnel

Similarly to Eq. (B5), the molar outflow rate of water vapour was formulated as

$$F_{\text{out,mol,w}} = \frac{F_{\text{out,v}} P_{\text{w,out}}}{R_{\text{mol}} T_{\text{out}}}. \tag{B8}$$

Equating Eq. (B8) with (B2), substituting Eq. (B5), and solving for $P_{\text{w,out}}$, we obtain an equation of the steady-state vapour pressure of the outgoing air as a function of the transpiration rate, the incoming air flow, its vapour pressure, and temperature:

$$P_{\text{w,out}} = \frac{E_{\text{l,mol}} L_{\text{A}} R_{\text{mol}} T_{\text{in}} T_{\text{out}} + F_{\text{in,v}} P_{\text{w,in}} T_{\text{out}}}{F_{\text{out,v}} T_{\text{in}}}. \tag{B9}$$

Due to changes in air temperature and humidity inside the wind tunnel, the volumetric outflow rate ($F_{\text{out,v}}$) does not necessarily equal the volumetric inflow rate ($F_{\text{in,v}}$). To calculate $F_{\text{out,v}}$, we used again the ideal gas law to obtain

$$F_{\text{out,v}} = \frac{\left(F_{\text{out,mol,a}} + F_{\text{out,mol,w}}\right) R_{\text{mol}} T_{\text{out}}}{P_{\text{a}}}. \tag{B10}$$

Considering that, at steady state, $F_{\text{out,mol,a}} = F_{\text{in,mol,a}}$, we substituted Eqs. (B2), (B4), and (B5) into Eq. (B10) to obtain $F_{\text{out,v}}$ as a function of directly measured quantities only:

$$F_{\text{out,v}} = \frac{1}{P_{\text{a}} T_{\text{r}}} \Big(\left(E_{\text{l,mol}} L_{\text{A}} + F_{\text{in,mol,w}} \right) R_{\text{mol}} T_{\text{out}} T_{\text{r}} + F_{\text{in,v,a,n}} P_{\text{r}} T_{\text{out}} \Big). \tag{B11}$$

Assuming well-mixed air inside the wind tunnel, we assumed that $P_{\text{wa}} = P_{\text{w,out}}$ in our simulations and used Eqs. (B9), (B7), and (B11) to calculate it. On some occasions, we attached an infrared gas analyser (LI6400XT, LI-COR, Lincoln, Nebraska, USA) to the outlet of the wind tunnel and verified results obtained from Eq. (B9), without detecting any significant discrepancy (Fig. B2).

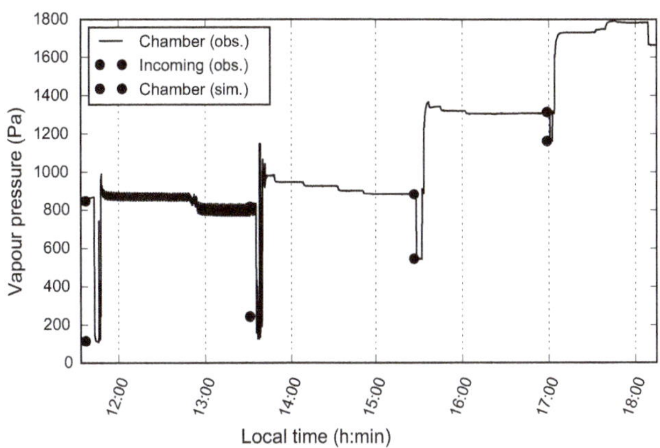

Figure B2. Time series of vapour pressure in the incoming and outgoing air. The experiment was conducted using an artificial leaf with a wet surface on the lower side, air temperature between 295.1 and 295.4 K, and 3.9 m s^{-1} wind speed. Outgoing air was passed through the LI-6400 XT IRGA and its vapour pressure was recorded continuously (blue line). At steady state, air flow was switched for a few minutes, so that the incoming air was passed through the IRGA (expressed as downwards steps in the blue line and marked by blue circles. The blue circles represent the average vapour pressures recorded for the incoming air; red circles represent the computed steady-state vapour pressure inside the chamber, using the chamber mass balance described in the main paper. The apparently thick line between 12:00 and 13:30 local time was the result of oscillations in vapour pressure caused by the control loop in the humidifier.

B3 Measurement of net radiation of artificial leaf

The leaf is exposed to down-welling global radiation (R_{d}) and up-welling global radiation (R_{u}). R_{d} is composed of shortwave irradiance entering through the upper window plus the longwave irradiance transmitted through the upper window and emitted by the chamber walls. R_{u} is composed of radiation transmitted through the lower window and longwave radiation emitted by the chamber walls. The leaf itself reflects some of the radiation in both directions and emits its own black-body longwave radiation. The sum of reflected and emitted radiation away from the leaf is denoted as R_{lu} and R_{ld} for the upper and lower sides respectively. We have three net radiation sensors in place, one above the leaf measuring a signal S_{a}, one below the leaf measuring S_{b}, and

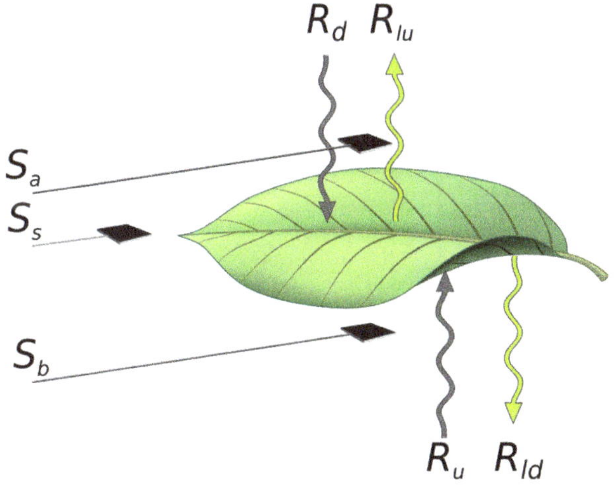

Figure B3. Leaf radiative energy exchange components and net radiation sensors in the radiative field. R_{d}: down-welling global radiation; R_{lu}: upwards emitted/reflected global radiation from leaf; R_{u}: up-welling global radiation; R_{ld}: downwards emitted/reflected global radiation from the leaf; S_{a}: sensor above the leaf; S_{s}: sensor beside the leaf; S_{b}: sensor below the leaf (Eqs. B12–B15). See Fig. 3 for a picture of the sensors.

one at the same level beside the leaf measuring S_s (Fig. B3). These sensors are expected to measure

$$S_a = R_d - R_{lu}, \quad \text{(B12)}$$

$$S_b = R_{ld} - R_u, \quad \text{(B13)}$$

and

$$S_s = R_d - R_u. \quad \text{(B14)}$$

This leaves us with three equations and four unknowns (R_d, R_{lu}, R_{ld}, and R_u), suggesting that this arrangement of sensors was not suited to capturing all components of the leaf radiative balance. In hindsight, the sensor beside the leaf (S_s) is not very useful, as it measures the *difference* between the down-welling and the up-welling global radiation, whereas the radiative load on the leaf is the *sum* of both. Furthermore, we found that the signal S_s was not reliable, as it responded strongly to leaf temperature and wind speed, suggesting that the measurement was likely affected by vertical air temperature gradients down-wind of the leaf. The sensors above and below the leaf, which were outside of the leaf boundary layer (1 cm away from the leaf surface), produced more consistent signals. Therefore, we only used S_a and S_b and assumed that the difference between these signals represents the net radiative energy absorbed by the leaf:

$$\alpha_l R_s - R_{ll} = S_a - S_b. \quad \text{(B15)}$$

The net radiation sensors above and below the artificial leaf were positioned at a distance (L_{ls}) of 1 cm from the artificial leaf. Depending on the distance between the leaf and the sensors, as well as their sizes, the sensors only absorb a fraction of the radiation emitted by the leaf. For two rectangular plates, parallel and centred, this fraction (F_s, "view factor") can be calculated as (Incropera et al., 2006, Table 13.1)

$$F_s = \frac{L_{ls}\left(\sqrt{\left(\frac{L_l}{L_{ls}} + \frac{L_s}{L_{ls}}\right)^2 + 4} - \sqrt{\left(\frac{L_l}{L_{ls}} - \frac{L_s}{L_{ls}}\right)^2 + 4}\right)}{2\,L_s}, \quad \text{(B16)}$$

where L_l and L_s are the widths of the leaf and the net radiation sensors respectively. For 1 cm wide sensors 1 cm away from a 3 cm wide leaf, this fraction amounts to 0.82.

Appendix C: Instruments and sensors

We used a range of commercial instruments and sensors in this study, some of which were modified and calibrated to fit our purpose. The use of these sensors is described in the main text, with reference to their type. In Table C1, we provide further details on the manufacturers and sensor specifications.

Table C1. Sensors and instruments used in this study.

Function	Type and manufacturer	Specifications
Liquid flow	SLI-0430, Sensirion AG, Staefa, Switzerland http://www.sensirion.com	Max. flow rate: $50\,\mu L\,min^{-1}$ Lowest calib. flow (LCF): $2\,\mu L\,min^{-1}$ Accuracy above LCF: 5.0 % of m.v. Accuracy below LCF: 0.2 % of f.s.
Wind speed (calibration)	EE75, E + E Elektronik GmbH, Engerwitzdorf, Germany http://www.epluse.com	Range: $0.15–10\,m\,s^{-1}$ Accuracy in air: $\pm 0.10\,m\,s^{-1}$ + 1 % of m.v.
Wind speed (monitoring)	FS5 Flowmodule, attached to IST evaluation board (P/N: 160.00001), Innovative Sensor Technology IST AG, Ebnat-Kappel, Switzerland http://www.ist-ag.com	Range: $0–100\,m\,s^{-1}$ Accuracy in air: < 3 % of m.v. Temperature sensitivity: $< 0.1\,\%\,K^{-1}$
Fan power supply	Programmable power supply 1786B, B&K Precision Corporation, Yorba Linda, CA 92887-4610, USA http://www.bkprecision.com	Resolution: 10 mV Accuracy: < 0.05 % + 10 mV
Fan power control	Bus-Powered Multifunction DAQ for USB, NI USB-6008 National Instruments Corporation, Austin, TX 78759-3504, USA, http://sine.ni.com	Inputs: eight analogue, at 12 bits, up to $48\,kS\,s^{-1}$ Outputs: two analogue, at 12 bits, software-timed.
Fan	MULTICOMP – MC35357 Farnell AG, 6300 Zug, Switzerland http://ch.farnell.com	Power: 12 V (DC) Dimensions: 50 × 50 × 15 mm Max. flow rate: $0.526\,m^3\,min^{-1}$
Net radiation (calibration)	Net radiometer NR Lite2, Kipp & Zonen, Delft, the Netherlands http://www.kippzonen.com	Spectral range: 200 to 100.000 nm Sensitivity: $10\,\mu V$ per $W\,m^{-2}$ energy flux Response time: < 20 s
Net radiation (monitoring)	gSKIN heat flux sensor, greenTEG, Zurich, Switzerland http://www.greenteg.com	Sensitivity: $1.9\,\mu V$ per $W\,m^{-2}$ heat flux Relative error: ±5 % Sensor range: -10 to $+10\,kW\,m^{-2}$ Resolution: $0.5\,W\,m^{-2}$
Temperature	T type thermocouples TG-T-30-SLE, TG-T-40-SLE, Omega Engineering GmbH 75392 Deckenpfronn, Germany http://www.omega.de	Temperature range: 0–350 °C Accuracy: ±0.5 K or 0.4 % Wire diameter: 0.25 mm (30); 0.08 mm (40)
Temperature imaging	FLIR SC6000 FLIR Systems®, Inc. Wilsonville, OR, USA http://www.flir.com	Image resolution: 640 × 512 pixel Detector type: QWIP Spectral range: $8–9.2\,\mu m$ Sensitivity (NEDT): < 5 mK
Controlled air supply	Humidifier P-10C-0A-1-0-000100-v7 Cellkraft AB SE-114 19 Stockholm, Sweden http://www.cellkraft.se	Air flow range: $0–10\,L\,min^{-1}$ Flow accuracy: ±1 % of f.s. Humidity range: 0–90 % RH Humidity accuracy: ±1.7 % of m.v. Temperature range: 4–300 °C Air supply: pressurized at 0–20 bar
Data logging	Data logger CR 1000 and 25-channel solid state multiplexer AM25T Campbell Scientific, Inc. Logan, UT 84321-1784, USA https://www.campbellsci.com	Analogue voltage accuracy: ±0.06 % of reading + offset Analogue resolution: $0.33\,\mu V$ AD converter resolution: 13 bits

m.v.: measured value; f.s.: full scale.

Appendix D: Additional experimental data

Figures D1 and D2 illustrate additional experimental results obtained in the absence of shortwave irradiance for different perforation densities and varying wind speed or vapour pressure. Experimental conditions are summarized in the figure captions.

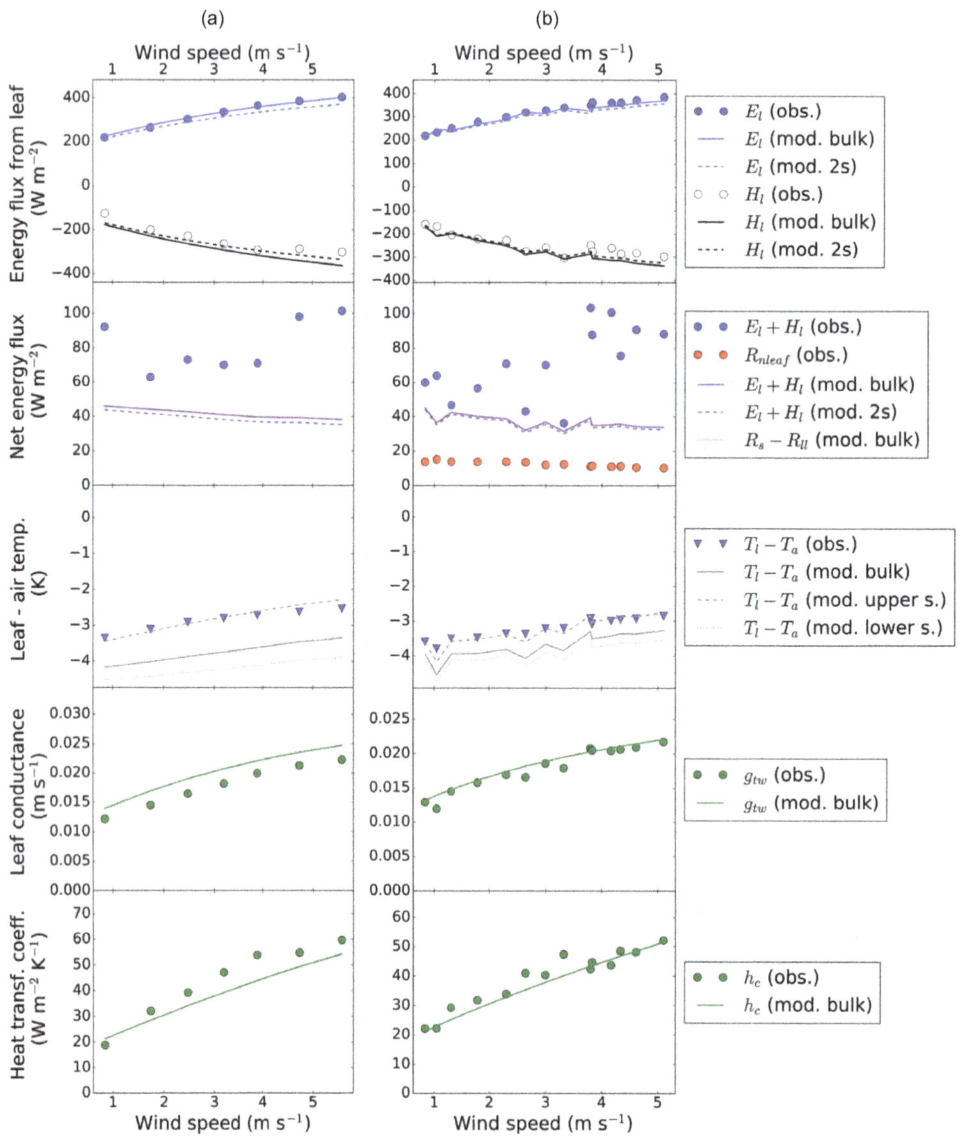

Figure D1. Artificial leaves with **(a)** 65 and **(b)** 35 pores mm^{-2} under varying wind speed. Numerical model results (lines) based on the same boundary conditions as observations (symbols). E_l: latent heat flux; H_l: sensible heat flux; $R_s - R_{ll}$: absorbed net radiation; $T_l - T_a$: leaf–air temperature difference; g_{tw}: total leaf conductance to water vapour; h_c: convective heat transfer coefficient; "mod. bulk": bulk leaf temperature model; "mod. 2s": model based on different leaf temperatures on both "upper" and "lower" leaf sides. Boundary conditions: $g_{sw} = 0.05\ m\,s^{-1}$ **(a)** and $0.042\ m\,s^{-1}$ **(b)**; $R_s = 0$; $T_a = 295.1$–295.3 K **(a)** and 295.0–296.5 **(b)**; $P_{wa} = 1200$–1340 Pa **(a)** and 1190–1280 **(b)**; $k_l = 0.03\ W\,K^{-1}\,m^{-1}$ **(a)** and $0.1\ W\,K^{-1}\,m^{-1}$ **(b)**; $z_l = 0.35$ mm **(a)** and 0.6 mm **(b)**.

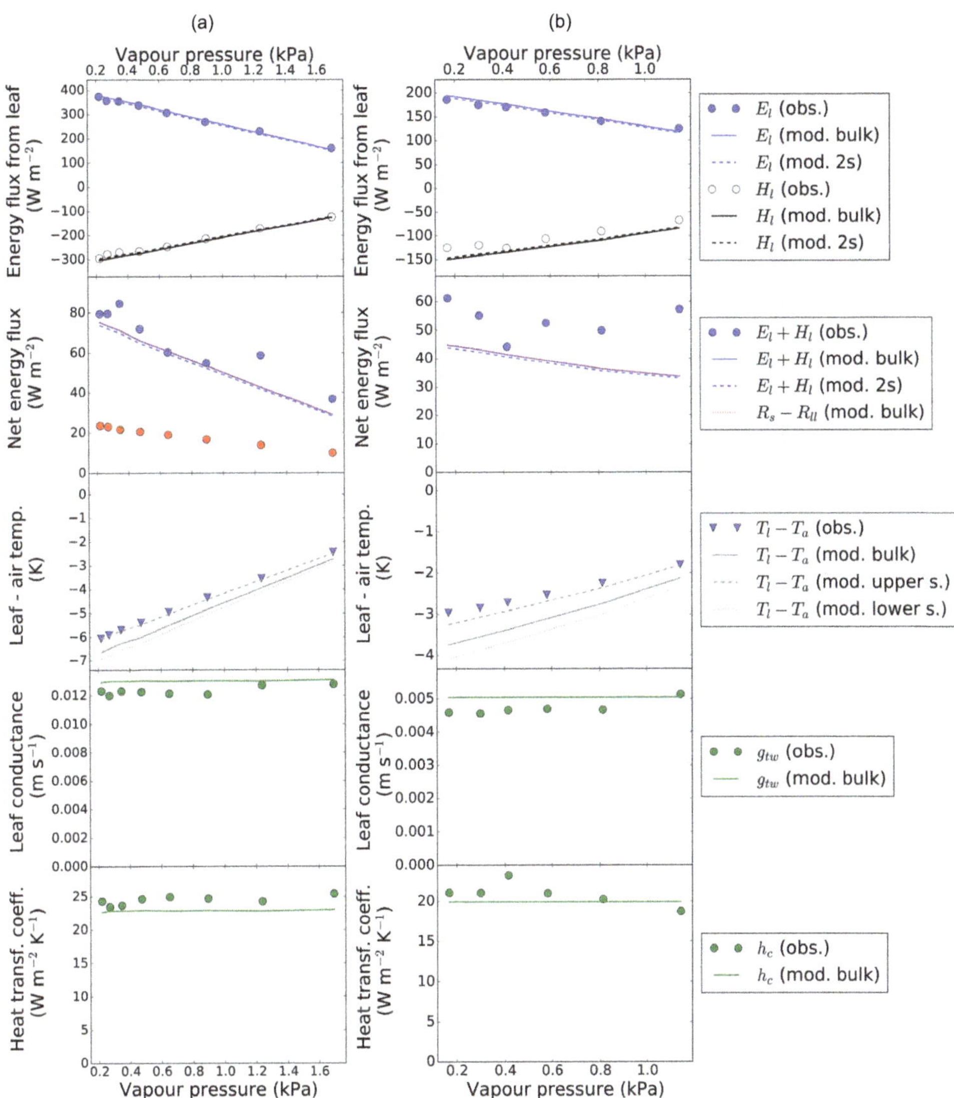

Figure D2. Artificial leaves with **(a)** 37 and **(b)** 7 pores $\mathrm{mm^{-2}}$ under varying vapour pressure. Numerical model results (lines) based on the same boundary conditions as observations (symbols). E_l: latent heat flux; H_l: sensible heat flux; $R_\mathrm{s} - R_\mathrm{ll}$: absorbed net radiation; $T_\mathrm{l} - T_\mathrm{a}$: leaf–air temperature difference; g_tw: total leaf conductance to water vapour; h_c: convective heat transfer coefficient; "obs.": observations; "mod. bulk": bulk leaf temperature model; "mod. 2s": model based on different leaf temperatures on both "upper" and "lower" leaf sides. Boundary conditions: $g_\mathrm{sw} = 0.035\,\mathrm{m\,s^{-1}}$ **(a)** and $0.007\,\mathrm{m\,s^{-1}}$ **(b)**; $R_\mathrm{s} = 0$; $T_\mathrm{a} = 295.7$–$296.0\,\mathrm{K}$ **(a)** and 296.1–$296.7\,\mathrm{K}$ **(b)**; $v_\mathrm{w} = 1\,\mathrm{m\,s^{-1}}$ **(a)** and $0.7\,\mathrm{m\,s^{-1}}$ **(b)**; $k_\mathrm{l} = 0.1\,\mathrm{W\,K^{-1}\,m^{-1}}$ **(a)** and $0.05\,\mathrm{W\,K^{-1}\,m^{-1}}$ **(b)**; $z_\mathrm{l} = 0.5\,\mathrm{mm}$.

Table D1. Table of symbols and standard values used in this paper. All area-related variables are expressed per unit leaf area.

Variable	Description (value)	Units
A_i	Conducting area of insulation material	m^2
A_p	Cross-sectional pore area	m^2
α_l	Leaf albedo, fraction of shortwave radiation reflected by the leaf	1
c_{pa}	Specific heat of dry air (1010)	$J\,K^{-1}\,kg^{-1}$
c_{pv}	Specific heat of water vapour at 300 K	$J\,K^{-1}\,kg^{-1}$
C_{wa}	Concentration of water in the free air	$mol\,m^{-3}$
C_{wl}	Concentration of water in the leaf air space	$mol\,m^{-3}$
d_p	Pore depth	m
D_{va}	Binary diffusion coefficient of water vapour in air	$m^2\,s^{-1}$
E_l	Latent heat flux from leaf	$J\,m^{-2}\,s^{-1}$
E_{ll}	Latent heat flux from lower side of leaf	$J\,m^{-2}\,s^{-1}$
$E_{l,mol}$	Transpiration rate in molar units	$mol\,m^{-2}\,s^{-1}$
E_{lu}	Latent heat flux from upper side of leaf	$J\,m^{-2}\,s^{-1}$
ϵ_l	Longwave emmissivity of the leaf surface (1.0)	1
$F_{in,mol,a}$	Molar flow rate of dry air into chamber	$mol\,s^{-1}$
$F_{in,mol,w}$	Molar flow rate of water vapour into chamber	$mol\,s^{-1}$
$F_{in,v}$	Volumetric flow rate into chamber	$m^3\,s^{-1}$
$F_{in,v,a,n}$	Volumetric inflow of dry air at 0 °C and 101 325 Pa	$m^3\,s^{-1}$
$F_{out,mol,a}$	Molar flow rate of dry air out of chamber	$mol\,s^{-1}$
$F_{out,mol,w}$	Molar flow rate of water vapour out of chamber	$mol\,s^{-1}$
$F_{out,v}$	Volumetric flow rate out of chamber	$m^3\,s^{-1}$
F_s	Fraction of radiation emitted by leaf, absorbed by sensor	1
g	Gravitational acceleration (9.81)	$m\,s^{-2}$
g_{bw}	Boundary layer conductance to water vapour	$m\,s^{-1}$
$g_{bw,mol}$	Boundary layer conductance to water vapour	$mol\,m^{-2}\,s^{-1}$
g_{sw}	Stomatal conductance to water vapour	$m\,s^{-1}$
g_{tw}	Total leaf conductance to water vapour	$m\,s^{-1}$
$g_{tw,mol}$	Total leaf layer conductance to water vapour	$mol\,m^{-2}\,s^{-1}$
h_c	Average one-sided convective transfer coefficient	$J\,K^{-1}\,m^{-2}\,s^{-1}$
H_l	Sensible heat flux from leaf	$J\,m^{-2}\,s^{-1}$
H_{l_l}	Sensible heat flux from lower side of leaf	$J\,m^{-2}\,s^{-1}$
H_{l_u}	Sensible heat flux from upper side of leaf	$J\,m^{-2}\,s^{-1}$
k_a	Thermal conductivity of dry air	$J\,K^{-1}\,m^{-1}\,s^{-1}$
k_{dv}	Ratio D_{va}/V_m	$mol\,m^{-1}\,s^{-1}$
k_l	Thermal conductivity of a fresh leaf	$J\,K^{-1}\,m^{-1}\,s^{-1}$
L_A	Leaf area	m^2
L_l	Characteristic length scale for convection (size of leaf)	m
L_{ls}	Distance between leaf and net radiation sensor	m
L_s	Width of net radiation sensor	m
λ_E	Latent heat of evaporation (2.45e6)	$J\,kg^{-1}$
M_{air}	Molar mass of air ($kg\,mol^{-1}$)	$kg\,mol^{-1}$
M_w	Molar mass of water (0.018)	$kg\,mol^{-1}$
N_{Gr_L}	Grashof number	1
N_{Le}	Lewis number	1
N_{Nu_L}	Nusselt number	1
n_p	Pore density	m^{-2}
N_{Re_L}	Reynolds number	1
N_{Re_c}	Critical Reynolds number for the onset of turbulence	1
N_{Sh_L}	Sherwood number	1
P_a	Air pressure	Pa
P_r	Reference pressure	Pa
$P_{w,in}$	Vapour pressure of incoming air	Pa
$P_{w,out}$	Vapour pressure of outgoing air	Pa
P_{wa}	Vapour pressure in the atmosphere	Pa
P_{wl}	Vapour pressure inside the leaf	Pa
N_{Pr}	Prandtl number (0.71)	1
Q_{in}	Internal heat sources, such as fan	$J\,s^{-1}$
Q_l	Conductive heat flux from upper to lower side of leaf	$J\,m^{-2}\,s^{-1}$
R_d	Down-welling global radiation	$J\,m^{-2}\,s^{-1}$
R_{ll}	Longwave radiation away from leaf	$J\,m^{-2}\,s^{-1}$
R_{ll_l}	Longwave heat flux from lower side of leaf	$J\,m^{-2}\,s^{-1}$
R_{ll_u}	Longwave heat flux from upper side of leaf	$J\,m^{-2}\,s^{-1}$

Table D1. Continued.

Variable	Description (value)	Units
R_{lu}	Upwards emitted/reflected global radiation from leaf	$J\,m^{-2}\,s^{-1}$
R_{mol}	Molar gas constant (8.314472)	$J\,K^{-1}\,mol^{-1}$
R_s	Solar shortwave flux	$J\,m^{-2}\,s^{-1}$
R_u	Up-welling global radiation	$J\,m^{-2}\,s^{-1}$
r_p	Pore radius (for ellipsoidal pores, half the pore width)	m
r_{sp}	Diffusive resistance of a stomatal pore	$s\,m^2\,mol^{-1}$
r_{vs}	Diffusive resistance of a stomatal vapour shell	$s\,m^2\,mol^{-1}$
S_a	Radiation sensor above leaf reading	$J\,m^{-2}\,s^{-1}$
S_b	Radiation sensor below leaf reading	$J\,m^{-2}\,s^{-1}$
S_s	Radiation sensor beside leaf reading	$J\,m^{-2}\,s^{-1}$
s_p	Spacing between stomata	m
σ	Stefan–Boltzmann constant (5.67e-8)	$J\,K^{-4}\,m^{-2}\,s^{-1}$
T_a	Air temperature	K
T_{in}	Temperature of incoming air	K
T_l	Leaf temperature	K
T_{l_l}	Leaf surface temperature of lower side	K
T_{l_u}	Leaf surface temperature of upper side	K
T_{out}	Temperature of outgoing air (= chamber T_a)	K
T_r	Reference temperature	K
T_w	Radiative temperature of objects surrounding the leaf	K
V_m	Molar volume of air	$m^3\,mol^{-1}$
v_w	Wind velocity	$m\,s^{-1}$
z_l	Leaf thickness (m)	m

Author contributions. SJS performed the mathematical derivations, designed and carried out the experiments, and wrote the paper. DB designed and constructed major parts of the wind tunnel and custom sensor set-up and contributed visuals and feedback for the manuscript. DO was involved in the design of the experimental set-up, interpretation of the results, and writing the paper.

Competing interests. The authors declare that they have no conflict of interest.

Acknowledgements. The authors are very grateful to Hans Wunderli for assistance in the lab, Stefan Meier and Joni Dehaspe for assistance in constructing artificial leaves, and Ralph Beglinger (Lasergraph AG, Würenlingen, Switzerland) for laser perforation services. We also wish to acknowledge funding by the Swiss National Science Foundation, Project 2000021 135077.

Edited by: Pierre Gentine

References

Ball, M., Cowan, I., and Farquhar, G.: Maintenance of Leaf Temperature and the Optimisation of Carbon Gain in Relation to Water Loss in a Tropical Mangrove Forest, Funct. Plant Biol., 15, 263–276, 1988.

Bange, G. G. J.: On the quantitative explanation of stomatal transpiration, Acta Bot. Neerl., 2, 255–296, 1953.

Brown, H. T. and Escombe, F.: Static Diffusion of Gases and Liquids in Relation to the Assimilation of Carbon and Translocation in Plants, Abstract, P. R. Soc. London, 67, 124–128, 1900.

Cannon, J. N., Krantz, W. B., Kreith, F., and Naot, D.: A study of transpiration from porous flat plates simulating plant leaves, International Journal of Heat and Mass Transfer, 22, 469–483, https://doi.org/10.1016/0017-9310(79)90013-9, 1979.

Field, C., Berry, J. A., and Mooney, H. A.: A portable system for measuring carbon dioxide and water vapour exchange of leaves, Plant Cell Environ., 5, 179–186, https://doi.org/10.1111/1365-3040.ep11571607, 1982.

Grace, J., Fasehun, F. E., and Dixon, M.: Boundary layer conductance of the leaves of some tropical timber trees, Plant Cell Environ., 3, 443–450, https://doi.org/10.1111/1365-3040.ep11586917, 1980.

Hays, R. L.: The thermal conductivity of leaves, Planta, 125, 281–287, https://doi.org/10.1007/BF00385604, 1975.

Incropera, F. P., DeWitt, D. P., Bergman, T. L., and Lavine, A. S.: Fundamentals of Heat and Mass Transfer, John Wiley & Sons, 6th Edn., 2006.

Kaiser, H. and Kappen, L.: In Situ Observations of Stomatal Movements in Different Light-Dark Regimes: The Influence of Endogenous Rhythmicity and Long-Term Adjustments, J. Exp. Bot., 48, 1583–1589, https://doi.org/10.1093/jxb/48.8.1583, 1997.

Kappen, L., Andresen, G., and Lösch, R.: In situ Observations of Stomatal Movements, J. Exp. Bot., 38, 126–141, https://doi.org/10.1093/jxb/38.1.126, 1987.

Kumar, A. and Barthakur, N.: Convective heat transfer measurements of plants in a wind tunnel, Bound.-Lay. Meteorol., 2, 218–227, 1971.

Lehmann, P. and Or, D.: Effects of stomata clustering on leaf gas exchange, New Phytol., 207, 1015–1025, https://doi.org/10.1111/nph.13442, 2015.

Monteith, J. and Unsworth, M.: Principles of Environmental Physics, Academic Press, Oxford, UK, 3rd Edn., 2007.

Morrison Jr., J. E. and Barfield, B. J.: Transpiring artificial leaves, Agricultural Meteorology, 24, 227–236, 1981.

Parkhurst, D. F., Duncan, P. R., Gates, D. M., and Kreith, F.: Wind-tunnel modelling of convection of heat between air and broad leaves of plants, Agr. Meteorol., 5, 33–47, 1968.

SageMath: Sage Mathematics Software, available at: https://zenodo.org/record/820864#.WVoYf2dQEb1 (last access: 3 July 2017), 2016.

Schuepp, P. H.: Studies of forced-convection heat and mass transfer of fluttering realistic leaf models, Bound.-Lay. Meteorol., 2, 263–274, 1972.

Schymanski, S. J.: Supporting information for Schymanski et al., HESSD, doi:10.5194/hess-2016-643, 2017, Zenodo, https://doi.org/10.5281/zenodo.241217, 2017.

Schymanski, S. J. and Or, D.: Wind increases leaf water use efficiency, Plant Cell Environ., 39, 1448–1459, https://doi.org/10.1111/pce.12700, 2016.

Schymanski, S. J. and Or, D.: Leaf-scale experiments reveal an important omission in the Penman–Monteith equation, Hydrol. Earth Syst. Sci., 21, 685–706, https://doi.org/10.5194/hess-21-685-2017, 2017.

Schymanski, S. J., Or, D., and Zwieniecki, M.: Stomatal Control and Leaf Thermal and Hydraulic Capacitances under Rapid Environmental Fluctuations, PLoS ONE, 8, e54231, https://doi.org/10.1371/journal.pone.0054231, 2013.

Sierp, H. and Seybold, A.: Weitere Untersuchungen über die Verdunstung aus multiperforaten Folien mit kleinsten Poren, Planta, 9, 246–269, https://doi.org/10.1007/BF01913327, 1929.

Thom, A. S.: The exchange of momentum, mass, and heat between an artificial leaf and the airflow in a wind-tunnel, Q. J. Roy. Meteor. Soc., 94, 44–55, https://doi.org/10.1002/qj.49709439906, 1968.

Ting, I. P. and Loomis, W. E.: Diffusion Through Stomates, American J. Bot., 50, 866–872, https://doi.org/10.2307/2439773, 1963.

Wigley, G. and Clark, J. A.: Heat transport coefficients for constant energy flux models of broad leaves, Bound.-Lay. Meteorol., 7, 139–150, https://doi.org/10.1007/BF00227909, 1974.

Wohlfahrt, G. and Widmoser, P.: Can an energy balance model provide additional constraints on how to close the energy imbalance?, Agr. Forest Meteorol., 169, 85–91, https://doi.org/10.1016/j.agrformet.2012.10.006, 2013.

Zwieniecki, M. A., Haaning, K. S., Boyce, C. K., and Jensen, K. H.: Stomatal design principles in synthetic and real leaves, J. R. Soc. Interface, 13, 20160535, https://doi.org/10.1098/rsif.2016.0535, 2016.

3

LiDAR measurement of seasonal snow accumulation along an elevation gradient in the southern Sierra Nevada, California

P. B. Kirchner[1,*], R. C. Bales[1], N. P. Molotch[2,3], J. Flanagan[1], and Q. Guo[1]

[1]Sierra Nevada Research Institute, UC Merced, Merced, CA, USA
[2]Department of Geography and the Institute of Arctic and Alpine Research, University of Colorado at Boulder, Boulder, CO, USA
[3]Jet Propulsion Laboratory, California Institute of Technology, Pasadena, CA, USA
[*]now at: Joint Institute for Regional Earth System Science and Engineering, UCLA, Los Angeles, CA, USA

Correspondence to: P. B. Kirchner (peter.b.kirchner@jpl.nasa.gov)

Abstract. We present results from snow-on and snow-off airborne-scanning LiDAR measurements over a 53 km^2 area in the southern Sierra Nevada. We found that snow depth as a function of elevation increased approximately 15 cm per 100 m, until reaching an elevation of 3300 m, where depth sharply decreased at a rate of 48 cm per 100 m. Departures from the 15 cm per 100 m trend, based on 1 m elevation-band means of regression residuals, showed slightly less steep increases below 2050 m; steeper increases between 2050 and 3300 m; and less steep increases above 3300 m. Although the study area is partly forested, only measurements in open areas were used. Below approximately 2050 m elevation, ablation and rainfall are the primary causes of departure from the orographic trend. From 2050 to 3300 m, greater snow depths than predicted were found on the steeper terrain of the northwest and the less steep northeast-facing slopes, suggesting that ablation, aspect, slope and wind redistribution all play a role in local snow-depth variability. At elevations above 3300 m, orographic processes mask the effect of wind deposition when averaging over large areas. Also, terrain in this basin becomes less steep above 3300 m. This suggests a reduction in precipitation from upslope lifting and/or the exhaustion of precipitable water from ascending air masses. Our results suggest a cumulative precipitation lapse rate for the 2100–3300 m range of about 6 cm per 100 m elevation for the accumulation period of 3 December 2009 to 23 March 2010. This is a higher gradient than the widely used PRISM (Parameter-elevation Relationships on Independent Slopes Model) precipitation products, but similar to that from reconstruction of snowmelt amounts from satellite snow-cover data. Our findings provide a unique characterization of the consistent, steep average increase in precipitation with elevation in snow-dominated terrain, using high-resolution, highly accurate data and highlighs the importance of solar radiation, wind redistribution and mid-winter melt with regard to snow distribution.

1 Introduction

In mountainous regions of the western United States, snowmelt is the dominant contributor to surface runoff, water use by vegetation, and groundwater recharge (Bales et al., 2006; Earman and Dettinger, 2011). Because of their importance and vulnerability of mountain snowpacks in a warmer climate, several researchers have recently developed scenarios for changes in annual and multiyear mountain water cycles, including trends in water storage and runoff, groundwater recharge, and feedbacks with vegetation (Peterson et al., 2000; Marks et al., 2001; Lundquist et al., 2005; Maxwell and Kollet, 2008; Barnett et al., 2008; Anderson and Goulden, 2011; Trujillo et al., 2012).

Given the challenges in measuring the spatial distribution of mountain precipitation, the processes controlling its distribution remain poorly understood. However, since the large majority of precipitation in the middle and upper elevations of the southern Sierra Nevada falls and accumulates as snow, with limited ablation through much of the winter, we can examine snow accumulation to assess processes governing the distribution of precipitation.

Snow accumulation across the mountains is primarily influenced by orographic processes, involving feedbacks between atmospheric circulation, terrain and the geomorphic processes of mountain uplift, erosion and glaciation on the Earth's surface (Roe, 2005; Roe and Baker, 2006; Pedersen et al., 2010; Kessler et al., 2006; Stolar et al., 2007; Galewsky, 2009; Mott et al., 2014). Orographic precipitation is well documented and central to determining the amount of snow water equivalent (SWE) in mountainous regions. The Sierra Nevada, a major barrier to land-falling storms from the Pacific, is ideally oriented to produce orographic precipitation, and exerts a strong influence on the upslope amplification of precipitation and the regional water budget (Pandey et al., 1999). Despite this well-developed conceptual understanding, our ability to apply this knowledge at spatial and temporal scales relevant to questions of regional climate and local water-supply forecasting are limited by a lack of accurate precipitation measurements across mountains (Viviroli et al., 2011). Additionally, long, narrow land-falling bands of extratropical Pacific water vapor, referred to as atmospheric rivers, frequently deposit large fluxes of orographic precipitation as they ascend over the Sierra Nevada (Neiman et al., 2008; Ralph and Dettinger, 2011). Atmospheric rivers deposit approximately 40 % of total winter snowfall in the Sierra Nevada, linking ocean–atmosphere interactions and the terrestrial water balance (Dettinger et al., 2011; Guan et al., 2012, 2013a).

Current mountain-basin operational SWE estimates are made with a limited set of snow-course and continuous in situ point measurements from snow pillows. Measurements at these index sites are used to develop statistically based runoff estimates for the subsequent spring and summer. While this approach has provided operationally robust predictions in years near the long-term normal, snow accumulation both varies from year to year and changes in response to long-term climatic conditions, and, in recent decades, has trended outside the statistical normal (Milly et al., 2008). Hence our current methods are becoming less reliable, and accurate predictions require a more comprehensive approach to understanding the processes affecting precipitation and the probabilities of extremes (Rahmstorf and Coumou, 2011).

Accurate estimates of the amount and spatial distribution of both precipitation and SWE are essential given the shift toward spatially distributed models for forecasts of runoff, moisture stress and other water-cycle components (Rice et al., 2011; Meromy et al., 2012). Current operational measurements for precipitation and SWE are limited by scale and by the heterogeneity of snow-accumulation processes, and do not provide spatially representative values (Viviroli et al., 2011; Bales et al., 2006; Grünewald and Lehning, 2011). Uncertainty in watershed-scale SWE and precipitation estimates result in part from the lack of measurements at both the rain–snow transition and highest elevations, as well as the lack of representative measurements across different slopes, aspects and canopy conditions (Molotch and Margulis, 2008).

Remotely sensed snow properties from satellites and aircraft are used in research, and on a limited basis in forecasts. In both cases these measurements can be blended using statistical or spatially explicit models to produce discharge forecasts (Rice et al., 2011; Molotch et al., 2004; Fassnacht et al., 2003; Bales et al., 2008; Kerkez et al., 2012). A recent review highlighted the promise of aircraft LiDAR measurements for snow-depth mapping at high spatial resolution and vertical accuracy, using repeat snow-on and snow-off LiDAR flights (Deems et al., 2013). The emergence of quality research data sets for snow mapping offers opportunities to assess LiDAR accuracy and coverage in complex, forested terrain, and also its potential for providing a much-needed spatial "ground truth" for watershed-scale snow depth (Harpold et al., 2014).

Research reported in this paper was aimed at determining the influences of terrain and orographic precipitation on patterns of seasonal snow accumulation along a 1650 m elevation gradient in the southern Sierra Nevada. Three questions were posed in this research: (i) what is the magnitude of the average elevation lapse rate for snow accumulation, (ii) what is the variability in snow accumulation at each elevation along an elevation gradient, and (iii) to what extent do local terrain and wind redistribution influence this pattern? It was also our aim to evaluate consistency between LiDAR-estimated SWE and prior model-based estimates of accumulated SWE and total precipitation.

2 Methods

Our approach involved analysis of (i) LiDAR-based snow-depth estimates derived from two LiDAR acquisitions, one when the ground was snow-free and one near peak snow accumulation on 23 March; (ii) continuous ground-based measurements of snow depth, SWE, wind speed and air temperature, plus operational bright-band radar observations; and (iii) model estimates of SWE and precipitation. The LiDAR data were used to estimate snow depth across the study area at a 1 m^2 spatial resolution in open areas without canopy cover. The ground measurements were used in interpreting the spatial patterns and in estimating SWE, and the bright-band radar in determining the rain–snow transition elevation for precipitation events, an important metric for interpreting snow depth and SWE along elevation gradients.

2.1 Location

Our study area is centered at approximately 36.5° N, 118.7° W and includes the 53.1 km^2 area covered by the two LiDAR flights in the southeastern part of the 135 km^2 Marble Fork of the Kaweah River watershed, located in Sequoia National Park in the southern Sierra Nevada, California

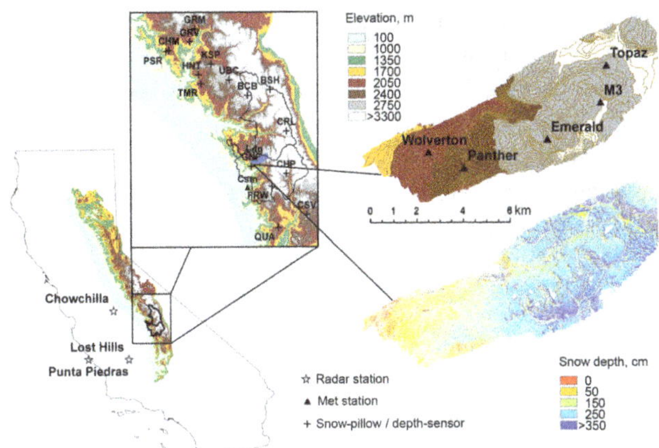

Figure 1. Study area and instrument locations. (Left) California with Sierra Nevada, outline of Sequoia and Kings Canyon National Parks, and location of radar stations. (Center) Location of snow sensors, and meteorological stations from north to south: Graveyard Meadow (GRV), Green Mountain (GRM), Chilkoot Meadow (CHM), Poison Ridge (PSR), Kaiser Pass (KSP), Huntington Lake (HNT), Upper Burnt Corral (UBC), Tamarack Summit (TMR), Bishop Pass (BSH), Black Cap Basin (BCB), Charlotte Lake (CRL), Lodgepole met (Ldg), Giant Forest (GNF), Chagoopa Plateau (CHP), Farewell Gap (FRW), Case Mountain met (Csm), Casa Vieja (CSV), and Quaking Aspen (QUA). (Upper right) Elevation and 50 m contour map with locations of meteorological stations in LiDAR footprint. (Bottom right) LiDAR-measured 1 m snow depth in areas free of vegetation.

(Fig. 1). Elevations of the LiDAR acquisition were 1850–3494 m, with aspects predominantly trending northwest, about orthogonal to the regions southwest prevailing storm tracks. The land features include glaciated lake basins, cirques and stepped plateaus at the highest elevations. Soils are characterized by moraine deposits and well-drained granitic soils at the lower elevations, and rock outcrops with pockets of course shallow soil at the higher elevations. The vegetation cover below 3000 m consists primarily of coniferous forests that transition with increasing elevation from a giant sequoia grove, to mixed-conifer forests, and to red fir forests. Above 3000 m there are increasing areas of bare rock with subalpine forests and alpine meadows in locations with soil (Fig. 2b).

2.2 LiDAR altimetry

Airborne-scanning LiDAR altimetry was collected by the National Center for Airborne Laser Mapping (NCALM) using an Optech Gemini® ALTM 1233 airborne-scanning laser (Zhang and Cui, 2007). The two campaigns were conducted in the 2010 water year: 23 March for snow-covered, and 15 August for snow-free conditions (Harpold et al., 2014). The instrument settings used for acquisition generated an average point density greater than 10 points per square meter, and a fine-scale beam-sampling footprint of approximately

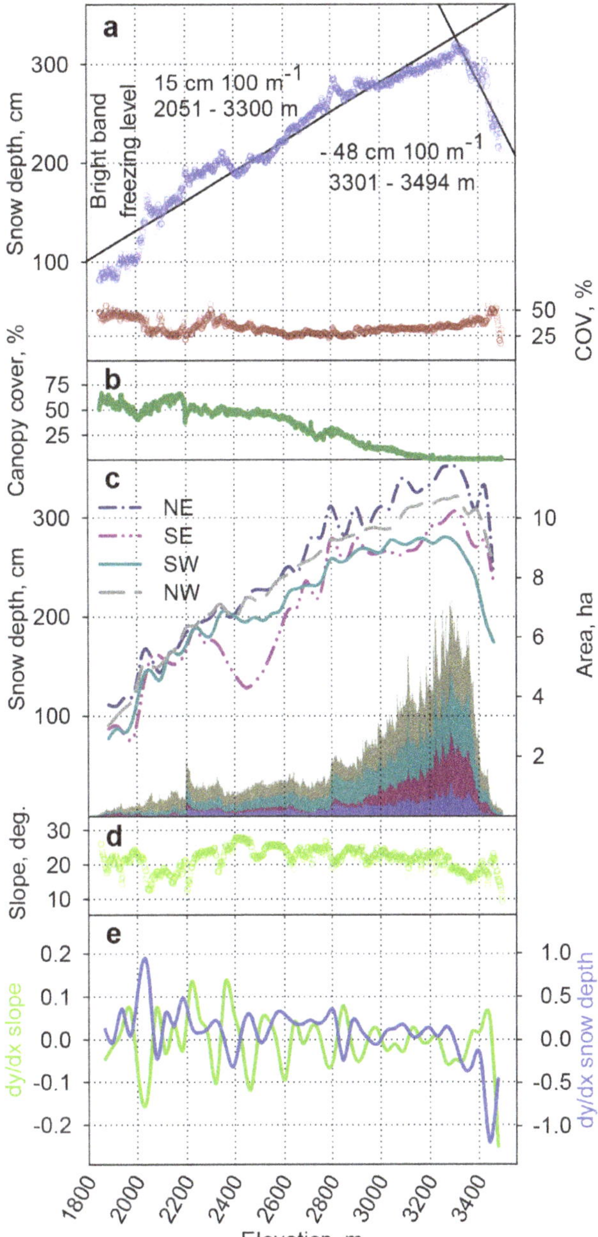

Figure 2. (a) Snow depth (blue) with regression lines and approximate 3 December–23 March bright-band radar freezing level noted, and snow depth percent coefficient of variation (dark red); **(b)** percent canopy cover; **(c)** 35 m running average of mean snow depth and stacked area by elevation for each 90° quadrant of aspect; **(d)** mean slope of 1 m elevation band; and **(e)** first derivative of mean slope (green) and snow depth (blue) over 35 m running average.

20 cm (Table 1). Ground points were classified by NCALM through iteratively building a triangulated surface model with discrete points classified as ground and non-ground (Shrestha et al., 2007; Slatton et al., 2007). The nominal horizontal and vertical accuracy for a single flight path are 0.5 and 0.11 m,

Table 1. Target parameters and attributes for LiDAR flights.

Flight parameters		Instrument attributes	
altitude AGL	600 m	wavelength	1064 nm
flight speed	$65\,m\,s^{-1}$	beam divergence	0.25 mrad
swath width	233.62 m	laser PRF	100 kHz
swath overlap	50 %	scan frequency	55 Hz
point density	$10.27\ m^2$	scan angle	+ 14°
cross track res.	0.233 m	scan cutoff	3°
down track res.	0.418 m	scan offset	0°

respectively, but higher accuracy was likely achieved, particularly where flight paths overlapped.

A digital surface model (DSM) was created by using first-return points and discarding outliers > 100 m (tallest trees are approximately 85 m) and returns below −0.1 m, where values in the range of −0.1 to 0 m were classified as 0. A continuous-coverage bare-earth digital elevation model (DEM) was created through kriging of ground points using a linear variogram with a nugget of 15 cm, a sill of 10 m, a range of 100 m, and a search radius of 100 m, where the minimum number of points was five (Guo et al., 2010). We used a $1\ m^2$ gridded model for representing our data, as this is the smallest footprint that most closely matches the expected beam sampling footprint and uncertainty in horizontal accuracy. After interpolation, digital models of mean elevation and point-return density grids were georegistered to a common grid for snow-on and snow-off flights. The average point-return densities were $8\ m^{-2}$ for the surface model and $3\ m^{-2}$ for the bare-earth model. Grids with no point returns in either flight, primarily under forest canopy, were not used.

The accuracy of the LiDAR altimetry was evaluated by using 352 georegistered 2.5 m × 2.5 m grid samples of the point cloud along the paved highway in the western part of the domain; because the highway is plowed regularly, surface heights do not change with snow accumulation. These samples had a bias of +0.05 m and a standard deviation of 0.07 m, which is below the estimated combined two-flight instrumental elevation error of 0.11 m (Xiaoye, 2008; Zhang and Cui, 2007). A possible explanation of the 0.05 m bias for the snow-on flight is that some sections of the road had a small amount of snow remaining after plowing.

A $1\ m^2$ gridded DSM of the vegetation canopy > 2 m was created by subtracting the DSM from the DEM. In order to accurately determine snow depth, values were further classified into two groups, where snow depth was either greater or less than the coincident vegetation height. This allowed us to consider, for further analysis, only snow from open slopes or where it had accumulated in the gaps between trees. To reduce the amount of error, we eliminated locations with slopes greater than 55°, warranted by the high number of erroneous values and known issues of vertical inaccuracies due to slope angle (Schaer et al., 2007; Deems et al., 2013). Additionally, we eliminated areas with rapid annual vegetation growth that had negative snow-depth values (e.g., areas within a wet meadow). Lastly, we filtered out areas with open water, buildings and parking lots where returns were not representative of local snow accumulation. Mean snow depth for each $1\ m^2$ elevation band with > $100\ m^2$ area was computed from the snow-depth grid. Additionally, a 5 m elevation model, aggregated from the $1\ m^2$ bare-earth model, was produced to remove scaling biases in the analysis of slope and aspect (Kienzle, 2004; Erskine et al., 2006).

2.3 Spatial analysis

To analyze possible correlations between terrain steepness and snow distribution, we calculated the first derivative of slope and snow depth, over distances of 5–100 m, using the $1\ m^2$ mean snow depths and the corresponding mean slope for each 1 m elevation band, computing the correlation at 5–100 m using 5 m steps. Using the derivatives identifies transition areas.

For quantifying the combined effect of slope and aspect on snow depth, we indexed aspect on a scale of 1 to −1 using methods adapted from Roberts (1986):

$$V_A = \cos(A - FA), \tag{1}$$

where V_A is the aspect value; A is the azimuth variable or direction for which the calculation is being indexed to; and FA is the focal aspect, e.g., FA = 0° is north and FA = 45° northeast. The aspect value was further scaled by the sine of the slope angle, yielding 0 in flat terrain and approaching 1 as the mean slope increases to 90°:

$$I_A = \sin(S) \cdot V_A, \tag{2}$$

where I_A is aspect intensity and S the slope angle. The method of scaling the cosine of aspect by sine of slope for $A = 0°$ is referred to as "northness" (Molotch et al., 2004).

2.4 Ground measurements

Meteorological data were obtained from six meteorological stations in the flight area for the period from the first significant snowfall on 3 December 2009 to the 23 March LiDAR acquisition date, henceforth referred to as the snow-accumulation period. At these stations, temperature was measured using Vaisala HMP-35 and Campbell T-108 sensors, with wind speed and direction measured using RM Young 5103 sensors. All meteorological stations measure hourly average wind speed, and two stations – Wolverton and Panther – recorded maximum wind gusts at 10 s scan intervals. The M3, Topaz, and Emerald Lake stations are managed by the University of California Santa Barbara; Giant Forest is operated by the California Air Resources Board (data available at http://mesowest.utah.edu/, 2014); and Case Mountain is managed by the Bureau of Land Management (data available at http://www.raws.dri.edu/, 2014). The Giant Forest station is located on an exposed shrub-covered slope; the

Case Mountain, Wolverton and Panther stations are in forest openings; Emerald Lake is an alpine cirque; and Topaz and M3 are in alpine fell fields.

Wind sensors are between 4.2 and 6.5 m above ground level, and we scaled wind speeds to 10 m using a logarithmic profile to estimate saltation thresholds:

$$V_{10} = V_z \left[\frac{\ln \frac{z_x}{k}}{\ln \frac{z_{10}}{k}} \right], \tag{3}$$

where V_{10} is wind velocity at 10 m, V_z is measured velocity, z is instrument height, and k the site specific roughness length.

To identify periods with the greatest potential for wind redistribution of snow, we filtered for times when temperature was below 0 °C and wind velocity was above the minimum saltation threshold of 6.7 m s^{-1} established by Li and Pomeroy (1997a).

Snow depth was measured continuously by 26 ultrasonic snow-depth sensors (manufactured by Judd Communications, Salt Lake City, UT) placed on meteorological stations and over or near snow pillows. These snow-depth sensors have an effective beam width of 22°, and were mounted up to 4.6 m above the ground on a steel arm extending from a vertical steel pipe anchored to a U-channel post. This arrangement provided a snow-depth observation area of up to 2.3 m^2 over flat, bare ground, with sampling area decreasing as snow depth increases.

The LiDAR measurements, plus ground-based snow-density measurements, were used to develop estimates of SWE versus elevation. Paired snow-depth and snow-pillow SWE measurements were part of the California Cooperative Snow Survey network, and data were acquired from the California Department of Water Resources (data available at http://cdec.water.ca.gov/, 2014) for all 16 operable stations on the western slope of the southern Sierra Nevada within 100 km north and 50 km south of the study area (Fig. 1). One snow pillow (GNF) is located 2.5 km west-southwest of the LiDAR acquisition area. Daily snow densities were estimated by dividing the daily mean SWE from the snow pillows by snow depth from the collocated ultrasonic depth sensors. To minimize the error from intermittent noise associated with snow pillows, we used the daily average SWE and did not consider measurements under a 20 cm SWE threshold. This procedure was necessary because complete snow coverage of the snow pillow is unlikely for shallow snow and the combined uncertainties of depth sensors and snow pillows can yield significant error in density measurements (Johnson and Schaefer, 2002). In addition, accumulated precipitation measurements from Alter-shielded Belfort gauges at Giant Forest (GNF), Quaking Aspen (QUA) and Charlotte Lake (CRL) and manually measured daily precipitation from Lodgepole ranger station were compared with SWE measurements to estimate total precipitation (data available at http://cdec.water.ca.gov/, 2014). All instrumental data were formatted, calibrated and gap filled by interpolation or correlation with other sensors and aggregated to daily means prior to analysis (Moffat et al., 2007). Under 1 % of the meteorological data required filtering or gap filling; snow-pillow data required slightly more (< 5 %) and snow depth required up to 20 %. Stations with data gaps >2 days with no nearby station for interpolation were not used in our analysis.

2.5 Bright-band radar

The transition elevation where hydrometeors turn from frozen to liquid, or freezing level, was determined from analysis of hourly Doppler radar data from wind profilers located upstream of the LiDAR-acquisition area. Radar reflectance is greatest, or brightest, in the altitude range where precipitation changes from snow to rain, owing to a difference in the dielectric factor for water and ice and the aggregation of hydrometeors (White et al., 2009; Ryzhkov and Zrnic, 1998). We analyzed bright-band altitudes and thus identified freezing levels from observations collected over the 2010 water year snow-accumulation period (3 December–23 March) from the three nearest upwind locations, i.e., Punta Piedras Blancas, Lost Hills, and Chowchilla, California (data available at http://www.esrl.noaa.gov/psd/data/, 2014) (Fig. 1).

2.6 Model reanalysis

We calculated spatial SWE from LiDAR snow-depth measurements using mean snow-density measurements from the 16 snow-pillow sites. These values were compared with two scales of the widely used PRISM (Parameter-elevation Relationships on Independent Slopes Model) precipitation estimates, plus SWE estimates from two different MODIS-based SWE reconstruction models (Daly et al., 2008, 1994; Rice and Bales, 2011; Rice et al., 2011). Using the available 4 km and 800 m PRISM model output, we summed precipitation for December–March at the spatial extent of the LiDAR acquisition. The 4 km data were monthly for the 2010 water year and the 800 m data were monthly 30-year mean climatology. We then calculated the cumulative precipitation for each 1 m elevation band across the elevation gradient of both data sets, and aggregated values to the resolution of the comparative data.

One reconstruction data set gives 2000–2009 accumulation-period means of the entire Kaweah River watershed, calculated at a 500 m resolution, based on 300 m elevation-bin averages of MODSCAG snow-cover data, local ground-based meteorological measurements, and a temperature-index snowmelt equation that was calibrated with snow-pillow data (Rice and Bales, 2013). Fractional snow-cover was adjusted for canopy using 2 standard deviations of the elevation-band mean.

The second reconstructed SWE data were developed using the algorithm developed by Molotch (2009) and applied to the Sierra Nevada as described in Guan et al. (2013b). Fractional snow cover was adjusted for canopy using vegetation data from the Global Forest Resource Assessment 2000.

The Guan et al. (2013b) values were a subset taken from a December–March Sierra Nevada-wide calculation. The primary difference between this method and the one developed by Rice and Bales (2013) is that the Guan et al. (2013b) method includes an explicit treatment of all radiative and turbulent fluxes, whereas the Rice and Bales (2013) method uses a degree-day melt-flux calculation.

3 Results

3.1 LiDAR-measured snow depth

Of the 53.1 km^2 planer footprint of the LiDAR survey, 0.8 km^2 was over water or in areas that exceeded filter thresholds of the DSM. An additional 0.01 km^2 of area with slope $> 55°$, roads and buildings, and rapidly growing meadow vegetation were also removed from the analysis. The total snow-covered area where both LiDAR and ground returns were available at a density $> 1\ m^{-2}$ was 40.2 km^2, and of this area 5.0 km^2 was under canopy and also eliminated from this analysis. This left an area of 35.2 km^2 remaining for analysis, and of this $< 0.2\ km^2$, mostly below 2300 m, was snow-free. Mean snow depth, measured by LiDAR, increased with altitude from 1850 to 3300 m elevation, with depths decreasing above 3300 m (Fig. 2a). Up to 3300 m, snow depth shows a strong correlation with elevation ($R^2 = 0.974$, $p < 0.001$), increasing at 15 cm per 100 m elevation, with a steep increase in snow depth at 2000–2050 m. Above 3300 m, snow depth sharply decreased at a rate of −48 cm per 100 m ($R^2 = 0.830$, $p < 0.001$). The mean "open" snow-covered area in each 1 m elevation band was 1.7 ha, with a range of 0.1 to 7.3 ha. Overall, 67 % of the study area (excluding water or developed areas) was free of canopy, including most of the 5.6 km^2 area above 3300 m. The increase in snow depth with elevation up to 3300 m is accompanied by a decrease in canopy cover with elevation. Canopy cover, based on the canopy-height model, is greater than 40 % below 2600 m, and near zero above 3200 m (Fig. 2b).

3.2 Wind and topographic effects

Hourly average wind speed at the six meteorological stations showed that the highest potential for redistributing snow was from the westerly directions, with a few periods of strong winds from the northeast at Topaz (Fig. 3). Winds were highest at the three stations above 2800 m and, to a lesser extent, at one lower-elevation station, Giant Forest, which is located in an exposed area free of upwind vegetation. Only five instantaneous gusts over 6.7 $m\,s^{-1}$ were recorded at Panther during the snow-accumulation period, and in one instance at Wolverton; no hourly averages at these sites were over 6.7 $m\,s^{-1}$.

Snow depths were lowest on the southwest- and southeast-facing slopes, and highest on the northwest- and northeast-facing slopes (Fig. 2c). This pattern was most pronounced at elevations above 2400 m, and depths were low especially in the southeast between 2300 and 2700 m, which is a small fraction of the area at this elevation (Fig. 2c). The aspect with the least overall area is northeast and the greatest areal aspect representation faces northwest.

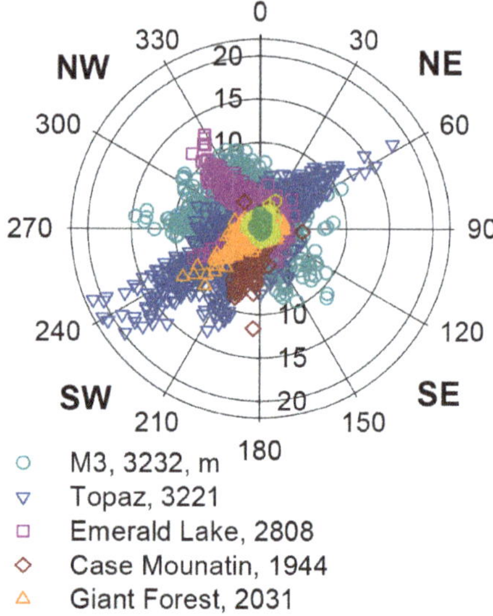

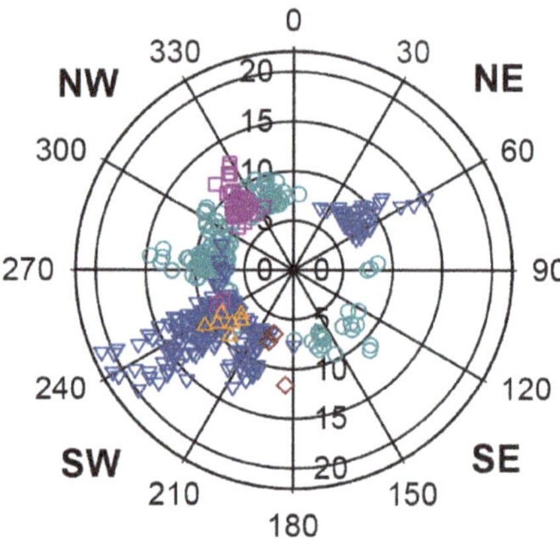

Figure 3. Top: hourly average wind speed and direction for 3 December–23 March accumulation period. Bottom: periods with highest probability of snow redistribution. Radius scale in $m\,s^{-1}$, azimuth in degrees, and north at 0°

The changes in mean snow depth and slope (Fig. 2a and d), over 5–100 m averaging lengths, show an (anti)correlation at −0.16 to −0.36, with the most negative correlation at 35 m (data not shown). The most-rapid changes in slope with elevation show the least increase in snow depth; this is most evident up to 3300 m, above which the terrain becomes flatter (Fig. 2e).

The combined effects of slope and aspect express the "aspect intensity" (I_A), where higher values represent more terrain at that aspect and/or greater slope angles (Fig. 4a).

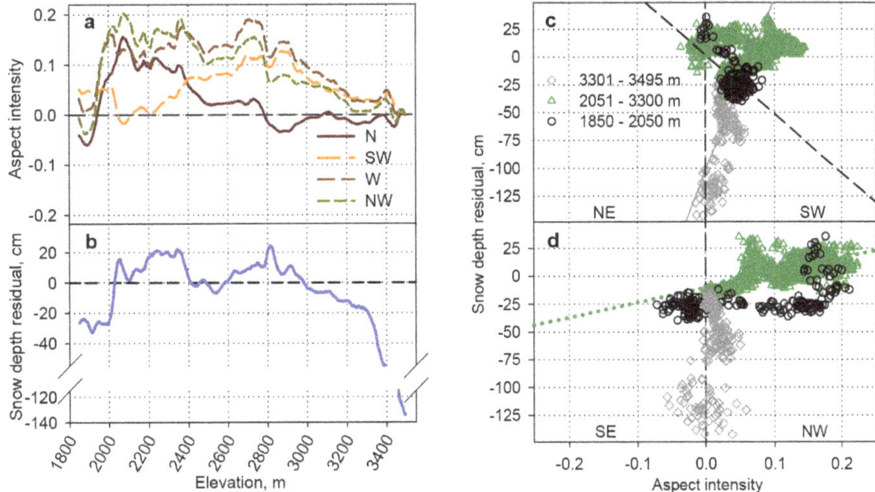

Figure 4. **(a)** Aspect intensity of LiDAR domain by elevation; **(b)** residuals of mean 23 March snow depth using model regression of slope, 1850–3300 m, from Fig. 2a; **(c)** regression of residuals for 1850–2050 m (black, dashed line) and 3301–3494 m (gray, solid line) showing departures from elevation trend for NE and SW aspect intensity; and **(d)** 2051–3300 m (green, dotted line) in SE and NW.

This analysis reveals the slope- and aspect-feature space of the study area, where the predominant sloped aspects of north, southwest, west and northwest have positive I_A values. Conversely, south, northeast, east and southeast have negative values closer to zero and are therefore less represented in the study area. At elevations < 2000 m, moderate east- and southeast-facing slopes, indicated by the slightly negative I_A values, quickly rise to steeper north, northwest and west slopes, as indicated by the higher and positive I_A values (Fig. 4a). Near 2400 m, southwest aspects become more predominant than north, as indicated by the crossover in I_A values, and at higher elevations aspect becomes equally represented by west, southwest and northwest, with some southerly aspects (negative north I_A values) above 2800 m (Fig. 4a).

To evaluate the terrain effects secondary to elevation, we applied a regression to all snow depths as a function of elevation using the slope (0.15) and intercept (-169) from the snow depth measured by LiDAR at 1850–3300 m (Fig. 4b). The residuals from this regression were then correlated with each of the predominant I_A values (Table 2, Fig. 4c and d). I_A snow-depth anomalies for the lowest elevations (1850–2051 m) were negatively correlated with the southeast, at the mid-elevations (2051–3301 m) most positively correlated with the northwest, and at the highest elevations (3300–3494 m) most positively correlated with the southwest (Fig. 4c and d).

3.3 Bright-band radar

The radar-sounding data include 8287 hourly observations (353 missing) from the 3 sites. While individual observations of freezing levels ranged from 200 to 2700 m, the 95th percentile values were in the range of 950–2550 m (Fig. 5). The greatest variability and highest mean freezing level occurred at the coastal station of Punta Piedras, with the lowest values at the furthest-inland station of Chowchilla. This decline in mean freezing levels going inland from the coast suggests that the snow level drops as the air mass moves inland. The third quartile of the freezing level of the Chowchilla station is 2063 m; this closely tracks the break in the coefficient of variation and correlation between snow depth and elevation observed from LiDAR at 2050 m (Figs. 2a and 5), as well as the steep increase in snow depth from 1950 to 2050 m elevation (Fig. 2d).

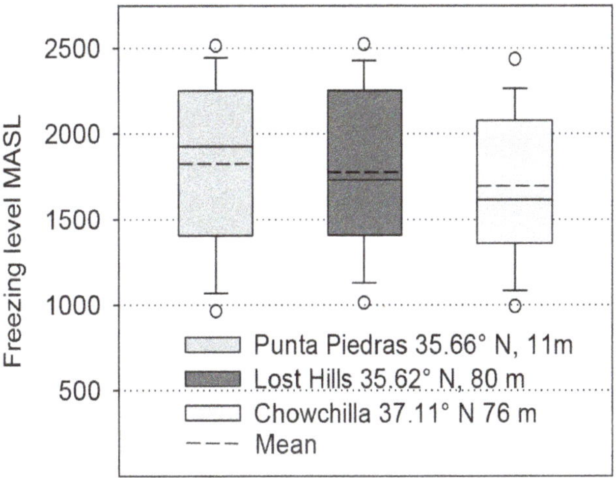

Figure 5. 3 December–23 March accumulation period hourly bright-band freezing level recorded at three wind-profiler stations upwind of the study area; locations shown in Fig. 1. Circles are 5th and 95th percentile.

Table 2. Regression of snow-depth residuals with aspect intensity (I_A).

	R^2/intercept/slope*			
Elevation, m	North	Northwest	Southwest	West
1850–2050	0.32/−23/124	0.22/−26.4/74.3	**0.34/2.0/−531.4**	0.14/−28/81
2051–3300	0.22/1/102	**0.42/−10/134**	0.00/3/10	0.37/−15/160
3301–3494	0/−68/−260	0.08/−72/594	**0.32/−105/1625**	0.25/−91/1028

* All $p < 0.001$, with exception of north at 3301–3494 m and southwest at 2051–3300 m. The three elevations and aspects with the highest R^2 values are in bold.

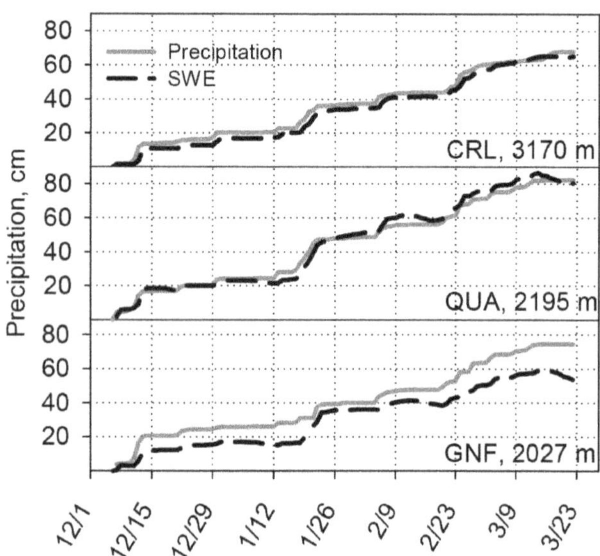

Figure 6. Accumulated gauge precipitation and snow-pillow SWE for the three sites with both measurements. Locations are shown in Fig. 1.

3.4 Ground measurements

Accumulated precipitation and SWE track each other closely at the two higher-elevation sites (CRL and QUA), but measured (GNF) SWE was up to 21 cm less than total precipitation at the lowest site, showing some melt prior to the LiDAR acquisition (Fig. 6). Total precipitation at the lowest total precipitation gauges was 75 cm at GNF (2027 m) and was 72 at Lodgepole (2053 m).

The LiDAR flights were 18 days after mean peak depth and three weeks before mean peak SWE (Fig. 7a, b). The mean and standard deviation (shown in paranthesis) of snow depth during LiDAR acquisition, recorded by the 26 depth sensors in the Wolverton and Panther areas, plus the 16 operational sensors co-located with snow pillows, was 210 (38) cm. This was 19 % less than the mean peak depth of 266 (44) cm recorded on 4 March. However, the mean SWE recorded by the 16 snow pillows during LiDAR acquisition was 82 (16) cm, 2 % less than the mean peak SWE of 83 (20) on 14 April. Two snow pillows, the lowest (GNF) at 2027 m and the most southerly (QUA) at 2195 m, reached peak SWE one week before acquisition, on 15 March, and had ablated 9 and 7 % SWE, respectively, prior to the time of the LiDAR acquisition (Fig. 7b). All other snow pillows either gained SWE or ablated < 5 % in the period prior to the snow LiDAR acquisition. Snow depths measured at the snow-pillow sites on the days of the LiDAR flights failed to show the elevation patterns apparent in the LiDAR depths (Fig. 8).

Daily density values calculated for the 16 snow pillows for 1 February to 30 April indicate a general trend of increasing density and consistent intrasite patterns of accumulation and densification corresponding to stormy and clear conditions (Fig. 7). Over the three month period, density decreased with each accumulation event and increased through densification as the snowpack settled, metamorphosed and integrated free water from melt or rain. At the time of the LiDAR flights the mean density was 384 kg m^{-3}, with a range of 83 kg m^{-3} and standard deviation of 42 kg m^{-3}, across the 1036 m elevation range represented in these data. The combined measurement error of snow-pillow and depth-sensor instruments used in the density calculation can be greater than the range of variability reported here (Johnson and Schaefer, 2002). We found low spatial variability in density that showed no significant relationship with elevation at our sites. This observation concurs with other studies of mountain snowpacks finding spatial consistency in the density of mountain snowpacks (Jonas et al., 2009; Mizukami and Perica, 2008).

3.5 Model reanalysis

The 4 km resolution PRISM data were comprised of seven grid elements in the study domain, whereas the 800 m product had approximately 4225 grid elements. Both PRISM data sets show a small upward trend in precipitation and elevation up to ∼ 300 m and a reversal of this trend at the higher elevations. The 4 km and 800 m PRISM data demonstrate similar magnitudes of increase in precipitation with elevation, 2–3 cm per 100 m, respectively. Because of this small precipitation lapse rate, the PRISM estimates diverge from the LiDAR values below about 2800 m. Total precipitation measured at two locations near the lower extent of the LiDAR footprint during the accumulation period was 72 cm at Lodgepole (2053 m) and 75 cm at Giant Forest (2027 m)

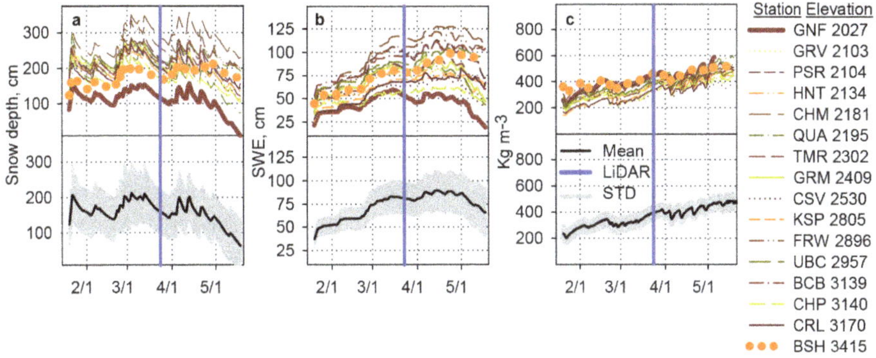

Figure 7. In situ measurements of **(a)** snow depth, **(b)** SWE and **(c)** density for all west-slope snow-pillow and depth sensors in sites located within 1° latitude of study area. Upper halves of the panels show data for individual stations, with highest and lowest elevations plotted in bold. Lower halves of the panels show mean in black, with +1 standard deviation shaded in gray; vertical blue line indicates 23 March LiDAR acquisition dates. Figure 1 shows station names and locations.

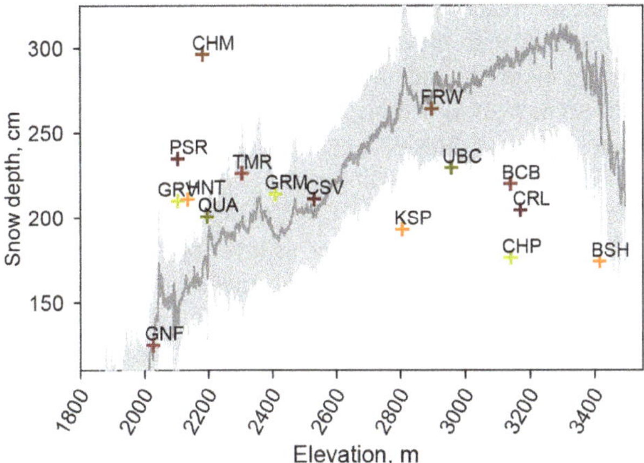

Figure 8. Observed snow depth on the 23 March LiDAR acquisition date for all west-slope snow-pillow sites equipped with depth sensors, plotted over mean LiDAR snow depth (dark gray) and 1 standard deviation (light gray). Giant Forest (GNF), Farewell Gap (FRW) and Chagoopa Plateau (CHP) are within 21 km of the measurement domain. Chilkoot Meadow (CHM) and Poison Ridge (PSR) are the sites furthest to the northwest. Locations shown in Fig. 1.

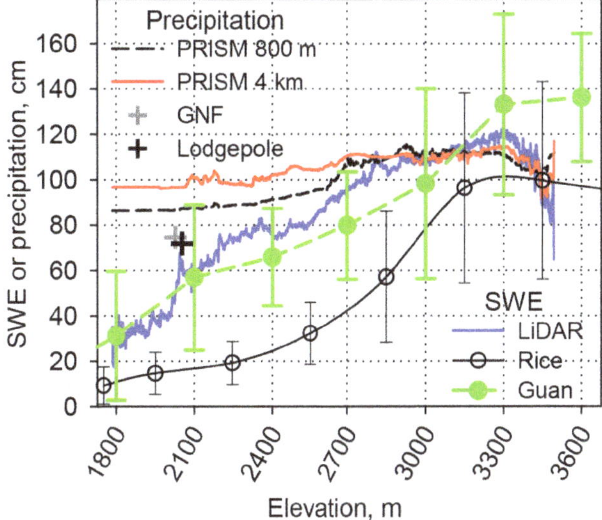

Figure 9. Elevation trend of cumulative precipitation for the Kaweah River watershed for two scales of PRISM and two reconstructions of SWE from daily snowmelt estimates, for December–March, with 23 March 2010 LiDAR SWE estimate and two cumulative precipitation gauge measurements.

(data available at http://cdec.water.ca.gov/, 2014) (Fig. 1). When compared with the LiDAR SWE estimates in Fig. 9, both stations show slightly more precipitation.

Total precipitation was also measured at two additional snow-pillow sensors: CRL (3170 m) and QUA (2195 m) (Fig. 1). From Fig. 6 it is apparent that snow does not account for all of the precipitation at elevations below 2200 m, but does above this elevation where rain had little influence for the accumulation period prior to the LiDAR flight. Thus the LiDAR data should reflect total precipitation above 2200 m.

The LiDAR SWE and the two reconstructed-snowmelt calculations have similar slopes of about 6 cm per 100 m (Fig. 9). The calculations from Guan et al. (2013b) most closely match the LiDAR estimates up to 3300 m, where those from Rice and Bales (2013) are offset by 20–40 cm but show a slight decrease in depth at the highest elevations. In contrast, the two PRISM precipitation models deviate from the LiDAR SWE estimates at elevations below 2800 m and have markedly different slopes.

4 Discussion

The overall increase in precipitation with elevation observed with airborne LiDAR is consistent with the orographic effect of mountains on precipitation (Roe, 2005; Roe and Baker,

2006). Variability of the snow accumulation along the elevation gradient and deviations from a regular increase with elevation can be attributed to the interactions of topography, wind and storm tracks. Deviations from a linear increase are apparent at the lower rain–snow-transition elevations, at higher elevations near the ridge, and at intermediate elevations that have a variety of aspect and steepness characteristics.

4.1 Variability of orographic trends

Elevations over 3300 m showed the greatest negative departure from the overall orographic trend, likely due to the southwest-to-northeast-trending terrain flattening out and no longer providing the necessary lift for the same rate of adiabatic cooling (Figs. 2 and 4). Above 3300 m, the reduced lift over flatter terrain, the exhaustion of precipitable water as storms rise less steeply, and the horizontal displacement of falling snow likely all contribute to declining precipitation at the higher elevations (Mott et al., 2014; Houze Jr., 2012). These processes have been approximated in the Sierra Nevada through simulations based on the convergence of the boundary layer and slope of the local terrain but, until now, have been difficult to observe (Alpert, 1986).

However, as other researchers have noted, it is also difficult to identify the effects of specific storms on snowpack ablation due to the variability of atmospheric conditions close to the Earth's surface (Lundquist et al., 2008). The extent to which high-altitude Sierra Nevada catchments receive more precipitation than adjacent low-altitude areas varies from storm to storm, and from year to year, from occasions during which nearly equal amounts of precipitation fall at high and low altitudes to occasions when 10 or more times as much precipitation falls at the higher altitudes (Dettinger et al., 2004). In the northern Sierra Nevada, the blocking and associated terrain-parallel southerly flow of air masses, referred to as the Sierra barrier jet, can enhance lower-elevation precipitation (Neiman et al., 2008). Conversely, in the central Sierra Nevada, it has been reported that, seasonally, the ratio between higher- versus lower-elevation annual winter-season total precipitation averages about 3, but in some years the ratio drops to as low as 1 (as in 1991) or rises to as much as 4 or 5 (Dettinger, et al., 2004). While the particular orographic patterns reported here could be unique to the 2010 water year, similar patterns have been observed in the mountains of Europe, and previous works have shown consistency in the interannual spatial patterns of snow accumulation (Grünewald et al., 2014; Sturm and Wagner, 2010a; Deems et al., 2008).

4.2 Wind redistribution and radiation effects

The high spatial resolution of LiDAR snow-depth measurements points to two possible controls on snow distribution. While wind affects snow accumulation during a storm, the combined effects of wind and radiation are apparent in post-depositional changes in snow depth. As the same topographic variables influence both wind and radiation, separating the effects based on an analysis of seasonal snow-depth is challenging.

While wind patterns from a single station may be a poor indicator of the wind fields influencing snow redistribution across the entire domain, we expect snow transport by wind to be coarsely defined by the consensus of the local station's wind direction when temperatures are below zero within 24 h of a snowstorm (Figs. 1 and 3). However, the Topaz Lake station, located in smooth terrain with limited upwind influence, may best represent the wind patterns of the free atmosphere and predominant southwest storm winds. We attribute the inconsistent wind direction of other stations to the terrain-induced turbulence of the free atmosphere upwind of the stations. The M3 and Emerald Lake sites have upwind obstacles, and the Wolverton and Panther stations have low wind speeds, reflecting the muting effect of tall forest cover on wind speed and consequently snow redistribution (Fig. 3).

Consistent with prevailing winds from the southwest, we observed more accumulation on the northeast slopes and less on the southwest; however, in our domain, northeast has the least total area of all aspect quadrants, and hence these areas may be underrepresented in the analysis (Fig. 2c).

The aspect intensity variable (I_A) combines the influences of slope and aspect, and serves as a proxy for several processes affecting snow depth, e.g., radiation, upslope orographic deposition, and potentially wind and gravitational redistribution. As a result, some local anomalies, such as deep-snow-patch development, are likely masked when considering topography and snow depth as elevation-band means.

Examining residuals from a linear orographic trend by I_A suggests that the steeper, northwest-facing slopes at the mid-elevations and northerly slopes at the lowest elevations show the greatest snow depths, likely due to the combined effects of wind deposition and lower radiation influx (Fig. 4c and d). Conversely, low- to mid-elevation slopes prone to the combined effects of ablation and wind erosion have the least snow. These findings suggest that departures from the overall orographic trend can be observed in the elevation profile using I_A, but there are limitations to the approach as used here.

It is also possible that there is limited utility in extrapolating prevailing winds from meteorological stations to predict effects of wind on snow redistribution because of the turbulence from local terrain. Research into the relationship between slope, aspect and wind has revealed that small-scale slope breaks and surface roughness have the most-significant effects on where snow accumulates locally (Li and Pomeroy, 1997b; Winstral et al., 2002; Fang and Pomeroy, 2009; Pomeroy and Li, 2000). While not part of this analysis, classification of downwind terrain has also been effective for identifying snow-patch development and persistence of localized wind deposition, offering a deterministic explanation for the spatial stationarity of snow (Winstral et al., 2002;

Schirmer et al., 2011). The I_A variable may also be effective for classifying locations where these processes are likely to occur.

4.3 Sublimation

Wind-driven sublimation may also play a role in the departure from the linear increase in snow depth at the higher elevations, where the highest wind velocities and thus greatest suspension of snow occur (Fig. 3).

In dry intercontinental locations, sublimation rates can be in excess of 50 %, but are much lower in the maritime climate of the Sierra Nevada and lowest during the accumulation period (Ellis et al., 2010; Essery and Pomeroy, 2001). Studies conducted at 2800 and 3100 m in the Emerald Lake basin, located in the center of our measurement domain, found net losses due to evaporation and sublimation of < 10 % for the period between 1 December and 1 April (Marks and Dozier, 1992; Marks et al., 1992). Consequently, we consider the 2010 water year cumulative loss due to sublimation and snowmelt to be limited (< 10 %) prior to the 23 March LiDAR acquisition at all elevation bands, with more melt occurring at the lowest elevations and on the southeast-facing slopes, as indicated by the loss of SWE measured at the low-elevation snow-pillow sites and reduced snow depths on the southeast mid-elevation slopes (Figs. 2c, 6, and 7).

4.4 Rain–snow transition

At lower elevations, below 2050 m, a mix of rain and snow precipitation appears to influence the amount of seasonal snow accumulation. Local SWE measurements are only available at one location below 2050 m (GNF), and this station does show a very small loss of SWE in mid-February as a result of a rain-on-snow event (Fig. 6). Nevertheless, given the expected large storm-to-storm variation in freezing level, the relatively sharp transition in slope of LiDAR-measured snow accumulation at about 2050 m suggests that most precipitation above this elevation fell as snow in the winter of 2010.

In addition, seasonal snow at the lowest elevations and on south-facing slopes has greater positive net energy exchange (from radiation or condensation), and is most susceptible to melt during the accumulation period. LiDAR snow-depth results show lower depths on south-facing versus greater depths on north-facing slopes (Fig. 2c).

4.5 Snow density

Our 23 March, calculations of snow density based on snow depth and snow pillow measurements are uncorrelated with depth or elevation but varied < 11 % from the mean, within the combined uncertainty of the sensors used to calculate them (Figs. 1 and 8). Elder et al. (1998), Anderton et al. (2004) and Anderson et al. (2014) found the variability of spring snow density to be insignificantly correlated with elevation in their studies, while Zhong et al. (2014) found negative correlations with elevation in a meta-study of densities in the former USSR. A range of results has also been reported for the snow-density correlation with depth showing both positive and negative correlations depending on the age of the snow and season (Arons and Colbeck, 1995; McCreight and Small, 2014).

These seemingly contradictory findings can be explained by the seasonal and climatic effects on snow depth and the snowpack energy balance and their affect on snow density. Snow depth is often positively correlated with elevation and the energy balance of the snowpack often negatively correlated with elevation; the magnitude of these effects depends on season and climate (Jonas et al., 2009; Sturm et al., 2010b). For example, in winter, when there are low levels of solar influx on low-albedo snowpacks, snow depth, which is positively correlated with elevation, has a greater influence on density. Conversely, in springtime, or in a warmer climate, a warming snowpack may reverse any previous correlation or be uncorrelated with elevation. Thus our assumption of uniform density may not be accurate for early winter but provides a reasonable estimate for spring snowpack conditions when the LiDAR snow-depth measurement was made.

4.6 Other measures of orographic trends

Although orographic precipitation is a well-documented first-order process, in the Sierra Nevada it is not well described at the watershed to basin scale owing to the very limited availability of ground-based precipitation measurements. Each set of comparative measurements used in this study provides a different index of orographic response: (i) LiDAR is a one-time snapshot of snow depth; (ii) point SWE data are small samples from highly variable spatial values; (iii) reconstructed snowmelt, or retrospective gridded SWE, reflects precipitation minus evaporation and sublimation; and (iv) PRISM is a retrospective precipitation estimate, based largely on lower-elevation stations. Nevertheless these complementary data offer spatially relevant indices of seasonally accumulated precipitation.

As Fig. 8 shows, snow depths from snow-pillow sites fail to capture the elevation patterns apparent in the LiDAR data. This pattern is also apparent in the SWE values from the same sites (Fig. 7b). While the shallowest depth is registered at the lowest elevation site (GNF, 2027 m), where a greater percentage of precipitation falls as rain, the other sites do not show a consistent increase in depth with elevation. Thus current operational measurements in the Sierra Nevada are insufficient to capture orographic trends in snow depth and precipitation.

The less steep increase in precipitation with elevation seen in the two PRISM profiles versus the LiDAR results are thought to be primarily due to the limited number of mountain stations used to calculate the PRISM trends. SWE loss from ablation and rain versus snowfall are important components of the observed LiDAR lapse rates at lower elevations, particularly below 2050 m; these processes should have only a small influence above that elevation. Evidence for this can be seen in three locations of coincident SWE and cumulative precipitation measurements (Fig. 6). The accumulated SWE and total precipitation at the two higher-elevation stations, CRL and QUA, are in close agreement, and the lowest station, GNF, shows 21 cm more total precipitation and a slight loss of SWE on 18–23 March, prior to the date of LiDAR acquisition, demonstrating that measurable rain and melt occurred at the site. In addition, precipitation station in the Kaweah Basin near the LiDAR footprint (LDG, 2053 m) had an accumulation-period total of 72 cm, higher than the LiDAR SWE estimate and lower than both PRISM estimates for the same time period (Fig. 9). The difference in annual precipitation at these sites versus annual SWE accumulation reflects in part the contribution of both rain and snow, as well as mid-winter melt, at this elevation. Thus, divergence of the PRISM and reconstructed SWE at elevations below 2200 m is expected. Temperature records in the area suggest only a small amount of winter melt at 2100 m, with very little winter melt and precipitation as rain above 2400 m (Rice and Bales, 2013).

The general pattern of SWE reconstructed from snowmelt by Guan et al. (2013b) compares with the LiDAR data, being somewhat higher at the highest elevations, lower in the mid-elevations, and similar at the lower elevations. Even though the reconstruction was based on energy-balance modeling, the match is somewhat surprising given the coarseness of the reconstruction model relative to the complex topography of the basin.

The Rice and Bales (2013) reconstruction, in which snowmelt was indexed to amounts and rates at the snow-pillow sites, has less SWE, particularly at the mid- to lower elevations. This offset may stem in part from the higher, 106 % of average 2010 seasonal precipitation versus the lower, 90 % of average, precipitation in the 2000–2009 snowmelt reconstruction period. Further, the reconstructed SWE estimates by Rice and Bales (2013) are based on a temperature-index calculation, versus a full energy-balance approach used by Guan et al. (2013b). Also, some offset in both reconstructed SWE estimates may reflect a bias in snow-covered-area estimates, which have a 500 m spatial resolution and are heavily influenced by canopy. In other words, the LiDAR data represent open areas, and the reconstructed SWE values represent the full domain, but are empirically corrected for canopy.

5 Conclusions

The current results show elevation as the primary determinant of snow depth near the time of peak accumulation over 1650 m on the west slope of the southern Sierra Nevada. LiDAR data reveal patterns potentially associated with orographic processes, mean freezing level, slope, terrain orientation and wind redistribution. Snow depth increased approximately 15 cm per 100 m elevation from snow line to about 3300 m, equivalent to approximately 6 cm SWE per 100 m elevation. This lapse rate is nearly equal to the SWE-reconstruction approach, but higher than the widely used PRISM precipitation data. Localized departures from this trend of +30 to −140 cm from the kilometer-scale pattern of linear increase with elevation are seen in an elevation profile of 1 m elevation bands. Interestingly, snow depth decreased by approximately 48 cm per 100 m elevation from 3300 to 3494 m elevation. Both PRISM and SWE reconstructions show a leveling-off or reductions in SWE at higher elevations as well.

The characterization of snow depth and SWE elevation lapse rates is unique given the high accuracy and high spatial resolution of these data. Moreover, the analysis of the residuals from this elevation trend reveals the role of aspect as a controlling factor, highlighting the importance of solar radiation and wind redistribution with regard to snow distribution. While previous works have come to similar conclusions, the use of LiDAR data reveals these signals in a spatially explicit manner. As LiDAR data become more available, the analyses performed here provide a framework for evaluating the sensitivity of snow-distribution patterns to variability in location and climate.

Acknowledgements. Research was supported by the National Science Foundation (NSF), through the Southern Sierra Critical Zone Observatory (EAR-0725097) and NSF grants (EAR 1141764, EAR 1032295, EAR 0922307), a fellowship for the first author from Southern California Edison, and a seed grant through Lawrence Livermore National Laboratory (6766). Supplemental support was also provided by T. Painter and NASA project NNX10A0976G. We acknowledge the helpful comments from an anonymous reviewer, G. Blöschel, J. Fernandez Diaz, M. Conklin, M. Goulden, R. Rice, C. Riebe and T. Harmon, and thank J. Sickman and J. Melack for providing meteorological data.

Edited by: R. Uijlenhoet

References

Alpert, P.: Mesoscale indexing of the distribution of orographic precipitation over high mountains, J. Clim. Appl. Meteorol., 25, 532–545, doi:10.1175/1520-0450(1986)025<0532:MIOTDO>2.0.CO;2, 1986.

Anderson, B. T., McNamara, J. P., Marshall, H. P., and Flores, A. N.: Insights into the physical processes controlling correlations between snow distribution and terrain properties, Water Resour. Res., 50, 4545–4563, doi:10.1002/2013wr013714, 2014.

Anderson, R. G. and Goulden, M. L.: Relationships between climate, vegetation, and energy exchange across a montane gradient, J. Geophys. Res.-Biogeosci., 116, G01026, doi:10.1029/2010jg001476, 2011.

Anderton, S. P., White, S. M., and Alvera, B.: Evaluation of spatial variability in snow water equivalent for a high mountain catchment, Hydrol. Process., 18, 435–453, doi:10.1002/hyp.1319, 2004.

Arons, E. M. and Colbeck, S. C.: Geometry of heat and mass-transfer in dry snow – a review of theory and experiment, Rev. Geophys., 33, 463–493, doi:10.1029/95rg02073, 1995.

Bales, R. C., Molotch, N. P., Painter, T. H., Dettinger, M. D., Rice, R., and Dozier, J.: Mountain hydrology of the western United States, Water Resour. Res., 42, W08432, doi:10.1029/2005WR004387, 2006.

Bales, R. C., Dressler, K. A., Imam, B., Fassnacht, S. R., and Lampkin, D.: Fractional snow cover in the Colorado and Rio Grande basins, 1995–2002, Water Resour. Res., 44, W01425, doi:10.1029/2006wr005377, 2008.

Barnett, T. P., Pierce, D. W., Hidalgo, H. G., Bonfils, C., Santer, B. D., Das, T., Bala, G., Wood, A. W., Nozawa, T., Mirin, A. A., Cayan, D. R., and Dettinger, M. D.: Human-induced changes in the hydrology of the western United States, Science, 319, 1080–1083, doi:10.1126/science.1152538, 2008.

Daly, C., Neilson, R. P., and Phillips, D. L.: A statistical topographic model for mapping climatological precipitation over mountainous terrain, J. Appl. Meteorol., 33, 140–158, doi:10.1175/1520-0450(1994)033<0140:ASTMFM>2.0.CO;2, 1994.

Daly, C., Halbleib, M., Smith, J. I., Gibson, W. P., Doggett, M. K., Taylor, G. H., Curtis, J., and Pasteris, P. P.: Physiographically sensitive mapping of climatological temperature and precipitation across the conterminous United States, Int. J. Climatol., 28, 2031–2064, doi:10.1002/joc.1688, 2008.

Deems, J. S., Fassnacht, S. R., and Elder, K. J.: Interannual Consistency in Fractal Snow Depth Patterns at Two Colorado Mountain Sites, J. Hydrometeorol., 9, 977–988, doi:10.1175/2008jhm901.1, 2008.

Deems, J. S., Painter, T. H., and Finnegan, D. C.: Lidar measurement of snow depth: a review, J. Glaciol., 59, 467–479, doi:10.3189/2013JoG12J154, 2013.

Dettinger, M., Redmond, K., and Cayan, D. R.: Winter orographic-precipitation ratios in the Sierra Nevada-Large-scale atmospheric circulations and hydrologic consequences, J. Hydrometeorol., 5, 1102–1116, 2004.

Dettinger, M. D., Ralph, F. M., Das, T., Neiman, P. J., and Cayan, D. R.: Atmospheric Rivers, Floods and the Water Resources of California, Water, 3, 445–478, 2011.

Earman, S. and Dettinger, M.: Potential impacts of climate change on groundwater resources – a global review, J. Water Clim. Change, 2, 213–229, 2011.

Elder, K., Rosenthal, W., and Davis, R. E.: Estimating the spatial distribution of snow water equivalence in a montane watershed, Hydrol. Process., 12, 1793–1808, 1998.

Ellis, C. R., Pomeroy, J. W., Brown, T., and MacDonald, J.: Simulation of snow accumulation and melt in needleleaf forest environments, Hydrol. Earth Syst. Sci., 14, 925–940, doi:10.5194/hess-14-925-2010, 2010.

Erskine, R. H., Green, T. R., Ramirez, J. A., and MacDonald, L. H.: Comparison of grid-based algorithms for computing upslope contributing area, Water Resour. Res., 42, 9, W09416, doi:10.1029/2005wr004648, 2006.

Essery, R. and Pomeroy, J. W.: Sublimation of snow intercepted by coniferous forest canopies in a climate model, in: Proceedings of the Sixth International Association of Hydrologic Sciences Assembly at Maastricht, the Netherlands, Soil-Vegetation-Atmosphere Transfer Schemes and Large-Scale Hydrological Models, IAHS Publication 270, 343–347, 2001.

Fang, X. and Pomeroy, J. W.: Modelling blowing snow redistribution to prairie wetlands, Hydrol. Process., 23, 2557–2569, doi:10.1002/hyp.7348, 2009.

Fassnacht, S. R., Dressler, K. A., and Bales, R. C.: Snow water equivalent interpolation for the Colorado River Basin from snow telemetry (SNOTEL) data, Water Resour. Res., 39, 1208, doi:10.1029/2002WR001512, 2003.

Galewsky, J.: Rain shadow development during the growth of mountain ranges: An atmospheric dynamics perspective, J. Geophys. Res.-Earth, 114, F01018, doi:10.1029/2008JF001085, 2009.

Grünewald, T. and Lehning, M.: Altitudinal dependency of snow amounts in two small alpine catchments: can catchment-wide snow amounts be estimated via single snow or precipitation stations?, 52, 153–158, doi:103189/172756411797252248, 2011.

Grünewald, T., Bühler, Y., and Lehning, M.: Elevation dependency of mountain snow depth, The Cryosphere Discuss., 8, 3665–3698, doi:10.5194/tcd-8-3665-2014, 2014.

Guan, B., Waliser, D. E., Molotch, N. P., Fetzer, E. J., and Neiman, P. J.: Does the Madden-Julian Oscillation Influence Wintertime Atmospheric Rivers and Snowpack in the Sierra Nevada?, Mon. Weather Rev., 140, 325–342, 2012.

Guan, B., Molotch, N. P., Waliser, D. E., Fetzer, E. J., and Neiman, P. J.: The 2010/2011 snow season in California's Sierra Nevada: Role of atmospheric rivers and modes of large-scale variability, Water Resour. Res., 49, 6731–6743, doi:10.1002/wrcr.20537, 2013a.

Guan, B., Molotch, N. P., Waliser, D. E., Jepsen, S. M., Painter, T. H., and Dozier, J.: Snow water equivalent in the Sierra Nevada: Blending snow sensor observations with snowmelt model simulations, Water Resour. Res., 49, 5029–5046, doi:10.1002/wrcr.20387, 2013b.

Guo, Q., Li, W., Yu, H., and Alvarez, O.: Effects of Topographic Variability and Lidar Sampling Density on Several DEM Interpolation Methods, Photogramm. Eng. Rem. S., 76, 701–712, 2010.

Harpold, A. A., Guo, Q., Molotch, N., Brooks, P. D., Bales, R., Fernandez-Diaz, J. C., Musselman, K. N., Swetnam, T. L., Kirchner, P., Meadows, M., Flanagan, J., and Lucas, R.: LiDAR-Derived Snowpack Datasets From Mixed Conifer Forests Across the Western U.S., Water Resour. Res., 50, 2749–2755, doi:10.1002/2013WR013935, 2014.

Houze Jr., R. A.: Orographic effects on precipitating clouds, Rev. Geophys., 50, RG1001, doi:10.1029/2011RG000365, 2012.

Johnson, J. B. and Schaefer, G. L.: The influence of thermal, hydrologic, and snow deformation mechanisms on snow water equivalent pressure sensor accuracy, Hydrol. Process., 16, 3529–3542, doi:10.1002/hyp.1236, 2002.

Jonas, T., Marty, C., and Magnusson, J.: Estimating the snow water equivalent from snow depth measurements in the Swiss Alps, J. Hydrol., 378, 161–167, doi:10.1016/j.jhydrol.2009.09.021, 2009.

Kerkez, B., Glaser, S. D., Bales, R. C., and Meadows, M. W.: Design and performance of a wireless sensor network for catchment-scale snow and soil moisture measurements, Water Resour. Res., 48, W09515, doi:10.1029/2011wr011214, 2012.

Kessler, M. A., Anderson, R. S., and Stock, G. M.: Modeling topographic and climatic control of east-west asymmetry in Sierra Nevada glacier length during the Last Glacial Maximum, J. Geophys. Res.-Earth, 111, F02002, doi:10.1029/2005jf000365, 2006.

Kienzle, S.: The Effect of DEM Raster Resolution on First Order, Second Order and Compound Terrain Derivatives, Trans. GIS, 8, 83–111, 2004.

Li, L. and Pomeroy, J. W.: Estimates of threshold wind speeds for snow transport using meteorological data, J. Appl. Meteorol., 36, 205–213, doi:10.1175/1520-0450(1997)036<0205:EOTWSF>2.0.CO;2, 1997a.

Li, L. and Pomeroy, J. W.: Probability of occurrence of blowing snow, J. Geophys. Res.-Atmos., 102, 21955–21964, doi:10.1029/97jd01522, 1997b.

Lundquist, J. D., Dettinger, M. D., and Cayan, D. R.: Snow-fed streamflow timing at different basin scales: Case study of the Tuolumne River above Hetch Hetchy, Yosemite, California, Water Resour. Res., 41, W07005, doi:10.1029/2004wr003933, 2005.

Lundquist, J. D., Neiman, P. J., Martner, B., White, A. B., Gottas, D. J., and Ralph, F. M.: Rain versus snow in the Sierra Nevada, California: Comparing Doppler profiling radar and surface observations of melting level, J. Hydrometeorol., 9, 194–211, doi:10.1175/2007jhm853.1, 2008.

Marks, D. and Dozier, J.: Climate and Energy Exchange at the Snow Surface in the Alpine Region of the Sierra-Nevada .2. Snow Cover Energy-Balance, Water Resour. Res., 28, 3043–3054, doi:10.1029/92wr01483, 1992.

Marks, D., Dozier, J., and Davis, R. E.: Climate and Energy Exchange at the Snow Surface in the Alpine Region of the Sierra-Nevada .1. Metrological Measurements and Monitoring, Water Resour. Res., 28, 3029–3042, doi:10.1029/92wr01482, 1992.

Marks, D., Link, T., Winstral, A., and Garen, D.: Simulating snowmelt processes during rain-on-snow over a semi-arid mountain basin, Ann. Glaciol., 32, 195–202, 2001.

Maxwell, R. M. and Kollet, S. J.: Interdependence of groundwater dynamics and land-energy feedbacks under climate change, Nat. Geosci., 1, 665–669, 2008.

McCreight, J. L. and Small, E. E.: Modeling bulk density and snow water equivalent using daily snow depth observations, The Cryosphere, 8, 521–536, doi:10.5194/tc-8-521-2014, 2014.

Meromy, L., Molotch, N. P., Link, T. E., Fassnacht, S. R., and Rice, R.: Subgrid variability of snow water equivalent at operational snow stations in the western USA, Hydrol. Process., 27, 2383–2400, 2012.

Milly, P. C. D., Betancourt, J., Falkenmark, M., Hirsch, R. M., Kundzewicz, Z. W., Lettenmaier, D. P., and Stouffer, R. J.: Climate Change: Stationarity Is Dead: Whither Water Management?, Science, 319, 573–574, doi:10.1126/science.1151915, 2008.

Mizukami, N. and Perica, S.: Spatiotemporal Characteristics of Snowpack Density in the Mountainous Regions of the Western United States, J. Hydrometeorol., 9, 1416–1426, doi:10.1175/2008jhm981.1, 2008.

Moffat, A. M., Papale, D., Reichstein, M., Hollinger, D. Y., Richardson, A. D., Barr, A. G., Beckstein, C., Braswell, B. H., Churkina, G., Desai, A. R., Falge, E., Gove, J. H., Heimann, M., Hui, D., Jarvis, A. J., Kattge, J., Noormets, A., and Stauch, V. J.: Comprehensive comparison of gap-filling techniques for eddy covariance net carbon fluxes, Agr. Forest. Meteorol., 147, 209–232, 2007.

Molotch, N. P., Painter, T. H., Bales, R. C., and Dozier, J.: Incorporating remotely-sensed snow albedo into a spatially-distributed snowmelt model, Geophys. Res. Lett., 31, L03501, doi:10.1029/2003GL019063, 2004.

Molotch, N. P. and Margulis, S. A.: Estimating the distribution of snow water equivalent using remotely sensed snow cover data and a spatially distributed snowmelt model: A multi-resolution, multi-sensor comparison, Adv. Water Resour., 31, 1503–1514, doi:10.1016/j.advwatres.2008.07.017, 2008.

Molotch, N. P.: Reconstructing snow water equivalent in the Rio Grande headwaters using remotely sensed snow cover data and a spatially distributed snowmelt model, Hydrol. Process., 23, 1076–1089, doi:10.1002/hyp.7206, 2009.

Mott, R., Scipión, D., Schneebeli, M., Dawes, N., Berne, A., and Lehning, M.: Orographic effects on snow deposition patterns in mountainous terrain, J. Geophys. Res.-Atmos., 119, 1419–1439, doi:10.1002/2013JD019880, 2014.

Neiman, P. J., Ralph, F. M., Wick, G. A., Lundquist, J. D., and Dettinger, M. D.: Meteorological characteristics and overland precipitation impacts of atmospheric rivers affecting the West Coast of North America based on eight years of SSM/I satellite observations, J. Hydrometeorol., 9, 22–47, doi:10.1175/2007jhm855.1, 2008.

Pandey, G. R., Cayan, D. R., and Georgakakos, K. P.: Precipitation structure in the Sierra Nevada of California during winter, J. Geophys. Res.-Atmos., 104, 12019–12030, doi:10.1029/1999JD900103, 1999.

Pedersen, V. K., Egholm, D. L., and Nielsen, S. B.: Alpine glacial topography and the rate of rock column uplift: a global perspective, Geomorphology, 122, 129–139, doi:10.1016/j.geomorph.2010.06.005, 2010.

Peterson, D. H., Smith, R. E., Dettinger, M. D., Cayan, D. R., and Riddle, L.: An organized signal in snowmelt runoff over the western United States, J. Am. Water Resour. Assoc., 36, 421–432, 2000.

Pomeroy, J. W. and Li, L.: Prairie and arctic areal snow cover mass balance using a blowing snow model, J. Geophys. Res.-Atmos., 105, 26619–26634, doi:10.1029/2000jd900149, 2000.

Rahmstorf, S. and Coumou, D.: Increase of extreme events in a warming world, Proc. Natl. Acad. Sci., 108, 17905–17909, doi:10.1073/pnas.1101766108, 2011.

Ralph, F. M. and Dettinger, M. D.: Storms, floods, and the science of atmospheric rivers, EOS Trans. AGU, 92, 2011.

Rice, R. and Bales, R. C.: An Assessment of Snowcover in 6 Major River Basins of Sierra Nevada and Potential Approaches for Long-term Monitoring, in: Fall Meeting of the American Geophysical Union, San Francisco, USA, December, 2011.

Rice, R., Bales, R. C., Painter, T. H., and Dozier, J.: Snow water equivalent along elevation gradients in the Merced and Tuolumne River basins of the Sierra Nevada, Water Resour. Res., 47, W08515, doi:10.1029/2010wr009278, 2011.

Rice, R. and Bales, R.: Water Qauntity: rain, snow, and temperature. Natural Resource Report. NPS/SEKI/NRR-2013/665.7a, National Park Service, Fort Collins, Colorado, 2013.

Roberts, D. W.: Ordination on the basis of fuzzy set theory, Vegetatio, 66, 123–131, 1986.

Roe, G. H.: Orographic precipitation, Annu. Rev. Earth Planet. Sci., 33, 645–671, doi:10.1146/annurev.earth.33.092203.122541, 2005.

Roe, G. H. and Baker, M. B.: Microphysical and geometrical controls on the pattern of orographic precipitation, J. Atmos., 63, 861–880, doi:10.1175/jas3619.1, 2006.

Ryzhkov, A. V. and Zrnic, D. S.: Discrimination between rain and snow with a polarimetric radar, J. Appl. Meteorol., 37, 1228–1240, doi:10.1175/1520-0450, 1998.

Schaer, P., Skaloud, J., Landtwing, S., and Legat, K.: Accuracy estimation for laser point cloud including accuracy estimation, Proceedings of the5th International Symposium on Mobile Mapping Technology, Padua, Italy, 2007.

Schirmer, M., Wirz, V., Clifton, A., and Lehning, M.: Persistence in intra-annual snow depth distribution: 1 Measurements and topographic control, Water Resour. Res., 47, W09516, doi:10.1029/2010WR009426, 2011.

Shrestha, R., Carter, W., Slatton, C., and Dietrich, W.: "Research-Quality" Airborne Laser Swath Mapping: The Defining Factors, The National Center for Airborne Laser Mapping, white paper, 25 pp., 2007.

Slatton, K. C., Carter, W. E., Shrestha, R. L., and Dietrich, W.: Airborne Laser Swath Mapping: Achieving the resolution and accuracy required for geosurficial research, Geophys. Res. Lett., 34, 5, L23s10, doi:10.1029/2007gl031939, 2007.

Stolar, D., Roe, G., and Willett, S.: Controls on the patterns of topography and erosion rate in a critical orogen, J. Geophys. Res., 112, F04002, doi:10.1029/2006JF000713, 2007.

Sturm, M. and Wagner, A. M.: Using repeated patterns in snow distribution modeling: An Arctic example, Water Resour. Res., 46, W12549, doi:10.1029/2010wr009434, 2010a.

Sturm, M., Taras, B., Liston, G. E., Derksen, C., Jonas, T., and Lea, J.: Estimating snow water equivalent using snow depth data and climate classes, J. Hydrometeorol., 11, 1380–1394, doi:10.1175/2010JHM1202.1, 2010b.

Trujillo, E., Molotch, N. P., Goulden, M. L., Kelly, A. E., and Bales, R. C.: Elevation-dependent influence of snow accumulation on forest greening, Nat. Geosci., 5, 705–709, 2012.

Viviroli, D., Archer, D. R., Buytaert, W., Fowler, H. J., Greenwood, G. B., Hamlet, A. F., Huang, Y., Koboltschnig, G., Litaor, M. I., Lopez-Moreno, J. I., Lorentz, S., Schadler, B., Schreier, H., Schwaiger, K., Vuille, M., and Woods, R.: Climate change and mountain water resources: overview and recommendations for research, management and policy, Hydrol. Earth Syst. Sc., 15, 471–504, doi:10.5194/hess-15-471-2011, 2011.

White, A. B., Gottas, D. J., Henkel, A. F., Neiman, P. J., Ralph, F. M., and Gutman, S. I.: Developing a performance measure for snow-level forecasts, J. Hydrometeorol., 11, 739–753, 2009.

Winstral, A., Elder, K., and Davis, R. E.: Spatial snow modeling of wind-redistributed snow using terrain-based parameters, J. Hydrometeorol., 3, 524–538, doi:10.1175/1525-7541(2002)003<0524:SSMOWR>2.0.CO;2, 2002.

Xiaoye, L.: Airborne LiDAR for DEM generation: some critical issues, Prog. Phys. Geogr., 32, 31–49, doi:10.1177/0309133308089496, 2008.

Zhang, K. and Cui, Z.: Airborne LiDAR data processing and analysis tools - ALDPAT 1.0., available at: http://lidar.ihrc.fiu.edu/lidartool.html (last access: March 2014), 2007.

Zhong, X., Zhang, T., and Wang, K.: Snow density climatology across the former USSR, The Cryosphere, 8, 785–799, doi:10.5194/tc-8-785-2014, 2014

4

Improving the precipitation accumulation analysis using lightning measurements and different integration periods

Erik Gregow[1], **Antti Pessi**[2], **Antti Mäkelä**[1], **and Elena Saltikoff**[1]

[1]Meteorological Research, Finnish Meteorological Institute, P.O. Box 503, 00101 Helsinki, Finland
[2]Applied Meteorology, Vaisala, 3 Lan Dr., Westford, MA 01886, USA

Correspondence to: Erik Gregow (erik.gregow@fmi.fi)

Abstract. The focus of this article is to improve the precipitation accumulation analysis, with special focus on the intense precipitation events. Two main objectives are addressed: (i) the assimilation of lightning observations together with radar and gauge measurements, and (ii) the analysis of the impact of different integration periods in the radar–gauge correction method. The article is a continuation of previous work by Gregow et al. (2013) in the same research field.

A new lightning data assimilation method has been implemented and validated within the Finnish Meteorological Institute – Local Analysis and Prediction System. Lightning data do improve the analysis when no radars are available, and even with radar data, lightning data have a positive impact on the results.

The radar–gauge assimilation method is highly dependent on statistical relationships between radar and gauges, when performing the correction to the precipitation accumulation field. Here, we investigate the usage of different time integration intervals: 1, 6, 12, 24 h and 7 days. This will change the amount of data used and affect the statistical calculation of the radar–gauge relations. Verification shows that the real-time analysis using the 1 h integration time length gives the best results.

1 Introduction

Accurate estimates of accumulated precipitation are needed for several applications such as flood protection, hydropower, road- and fire-weather models. In Finland, one of the most economically relevant users of precipitation is the hydropower industry. Between 10 and 20 % of Finnish annual electric power production comes from hydropower, depending on the amount of precipitation and water levels in dams and water reservoirs. In order to maintain correct calculation of the energy supplied to customers and to avoid (or at least minimize) the environmental risks and economical losses during extreme precipitation and flooding events, a profound analysis of the expected water amounts in dams and reservoirs from catchment areas is needed. The current hydropower strategy of Finland is to increase capacity by improving the efficiency of existing plants through technical adjustments. The maintenance and planning of proper dam structures need the most up-to-date information about the rain rates to be able to adjust the regulation functions of the dams, both for the current and the changing climatic conditions (IPCC-AR5, 2013).

Often, the accumulated precipitation values are based on pure radar analysis, unless there exists a surface gauge observation in the immediate surroundings. Radar echoes are related to rainfall rate and thereafter transformed into accumulation values. However, such conversions are based on general empirical relations which are not suitable for all meteorological cases (e.g., depending on precipitation type; Koistinen and Michelson, 2002). Radar reflectivity can, in some cases, suffer from poor quality, resulting from electronic miscalibration, beam blocking, clutter, attenuation and overhanging precipitation (Saltikoff et al., 2010), which results in poor estimations of the precipitation accumulation. In some cases, the radar can even be missing, e.g., during maintenance, upgrading or due to technical problems. Especially during thunderstorms, there is a potential of radar disturbances, either in the form of missing data due to interruptions in electricity and telecommunication systems, or in the form of quality issues such as attenuation, due to intervening heavy precipitation.

The research of combining radar and surface observations, to perform corrections to precipitation accumulation, is well explored. Many have made developments in this field and much literature is available, for example, Sideris et al. (2014), Schiemann et al. (2011) and Goudenhoofdt and Delobbe (2009). In general, combining radar and rain gauge data is very difficult in the vicinity of heavy local rain cells (Einfalt et al., 2005). Recently, Jewell and Gaussiat (2015) compared performances of different merging schemas and noted a large difference between convective and stratiform situations. In their study, the nonparametric kriging with external drift outperformed other methods in an accumulation period of 60 min. Wang et al. (2015) developed a sophisticated method for urban hydrology, which preserves the nonnormal characteristics of the precipitation field. They also noticed that common methods have a tendency to smooth out the important but spatially limited extremes of precipitation.

Comparing radars and gauges, an additional challenge arises from the different sampling sizes of the instruments. Radar measurement volume can be several kilometers wide and thick (a 1° beam is approximately 5 km wide at 250 km), while the measurement area of a gauge is 400 cm^2 (weighing gauges) or 100 cm^3 (optical instruments). Part of the disparateness of radar and gauge measurements is due to variability of the raindrop size distribution within the area of a single radar pixel. Jaffrain and Berne (2012) have observed variability up to 15 % of the rain rate in a 1×1 km pixel, with time steps of 1 min.

Lightning is associated with convective precipitation, but in areas where a large portion of precipitation is stratiform, lightning data alone are not adequate for precipitation estimation. Although convective events contribute only a fraction of the annual precipitation amount, they might be important during flooding events. However, lightning has been used to complement and improve other datasets. Morales and Agnastou (2003) combined lightning with satellite-based measurements to distinguish between convective and stratiform precipitation area and achieved a remarkable 31 % bias reduction, compared to satellite-only techniques. Lightning has also been assimilated to numerical weather prediction (NWP) models, using nudging techniques, or improving the initialization process of the model. This can be done by blending them with other remote sensing data to create heating profiles (e.g., estimating the latent heat release when precipitation is condensed). Papadopulos et al. (2005) used lightning data to identify convective areas and then modified the model humidity profiles, allowing the model to produce convection and release latent heat using its own convective parameterization scheme. They combined lightning with 6-hourly gauge data, within a mesoscale model in the Mediterranean area, and showed improvement in forecasts up to 12 h lead time. Pessi and Businger (2009) derived a lightning–convective rainfall relationship over the North Pacific Ocean and used it for latent heat nudging method in an NWP model. They were able to improve the pressure forecast of a North Pacific winter storm significantly.

Our situation is different from the above-mentioned experiments because lightning activity is usually low in Finland, compared to warmer climates (Mäkelä et al., 2011). Also, our analysis area already has a good radar coverage and a relatively evenly distributed network of 1 h gauge measurements. However, if we want to enlarge the analysis area, we will soon go to either sea areas or neighboring countries where availability of radar data and frequent gauge measurements is low. We also anticipate the usefulness of lightning data as a backup plan in the occasions when radar data are either missing or of deteriorated quality. Even though these occasions are rare, they often occur on days when detailed precipitation estimates are of great interest. Thunderstorms producing heavy localized rainfall are also often producing heavy winds, causing unavailability of radar data due to breaks on electricity and data communications. Our principal goal is to have as good an analysis as possible, which is different from having a best analysis to start a model.

Gregow et al. (2013) have demonstrated the benefit of assimilating different data sources (radars and gauges) in precipitation estimation. The largest uncertainties were observed during heavy convective rainfall. These are the situations when lightning occurs. The accumulation process is based on the radar reflectivity field, where gauges correct the initial field; e.g., if there is no reflectivity field, there is no accumulation (gauges are not used alone). To improve the spatially accurate real-time precipitation analysis, new methods are adopted by fusion of weather radar, lightning observations and rain gauge information in novel ways. This leads to better possibilities in estimating convective rainfall events (i.e., $> 5\,mm\,h^{-1}$) and the accumulated precipitation for the benefit of hydropower management and other related application areas. The work reported here has been performed using the Local Analysis and Prediction System (LAPS), which is used operationally in the weather service of the Finnish Meteorological Institute (FMI). Testing new approaches in an operational system has its challenges. For example, it is not possible to exclude a large amount of independent reference stations. Also, the possibilities to rerun cases with different settings have been limited. The major benefit of working in an operational environment is that we can be sure that we only use data and methods which are operationally available and feasible.

In this article, the observational datasets are described in Sect. 2. New methods on how to calculate the precipitation accumulation are handled in Sect. 3, and the results and discussion are shown in Sects. 4 and 5, respectively.

2 Observations and instrumentation

Here, we describe the three data sources employed in this study (rain gauge, radar and lightning observations) and the verification periods used in this study.

2.1 Rain gauge observations

Rain gauges provide point observations of the accumulation. They are usually considered more accurate than radar as point values and are frequently used to correct the radar field (Wilson and Brandes, 1979). The surface precipitation network (in total, 472 stations) consists of standard weighting gauges and optical sensors mounted on road-weather masts. Since 2015, FMI has managed 102 stations instrumented with the weighting gauge OTT Messtechnik Pluvio2. The Finnish Transport Agency (FTA) runs 370 road-weather stations with optical sensor measurements (Vaisala Present Weather Detectors models PWD22 and, to some extent, PWD11). The precipitation intensity is measured in different time intervals which are summed up to 1 h precipitation accumulation information. Uncertainties and more detailed information can be found in Gregow et al. (2013). If measurements consistently indicate poor data quality, either manually identified from station error logs or by inspecting the data, those stations are blacklisted within the LAPS process and do not contribute to the precipitation accumulation analysis. Hereafter, in this article, the weighting gauges and road-weather measurements are indistinctly called gauges and their placement in Finland is shown in Fig. 1a.

2.2 The radar data

As of summer 2016, FMI operates 10 C-band Doppler radars (with the newest one operational since late 2015). All but one station (VIM in western Finland; see Fig. 1b) are dual-polarization radars. At the moment, the quantitative precipitation estimation based on dual polarization is not used operationally in FMI, but the polarimetric properties contribute to the improved clutter cancellation (i.e., removal of non-meteorological echoes, especially sea clutter, birds and insects). In southern Finland, the distance between radars is 140–200 km, but in the north, the distance between stations LUO and UTA is 260 km. The location of the radars and the coverage is shown in Fig. 1b. As Finland has no high mountains, the horizon of all the radars is near zero elevation with no major beam blockage, and, in general, the radar coverage is very good except in the most northern part of the country. The Finnish radar network does have a very high system utilization rate (e.g., no interruption). During the years 2014 and 2015, the utilization rate was >99 %. Further details of the FMI radar network and processing routines are described in Saltikoff et al. (2010).

The basic radar volume scan consists of 13 plan position indicator (PPI) sweeps. The FMI-operated LAPS version (hereafter FMI-LAPS) is using the six lowest elevations: 0.3 (alternative 0.1 or 0.5, depending on site location), 0.7, 1.5, 3.0, 5.0 and 9.0, which are scanned out to 250 km, and repeated every 5 min. These radar volume scans are further used in LAPS routines for the rain rate calculations but also as proxy data to the lightning data assimilation (LDA) method (see Sect. 3.2).

2.3 The Lightning Location System (LLS)

The Lightning Location System (LLS) of FMI is part of the Nordic Lightning Information System (NORDLIS). The system detects cloud-to-ground (CG) and intracloud (IC) strokes in the low-frequency (LF) domain. Finland is situated between 60–70° N and 19–32° E, and thunderstorm season begins usually in May and lasts until September. During the period 1960–2007, on average, 140 000 ground flashes occurred during approximately 100 days per year (Tuomi and Mäkelä, 2008). The present modern LLS was installed in summer 1997 (Tuomi and Mäkelä, 2007; Mäkelä et al., 2010, 2016). The system consists of Vaisala Inc. sensors of various generations, and the sensor locations in 2015 and the efficient network coverage area can be seen in Fig. 2. Lightning location sensors detect the electromagnetic (EM) signals emitted by lightning return strokes, and measure the signal azimuth and exact time (GPS). Sensors send this information to the central processing computer in real time which combines them, optimizes the most probable strike point and outputs this information to the end user. More detailed information of LLS principles is described in Cummins et al. (1998).

2.4 Verification periods

The verification periods are limited to summer season (the active convective season in Finland) where two long periods were included in the verification: (a) 1 April to 1 September 2015 and (b) 1 May to 26 July 2016. These long verification periods include many cases of stratiform precipitation with no lightning, and therefore the effective impact by lightning is diluted (e.g., no influence by the LDA method). Hence, two subsets of two lightning intensive cases (e.g., situations with heavy rain and strong convection), datasets (c) and (d), were used to explicitly find the lightning-induced impacts. The dataset (c) includes full days (24 h periods) with more than 100 CG strokes per day. The dataset (d) includes only the stations and time intervals affected by lightning (defined as stations with maximum distance of 30 km to the lightning position and within the 1 h accumulation time interval, hereafter called the scaled dataset). An early dataset from 2014, dataset (e), consists of 4 days (3, 23, 24 and 30 July 2014) with more than 100 CG strokes per day. This dataset was used to perform several autonomous experiments with the FMI-LAPS LDA system in the early stage of the development of the LDA method.

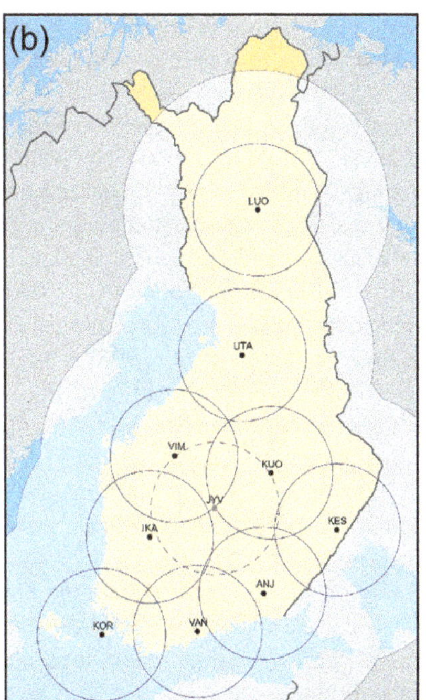

Figure 1. In panel **(a)**, the Finnish surface gauge stations are shown (as dots on the map); these are used to measure the hourly precipitation accumulation. The red dots indicate the position of the seven independent stations used for the verification. In panel **(b)**, the outer rectangular frame of the map depicts the LAPS analysis domain. The black dots represent the 10 Finnish radar stations and the outer black curved lines display their coverage. The thin circles surrounding each radar represent the areas where measurements are performed below 2 km height. The dashed circle indicates radar station JYV, which was not included in the radar network during summer 2015.

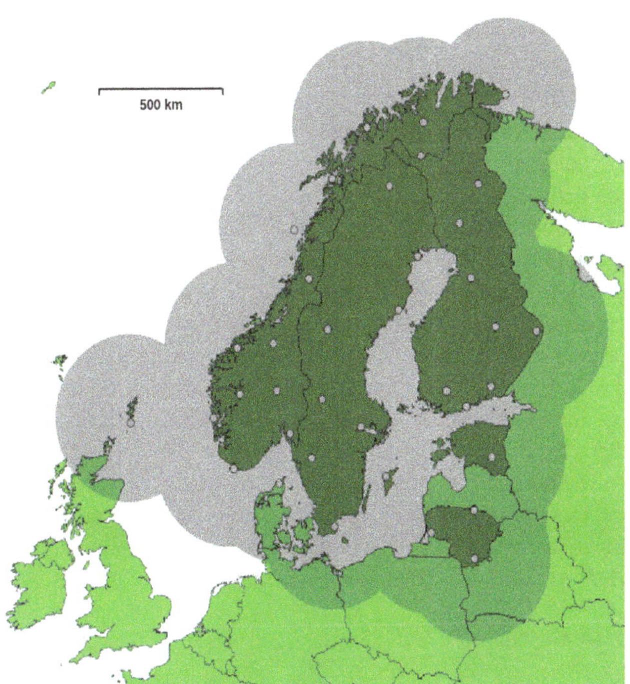

Figure 2. The LLS sensor locations (white dots) and coverage (grey circular areas) as of the year 2015.

3 Methods

The systems used to assimilate radar, gauge and lightning measurements are described in Sect. 3.1 and 3.2. The impact of different integration time periods on the regression and Barnes (RandB) method is shown in Sect. 3.3 and 3.4 and the verification methods in Sect. 3.5.

3.1 The Local Analysis and Prediction System (LAPS)

The LAPS produces 3-D analysis fields of several different weather parameters (Albers et al., 1996). LAPS performs a high-resolution spatial analysis where observational input from several sources is fitted to a coarser background model first-guess field (e.g., ECMWF forecast model). Additionally, high-resolution topographical data are used when creating the final analysis fields. The FMI-LAPS products are mainly used for nowcasting purposes (i.e., what is currently happening and what will happen in the next few hours), which is of critical interest for end users who demand near-real-time products.

The FMI-LAPS use a pressure coordinate system including 44 vertical levels distributed with a higher resolution (e.g., 10 hPa) at lower altitudes and decreasing with height. The horizontal resolution is 3 km and the temporal resolution is 1 h. The domain used in this article covers the whole country of Finland and some parts of the neighboring countries (Fig. 1b). LAPS highly relies on the existence of high-resolution observational network, in both space and time, and especially on remote sensing data. The FMI-LAPS is able to process several types of in situ and remotely sensed observations (Koskinen et al., 2011), among which radar reflectivity, weighting gauges and road-weather observations are used for calculating the precipitation accumulation. The Finnish radar volume scans are read into LAPS as NetCDF format files; thereafter, the data are remapped to the LAPS internal Cartesian grid and the mosaic process combines data of the different radar stations (Albers et al., 1996). The rain rates are calculated from the lowest levels of the LAPS 3-D radar mosaic data via the standard Z-R formula (Marshall and Palmer, 1948), which is then used for precipitation accumulation calculations (see Sect. 3.2). Other information on observational usage, first-guess fields, the coordinate system etc. is described in Gregow et al. (2013).

In this study, the lightning data are ingested into the FMI-LAPS. Modifications have been made to the software in order to use it together with FMI operational radar input data and the new lightning algorithms.

3.2 The LAPS lightning data assimilation (LDA) method

A lightning data assimilation (hereafter LDA) system has been developed by Vaisala and distributed as open and free software (Pessi and Albers, 2014). The LDA method is constructed to build up statistical relationships between radar and lightning measurements. The lightning information used for the LAPS LDA method is the location data (e.g., time, longitude and latitude) for each CG lightning stroke. LDA counts the amount of CG lightning strokes and converts lightning rates into vertical radar reflectivity profiles within each LAPS grid cell. The radar reflectivity–lightning (hereafter Rad-Lig) relationship profiles may differ depending on the local geographical regime and climate. A set of default profiles are included within the LDA package, which were derived over the eastern United States with the use of radar data from NEXRAD network and lightning data from the GLD360 network (Pessi, 2013; Said et al., 2010). These profiles can be used as a first guess if profiles for the local climate are not available.

For this study over Finland, climatological Rad-Lig reflectivity relationship profiles were estimated using NORDLIS-LLS lightning information and operational radar volume data from the Finland area during summer 2014. A total of approximately 220 000 lightning strokes were used for this calibration. The FMI-LAPS LDA used a 5 min interval of lightning and radar data, within a LAPS grid box of 3 × 3 km resolution. The collected strokes are divided into binned categories using an exponential division (i.e., 2^n ... 2^{n+1}), following the same method used in Pessi (2013). This results in six different lightning categories (e.g., with 1, 2–3, 4–7, 8–15, 16–31 and 32–63 strokes) for the NORDLIS-LLS dataset. For each of these six categories, the average reflectivity is calculated at each grid point for each level and gives the average Rad-Lig profiles (Fig. 3a), which is the baseline method. There is a good correlation ($R^2 = 0.95$) between the maximum reflectivity of profile and number of lightning strokes (Fig. 3b; results shown for the average Rad-Lig profiles). We extend this method to also calculate the third quartile (i.e., 75 % percentile) and the variable quartile Rad-Lig profiles, for each category. The variable quartile method uses a range between the 50 % percentile (for the lower dBZ values) and the 95 % percentile (for the highest dBZ values). The specific percentiles used for the six categories are the 50, 50, 60, 75, 90 and 95 % percentiles, respectively. The reasoning is to take into account the uncertainties in the low categories (due to larger spread and bias in the collected datasets) and, on the other hand, rely on the high percentiles for the high categories (since these have less spread). The profiles from the two categories with largest amount of strokes have the least data, because they are the rarest categories. All datasets suffer from missing data at some height levels, but these two categories are more sensitive due to the overall small data amounts. This can sometimes create artificial peaks of reflectivity values that are too low. This was especially seen at high altitudes, which can partly be explained by the radar measurement geometry. Therefore, these two reflectivity profiles have been manually smoothed to have the same shape as the other profiles.

The Rad-Lig reflectivity profiles can be used either independently or merged with the radar data in the LAPS accumulation analysis. When merging the two sources, radar and lightning reflectivity values are compared at each grid point both horizontally and vertically. The data source giving the highest reflectivity value will be used in that LAPS grid point. The logic behind this is that the radars are more likely to underestimate than overestimate the precipitation (due to attenuation, beam blocking or the nearest radar missing from the network; e.g., Battan, 1973; Germann, 1999), especially in thunderstorm situations. This is an approximation, aiming to compensate for the most serious radar error sources, which could be a subject for further improvement in future developments (especially if independent quality estimates of the radar data become available). LAPS then uses the generated 3-D volume reflectivity field in a similar manner, as it would use the regular volume radar data, for example, to adjust hydrometeor fields and rainfall.

The reflectivity (Z; $\mathrm{mm^6\,m^{-3}}$) parameter, measured by the radar or estimated by LDA method, is converted to precipitation intensity (R; $\mathrm{mm\,h^{-1}}$) within LAPS, using a pre-selected Z-R equation (Marshall and Palmer, 1948) as of the type

$$Z = A \cdot R^b, \tag{1}$$

where A and b are empirical factors describing the shape and size distribution of the hydrometeors. In FMI-LAPS's implementation, $A = 315$ and $b = 1.5$ for liquid precipitation,

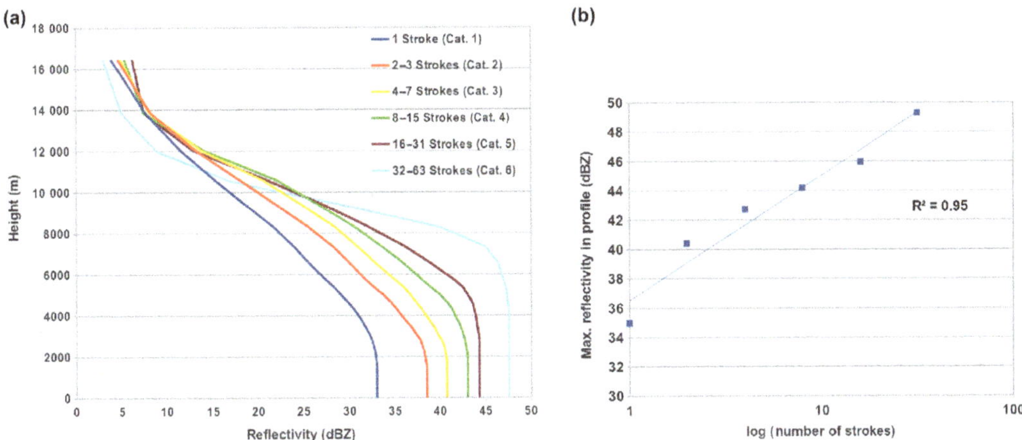

Figure 3. Panel **(a)** shows Rad-Lig relationship profiles (smoothed) from Finland NORDLIS-LLS, calculated using the dataset from summer 2014. Profiles are divided into binned categories of strokes, with a temporal resolution of 5 min and spatial resolution of 3 km. Panel **(b)** shows profiles' max reflectivity values vs. lightning rate (logarithmic scale of bins).

which is relevant in this study since it is carried out during the summer period. These static values introduce a gross simplification, since the drop size and particle shapes vary according to the weather situation (drizzle/convective, wet snow/snow grain). Challenging situations include both convective showers, with heavy rainfall, and the opposite event of drizzle, with little precipitation (Uijlenhoet, 2001). On the other hand, the same static factors have been used for many years in FMI's other operational radar products, and looking at long-term averages, the radar accumulation data do match the gauge accumulation values within reasonable accuracy (Aaltonen et al., 2008). The intensity field (R; Eq. 1) is calculated at every 5 min, and the 1 h accumulation is thereafter obtained by accumulating 5 min intervals. Gires et al. (2014) have shown that the scale difference has an effect on verification measures (such as normalized bias, e.g., RMSE) but it decreases with growing accumulation time (e.g., from 5 to 60 min). In our study, the 60 min accumulation period is smoothing some of the differences.

The following FMI-LAPS precipitation accumulation products are calculated based on radar (hereafter Rad_Accum), LDA (hereafter LDA_Accum) and the combined radar and LDA (hereafter Rad_LDA_Accum) precipitation accumulation.

3.3 The FMI-LAPS regression and Barnes (RandB) analysis method

The FMI-LAPS RandB method corrects the precipitation accumulation estimates using radar and gauge datasets. The first step in this method is to make the radar–gauge correction using the regression method. Data of hourly accumulation values are derived from the radar–gauge pairs within the LAPS grid (i.e., from the same location and time), and from this a linear regression function can be established. The corrections from the regression method are applied to the whole radar accumulation field and thereafter used as input for the second step, the Barnes analysis. Within LAPS routines, the Barnes interpolation converge the radar field towards gauge accumulation measurements at smaller areas (i.e., for gauge station surroundings). Several iterative correction steps are performed within the Barnes analysis, adjusting the final accumulation. The FMI-LAPS RandB method is described in more details in Gregow et al. (2013).

In this article, the RandB method is used to calculate the precipitation accumulation with the use of radar, gauges, lightning and the combination of radar–lightning. This gives the additional three FMI-LAPS accumulation products: Rad_RandB, LDA_RandB and Rad_LDA_RandB, respectively.

3.4 RandB method and the integration time period

The original FMI-LAPS RandB method uses radar and gauge data from the recent hour. Using only the latest hour, the gauge observational dataset can suffer from too few observations and thereby affect the quality and robustness of the regression and Barnes calculations. As a further investigation in this article, we use a selection of longer time periods (e.g., the previous 6, 12, 24 h and 7 days of data) in order to build up a larger radar–gauge dataset. These datasets are thereafter used to make the correction within the RandB method.

We have limited our studies to compare how the occurring synoptic weather situation, i.e., frontal or convective situation (1 to 12 h), and the medium-time-range information (24 h to 7 days) impact the accumulation analysis. The longer the integration time, the less information on the situational weather occurring at analysis time; i.e., the dataset is getting more smoothed and extremes might disappear.

Verification was done for the summer 2015 period using the input from radar and lightning, and gives the following resulting accumulation products: Rad_LDA_RandB (i.e., dataset collected within the last 1 h), Rad_LDA_RandB_6hr, Rad_LDA_RandB_12hr, Rad_LDA_RandB_24hr and Rad_LDA_RandB_7d, respectively.

3.5 Verification methods

The hourly accumulation results have been verified against surface gauge observations, both dependent and independent stations. The dependent station data are included in the FMI-LAPS analysis calculating the 1 h precipitation accumulation; i.e., the analysis is depending on the station information used as input. There are seven independent stations which are excluded from the LAPS analysis. Note that, in the Rad_Accum and Rad_LDA_Accum products, the gauge data have not been used; therefore, all gauge stations are independent references for their verification. In this study, we apply a filter to the verification datasets where hourly accumulation data less than 0.3 mm are discarded (due to the lowest threshold value of surface gauge measurements from the FMI database). In a separate verification exercise for the 2016 data, only stations located more than 100 km and more than 150 km from the nearest radar station were used to demonstrate the potentially deteriorating quality of radar data with distance to the radar due to, e.g., attenuation and beam broadening (a 1° beam is 5 km wide at a distance of 250 km).

The validation of the different analysis methods is based on the logarithmic standard deviation (SD; Eq. 2), root mean square deviation (RMSE; Eq. 3) and Pearson's correlation coefficient (CORR; Eq. 4):

$$\mathrm{SD} = \frac{1}{N-1}\sum_{i=1}^{N}\left(\log\left(\frac{\text{Analysis}}{\text{Gauge}}\right)_i - \overline{\log\left(\frac{\text{Analysis}}{\text{Gauge}}\right)}\right)^2 \quad (2)$$

$$\mathrm{RMSE} = \sqrt{\frac{\sum_{i=1}^{N}\left((\text{Analysis} - \text{Gauge})_i\right)^2}{N-1}} \quad (3)$$

$$\mathrm{CORR} = \frac{\sum_i\left(\left(\text{Gauge}_i - \overline{\text{Gauge}}\right)\left(\text{Analysis}_i - \overline{\text{Analysis}}\right)\right)}{\sqrt{\sum_i\left(\text{Gauge}_i - \overline{\text{Gauge}}\right)^2\sum_i\left(\text{Analysis}_i - \overline{\text{Analysis}}\right)^2}}. \quad (4)$$

SD quantifies the amount of variation (i.e., spread) of a dataset. A low SD indicates that the data points tend to be close to the mean value of the dataset. Here, we use the logarithm of the quotients, in order to get the datasets closer to be normally distributed. RMSE is a quadratic scoring rule which measures the average magnitude of the error. Since the errors are squared before they are averaged, RMSE gives a relatively high weight to large errors. CORR gives a measure of the linear relationship (both strength and direction) between two quantities.

4 Results

Verification results using lightning data are presented in Sect. 4.1 and the impact from different integration time intervals in Sect. 4.2.

4.1 FMI-LAPS LDA results

The verification for the entire summer of 2015, i.e., using verification dataset (a) including days with no thunderstorms, assures that introducing lightning data has no significant impact on the overall performance of the system. The impact of using the LDA method for estimating the precipitation accumulation is neutral for this long verification period (shown in Fig. 4, where the data are from dependent stations). The same result is seen in the scores of RMSE, SD and CORR values (not included here). Since the data have been much influenced by weather situations not related to lightning, the focus will be on the subsets, i.e., datasets (c) and (d), the 25-day periods of intense lightning days of both 2015 and 2016, respectively.

The 25-day period with frequent thunderstorms during summer 2015, verification dataset (c), for which we used the average method to calculate the Rad_Lig profiles, shows an inconsistent result using lightning data (see Table 1, left column). For the independent dataset, the Rad_LDA_Accum has a slightly improved result (lower RMSE value) when compared with Rad_Accum. On the other hand, Rad_LDA_RandB gets worse results, as can be seen from the RMSE and CORR. The dependent data show almost neutral impact (RMSE is slightly better for Rad_LDA_RandB) with the use of the LDA method and average calculated Rad-Lig profiles.

Figure 5 shows the results using verification dataset (e), where different Rad-Lig profiles are compared (e.g., average, third quartile and variable quartile profiles) and validated against Rad_Accum. The precipitation accumulation estimates are improved at high accumulation values (> 5 mm) using either third or variable quartile profiles. Simultaneously, they both add to the overestimate in low accumulation values (< 5 mm). The third quartile profiles give the largest overestimate over the whole accumulation scale. The variable quartile gives the overall best result, with improved estimates for high accumulation values and only slight overestimation at low values.

The results, from the scaled dataset (d) and the dependency of distance to radar location, reveal the positive impact of using the lightning data as input for the LAPS-LDA model. Hence, using the variable quartile profiles in the accumulation analysis for the 25-day dataset of summer 2016 has a positive impact on the accumulation estimates (see Table 1, right column). Even if the improved scores are relatively small (the largest reduction in RMSE being 6.3 %), the LDA method shows a consistent correction of the results. The independent verification gives decreased RMSE

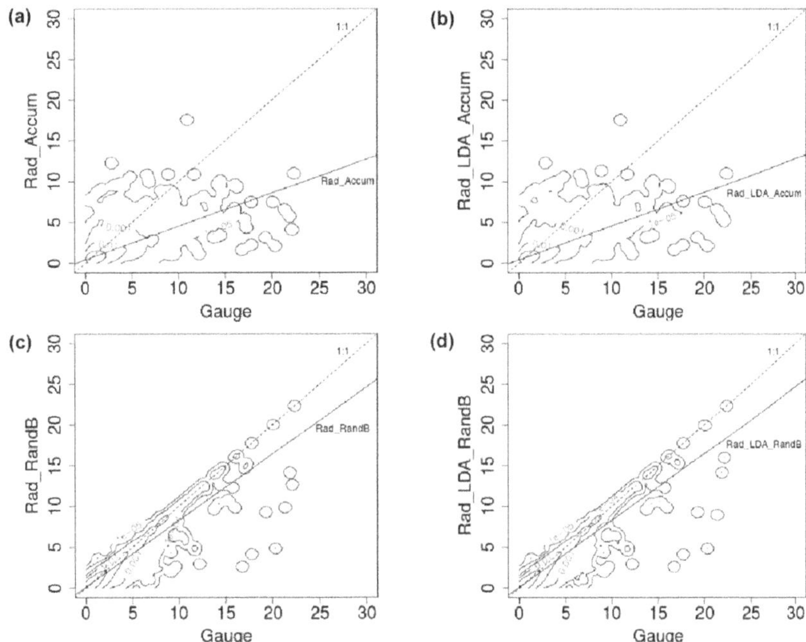

Figure 4. The FMI-LAPS precipitation accumulation (described in plots with density isolines of hourly accumulation values in millimeters) calculated using four different methods. Fit in solid line (see regression equations), the perfect solution would align on the 1 : 1 dashed line: **(a)** Rad_Accum ($y = 0.410x + 0.398$), **(b)** Rad_LDA_Accum ($y = 0.413x + 0.396$), **(c)** Rad_RandB ($y = 0.817x + 0.093$) and **(d)** Rad_LDA_RandB ($y = 0.819x + 0.091$). Results are from the dependent gauge dataset during summer 2015, i.e., verification dataset (a).

Table 1. Precipitation accumulation results from summer of 2015 (i.e., dataset c, left column) and 2016 (i.e., dataset d, right column), for periods of the 25 intensive lightning days (e.g., > 100 CG strokes per day) during both years. Precipitation results are shown for radar (Rad_Accum) and radar merged with lightning data (Rad_LDA_Accum), together with and without gauge measurements included with the RandB method (Rad_RandB and Rad_LDA_RandB, respectively). In the lowest panels, only data from more than 100 or 150 km from the nearest radar are used. Verification is performed against both independent and dependent stations, i.e., those used or left out from the gauge analysis.

Summer 2015 (average scheme)					Summer 2016 (variable quartile scheme)				
Independent	Rad_ Accum	Rad_LDA_ Accum	Rad_ RandB	Rad_LDA_ RandB	Independent	Rad_ Accum	Rad_LDA_ Accum	Rad_ RandB	Rad_LDA_ RandB
No. obs	3206	3332	256	256	No. obs	1320	1333	74	74
SD	0.27	0.27	0.11	0.11	SD	0.32	0.32	0.12	0.11
RMSE	1.66	1.64	0.58	0.70	RMSE	2.62	2.60	0.92	0.89
CORR	0.67	0.67	0.97	0.96	CORR	0.64	0.65	0.96	0.96
Dependent					Dependent				
No. obs			3566	3567	No. obs			1364	1376
SD			0.12	0.12	SD			0.14	0.13
RMSE			0.77	0.76	RMSE			1.27	1.19
CORR			0.93	0.93	CORR			0.93	0.94
> 100 km					> 100 km				
No. obs					No. obs	656	656	694	698
SD					SD	0.34	0.34	0.15	0.15
RMSE					RMSE	2.44	2.39	1.03	1.01
CORR					CORR	0.66	0.67	0.95	0.95
> 150 km					> 150 km				
No. obs					No. obs	153	153	168	171
SD					SD	0.39	0.39	0.20	0.20
RMSE					RMSE	2.46	2.42	1.47	1.43
CORR					CORR	0.33	0.35	0.80	0.81

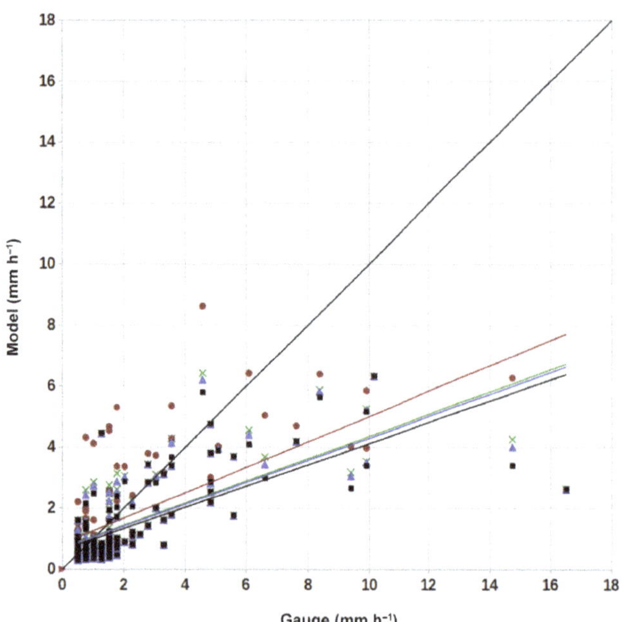

Figure 5. Verification of hourly accumulation values for Rad_Accum (black squares, with regression line equation $y = 0.349x + 0.638$) and LDA_Accum (triangle, cross and circular markers), using three different methods to calculate the relationship profiles: average (blue triangles, $y = 0.360x + 0.691$), third quartile (red circles, $y = 0.417x + 0.844$) and the variable quartile (green crosses, $y = 0.365x + 0.710$) accumulation estimates. The corresponding regression lines (see equations) are represented with same color as the markers for each method. Data are for the 4-day period in summer 2014, i.e., verification dataset (e). The best-fit curve (i.e., the 1 : 1 fit) is shown as a black solid line.

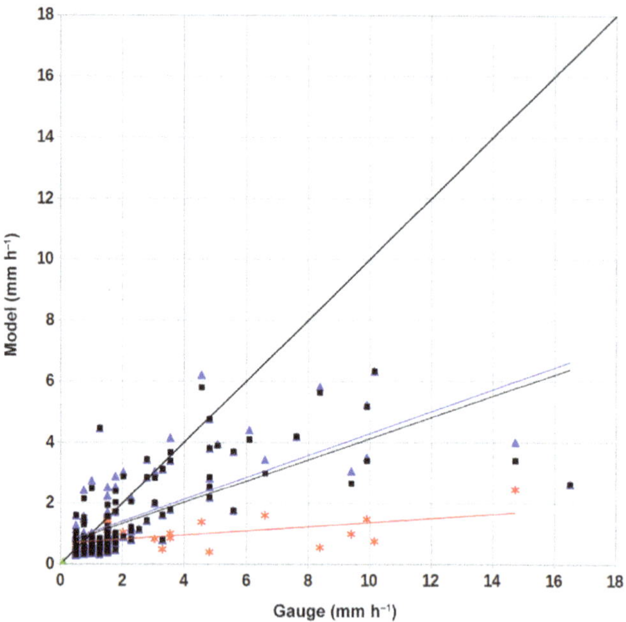

Figure 6. Verification of hourly accumulation values for LDA_Accum (red stars, with regression line equation $y = 0.068x + 0.685$) and the merged Rad_LDA_Accum (blue triangles, $y = 0.360x + 0.691$), compared to Rad_Accum (black boxes, $y = 0.349x + 0.638$). The corresponding regression lines (see equations) are represented with same color as the markers for each method. Data are for the 4-day period in summer 2014, i.e., verification dataset (e). The black solid line is the best-fit line (1 : 1 fit).

and increased CORR values for Rad_LDA_Accum compared to Rad_Accum. Also, Rad_LDA_RandB gets smaller errors than Rad_RandB (see SD and RMSE in Table 1, most upper-right panel). For the dependent stations, all scores are improved using the LDA method, especially the RMSE (as seen in Table 1, right column, second panel). The verification of distance dependencies, i.e., for observations further away than 100 and 150 km from the nearest radar stations, shows improved accumulation estimates when using the LDA method (see Table 1, right column, two last panels). The RMSE and CORR scores for Rad_LDA_Accum and Rad_LDA_RandB are better than Rad_Accum and Rad_RandB, respectively. Here, only dependent gauges are available for verification.

Comparing accumulation results from the 4-day period, i.e., verification dataset (e), for radar alone (Rad_Accum; black markers in Fig. 6) and lightning alone (LDA_Accum; red markers in Fig. 6), it is clear that the use of LDA_Accum is less accurate than Radar_Accum results. Figure 6 also shows that the Rad_LDA_Accum estimates (using the baseline method, with average Rad-Lig profiles) are amplified over the whole range of precipitation values, compared to Rad_Accum (Fig. 6; compare the blue with the black markers). For the high accumulation values ($>5\,\mathrm{mm\,h^{-1}}$), this is a positive effect, while in the lower range ($<5\,\mathrm{mm\,h^{-1}}$) there is an overestimation of the results.

4.2 RandB method and impact from different integration periods

The plotted results of different time sampling periods are seen in Fig. 7, where the density of points are drawn as isolines in the scatter plot, with verification against the independent stations from verification dataset (a). The Rad_LDA_RandB (i.e., using observations from the latest 1 h) does give the best result, when compared to Rad_LDA_Accum, Rad_LDA_RandB, Rad_LDA_RandB_6hr, Rad_LDA_RandB_12hr, Rad_LDA_RandB_24hr and the Rad_LDA_RandB_7d output. The statistical scores shown in Table 2 also imply the same result. The Rad_LDA_Accum (e.g., a method not using RandB) is included as a reference when comparing the results of different integration periods.

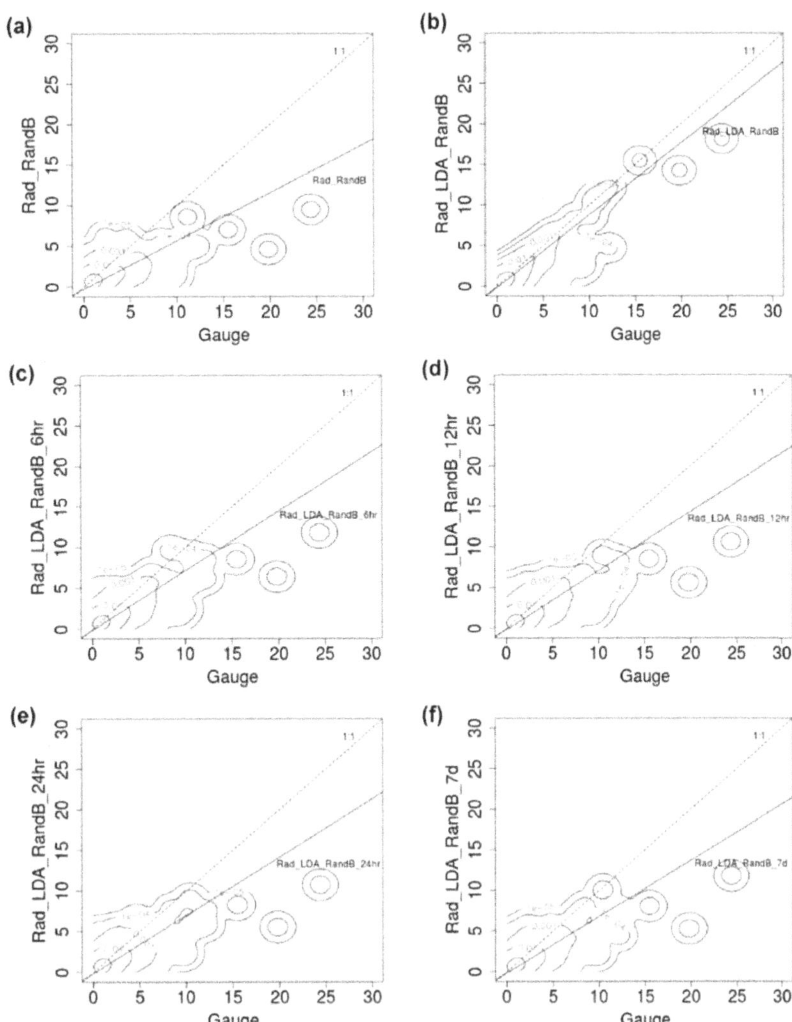

Figure 7. Impact of changing the integration time length, with verification for the independent gauges, using verification dataset (a) from summer 2015. Accumulation plots with density isolines of hourly values in millimeters: **(a)** Rad_LDA_Accum (with regression line equation $y = 0.594x - 0.312$), **(b)** Rad_LDA_RandB ($y = 0.891x - 0.147$), **(c)** Rad_LDA_RandB_6hr ($y = 0.732x - 0.160$), **(d)** Rad_LDA_RandB_12hr ($y = 0.725x - 0.169$), **(e)** Rad_LDA_RandB_24hr ($y = 0.715x - 0.167$) and **(f)** Rad_LDA_RandB_7d ($y = 0.692x - 0.166$). The fit is shown in solid lines (see regression equations); the perfect solution would align on the 1 : 1 dashed line.

Table 2. Impact of the integration time length on the RandB method for the dependent and independent stations datasets during summer 2015, i.e., dataset (a). The Rad_LDA_Accum (e.g., a method not using RandB) is included as a reference.

Dependent	Rad_LDA_ Accum	Rad_LDA_ RandB_1hr	Rad_LDA_ RandB_6hr	Rad_LDA_ RandB_12hr	Rad_LDA_ RandB_24hr	Rad_LDA_ RandB_7d
No. of observations	13 200	16 311	10 956	10 917	10 915	11 033
SD ($\log(R/G)$)	0.25	0.13	0.13	0.13	0.14	0.14
RMSE	1.20	0.52	0.67	0.71	0.72	0.72
CORR	0.64	0.93	0.91	0.90	0.89	0.89
Independent						
No. of observations	1177	1492	1028	1013	1005	1014
SD ($\log(R/G)$)	0.25	0.15	0.22	0.22	0.22	0.22
RMSE	1.38	0.68	1.16	1.23	1.24	1.24
CORR	0.39	0.92	0.79	0.77	0.77	0.77

5 Discussions and conclusions

The aim of this article is to describe new methods on how to improve the hourly precipitation accumulation estimates, especially for heavy rainfall events (> 5 mm) and as much as possible for the low-valued ranges (< 5 mm).

The strength of the LDA method is that the radar and lightning information can be merged and complement each other. This is especially important in areas of poor or even non-existent radar coverage, where the lightning information will improve the reflectivity field and thereby the hourly precipitation accumulation analysis. It is important to recall that, in the LAPS accumulation process, the reflectivity field is the first step, which is then corrected with gauges (e.g., if there is no reflectivity field, gauges will not be used and there will be no accumulation field). The results in this article are limited to Finland but should this area be extended to include Scandinavia, the LDA method will become even more useful. There are also other LAPS users in other parts of the world, whom we want to encourage to continue this work.

The whole summer periods of 2015 and 2016 show neutral impact on the results using the LDA method; scores are not included here but Fig. 4 shows the graphs for verification dataset (a). It is important to make long-term verification in order to see that the system is robust and does not generate any bad data during any weather situation, i.e., perform a sanity check of the system. However, in order to narrow down our analysis to areas and times where lightning did occur (i.e., exclude stratiform precipitation), we focused our results on the subset of 25 lightning intensive days for both 2015 and 2016, datasets (c) and (d), respectively. The subset of 2015, using the average method, gave inconsistent results and no unambiguous conclusions could be drawn (Table 1, left column).

New methods to calculate the Rad-Lig profiles were tested and reveal that the variable quartile method improves the estimates for the large accumulation (i.e., > 5 mm), though with some overestimation in low accumulation (Fig. 5). The third quartile approach has the highest impact on the whole accumulation field, which results in large overestimates for the low accumulation values (i.e., 0–5 mm). The average method smoothes out the small-scale variances, which are observed in heavy convection. Hence, the collected radar reflectivity profiles are less representative, and therefore the calculated Rad-Lig profiles will have values that are too low in these cases. As a result, the average method will have a low impact on the final precipitation accumulation estimates, compared to the use of the third quartile and variable quartile methods (Fig. 5). One should also mention that there is an overall uncertainty due to instrumental errors and the collocation between observations within the LDA method. This could potentially result in dislocation and bad quality of the received radar and lightning measurements, which would affect the calculated Rad-Lig profiles (for example, in the event of radar attenuation, where strong rainfall weakens some part of the reflectivity field). Here, the collected radar profiles will have reflectivity values that are too low and give underestimated Rad-Lig profiles, especially when using the average method.

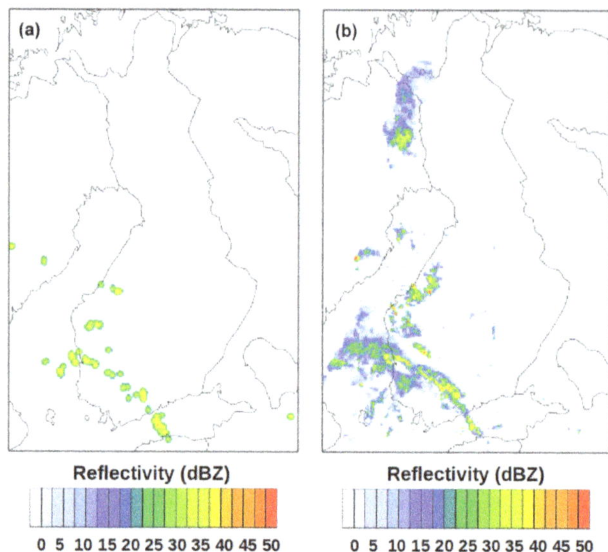

Figure 8. Reflectivity field simulated from lightning data alone (left) and, for verification, from radar data alone (right) 30 July 2014 at 16:00 UTC. The reflectivity color scale is shown below plots.

The newest results from 2016 and the 25-day subset show that there is a benefit to using the LDA (variable quartile) method. Mainly, all scores are becoming better and few are unchanged when lightning information is used to estimate the precipitation accumulation (see Table 1, right column). Verifying the dataset with distance to radar stations (i.e., gauges situated further away than 100 and 150 km) also shows the same results; the accumulation product is improved with the LDA method. The impact on scores is mainly in the second decimal, but they are consistent, and clearly show the tendency of improvement by using the LDA method with the variable quartile profiles. One reason we do not see a larger impact by the LDA method could be that the Finnish radar network does have a very high quality and system utilization rate and therefore is less impacted by the LDA method. In an upcoming version of FMI-LAPS, the verification will be focusing on including areas with poor (or non-existent) radar coverage where gauges are available.

The accumulation products generated from the RandB method are corrected using gauge information. This process influences the final accumulation results much more than the contribution from the LDA method (seen in Fig. 4 results from the dependent dataset, where a, c and b, d panels, respectively, are almost identical). The same result was seen for the independent dataset (not shown here). Nonetheless, we have proven that if there were no radar data (for example, if the radar is malfunctioning), precipitation accumulation information would be available from lightning data and add value to the final product. This is shown in Fig. 6, where accumulation would be generated from the LDA method (as seen in Fig. 6; red markers) and also visualized through the example in Fig. 8, where the radar and Rad-Lig lowest reflectivity fields are plotted for one analysis time: 16:00 UTC, 30 July 2014. This case study also demonstrates how the LDA method can reconstruct the highest reflectivities, but areas with weak precipitation are missing.

In the RandB method, the regression is used to correct for large-scale multiplicative biases between radar and gauge data. In this article, we introduce lightning into the RandB method as an additional data source. However, lightning errors are likely to be different from those of radar and gauges, and this could have an effect on the methodology used here. In future developments, after collecting longer time series to quantify the nature of uncertainty of lightning-based precipitation estimates, we intend to improve the analysis in this direction.

In the present analysis area, we mainly anticipate the usefulness of lightning data as a backup plan of rare but significant cases. Due to the rare nature of such events, it is not possible to collect a statistically representative dataset in a few years; even though attenuation of radar signals or completely missing data are observed several times a summer, it is not so often that such events happen just over a rain gauge station. However, our overall analysis shows that when we include the lightning data every day at every point, they make, on average, a small improvement, and they are there as a safety network waiting for the cases where radars fail.

For the near-real-time accumulation product, data used from the recent hour of analysis time do give the best precipitation accumulation result (Table 2 and Fig. 7). We see correlation peaking at the 1 h integration period and decreasing already for the 6 h period. Therefore, according to the results in this study, the use of long time integration periods for the RandB method (until 7 days in this case) does not improve the hourly precipitation accumulation analysis. Berndt et al. (2014) compared data resolutions from 10 min to 6 h and reported a large improvement in the correlation (from 10 min to 1 h, the correlation increased 0.37 to 0.57). From 1 to 6 h, the corresponding increase was 0.57 to 0.62, respectively. In Norway, Abdella and Alfredsen (2010) have shown that the use of average monthly adjustment factors leads to less than optimal results. One could speculate that there is an intermediate choice of temporal resolution that would improve the results in this article. For example, there could be better results using periods of 2 to 5 h. This has not been investigated in this article but will be considered in future studies.

Acknowledgements. We thank NOAA ESRL/GSD and Vaisala for their support of LAPS-LDA developments, Marco Gabella for his encouraging words and Asko Huuskonen for helping in the final and critical stage of evaluating the results.

Edited by: P. Molnar

References

Aaltonen, J., Hohti, H., Jylhä, K., Karvonen, T., Kilpeläinen, T., Koistinen, J., Kotro, J., Kuitunen, T., Ollila, M., Parvio, A., Pulkkinen, S., Silander, J., Tiihonen, T., Tuomenvirta, H., and Vajda, A.: Strong precipitation and urban floods (Rankkasateet ja taajamatulvat RATU), Finnish Environment Institute (Suomen ympäristókeskus), Helsinki, Finland, 80 pp., 2008.

Abdella, Y. and Alfredsen, K.: Long-term evaluation of gauge-adjusted precipitation estimates from a radar in Norway, Hydrol. Res., 41, 171–192, 2010.

Albers, S. C., McGinley, J. A., Birkenheuer, D. L., and Smart, J. R.: The local analysis and prediction system (LAPS): Analyses of clouds, precipitation, and temperature, Weather Forecast., 11, 273–287, 1996.

Battan, L. J.: Radar Observation of the Atmosphere, University of Chicago Press, Chicago, USA, 1973.

Berndt, C., Rabiei, E., and Haberlandt, U.: Geostatistical merging of rain gauge and radar data for high temporal resolutions and various station density scenarios, J. Hydrology, 508, 88–101, 2014.

Cummins, K. L., Murphy, M. J., Bardo, E. A., Hiscox, W. L., Pyle, R. B., and Pifer, A. E.: A combined TOA/MDF technology upgrade of the U.S. National Lightning Detection Network, J. Geophys. Res., 103, 9035–9044, doi:10.1029/98JD00153, 1998.

Einfalt, T., Jessen, M., and Mehlig, B.: Comparison of radar and raingauge measurements during heavy rainfall, Water Sci. Technol., 51, 195–201, 2005.

Germann, U.: Radome attenuation – a serious limiting factor for quantitative radar measurements?, Meteorol. Z., 8, 85–90, 1999.

Gires, A., Tchiguirinskaia, I., Schertzer, D., Schellart, A., Berne, A., and Lovejoy, S.: Influence of small scale rainfall variability on standard comparison tools between radar and rain gauge data, Atmos. Res., 138, 125–138, doi:10.1016/j.atmosres.2013.11.008, 2014.

Goudenhoofdt, E. and Delobbe, L.: Evaluation of radar-gauge merging methods for quantitative precipitation estimates, Hydrol. Earth Syst. Sci., 13, 195–203, doi:10.5194/hess-13-195-2009, 2009.

Gregow, E., Saltikoff, E., Albers, S., and Hohti, H.: Precipitation accumulation analysis – assimilation of radar-gauge measurements and validation of different methods, Hydrol. Earth Syst. Sci., 17, 4109–4120, doi:10.5194/hess-17-4109-2013, 2013.

IPCC-AR5: Climate change 2013, available at: http://www.ipcc.ch/report/ar5/wg1/ (last access: 9 January 2017), 2013.

Jaffrain, J. and Berne, A.: Influence of the subgrid variability of the raindrop size distribution on radar rainfall estimators, J. Appl. Meteorol. Clim., 51, 780–785, doi:10.1175/JAMC-D-11-0185.1, 2012.

Jewell, S. A. and Gaussiat, N.: An assessment of kriging-based rain-gauge-radar merging techniques, Q. J. Roy. Meteor. Soc., 141, 2300–2313, doi:10.1002/qj.2522, 2015.

Koistinen, J. and Michelson, D. B.: BALTEX weather radar-based precipitation products and their accuracies, Boreal Environ. Res., 7, 253–263, 2002.

Koskinen, J. T., Poutiainen, J., Schultz, D. M., Joffre, S., Koistinen, J., Saltikoff, E., Gregow, E., Turtiainen, H., Dabberdt, W. F., Damski, J., Eresmaa, N., Göke, S., Hyvärinen, O., Järvi, L.,

Karppinen, A., Kotro, J., Kuitunen, T., Kukkonen, J., Kulmala, M., Moisseev, D., Nurmi, P., Pohjola, H., Pylkkö, P., Vesala, T., and Viisanen, Y.: The Helsinki Testbed: A mesoscale measurement, research, and service platform, B. Am. Meteorol. Soc., 92, 325–342, doi:10.1175/2010BAMS2878.1, 2011.

Mäkelä, A., Tuomi, T. J., and Haapalainen, J.: A decade of high-latitude lightning location: Effects of the evolving location network in Finland, J. Geophys. Res., 115, D21124, doi:10.1029/2009JD012183, 2010.

Mäkelä, A., Rossi, P., and Schultz, D. M.: The daily cloud-to-ground lightning flash density in the contiguous United States and Finland, Mon. Weather Rev., 139, 1323–1337, doi:10.1175/2010MWR3517.1, 2011.

Mäkelä, A., Mäkelä, J., Haapalainen, J., and Porjo, N.: The verification of lightning location accuracy in Finland deduced from lightning strikes to trees, Atmos. Res., 172, 1–7, 2016.

Marshall, J. S. and Palmer, W. M.: The Distribution of raindrops with size, J. Meteorol., 5, 165–166, 1948.

Morales, C. A. and Anagnostou, E. N.: Extending the capabilities of high-frequency rainfall estimation from geostationary-based satellite infrared via a network of long-range lightning observations, J. Hydrometeorol., 4, 141–159, 2003.

NOAA: Latest release of LAPS software, available at: http://laps.noaa.gov/cgi/LAPS_SOFTWARE.cgi, last access: 11 January 2017.

Papadopoulos, A., Chronis, T. G., and Anagnostou, E. N.: Improving convective precipitation forecasting through assimilation of regional lightning measurements in a mesoscale model, Mon. Weather Rev., 133, 1961–1977, 2005.

Pessi, A.: Characteristics of Lightning and Radar Reflectivity in Continental and Oceanic Thunderstorms, 93th Annual American Meteorological Society Meeting, Austin, Texas, USA, 6–10 January 2013, availabe at: https://ams.confex.com/ams/93Annual/webprogram/Paper215562.html (last access: 9 January 2017), 2013.

Pessi, A. and Albers, S.: A Lightning Data Assimilation Method for the Local Analysis and Prediction System (LAPS): Impact on Modeling Extreme Events, 94th Annual American Meteorological Society Meeting, Atlanta, Georgia, USA, 1–6 February, 2014, available at: https://ams.confex.com/ams/94Annual/webprogram/Paper238715.html (last access: 9 January 2017), 2014.

Pessi, A. and Businger, S.: The Impact of Lightning Data Assimilation on a Winter Storm Simulation over the North Pacific Ocean, Mon. Weather Rev., 137, 3177–3195, 2009.

Said, R. K., Inan, U. S., and Cummins, K. L.: Long-range lightning geolocation using a VLF radio atmospheric waveform bank, J. Geophys. Res., 115, D23108, doi:10.1029/2010JD013863, 2010.

Saltikoff, E., Huuskonen, A., Hohti, H., Koistinen, J., and Järvinen, H.: Quality assurance in the FMI Doppler Weather radar network, Boreal Environ. Res., 15, 579–594, 2010.

Schiemann, R., Erdin, R., Willi, M., Frei, C., Berenguer, M., and Sempere-Torres, D.: Geostatistical radar-raingauge combination with nonparametric correlograms: methodological considerations and application in Switzerland, Hydrol. Earth Syst. Sci., 15, 1515–1536, doi:10.5194/hess-15-1515-2011, 2011.

Sideris, I. V., Gabella, M., Erdin, R., and Germann, U.: Real-time radar-raingauge merging using spatiotemporal co-kriging with external drift in the alpine terrain of Switzerland, Q. J. Roy. Meteor. Soc., 140, 1097–1111, 2014.

Tuomi, T. J. and Mäkelä, A.: Lightning observations in Finland, Reports, Finnish Meteorological Institute, Helsinki, Finland, 2007:5, 49 pp, 2007.

Tuomi, T. J. and Mäkelä, A.: Thunderstorm climate of Finland 1998–2007, Geophysica, 44, 29–42, 2008.

Uijlenhoet, R.: Raindrop size distributions and radar reflectivity–rain rate relationships for radar hydrology, Hydrol. Earth Syst. Sci., 5, 615–628, doi:10.5194/hess-5-615-2001, 2001.

Wang, L.-P., Ochoa-Rodríguez, S., Onof, C., and Willems, P.: Singularity-sensitive gauge-based radar rainfall adjustment methods for urban hydrological applications, Hydrol. Earth Syst. Sci., 19, 4001–4021, doi:10.5194/hess-19-4001-2015, 2015.

Wilson, J. W. and Brandes, E. A.: Radar Measurement of Rainfall – A Summary, B. Am. Meteorol. Soc., 60, 1048–1058, 1979.

5

Statistical analysis of hydrological response in urbanising catchments based on adaptive sampling using inter-amount times

Marie-Claire ten Veldhuis[1,3] **and Marc Schleiss**[2,3]

[1]Delft University of Technology, Water Management Department, Delft, the Netherlands
[2]Delft University of Technology, Geosciences and Remote Sensing Department, Delft, the Netherlands
[3]Princeton University, Hydrometeorology Group, Princeton, USA

Correspondence to: Marie-Claire ten Veldhuis (j.a.e.tenveldhuis@tudelft.nl)

Abstract. Urban catchments are typically characterised by a more flashy nature of the hydrological response compared to natural catchments. Predicting flow changes associated with urbanisation is not straightforward, as they are influenced by interactions between impervious cover, basin size, drainage connectivity and stormwater management infrastructure. In this study, we present an alternative approach to statistical analysis of hydrological response variability and basin flashiness, based on the distribution of inter-amount times. We analyse inter-amount time distributions of high-resolution streamflow time series for 17 (semi-)urbanised basins in North Carolina, USA, ranging from 13 to 238 km^2 in size. We show that in the inter-amount-time framework, sampling frequency is tuned to the local variability of the flow pattern, resulting in a different representation and weighting of high and low flow periods in the statistical distribution. This leads to important differences in the way the distribution quantiles, mean, coefficient of variation and skewness vary across scales and results in lower mean intermittency and improved scaling. Moreover, we show that inter-amount-time distributions can be used to detect regulation effects on flow patterns, identify critical sampling scales and characterise flashiness of hydrological response. The possibility to use both the classical approach and the inter-amount-time framework to identify minimum observable scales and analyse flow data opens up interesting areas for future research.

1 Introduction

Hydrological response in urban catchments tends to be more flashy compared to natural ones as a result of their higher degree of imperviousness. Increases in flashiness are typically characterised by shorter response times to rainfall, higher run-off ratios and higher peak flows (Berne et al., 2004; Smith et al., 2005). On the other hand, high impervious degrees may reduce base flows and lead to intermittent flow during dry periods. At the same time, urbanisation is usually tied to development of urban drainage infrastructure, associated with artificial flow control as well as higher peak flows due to increased drainage connectivity. Predicting the degree of flashiness or base flow reduction associated with urbanisation is not straightforward, as it depends on the interplay of impervious cover, basin size and shape, soil properties, basin slope, drainage connectivity, and control structures such as detention ponds, weirs and pumps (Emmanuel et al., 2012; Fletcher et al., 2013; Smith et al., 2013). Traditional analyses of flow time series tend to focus on specific aspects and flow characteristics, aiming for example at predicting low flow durations or peak flow magnitudes. For analysis of change in hydrological response, it may be beneficial to combine both peak flow and low flow statistics into a single framework. This applies in particular to the context of urban hydrology where urbanisation and human intervention alter both high flow and low flow characteristics of the hydrological response. Combining both aspects in a single analysis is difficult, as flow distributions are highly skewed and frequencies of low and high flow values are very different. In this paper, we show how alternative sampling of flow time series based on inter-amount times leads to more balanced statistical distributions, better representation of both high and low flows in a single framework, and more robust behaviour of statistical distributions across scales.

1.1 Statistical analysis of hydrological response

Many authors have investigated methods for characterising hydrological response and changes therein, including univariate analysis and multivariate statistics, combining several hydrograph properties such as flood peak, flood volume and flood duration (e.g. Salvadori and De Michele, 2004; Favre et al., 2004; Grimaldi and Serinaldi, 2006; Vittal et al., 2015). Traditional statistical analysis techniques tend to focus on either left or right tail properties of statistical distributions, but not necessarily using the same statistical framework. Low flow analyses for example are primarily concerned with the total time the flow stays below a critical threshold (see for example Smakhtin, 2001, for an extensive review). By contrast, peak flow analysis puts more weight on total accumulated flows at a given timescale using annual flow maxima or peak-over-threshold values to derive extreme value statistics and establish flood frequency curves (e.g. Stedinger, 1983; Lang et al., 1999; Villarini et al., 2009; Smith and Smith, 2015). Both approaches are valid and solidly rooted in the context of extreme event analysis, with numerous applications in drought and flood risk analysis. However, the statistical frameworks they rely on are not necessarily the same. Low flow analysis favours "time" as a random variable. Peak flow analysis on the other hand treats the "flow amount" over a fixed time interval as the main random quantity. This might seem more intuitive to many but there is no strong compelling reason to prefer one approach over the other a priori. For example, one might as well adopt an alternative framework in which the unknown random variable is the "time" necessary to cumulate a fixed, critical amount of flow. By doing so, both low flows and peak flows can be analysed using the same statistical framework. This approach is known as the inter-amount time (IAT) method (Schleiss and Smith, 2016) and has been previously proposed to analyse the properties of intermittent rainfall time series. An important goal of this paper is to derive properties of statistical distributions obtained by applying the IAT formalism to flow time series and to compare the results to the ones obtained using the classical fixed-time framework.

1.2 Change in hydrological response, basin flashiness

An important characteristic that has been used to analyse change in hydrological response is basin flashiness, qualitatively described by Poff (2002) as one of the indicators characterising change in natural flow regimes and how this affects the ecological integrity of river ecosystems. Richter (1996) developed a set of 33 indices, the Indicators of Hydrological Alteration (IHA), including indicators for conditions associated with flashiness, such as frequency and duration of high and low pulses, and rate and frequency of change in flow conditions. Smith and Smith (2015) quantified flashiness of 5436 catchments in the contiguous United States based on peak flows exceeding $1\,\mathrm{m}^3\,\mathrm{s}^{-1}\,\mathrm{km}^{-2}$ normalised flows (i.e. flows normalised by basin area). A frequently used index in the literature is the Richards–Baker (R–B) flashiness index (Baker et al., 2004), based on the Richards pathlength (Gustafson et al., 2004). The R–B index is defined as the sum of absolute values of changes in flow values divided by the total cumulative flow, and is usually computed at the daily timescale. Similar to the coefficient of variation, it measures the relative dispersion of the flow at a given scale. A downside of the R–B index is that it highly sensitive to the scale of analysis. Baker et al. (2004) argued that for smaller basins ($< 50\,\mathrm{km}^2$) the use of hourly instead of daily flow data should be considered to compute the R–B flashiness index, but also found that R–B flashiness values computed at hourly scale are highly sensitive to diurnal or other sub-daily low flow fluctuations. An important and still unanswered question remains how to overcome scale sensitivity of flashiness indicators in different hydrological basins. This is crucial for establishing how urbanisation impacts flashiness and how changes relate to basin characteristics such as size, slope, imperviousness degree, and whether urbanisation thresholds can be identified, at a value above which basin response is characteristically urban (Praskievicz and Chang, 2009).

1.3 Scaling analysis of hydrological flows

Scaling behaviour of river flows has been investigated by various authors, aiming to identify characteristic length and timescales and to detect scale dependence of hydrological response processes. Among the various statistical methods that have been proposed to investigate scaling, fractals and multifractals are among the most popular and powerful. Approaches for fractal analysis include spectral analysis based on second-order properties and trace moment analysis based on a wider range of statistical moments, typically between 0.1 and 4. The universal multifractal framework is based on the identification of scaling exponents summarising the changes in flow distributions across a given range of scales, (see Schertzer and Lovejoy, 1987 and Schertzer and Lovejoy, 2011 for a review). One important drawback of multifractal analyses is that scaling of hydrological flow time series only holds in approximation and only over a limited range of scales. Many studies report the existence of "scale breaks" at which scaling parameters change and significant departures from (multi)fractality can be observed. Table 1 summarises findings from selected scaling analyses of flow time series in the literature. It shows that the number and location of the scale breaks, as well as the values of the multifractal parameters, are sensitive to the method applied to estimate them and the resolution of the data used to conduct the analysis. For example, Labat et al. (2013) performed spectral analysis and

Table 1. Summary of results reported on the literature for (multi)fractal analysis of hydrological flows. MA: moment analysis, MFA: multifractal analysis, SA: spectral analysis, TMA: trace moment analysis.

Reference	Method	Sampling scale	Basins	Time series length	Scale break	Value C_1	Value alpha
Tessier et al. (1996)	MFA	day	30 basins in FR	11–30 years	16 days	1–16 days: 0.2 ± 0.1	1–16 days: 1.45 ± 0.25
			40–200 km^2		16 days	30–4096 days: 0.2 ± 0.1	30–4096 days: 1.45 ± 0.2
Sauquet et al. (2008)	SA	Hour	34 basins in FR	16–37 years	8.7 h–7 days	–	–
Sauquet et al. (2008)	MA	Hour	12.7–703 km^2	16–37 years	10 h-6.25 days*	–	–
Sauquet et al. (2008)	SA	Day	Idem	Idem	12 days	–	–
Pandey et al. (1998)	SA	Day	19 basins USA	9–73 years	8 days	1–8 days: 0.2 ± 0.1	1–8 days: 1.65 ± 0.12,
			$5–1.8 \times 10^6$ km^2	9–73 years	8 days	1–8 days: 0.2 ± 0.1	1–8 days: 1.65 ± 0.12
Labat et al. (2013)	SA	30 min	3 basins in FR	–	1 day	–	–
Labat et al. (2013)	TMA	30 min	ca. 13 km^2	–	16 h	30 min–16 h: 0.22 > 16 h: 0.35	30 min–16 h: 1.18 >16 h: 0.79

* only for higher-order moments.

trace moment analysis for 30 min flow time series and identified different flow regimes with scale breaks at 1 day for spectral and 16 h for trace moment analysis. But when they performed the same analysis at daily and at 3 min resolution, they identified different scaling regimes, with scale breaks at 16 days and 1 h for daily and 3 min resolution, respectively. Similarly, Sauquet et al. (2008) found different scaling regimes in their scaling analysis of flows for 34 basins, with scale breaks at 12 days for daily resolution and scale breaks varying between 8.7 h and 7 days across basins when using hourly data resolution, based on spectral analysis. When they applied trace moment analysis for the same time series at hourly resolution, they found no scale breaks for the lower-order moments and scale breaks between 10 and 150 h for higher-order moments. This shows that while most flows exhibit some sort of scaling behaviour, the identified scaling laws are not very robust or consistent, as they are dependent on analysis methods and data resolution.

1.4 Statistical analysis of hydrological response based on adaptive sampling using inter-amount times

In this paper, the IAT formalism is applied to flow time series and statistical distributions, and scaling properties are compared to the ones obtained using the classical fixed-time framework. To do this, we use flow observations collected in 17 hydrological basins in Charlotte, North Carolina. We aim to investigate what effects an adaptive sampling strategy such as IAT sampling has on statistical properties of the time series, in particular on the tails of the statistical distributions associated with peak flow and low flow extremes. The main problem with a fixed sampling rate, as in traditional flow time series analysis, is that it can only accurately represent frequencies of variations at timescales larger than a certain threshold. When frequencies higher than that exist, errors are introduced as information about the higher frequency variability is lost (Dippe and Wold, 1985). Increasing the sampling resolutions solves this problem, but results in oversampling of base flow values with respect to peak flows. An alternative consists of adopting an adaptive sampling strategy, i.e. one that adapts the sampling rate to the variability of the signal itself (e.g. Feizi et al., 2011). This makes sense for processes that are very unevenly distributed in time (such as rainfall and hydrological flows), and means taking more samples during periods of high activity (e.g. peak flows following storm events) and fewer during lower activity (e.g. periods of base flow). A well-designed adaptive sampling technique lowers the probability of missing an interesting feature like peak flow and avoids oversampling during periods of small flow variations. We examine to what extent IATs influence the variance, skewness and shape of the sample distributions and how they can be used to better characterise basin flashiness and derive more robust scaling laws. Our results show that because IATs give more weight to rare peak flows compared to common base flows, they can provide different insights into flow properties and complement traditional flow time series analyses and metrics. Advantages of IAT sampling compared to conventional time series analysis are that IAT time series contain more information about peak flows and evolve in a more predictable way across ranges of smaller to larger scales. This makes them a more robust and reliable source of information to make predictions about flow characteristics at small, unobserved scales, including crucial information about rapidly evolving peak flows.

This paper is organised as follows. In Sect. 2 we present the flow datasets and methods used for analysis. We explain the methodology for deriving normalised IATs and introduce metrics we used to compare properties of flows and IAT time series, to characterise hydrological response and compare response across basins. In Sect. 3, results of the analyses are presented and discussed, first based on results obtained using a daily sampling scale, and followed by results obtained a range of sampling scales, from hourly up to seasonal sampling scale. Conclusions and suggestions for future work are summarised in Sect. 4.

Table 2. Summary of hydrological basins in the Charlotte area: basin area (km^2), imperviousness (%), average 24 h flow (m^3), average 24 h flow normalized by basin area (mm) and length of observation in years.

ID	Name	Area	Imperv.	Dams	Mean flow	Mean norm. flow	N years
825	UBriar	13.3	24.0	22	12 275	0.92	17.4
315	Taggart	13.6	35.0	3	13 559	1.00	17.2
562	Campbell	15.3	28.0	48	13 567	0.89	16.2
175	Steele	17.9	32.0	21	17 838	1.00	17.4
700	McMullen	18.3	21.0	15	20 348	1.11	29.0
255	UMcAlpine	18.9	18.1	100	15 061	0.80	16.3
975	Irvins	21.8	8.0	62	14 821	0.68	16.3
970	Stewart	23.4	33.0	55	38 800	1.66	15.3
348	Coffey	23.8	25.0	72	24 104	1.01	17.0
409	LSugarM	31.7	48.0	2	46 775	1.48	21.0
022	LBriar	48.5	25.0	17	53 246	1.10	19.8
800	SixMile	52.6	15.0	−99	38 914	0.74	8.0
300	UIrwin	78.1	34.0	39	107 119	1.37	29.0
600	MMcAlpine	100.2	20.0	51	105 640	1.05	29.0
507	LSugarA	111.1	32.0	24	199 002	1.79	29.0
530	LSugarP	127.4	26.0	−99	205 202	1.61	18.3
750	LMcAlpine	238.4	19.4	−99	269 534	1.13	29.0

2 Data and methods

2.1 Flow datasets

The data used in the study were collected at 17 USGS stream gauging stations in Charlotte–Mecklenburg county, North Carolina. Gauging stations are located at the outlet of hydrological basins that range from 13 to 238 km^2 in size. The area is largely covered by low to high intensity urban development, covering 60 to 100 % of basin areas. Percentage of impervious cover varies from 8 % in the least developed to 48 % in the most urbanised basin covering the city centre of Charlotte. Figure 1 shows a map with the location of the area, boundaries of hydrological basins and location of stream gauges used in the analysis. Table 2 summarises the main characteristics of the 17 basins.

Stream gage data were collected at 5 to 15 min intervals over the period 1986–2011. Table 2 summarises the characteristics of the basins associated with each basin as well as the time period covered by the data. The temporal scale of observations changed from 15 to 5 min between 2010 and 2014, at different times for each gauge; overall 20–30 % of the total observation record was covered by 5 min intervals. Gauges measure water depth using pressure transducers and flow is derived using stage–discharge curves. These curves were established based on protocols developed by USGS and include manual flow measurements during site visits performed by USGS staff. As part of this procedure, stage–discharge curves are checked and recalibrated during site visits several times per year (https://waterdata.usgs.gov/nwis/measurements). The percentage of missing flow data was smaller than 5 % for all gauges included in the analysis; missing data were treated like zeros. The effect of missing data on IATs is difficult to predict as this depends on the pattern of missing values and whether or not they occur during a period of low or peak flow. Sensitivity studies by Schleiss and Smith (2016) have shown that the general effect of replacing missing values by zeros is that a few sample IATs will be overestimated. This mostly affects the right tail of the distribution and tends to have limited impact on peak flow

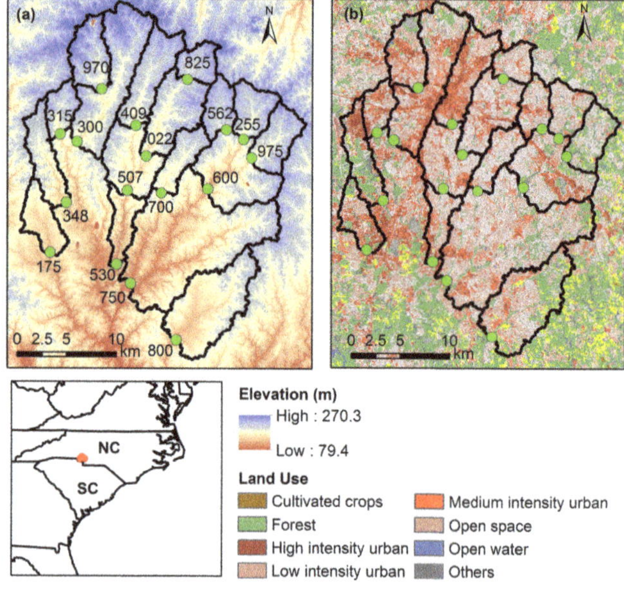

Figure 1. Map with the location of the area, boundaries of hydrological basins and location of stream gauges used in the analysis. NC: North Carolina. SC: South Carolina.

characteristics. Another strategy would be to replace missing values by mean or median flow value, which may slightly reduce the overestimation of IATs in case several missing values occur in row. However, in this paper only the worst-case scenario will be considered, i.e. missing values were replaced by zeros.

2.2 Definition of inter-amount times

In this paper we analyse hydrological flow variability, based on the distribution of IATs. We use the following definition of IATs, based on Schleiss and Smith (2016), when $\Delta q > 0$ denote a fixed flow amount: the series of IATs $\tau_n(\Delta q)$ with respect to Δq is defined as follows.

$$\tau_n(\Delta q) = t_n(\Delta q) - t_{n-1}(\Delta q), \tag{1}$$

where $t_n(\Delta q)$ denotes the time at which the cumulative flow amount first exceeded n times (Δq):

$$t_n(\Delta q) = \inf\{u : Q(u) \geq n \cdot \Delta q\}, \tag{2}$$

where $Q(u)$ denotes the cumulated flow at time u, $Q(0) = 0$, and inf stands for infimum, also known as the greatest lower bound in a set.

A steady flow pattern with constant flow has equal IATs for all values of Δq. A variable flow pattern, on the other hand, is characterized by a more variable IAT distribution.

2.3 Normalized inter-amounts

Flow magnitudes strongly vary from one gauge to another. To overcome this scale dependence and compare flow IATs across basins with different sizes and flow amounts, one needs to normalize IATs with respect to a common timescale. A possible way to do this is to fix an average IAT $\overline{\tau}$ (e.g. 24 h) and determine the inter-amount $\Delta q_{\overline{\tau}}$ at this timescale:

$$\Delta q_{\overline{\tau}} = \overline{\tau}\frac{Q_N}{T}, \tag{3}$$

where Q_N denotes the total cumulative flow amount at the considered location and T is the length of the studied time period. In other words, instead of comparing IATs for a fixed accumulation, we choose the mean IAT $\overline{\tau}$ and compute $(\Delta q)_{\overline{\tau}}$ such that the series of IATs $\{\tau_n(\Delta q_{\overline{\tau}}) : n = 1, \ldots, N\}$ has mean $\overline{\tau}$. Two locations with different cumulative flow amounts over a given period of time, e.g. over a year, therefore have different normalized inter-amounts.

2.4 Sample estimates and minimum inter-amount scale

Inter-amount times can be estimated from a sample flow time series $q_1, .., q_N$ with temporal observation scale Δt that may vary in time. But for simplicity, only the case with fixed temporal resolution Δt will be considered below. A key step in this procedure is the determination of the first passage times $t_1, .., t_n$ in Eq. (2). This is done by considering the sample accumulated flow amounts $Q_1 < .. < Q_N$ at times $t_n = t_0 + n\Delta t$:

$$Q_n = \sum_{i=1}^{n} q_i n = 1, \ldots, N. \tag{4}$$

The exact first passage times $t_1, .., t_n$ for a fixed flow amount $\Delta q > 0$ are likely to be unknown due to the limited temporal resolution of the data. But we can approximate them based on linear interpolation:

$$\hat{t}_n(n\Delta q) = \Delta t\left(i_{n\Delta q} - \frac{Q_{i_{n\Delta q}} - n\Delta q}{q_{i_{n\Delta q}}}\right) n = 1, \ldots, N, \tag{5}$$

where $\hat{t}_n$ are the estimated passage times and $i_{n\Delta q}$ denotes the index (in the sample) at which the total cumulated flow first exceeded n times (Δq):

$$i_{n\Delta q} = \min\{i \in \mathbb{N} | Q_i \geq n\Delta q\} n = 1, \ldots, N. \tag{6}$$

The sample IAT estimates are then given by the following:

$$\hat{\tau}_n(\Delta q) = \hat{t}(n\Delta q) - \hat{t}(n\Delta q - \Delta q). \tag{7}$$

Because of the linear interpolation in Eq. (5), each sample IAT estimate, regardless of its length and the scale of analysis, will be affected by a small interpolation error $\varepsilon_n(\Delta q) < \Delta t$. This error is random and has little influence on key statistics as long as IATs remain much larger than Δt, as is usually the case for large enough values of Δq and during periods of low to moderate flow. Most of the interpolation errors happen during peak flows, when large flow amounts are accumulated over small periods of time. It is therefore important, for any given gauge, to identify the values of Δq above which reliable IAT estimates can be derived. To identify the range of scales over which IATs can be reliably estimated, we consider the worst-case scenario in which all interpolation errors are equal to $\pm\Delta t$. In this case, the maximum relative error affecting IAT estimates is given by the following:

$$\varepsilon_n(\Delta q) = \frac{\Delta t}{\hat{\tau}_n(\Delta q)}. \tag{8}$$

The minimum value of Δq for which IATs can be reliably estimated depends on how strictly we want to control the estimation errors in Eq. (8). In our analysis, we set the mean of absolute relative errors to be smaller than 50 %. This is a rather conservative approach as the estimation errors in Eq. (8) represent the worst-case scenario and actual errors are likely to be much smaller than that. This leads to the following rule for determination of minimum inter-amounts Δq that can be used for analysis:

$$\Delta q_{\min} = \min\{\Delta q > 0 : \overline{\varepsilon}_{\Delta q} < 0.5\}, \tag{9}$$

where $\overline{\varepsilon}_{\Delta q}$ represents the arithmetic mean of the maximum relative errors in Eq. (8).

In addition to the lower bound, we also impose an upper bound on the inter-amounts used in our analysis. This is necessary to ensure IAT time series are long enough to compute relevant statistical moments. Typically, there should be at least 100 consecutive IATs, which yields the following upper bound for inter-amount Δq:

$$\Delta q_{\max} = \lfloor \frac{Q_N}{100} \rfloor, \tag{10}$$

where $\lfloor\rfloor$ denotes the lower integer part and Q_N is the total cumulative flow for the considered time series.

It is worth pointing out that the lower bound on the inter-amount in Eq. (9) also provides an indication of the left-tail properties of IATs, and thus of the degree of flashiness of the hydrological response, i.e. the smallest scale at which flow variations can be studied given a fixed temporal observational resolution. We will elaborate on this in Sect. 2.7, where we discuss this property in relation to basin flashiness. More generally, the left tail properties of IAT distributions provide a good indication of what observational resolution is necessary to adequately capture the most extreme flow variations. For more details on this important point, the reader is referred to the results section.

Note also that analyses of IATs were conducted for all gauges over the entire period of available data, without distinguishing between year, season or hour of the day. This was necessary as time series would otherwise be too short to study IATs across different scales. This means we mostly focus on average characteristics of IAT and flow distributions with respect to area size and imperviousness degree and potential influence of flow regulation and stormwater detention facilities, as far as this information is available for the 17 basins. We refrain from investigating long-term trends, as our time series are restricted to maximum 30 years and because a recent study by Villarini (2016) showed no signs of long-term trends at 7506 gauges in the contiguous USA in the last 30 years. Indeed, our own analyses revealed no significant long-term trend in mean IAT or flow variability over the considered time period.

2.5 Distribution of inter-amount times versus flows

Sample histograms of IATs and flows were analysed to investigate what different insights they provide into characteristics of the flow regimes. We plotted sample histograms for all gauges; appropriate bin widths were determined based on Scott's rule (Scott, 1979). We computed the coefficient of variation (CV), defined as the standard deviation divided by the mean, as an indicator for relative spread around the mean. Values of skewness and "medcouple" (Brys et al., 2004), a more robust skewness metric based on ordered statistics instead of statistical moments, were computed to investigate asymmetry of the histograms and influence of outliers. We compared coefficient of variation, skewness and medcouple values for IATs with those for traditional flow time series and investigated relationships of the three statistics with basin area and imperviousness degree.

2.6 Distribution of changes in inter-amount times

First-order differences of IATs and flows were computed to look into characteristics of the rising and falling limbs of hydrographs. Because IATs are measured on an inverted scale, positive differences are associated with the falling limb of the hydrograph and negative differences with the rising limb of the hydrograph. Narrow ranges of histogram values for IAT differences indicate slowly varying flow; wide range histograms indicate more flashy behaviour. Positively skewed histograms for IAT differences indicate that the distribution is dominated by values on the rising limb and short recession limbs, while negatively skewed histograms indicate a larger part of the flow is associated with flow recession, i.e. long, slowly receding hydrographs, for instance, induced by a strong groundwater flow component. Differences were computed at the 24 h timescale, imposed by the minimum inter-amount scale rule. Similarly to the other histograms, bin widths were chosen based on Scott's rule.

2.7 Flashiness indicator and minimum observable scale

As mentioned earlier, the lower bound on the inter-amount provides an indication of left-tail properties of IAT distributions (i.e. short waiting times) and can therefore be used to characterise the degree of flashiness of the hydrological response. In flashier catchments, the flow can rise quicker, resulting in lower IATs during times of heavy rain. The minimum observable inter-amount represents the smallest scale at which flow variations can be studied with acceptable interpolation errors, given a fixed temporal observational resolution. By extension, the lower tail of the IAT distribution provides a good indication of what observational resolution is necessary to adequately capture the most extreme flow variations. The IAT flashiness indicator used in this paper is defined as the mean scale μ (expressed in hours) at which the 1 % quantile of the IAT distribution equals the observational scale Δt (15 min in our case). That is, the IAT flashiness indicates the average time needed to accumulate the amount of flow that can be accumulated in 15 min or less, 1 % of the time. The larger the flashiness, the more flow can be accumulated over short amounts of time. To better interpret results, we compared the IAT flashiness index with the frequently used R–B flashiness index defined in Baker et al. (2004):

$$\mathrm{R-B\ index} = \frac{\sum_{i=1}^{N} |q_i - q_{i-1}|}{\sum_{i=1}^{N} q_i}, \tag{11}$$

where q_i denotes the flow at time step i. The R–B flashiness index is dimensionless and can vary between 0 and 2. It is 0 for constant flow and 2 for highly variable and continuously changing flow. Its value is independent of the units chosen to represent flow (Baker et al., 2004). However, index values do depend on the timescale at which they are computed, as will be discussed later in the results section. In our analysis, we computed R–B flashiness indices on daily aggregated flow values.

2.8 Scaling of inter-amount times

Multifractal analysis techniques were applied to investigate the scaling behaviour of IAT time series across different inter-amount scales. Multifractal analyses are based on the assumption of generalised scale invariance, in which the statistical moments or order $q > 0$ of a stochastic process X_λ at scale ratio λ are related by a power law:

$$\langle X_\lambda^q \rangle = C(q)\lambda^{K(q)}, \tag{12}$$

where $\langle X_\lambda^q \rangle$ denote the moments of order q of X measured at a scale ratio λ, $C(q)$ is a constant (for each q) and $K(q)$ is called the moment scaling function. Within the universal multifractal framework, $K(q)$ is characterised with the help of only three parameters, α, C_1 and H (Schertzer and Lovejoy, 1987, 2011):

$$K(q) = \begin{cases} \dfrac{C_1}{\alpha - 1}(q^\alpha - q) - qH & \text{if } \alpha \neq 1 \\ C_1 q \ln(q) - qH & \text{if } \alpha = 1 \end{cases}. \tag{13}$$

The parameter C_1 is referred to as the intermittency and characterises the clustering of the time series at smaller and smaller scales. $C_1 = 0$ for a homogeneous field that fills the embedded space and approaches 1 for an extremely concentrated field. The parameter α is called the multifractality index ($0 < \alpha < 2$) and it controls how the moments change when going from one scale to another. Finally, $H = -K(1)$ is called the Hurst exponent. Note that in the case of IATs, the mean inter-amount time $\overline{\tau}$ and scaling ratio λ are inversely proportional to each other (i.e. $\Delta q_{\overline{\tau}} \sim \lambda^{-1}$). So either of them can be used here as a measure of scale. The only difference will be the value of the constant $C(q)$ and the sign of the exponent in Eq. (12).

The scaling quality is assessed by noting that if Eq. (12) is true, the log moments for fixed values of q should be a linear function of the log-scale of $\ln(\lambda)$:

$$\ln(\langle X_\lambda^q \rangle) = K(q)\ln(\lambda) + \ln(C(q)). \tag{14}$$

The extent to which this equality holds can be assessed by fitting a linear regression model and computing the R^2 values, i.e. the coefficient of determination of the log moments versus the log-scale for each value of q. A R^2 of 1 indicates perfect scaling. The lower the coefficient of determination, the larger the deviations from scale-invariance. The approach was repeated for different values of q and the mean or minimum value of R^2 were chosen as a way to assess the overall quality of the scaling. Based on recommendations by Lombardo et al. (2014), we refrained from using too low- or high-order moments and only considered values of q between 0.4 and 2.5, with an equal number of moments above and below 1 to avoid favouring one tail of the distribution over the other. The range of IAT scales that was used for the analysis was constrained by the length of the time series and the minimum and maximum inter-amounts defined in Eqs. (9) and (10). The corresponding scales varied from 0.1 to 0.6 days up to 28 to 100 days for the longest time series.

3 Results

In the following sections we compare statistical properties of flow and IAT time series and highlight differences that result from the different sampling strategies. Analyses are first conducted at the 24 h timescale and associated mean inter-amount sampling scale. In the second part of this section, we analyse how statistical properties of flow and IAT time series vary across scales, and quantify flashiness and scaling behaviour of both time series.

3.1 Time series and variability analysis of inter-amount times and flow values

Figure 2 shows an example of times series for flows and for IATs for the gauge at Taggart Creek, a 13.6 km^2 basin in the Charlotte catchment, at 24 h sampling scale. The two graphs bring out different aspects of flow variability: flow time series have most of their data points concentrated in the low flow region, with intermittent peak flows characterising rain events. For IATs, peak flows appear as minima, while periods of low flow show up as maxima in the time series. The graph illustrates how IAT samples are more evenly distributed across high and low values in the time series compared to flows. The mean inter-amount for Taggart Creek at 24 h sampling scale is 13 559 m^3, equivalent to 0.998 mm when normalised by basin area. Hence, in IAT analysis, the time series is sampled each time 0.998 mm of normalised flow has been accumulated, which amounts to frequent samples during high flows and fewer samples during low flow periods. For instance, a high concentration of IAT samples is clearly visible for the wet year 2003: this year is represented by 802 IAT samples compared to the 365 samples per year we have on average.

Figure 3 illustrates the adaptive sampling strategy based on flow amounts as the sampling unit, instead of fixed time steps. Figure 3b shows cumulative flow over a week, where a storm event occurred on 7 August. In conventional flow time series analysis, flow is sampled daily (in this example), resulting in one sample representing the peak period of the event (i.e. on 7 August). In IAT analysis, flow accumulation determines the sampling frequency, so periods of low flow are sparsely sampled, while the storm event is represented by eight samples. This illustrates how, even for 24 h mean

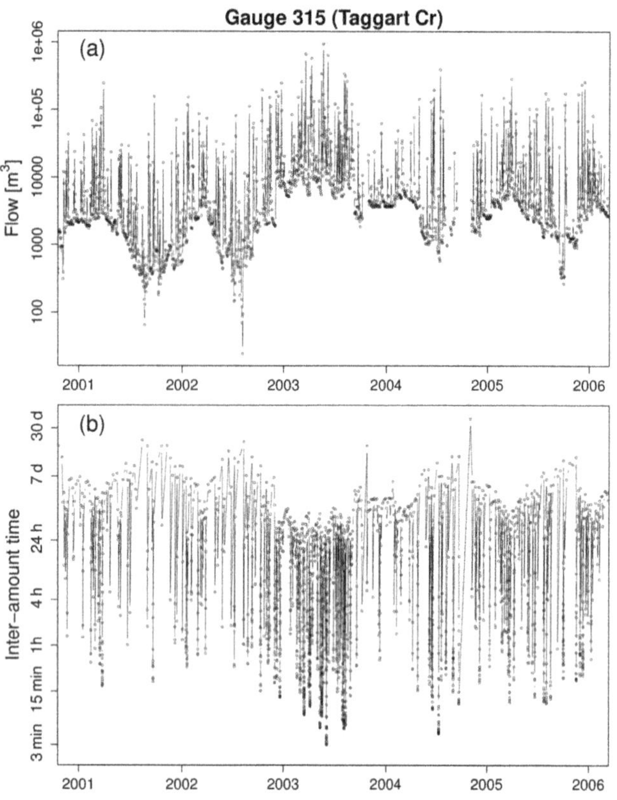

Figure 2. Example of times series for flow **(a)** and for associated inter-amount times **(b)** for the flow gauge at Taggart Creek, a 13.6 km^2 basin in the Charlotte catchment.

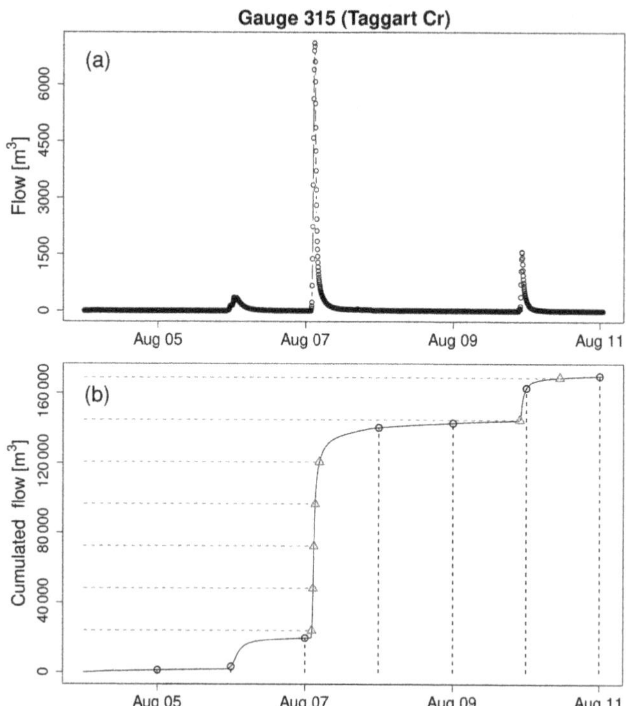

Figure 3. Illustration of inter-amount data sampling for cumulative flow over a period of 7 days, for Taggart Creek. **(a)** Flow data series at original 15 min observational resolution; **(b)** cumulative graph for flows and IATs at the same mean sampling resolution, illustrating how adaptive sampling based on IATs differs from classical fixed-time sampling.

inter-amounts, sampling frequency can be much higher during periods of peak flow.

Histograms of flow time series and IATs at daily timescale are plotted in Fig. 4, for two basins, Taggart Creek (13.6 km^2) and LSugarA (111 km^2). The corresponding inter-amounts are 1 and 1.8 mm of normalised flow (for Taggart and LSugarA, respectively). Histograms for the other 15 basins are available in the Supplement to this paper. Figure 4 shows that both histograms of flows and IATs are positively skewed. In both cases however, left and right tails represent very different flow characteristics. The left tail of the flow's histogram essentially features common base flow values while the right tail captures rare peak flow events. By contrast, the left tail of IAT distributions, which makes up most of the values, predominantly features short IAT values associated with periods of high flow. The rare samples that make up for the right tail represent long waiting times associated with extended periods of low flow. The low density of the first bin in the flow histogram for LSugarA reflects the effect of low flow regulation for this basin. The same effect is reflected in the bimodal shape of the IATs histogram. Note that the low density 0–0.5 bin in the flow histogram for LSugarA corresponds to the > 3.5 day bins in the IAT histogram.

Tables 2 (6th and 7th columns) and 3 summarise statistics of flow and IAT time series, at 24 h sampling scale. The results show that mean inter-amounts vary from 12 275 m^3 for the smallest to 269 534 m^3 for the largest basin in size. Mean normalised inter-amounts vary from 0.68 mm for Irvins Creek, the least-urbanised basin (8.2 % imperviousness) to 1.79 mm for Little Sugar Creek at Archdale, one of the largest basins with a high degree of imperviousness (32 %). Coefficients of variation at the daily scale are consistently higher for flows than for IATs (e.g. 1.7 times higher on average), which highlights the more balanced nature of IAT distributions. Skewness values at the daily timescale are 3.6 times higher for flows than for IATs, on average, and even up to a factor of 15 higher for Stewart Creek. By contrast, medcouple values for flows are lower than for IATs by a factor of 2.1 on average. This shows that statistical distributions of flows are strongly influenced by the presence of a few very large outliers. Most of the weight, however, lies close to the median (low medcouple). The IAT sampling gives more weight to rare peak flow values and less to common base flow, therefore producing distributions with lower skewness and more information about peak flow values. The larger medcouple values mean that IATs above the median value tend to be much further away from the median than

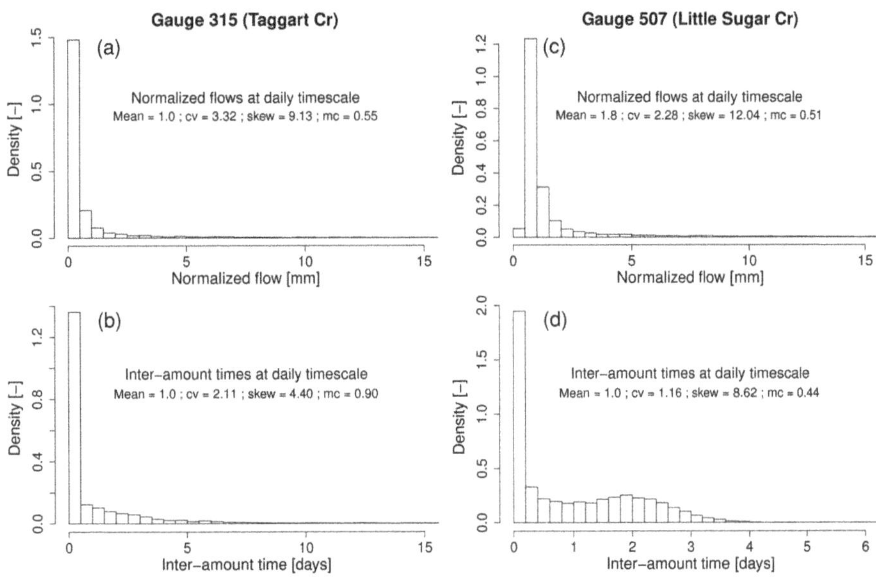

Figure 4. Histograms of flow time series **(a)** and time series of inter-amount times **(b)** for Taggart Creek and Little Sugar Creek at Archdale (LSugarA), for 24 h scale.

Table 3. Summary statistics of time series for flows and inter-amount times, at 24 h sampling scale: coefficient of variation (CV), skewness (Skew) and medcouple (Mc).

name	CV IAT	CV flow	Skew IAT	Skew flow	Mc IAT	Mc flow	Skew dIAT	Skew dflow	Mc dIAT	Mc dflow
UBriar	1.95	3.69	4.91	14.79	0.84	0.41	0.39	−2.81	0.51	−0.30
Taggart	2.11	3.32	4.40	9.13	0.90	0.55	0.00	−0.16	0.57	−0.46
Campbell	2.02	3.25	4.26	10.40	0.84	0.51	−0.02	0.15	0.66	−0.41
Steele	2.24	3.65	4.39	10.21	0.86	0.58	−0.43	−0.41	0.74	−0.46
McMullen	2.22	3.37	5.35	10.10	0.90	0.56	−0.61	−0.40	0.61	−0.35
UMcAlpine	2.04	3.55	5.53	13.51	0.79	0.42	−2.48	2.17	0.63	−0.38
Irvins	2.52	4.32	8.37	11.74	0.89	0.42	−3.84	0.07	0.78	−0.41
Stewart	0.96	2.47	0.84	12.90	0.12	0.37	−0.23	−0.25	0.26	−0.02
Coffey	2.15	2.94	7.34	8.44	0.85	0.54	−1.05	0.21	0.64	−0.41
LSugarM	1.57	2.95	2.06	11.55	0.90	0.55	−0.43	0.73	0.37	−0.31
LBriar	1.74	3.30	3.13	13.77	0.87	0.51	−0.87	1.13	0.56	−0.32
SixMile	2.23	2.59	6.29	6.42	0.82	0.38	−1.34	1.23	0.69	−0.31
UIrwin	1.36	2.70	2.65	14.43	0.69	0.53	−0.32	1.77	0.45	−0.22
MMcAlpine	2.00	3.19	5.42	10.30	0.84	0.50	−1.51	0.66	0.68	−0.38
LSugarA	1.16	2.28	8.62	12.04	0.44	0.51	0.77	0.40	0.33	−0.26
LSugarP	1.04	2.10	1.52	9.20	0.49	0.58	−0.71	−1.04	0.33	−0.33
LMcAlpine	2.16	2.84	6.56	7.65	0.88	0.50	−1.63	0.27	0.50	−0.32

values below the median. In other words, the right part of the distribution, which features long waiting times during low flow conditions, can be very stretched.

These results show that adaptive sampling based on inter-amounts leads to more balanced representation of high and low flows, resulting in lower coefficients of variation reflecting more stable statistical variance compared to traditional flow time series sampling. We would like to point out that these results were obtained at the 24 h sampling scales. In Sect. 3.4, behaviour of the statistical distributions of flows and IATs, as well as associated CV, skewness and medcouple values, will be analysed across a range of sub-daily to seasonal scales.

3.2 Statistical distribution properties comparison across different hydrological basins

Subsequently, we compared properties of IAT and flow distributions across the 17 basins in relation to basin characteristics. Figure 5 shows scatter plots of mean normalised inter-amounts, CV, skewness and medcouple values for flows and IATs as a function of basin area and imperviousness de-

Table 4. Summary statistics of time series for flows and inter-amount times, at 24 h sampling scale: coefficient of variation (CV), skewness (Skew) and medcouple (Mc), for three sets of connected sub-basins in the Charlotte catchments: Irwin, Little Sugar and McAlpine.

ID	Name	CV IAT	CV flow	Skew IAT	Skew flow	Mc IAT	Mc flow	Skew dIAT	Skew dflow	Mc dIAT	Mc dflow
970	Stewart	0.96	2.47	0.84	12.90	0.12	0.37	−0.23	−0.25	0.26	−0.02
300	UIrwin	1.36	2.70	2.65	14.43	0.69	0.53	−0.32	1.77	0.45	−0.22
825	UBriar	1.95	3.69	4.91	14.79	0.84	0.41	0.39	−2.81	0.51	−0.30
022	LBriar	1.74	3.30	3.13	13.77	0.87	0.51	−0.87	1.13	0.56	−0.32
409	LSugarM	1.57	2.95	2.06	11.55	0.90	0.55	−0.43	0.73	0.37	−0.31
507	LSugarA	1.16	2.28	8.62	12.04	0.44	0.51	0.77	0.40	0.33	−0.26
530	LSugarP	1.04	2.10	1.52	9.20	0.49	0.58	−0.71	−1.04	0.33	−0.33
562	Campbell	2.02	3.25	4.26	10.40	0.84	0.51	−0.02	0.15	0.66	−0.41
255	UMcAlpine	2.04	3.55	5.53	13.51	0.79	0.42	−2.48	2.17	0.63	−0.38
975	Irvins	2.52	4.32	8.37	11.74	0.89	0.42	−3.84	0.07	0.78	−0.41
600	MMcAlpine	2.00	3.19	5.42	10.30	0.84	0.50	−1.51	0.66	0.68	−0.38
750	LMcAlpine	2.16	2.84	6.56	7.65	0.88	0.50	−1.63	0.27	0.50	−0.32

gree. The results show a positive correlation of 24 h mean normalised flows or inter-amounts with basin size (Spearman correlation 0.55). This is mainly explained by a lower likelihood of low flows that have a large influence at this scale (24 h). Mean normalised flows correlate positively with imperviousness degree (Spearman correlation 0.58), which is likely to be explained by a generally growing importance of flow regulation, resulting in maintenance of higher mean base flows in urbanised basins.

Looking at CV values across all basins (Fig. 5c, d), we found that CV values for both flows and IATs generally decrease with basin size and with imperviousness degree. CV values are significantly negatively correlated with basin size for flows (Spearman rank correlation −0.75). This can be explained by an increased smoothing effect on flow variation, in particular a lower likelihood of low flow extremes during dry periods for larger basins. CV values for IAT distributions do not show a significant correlation with basin size, while they are significantly negatively correlated with imperviousness (Spearman rank correlation −0.57). Since IAT distributions put more weight on high flows compared to low flows as a result of their adaptive sampling strategy, this probably indicates stronger influence of flow regulation in urbanised basins resulting in more uniform run-off during rainy periods. IATs during these periods concentrate relatively more closely to the mean and show fewer extremes (this is clearly visible for the most urbanised basin, LSugarM, gauge 409). The effect of urbanisation as reflected by imperviousness degree on IAT statistics appears to be more important than basin size.

Scatter plots for skewness and medcouple values (Fig. 5e, f, g, h) show generally weak correlation with basin area (Spearman correlations not significant at the 5 % level). Skewness of IAT distributions is significantly negatively correlated with imperviousness (Spearman rank correlation −0.63). Similar to CV values, this probably indicates stronger influence of flow regulation on flows in urbanised basins. Medcouple values for IATs clearly show three low-value outliers: for Stewart Creek (970), LSugarP (530) and LSugarA (507). In these basins, active low flow control is applied[1] preventing occurrence of low flow extremes and high IAT extremes. The effect shows up more clearly for IAT medcouple values, as a result of the adaptive sampling strategy that gives more weight to peak flows, leading to generally higher medcouple values, but also reflecting more clearly the absence of low flow extremes. Some of the basins in this study are sub-basins of each other, which implies that flows can be correlated. Table 4 summarises CV, skewness and medcouple values for three sets of sub-basins in the Charlotte catchment. The results show that variability in skewness and medcouple values is unrelated to inter-basin connections. The same applies for flow CV values, while CV values for IATs seem to be clustered by group of sub-basins, indicating that inter-basin correlation plays a role in explaining IAT second-order variability. The fact that the effect is only visible for IAT, not for flows, indicates that correlation is mainly associated with occurrence of peak flows, which receive more weight in IAT than in flow statistics.

In this section we discussed distributions of IATs and flows at the 24 h scale. Results showed that larger basins are generally characterised by stronger smoothing of flows, resulting in higher mean flow, lower CV and lower skewness of the flow histograms. Flow variability is clearly correlated with basin size, which is mainly a result of smoothing of low flows, in the left tail of the flow histogram. This confirms results previously reported in the literature on scaling between flows and basin area (e.g. Goodrich et al., 1997; Smith, 1992) and specifically between CV of flows and basin area (Bloeschl and Sivapalan, 1997). These authors also refer to complexities in hydrological response resulting in deviations from this general relationship. The same applies for the basins in our study, where basin area only explains part of the flow variability, especially for smaller basins. Results

[1] USGS, water year reports

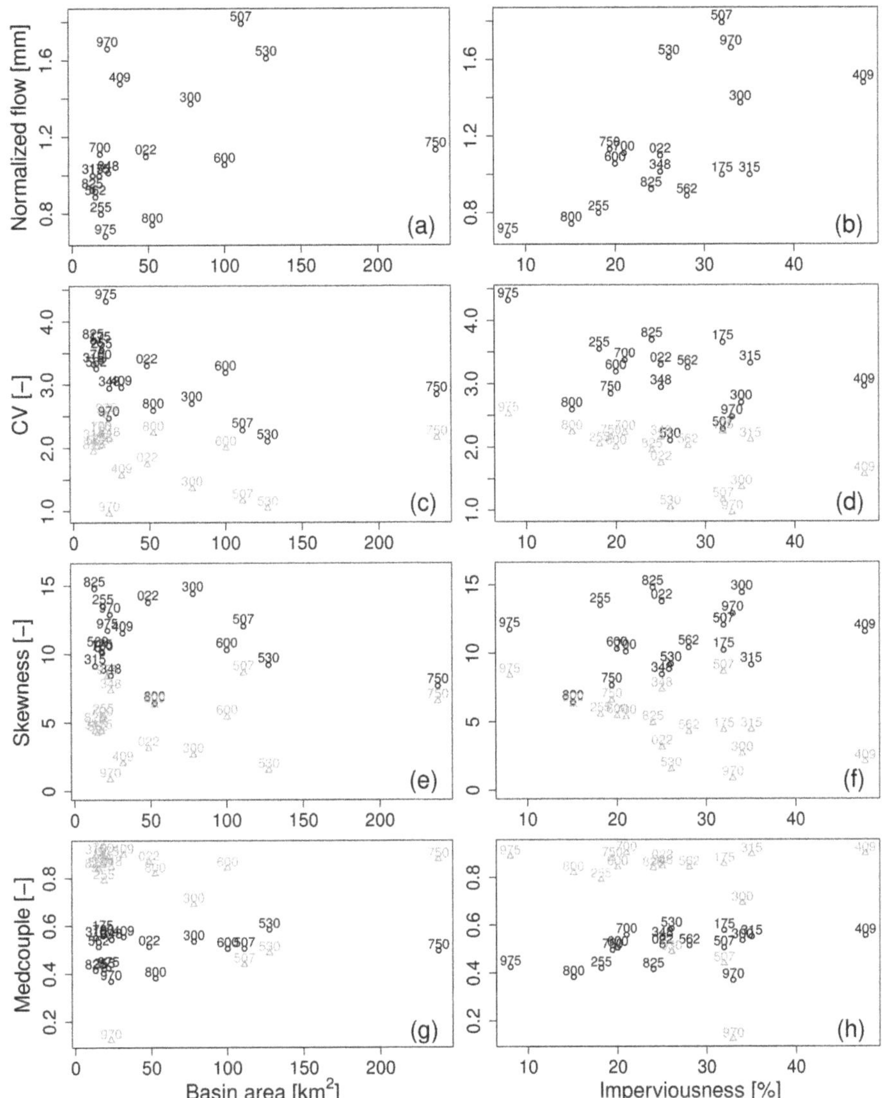

Figure 5. Scatter plots for mean normalised flow inter-amounts **(a, b)**, coefficient of variation **(c, d)** and medcouple values **(e, f)** for flows and inter-amount times versus basin area and imperviousness degree. Grey triangle symbols represent inter-amount times, black circles represent flows.

showed that larger imperviousness is associated with higher mean flows and significantly lower CV values for IATs, while there was not significant correlation between CV values for flows and imperviousness. This is probably explained by urbanisation being mainly associated with stronger flow regulation by detention and capacity constraints in the drainage system. Since IATs are relatively more sensitive to high flows, this effect showed up more clearly in CV values for IATs than for flows. CV and skewness values are much higher for flows than for IATs, while medcouple values are lower for flows, indicating strong asymmetry of the flow distributions and low representation of high flow extremes in the statistical distribution. While Kjeldsen (2010) reported a decrease in CV and skewness associated with urbanization for basins in the UK, we did not find significant correlations based on CV and skewness indicators for flows. Skewness for IATs was significantly negatively correlated with imperviousness; as stated before, this is probably associated with IAT statistics being more sensitive to variability in high flows than conventional flow statistics.

3.3 Distribution of changes in inter-amount times

Figure 6 shows histograms of first-order differences in IATs and flows at the 24 h analysis scale, for Irvins Creek (the least-urbanised basin), LSugarM (the most impervious basin), Stewart Creek (a basin with low flow regulation) and McAlpine (the largest of all studied basins). In the flow histograms, negative differences are associated with recession,

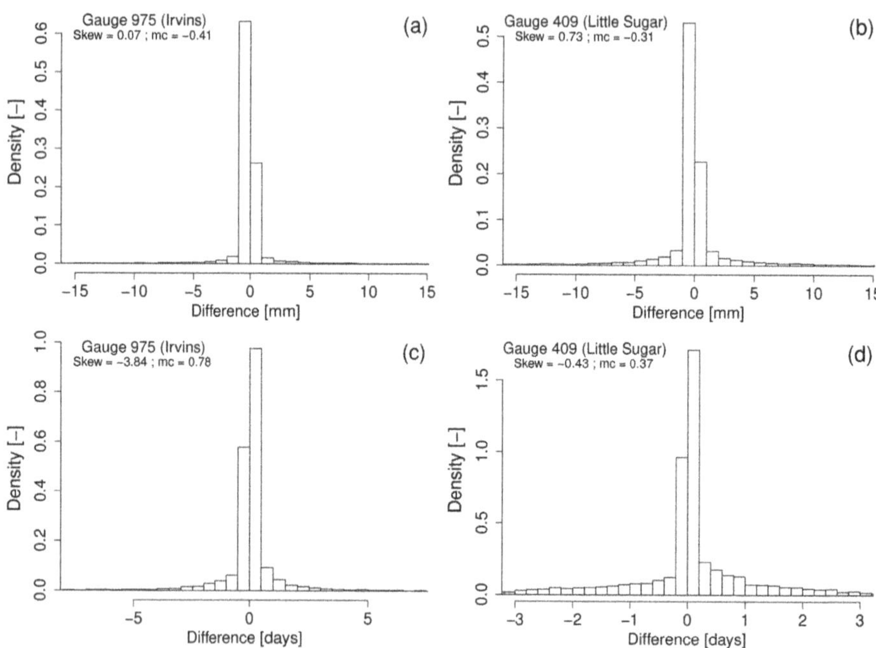

Figure 6. histograms of first-order differences in flows **(a, b)** and inter-amount times **(c, d)**, at 24 h analysis sampling scale, for Irvins Creek and LSugarM Creek.

positive differences with flow rise. Conversely, negative differences in IATs occur during flow rise, positive differences during flow recession. Most flow differences are concentrated in the 0 to -0.5 mm bin, associated with slow flow recession of 0.5 mm day^{-1}. Most IAT differences are concentrated in the 0 to 0.1 or 0.2 day bin, associated with steeper flow recession of approximately 5 to 10 mm day^{-1}. This reflects the relatively higher sampling of rapid flow response for IATs compared to conventional flow sampling. Skewness and medcouple values of the histograms provide indications of hydrograph shape, in particular of the steepness of the hydrograph recession limb: higher skewness, and thus more weight of the distribution concentrated in one of the tails, indicates slow flow recession compared to relatively rapid flow rise. Figure 7 shows scatter plots for skewness and medcouple values versus basin size and imperviousness, for all basins. The three basins with low flow regulation (970, 530, 507) can be recognised by their low medcouple values for IAT difference indicating near-symmetrical histograms, i.e. flow rise and recession occur at similar rates. Most IAT difference histograms are negatively skewed, with a longer left tail than right tail, i.e. IATs generally decrease quicker (flow rise) than they increase (flow recession). The strongest negative skewness for IAT differences was found for the least-urbanised basin (Irvins Creek, gauge 975), indicative of steep flow rise occurring in this basin. Significant positive correlation was found between skewness of IAT difference histograms and imperviousness (Spearman correlations 0.75), indicating lower probability of steep flow rise in higher urbanised basins. Negative correlation was found between medcouple and imperviousness (Spearman correlation -0.55); thus relatively more symmetrical hydrographs with flow rise and recession at similar rates occur for urbanised basins. Here, sub-basin correlation appears to play role: medcouple values are higher overall in the McAlpine sub-basins than in Little Sugar Creek and Irwin sub-basins (see Table 4). Significant correlations of IAT difference skewness and medcouple with imperviousness show that urbanisation is associated with more regulated flows, confirming findings in Sect. 3.1.

3.4 Inter-amount-time variability across scales, from sub-daily to seasonal sampling scale

In this section we analyse the variability of IATs and flows across a wide range of sampling scales. We investigate how the statistical distributions and hydrological response characteristics change when moving from inter-event (multiple days) to intra-event (sub-daily) scales. Figure 8 shows quantile plots for normalised flows and IATs at scales between 12 h and 64 days, for Taggart Creek. On the horizontal axis is the sampling scale, i.e. fixed sampling time for conventional flow statistics or, equivalently, mean inter-amounts for IAT statistics. Note that for the IAT analysis, mean inter-amounts are normalised by basin area size and reported in millimetres to allow easier interpretation of flow magnitudes and to allow easier comparison between basins. For instance, the normalised inter-amount Δq for Taggart Creek at the daily scale is 0.998 mm. The vertical axis shows quantiles of normalised flows and IATs corresponding to the sampling

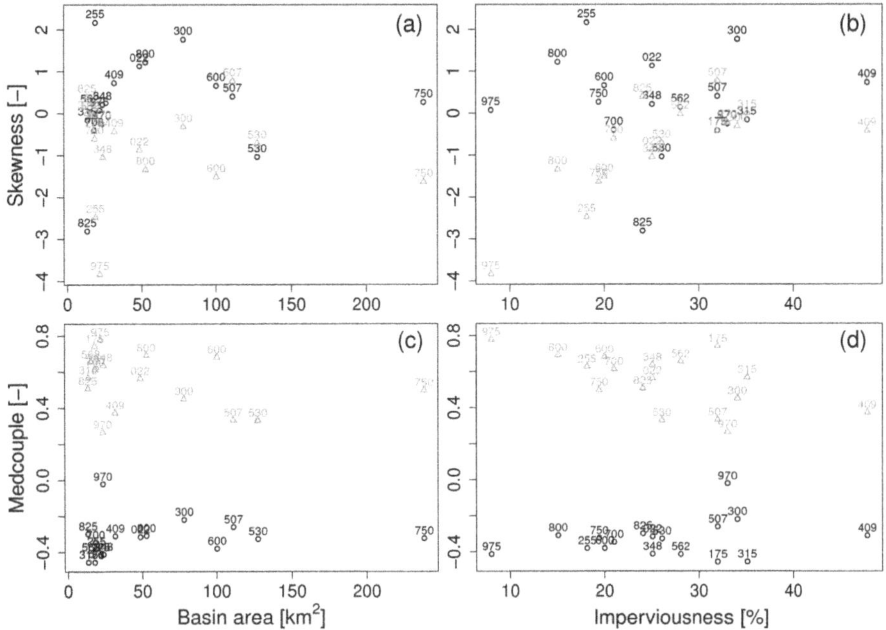

Figure 7. Scatter plots of skewness **(a, b)** and medcouple values **(c, d)** of histograms for differences in flows and inter-amount times, plotted versus basin size and imperviousness degree. Grey triangle symbols represent inter-amount times, black circles represent flows.

scale in time or Δq. Values on both x and y axes are plotted on log scales to allow easier visualisation of quantile values that vary by 2 to 4 orders of magnitude. The bold black line denotes the mean, and the dotted black line shows median values. The central part of the quantile plots represents the 25–75 percentile range, upper and lower whiskers 10–90 percentiles and crosses the 1 and 99 percentiles.

We can see that mean values of normalised flows and IATs decrease log-linearly with sampling scale, as indicated by a straight line in the log–log plot, i.e. the sampling mean follows power-law scaling. As histogram analysis at the 24 h scale already showed, statistical distributions of both flows and IATs are highly skewed. Moreover, skewness increases at smaller scales, as indicated by an increasing distance between mean and median values. Median values for flows follow close to log-linear scaling (albeit steeper compared to the mean) but exhibit stronger departures from log-linear scaling for IATs. In particular, the median of IATs shifts from close to log-linear scaling between 16 and 64 mm (associated with about 16 to 64 days) to non-log-linear scaling between 1 and 14 mm scales (1–14 days) and again to near-log-linear scaling below 1 mm. Coincidentally, these transitions correspond to the range of scales over which IATs generally transition from being inter-event to intra-event dominated. Indeed, IATs at coarser scales mostly combine the properties of multiple storms, resulting in a more symmetric distribution. This effect is much stronger in IAT than in flow distributions, because it is mainly associated with changes in sampling of peak flows which are more frequently sampled in the IAT framework than in the conventional fixed time approach.

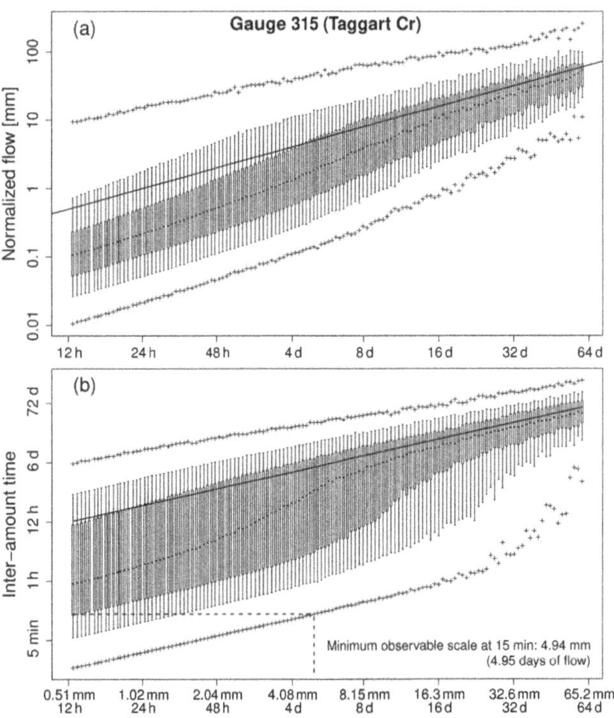

Figure 8. Quantile plots of flows **(a)** and inter-amount times **(b)** for Taggart Creek for a range of scales, from 12 h to 60 days. The bold black line denotes the mean values. The dotted black line shows median values. The central part of box plots represents the 25–75 percentile range, upper and lower whiskers the 10–90 percentile range, crosses the 1–99 percentile range.

Comparing the 10–90 and 1–99 percentile ranges in Fig. 8a and b we see that the 10–90 percentile range of IATs gradually increases towards smaller scales. For flows, the 10–90 percentile range remains approximately constant; however, distance between 90 and 99 percentile values rapidly increases towards smaller scales. This reflects the highly skewed nature of flow distributions caused by oversampling of low flows compared to high flows; an effect that increases progressively towards smaller scales. By contrast, 10–90 and 1–99 percentile ranges for IATs increase more or less similarly with scale, for sampling scales ranging from 0.51 mm to approximately 10–16 mm. This indicates that the tails of IAT distributions are more or less equally sampled, at least up to the 1 and 99 percentiles. The upper 75, 90 and 99 IAT percentiles of IATs, associated with low flow periods, change approximately log-linearly with scale, showing that upper tail percentiles of IAT values refer to the same low flow periods across all scales, up to 8–16 mm scale. Associated low flows are approximately 0.1 mm day^{-1}. The 1 percentiles for flows are associated with approximately 0.02 mm day^{-1}, for the 12 h to 4-day scale, showing that the distribution tail associated with low flows captures lower flow extremes in conventional sampling than in IAT sampling. This is a result of the relatively high frequency at which low flows are sampled. Conversely, peak flows, associated with the right tail of the flow distribution, are sampled less frequently in conventional flow sampling: the 99 percentiles are associated with peak flows of 0.78 to 0.38 mm h^{-1} for 12 h to 4-day scale. The 1 percentiles of IATs are associated with peak flows of about 20 mm h^{-1}, at the 0.5 to 4 mm inter-amount scale, associated with mean IATs of 12 h to 4 days. This shows that the IAT distribution captures more extreme peak flow values than conventional flow sampling, at the same sampling scale.

Quantile plots of inter-amounts over a range of scales were created for all 17 gauges included in our analysis (results are added as a Supplement to this paper). This allowed us to compare transition ranges between inter-event-dominated and intra-event-dominated IAT distributions for all basins. Results show that for 10 % IAT quantiles, the lower end of the transition range, where intra-event characteristics start to be mixed with inter-event phenomena, lies roughly between 10 and 25 mm mean inter-amounts, being accumulated in about 1 h in most of the basins. Lower values are found for basins with higher urbanisation degree and for basins where low flow control is applied, reflecting the smoothing influence of flow control measures on peak flows. Similarly, one can compare the amount of flow that is being generated in an hour, compared to the mean flow. This can be derived from the IAT quantile plots by looking at the scale at which a given IAT quantile, for instance 10 % or 1 %, equals 1 h. For Taggart Creek, the IAT 1 percentile equals 1 h at sampling scale of 18 mm of mean normalised flow or, equivalently, 18 days of mean IAT. This means there is a 1 % probability of exceeding 18 mm of flow accumulation in 1 h or less, or, in terms of time, it implies that there is a 1 % chance to accumulate the amount of flow measured on average over a period of 18 days in 1 h or less. Thus, higher values of 1 h, 1 percentiles indicate stronger flashiness of basin response. Comparing values across basins, we found that higher values of 1 %, 1 h accumulations were strongly correlated with basin area, while no significant correlation with imperviousness was observed.

Subsequently, we investigated scaling behaviour from the perspective of statistical moments, by looking at coefficients of variation for flows and IATs across scales. For the purpose of statistical analysis and downscaling applications, it is important to have a robust scaling model that predicts how distributions change when going from one scale to another. Scale invariance means that a distribution can be derived at any scale, especially small scales, by shifting and scaling the distribution at larger scales. One way to assess the property of scale invariance is to check if the statistical moments of distributions follow a power law of scale. Figure 9 shows coefficients of variation, computed as the ratio of the second- over the first-order moment, for four gauges, across a range of sub-daily (3 to 12 h) up to bi-monthly (60–68 days) scales. Results show that coefficients of variation for flows vary non-linearly with scale, while they approximately follow a power law with scale for IATs. For Irvins Creek, the most natural basin in this study (8.2 % imperviousness, Fig. 9a), CV values of IATs and flows are similar over a range of 10 to 50 days. At smaller scales, CV values for flows increase more rapidly than for IATs, indicating that IAT variance remains more stable at smaller scales, while variance rapidly increases at small scales for flows, as a result of growing skewness of the statistical distribution, caused by relative oversampling of low flows, or conversely, undersampling of high flows. CV values for Upper LSugar Creek, the most urbanised basin are lower than for Irvins Creek, especially at smaller scales (Fig. 9b). This is explained by the influence of flow control measures in this basin, as flows are constrained by the stormwater drainage system. The difference is more pronounced for IATs, because IAT variance is more sensitive to peak flows as a result of the adaptive sampling strategy. Figure 9c shows that for LMcAlpine, the largest basin (238.4 km^2), CV values for flow are more or less stable between 3 and 24 h scale, due to strong smoothing of peak flows at this intra-event scale. In contrast, CV values for IATs increase over this range, due to scale sensitivity of the upper tail of the IAT distribution, where long IATs at this small scale (0.1 to 1.1 mm for 3 to 24 h) are broken up more unevenly, creating increased CV and skewness. This shows that for analysis of low flows, especially in basins characterised by strongly smoothed flow variability, IAT analysis offers little advantage and conventional flow statistics are more suitable. CV values for Stewart Creek in Fig. 9d show very low CV values for IATs that vary little with scale, while CV values for flows are much higher and strongly sensitive to scale. Stewart Creek is a small, semi-urbanised basin (33 % imperviousness) where active low flow control is applied. This results in very low variability in IATs across the entire range

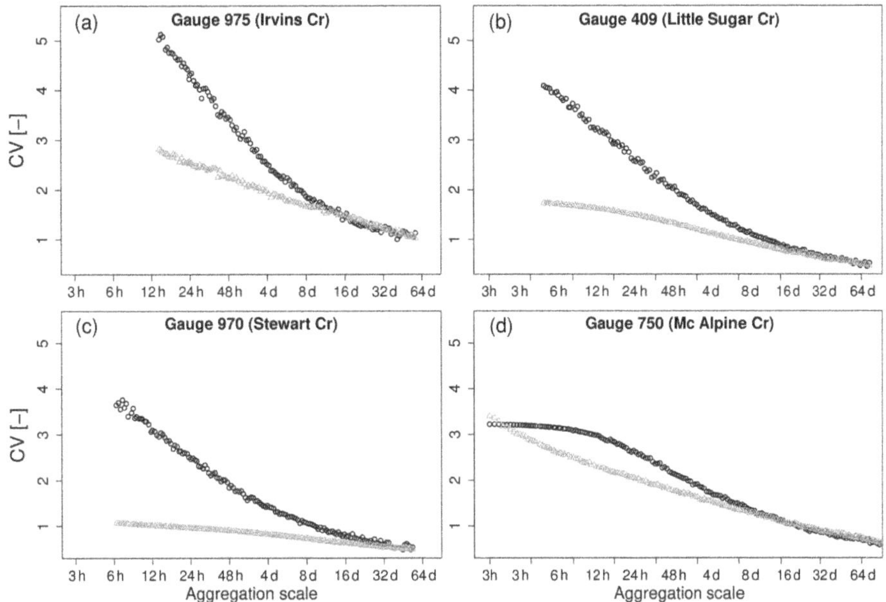

Figure 9. Coefficients of variation for flows and inter-amount-time scales across a range of sub-daily (3 to 12 h) up to bi-monthly (60–68 days) scale, for Irvins Creek, LSugarM, Stewart Creek and McAlpine. Grey triangle symbols represent inter-amount times, black circles represent flows.

of scales, while CV values for flows are lower than those for similar basins, but highly sensitive to scale, probably due to unbalanced sampling of peak flows compared to very stable low flows.

In Sect. 3.1 we analysed skewness and medcouple values of flow and IAT distributions at the 24 h scale and found that skewness values were lower and medcouple higher for IATs than for flows. This was explained by the sensitivity of flow distributions to rare peak flows compared to frequently sampled low flows. Initial analyses of skewness and medcouple values across scales showed that results are highly sensitive to the sampling scale. While CV values show a stable pattern across scales, results for skewness and medcouple are much more variable, across scales and across basins. Explanation of this scale sensitivity of skewness metrics and what information can be derived from this about the tails of the distributions requires deeper analysis that will be part of future work.

3.5 Flashiness indicators and minimum observable scale

Two flashiness indicators were computed, as explained in Sect. 2: the classical R–B flashiness index and an IAT flashiness indicator based on characteristics of the IAT distribution. Table 3 summarises flashiness values for all gauges, as well as minimum and maximum observable inter-amounts, as defined in Eqs. (9) and (10). IAT flashiness indicators vary between 12.5 and 165 h; higher values are generally associated with smaller basins. R–B flashiness values vary between 0.8 and 1.3, indicative of moderately variable flows (R–B flashiness can vary between 0 and 2). Values are in the same range as those reported by Baker et al. (2004) for smaller basins: they found R–B flashiness values larger than 1 for basins smaller than 50 km^2. R–B flashiness is strongly correlated with CV values (Fig. 10c, Spearman correlation 0.77); this confirms that R–B flashiness is essentially a metric of flow variability. Figure 10a shows that IAT-based flashiness and R–B flashiness are moderately correlated (Spearman rank correlation 0.55), yet there are some striking differences. The three low-flow-regulated basins have very low R–B flashiness values, while IAT flashiness values are in line with values for other basins. This is explained by R–B flashiness being strongly sensitive to low flow variability, while IAT flashiness is more sensitive to occurrence of peak flow values. For instance, the McAlpine basin (gauge 255) has a very high IAT flashiness as a result of high occurrence of peak flows. On the other hand LSugarM (gauge 409), the most urbanised basin, has low IAT flashiness as a result of peak values being capped by maximum capacity of pipes in the drainage network.

Figure 10b and d show scatter plots of IAT flashiness (left y axis) and R–B flashiness (right y axis) versus basin area and imperviousness, for all gauges. They show a clear relationship between flashiness and basin area (Spearman correlation −0.83 for IAT, −0.71 for R–B flashiness), with a large range of flashiness values for the smallest basins (< approx. 30 km^2). Here,other processes than basin size clearly play a role in explaining flashiness. Correlations between R–B and IAT flashiness versus imperviousness degree

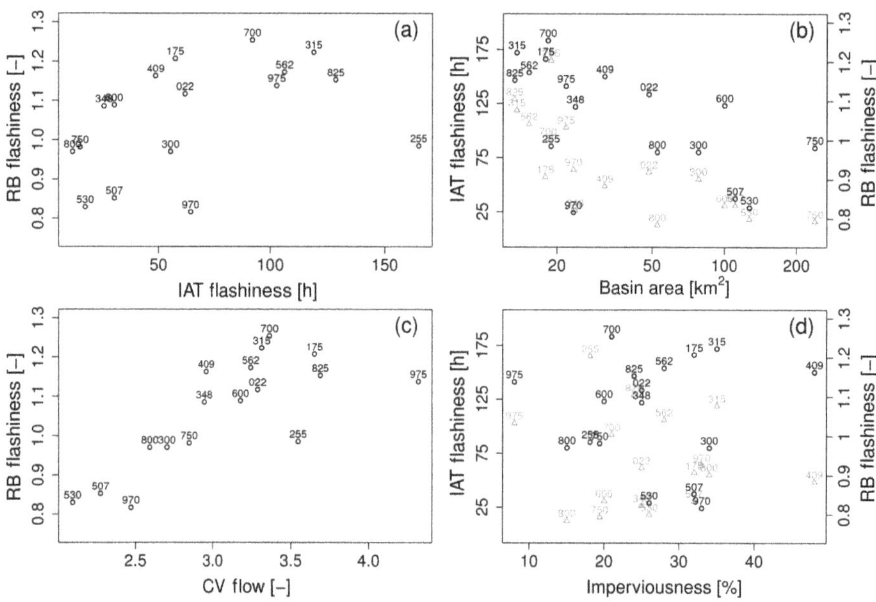

Figure 10. Scatter plots of flashiness versus basin area and imperviousness, for all gauges. Grey triangle symbols represent inter-amount times, black circles represent flows.

are not significant at the 5 % level. For R–B flashiness, the most pervious and the most impervious basins (gauges 975 and 409 respectively) are both in the high range of flashiness values, showing that other influences, such as basin size and presence or absence of low flow regulation play a more important role than imperviousness degree. IAT flashiness tends to decrease for a combination of higher imperviousness and larger basins, basin size playing a stronger role than urbanisation. The most urbanised basin, LSugarM (gauge 409, 31.7 km^2, 48 % imperviousness) has a relatively low flashiness value of 48.8 h, while the least impervious basin, Irvins Creek (gauge 975, 21.8 km^2, 8 % imperviousness) has a high flashiness value of 102.8 h. As discussed in Sect. 3.1, the effect of urbanisation on flow patterns for the basins in the study area seems to be mainly determined by increased flow regulation associated with introduction of dams, stormwater detention basins and stormwater drains with capacity limitations. While higher imperviousness leads to higher mean runoff flows (for instance, 1.5 mm for LSugarM versus 0.68 mm for Irvins Creek, at 24 h scale), rainfall tends to run off relatively more uniformly in impervious basins, without rapid flow rise or sharp flow peaks, depending on the degree of flow regulation. The leads to a mixed effect of basin size, imperviousness and flow regulation on IAT flashiness and peak flows. In this study, IAT flashiness values were defined as the time that is needed on average to accumulate the amount of flow that is accumulated in 15 min or less, 1 % of the time.

R–B flashiness indices were computed at the daily scale, to allow comparison with results obtained by Baker et al. (2004). For a fair comparison, both flashiness indices should be computed at similar scales, as far as possible, given that definitions used in the two approaches are different. We aimed to compute both indices at hourly scale, as this is an appropriate scale in relation to the size of most of the basins in our analysis and a reasonable compromise between the 15 min and 24 h timescales used for IAT flashiness and R–B flashiness index respectively. Note that Baker et al. (2004) stated that the hourly scale would be more suitable for smaller basins (< 30 km^2), but never computed R–B flashiness values at this scale, only Richard's pathlengths. When we computed R–B flashiness indices at the hourly scale, using the same definition, we found lower flashiness than at the daily scale, which is rather counterintuitive, as one would expect higher flashiness at smaller scales due to the fact that Richard's pathlengths increase from daily to hourly scales. However, R–B flashiness is based on absolute differences of flow values, not gradients (i.e. differences per unit of time). And since flow differences decrease when moving toward smaller scales, R–B index also decreases. Alternatively, one could use discharges instead of flow amounts, but then values could grow much larger than 2. Regardless of the approach used, R–B flashiness index appears to be rather sensitive to the scale of analysis. By contrast, the IAT flashiness index proposed in this paper tends to be much more robust. Additional sensitivity analyses (not shown) revealed almost no changes in IAT flashiness estimates for 15 min to 3–6 h aggregation scales. Beyond that, significant underestimation started to occur as the resolution is not sufficient anymore to correctly capture peak flow variability. For data aggregated at 24 h resolution (instead of the original 15 min), IAT flashiness values were underestimated by 20–80 %, depending on the considered gauge.

Table 5. Minimum and maximum observable scales (in hours), flashiness index for 15 min observation time (in hours) and fitted multifractal parameters α and C_1 for inter-amount time flows.

ID	Min. scale	Max. scale	Flash	RB	Min. R2 IAT	Min. R^2 flow	Alpha IAT	Alpha flow	C_1 IAT	C_1 flow
UBriar	13.75	1462	128.75	1.15	0.999	0.994	1.05	1.53	0.21	0.35
Taggart	12.50	1443	118.75	1.22	0.999	0.993	0.88	1.30	0.26	0.36
Campbell	9.25	1360	106.00	1.17	1.000	0.993	1.01	1.45	0.24	0.33
Steele	9.50	1457	57.25	1.21	1.000	0.991	0.86	1.30	0.25	0.36
McMullen	11.00	2420	92.25	1.25	0.999	0.992	0.94	1.32	0.26	0.32
UMcAlpine	10.00	1367	165.00	0.99	1.000	0.990	1.24	1.59	0.19	0.33
Irvins	13.75	1367	102.75	1.14	0.999	0.991	1.25	1.40	0.22	0.35
Stewart	6.25	1284	64.00	0.82	1.000	0.994	0.72	2.06	0.09	0.24
Coffey	4.75	1422	26.25	1.09	0.999	0.997	1.53	1.37	0.21	0.28
LSugarM	7.50	1752	48.75	1.16	1.000	0.996	0.66	1.48	0.20	0.33
LBriar	6.75	1658	61.50	1.12	1.000	0.996	0.88	1.51	0.20	0.31
SixMile	3.00	672	12.50	0.97	0.999	0.995	1.64	1.36	0.21	0.26
UIrwin	5.00	2420	55.25	0.97	1.000	0.995	1.14	1.81	0.14	0.25
MMcAlpine	5.50	2420	30.75	1.09	1.000	0.996	1.28	1.46	0.20	0.28
LSugarA	3.50	2420	30.75	0.85	0.995	0.996	2.89	1.89	0.07	0.22
LSugarP	2.75	1532	18.00	0.83	1.000	0.996	1.37	1.87	0.09	0.20
LMcAlpine	3.00	2420	15.75	0.98	0.999	0.997	1.64	1.32	0.19	0.24

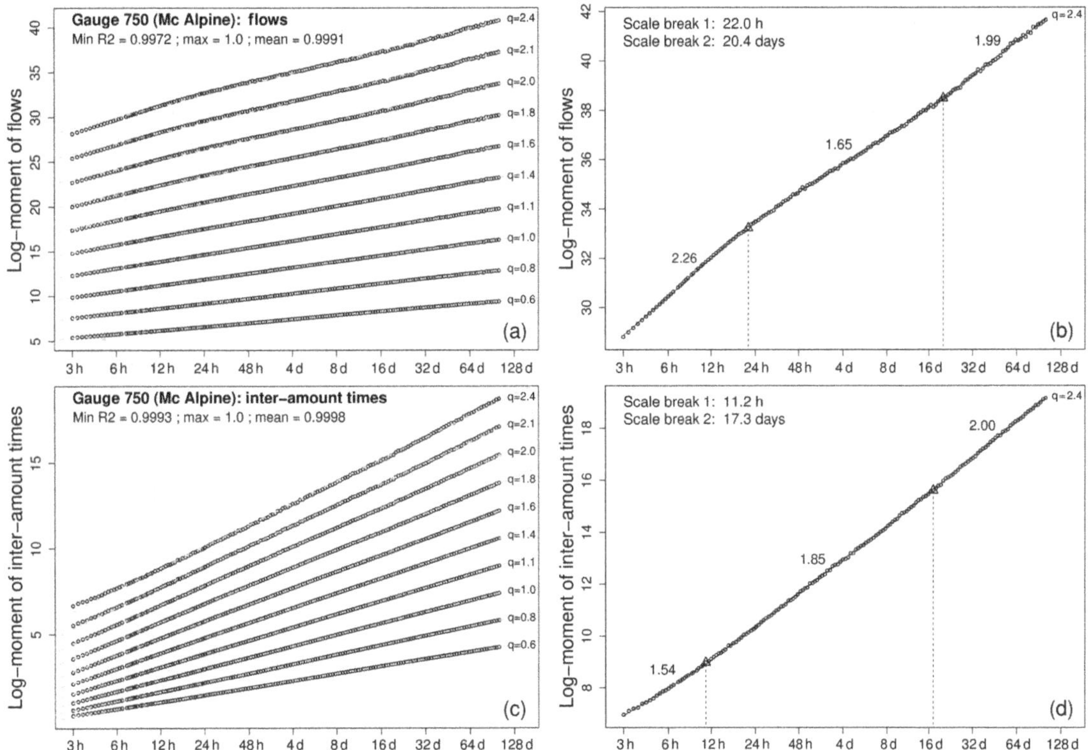

Figure 11. Example of log–log plots for flows and inter-amount times **(a, b)**, for Mc Alpine Creek, illustrating departures from linearity at high-order moments. Reported R^2 values are for the entire range of results, without scale breaks. Log–log curve for moment $q = 2.4$ illustrating scale breaks for flows and inter-amount times **(c, d)**.

Quantile plots of IAT distributions furthermore provide information about the minimum observable scale at a given observational resolution (15 min, in the data series used in our analysis), i.e. the degree of flow variability that occurs at scales smaller than the observation scale. When moving towards smaller sampling scales, a growing percentage of flow accumulations occurs in less than 15 min, and hence cannot be analysed at the given observational resolution. This typically coincides with peak flows and implies that during peak events, the observational resolution is too low to measure flow variability. IAT analysis can thus be used to identify a critical resolution for flow observations, if a given peak flow accumulation is of interest. This could be associated with, for instance, the capacity of detention ponds or flooding caused by exceedance of stormwater drainage capacity. For the example of Taggart Creek (Fig. 8b), the scale at which 1 % of flow accumulations occurs in less than 15 min is associated with an inter-amount sampling scale of 4.76 mm. This implies that flows that exceed 4.76 mm in 15 min, i.e. peak flows above 19.0 mm h^{-1}, cannot be observed 1 % of the time. If correct observation of peak flows of this magnitude or larger is important, flow data need to be collected at a higher than 15 min resolution during times of peak flows. This is typically the case in urban basins, where stormwater drainage systems are often designed for peak flows associated with 10- to 50-year return periods.

3.6 Scaling of inter-amount times across scales: multifractal analysis

As explained in Sect. 2, log–log plots of statistical moments versus sampling scale can be used to study scaling behaviour of time series. In the following, we plotted the moments $\langle X_\lambda^q \rangle$ of order q of IATs as a function of mean inter-amount scale Δq (proportional to the inverse of the scaling ratio λ), on a log–log scale, for moments of order 0.6 to 2.4. We applied the same procedure for flow time series over the same range of equivalent scales. Figure 11 shows examples of log–log plots for flow volumes and IATs for McAlpine Creek (gauge 750). They show that log-linear fits are better for IATs than for flows, especially for higher-order moments; minimum R^2 values, that are associated with fits for higher-order moments, are 0.9972 and 0.9993 for flows and IAT respectively.

Plots in Fig. 11 show stronger departures from linearity in the log–log plots for flows than for IATs, especially for higher-order moments. Figure 11c and d illustrate this for log–log curves of moment $q = 2.4$, where a scale break was detected at 22 h for flows. Based on a Davies test (Davies, 2002), two breakpoints were significant for flows (p value 0.001). For IATs, there was at least one significant breakpoint, but the test for two breakpoints returned a p value of 0.071. This shows that scaling is slightly better for IATs than for flows. Similar analyses were conducted for all gauges, Table 5 summarises minimum R^2 values for log-moment fits for flows and IATs. Log moments for IATs show near perfect fits for all gauges, with minimum R^2 values between 0.995 and 1.000. Quality of log moments is consistently lower for flows, for all basins; minimum R^2 values are between 0.990 and 0.997, lower quality fits generally occurring for smaller basins. Investigation of departures from linearity showed that for flows, most gauges exhibited a scale break between 8 and 20 days. Similar scale breaks, between timescales of 8 to 16 days, were found in scaling analyses of flow data by other authors based on flow data at daily resolution (Tessier et al., 1996; Labat et al., 2002; Sauquet et al., 2008). Labat et al. (2013) and Sauquet et al. (2008) found scale breaks in the range of 16 to 27 h, respectively, for 30 min hourly resolution. We did not detect any strong departures from linearity in the IAT framework except for the three gauges where low flow regulation is applied (LSugarA, 507, LSugarP, 530, Stewart Creek, 750).

Using the empirical log moments, we fitted the multifractal parameters C_1 and α for IATs and flow amounts. Table 5 summarises C_1 and α values for all basins, for flows and for IATs. Results show that C_1-values, characterising intermittency of the time series, are lower for IATs than for flows. This makes sense and can be explained by the adaptive sampling strategy of IATs, especially the fact that low flows are sampled less often than in the classical fixed-time framework. Values of the multifractality index α are generally lower for IATs, with the exception of four basins. Two of these basins are characterised by low flow regulation; one basin has anomalous land-use distribution with a high concentration of imperviousness in the upper part of the basin. Time series of the fourth basin is short (8 years), which might influence outcomes of the scaling analysis. C_1 and α values for flows are in the range of values found by other authors. Figure 12 shows scatter plots of values for C_1 and α for flow and for IATs versus basin size and imperviousness. C_1 values are clearly negatively correlated with basin area. Rank correlations for IATs are -0.67 and -0.85 for flows. No significant correlation of C_1 with imperviousness was found, but the three basins with low flow control stand out with lower-than-average C_1 values. This shows up both in the IAT analyses and in the classical approach based on flows. The α values for IATs are positively correlated with area (0.6) and negatively with imperviousness (-0.56). No significant correlation with area or imperviousness was detected. For IATs, negative correlation of α with imperviousness comes from the fact that IATs in highly impervious basins are redistributed more evenly when moving from large to small scales (due to high imperviousness).

4 Summary and conclusions

In this study, we introduced an alternative approach for analysis of hydrological flow time series, using an adaptive sampling framework based on inter-amount times (IATs). The main difference between flow time series and time series for IATs is the rate at which low and high flows are sampled; the unit of analysis for inter-amount times is a fixed flow amount, instead of a fixed time window. Thus, in IAT analysis, sampling rate is adapted according to the local variability in flow time series, as opposed to time series sampling using fixed time steps. We aimed to investigate the effect of adaptive IAT sampling on flow statistics, especially on the tails of the statistical distributions associated with peak flow and low flow extremes. We analysed and compared statistical distributions

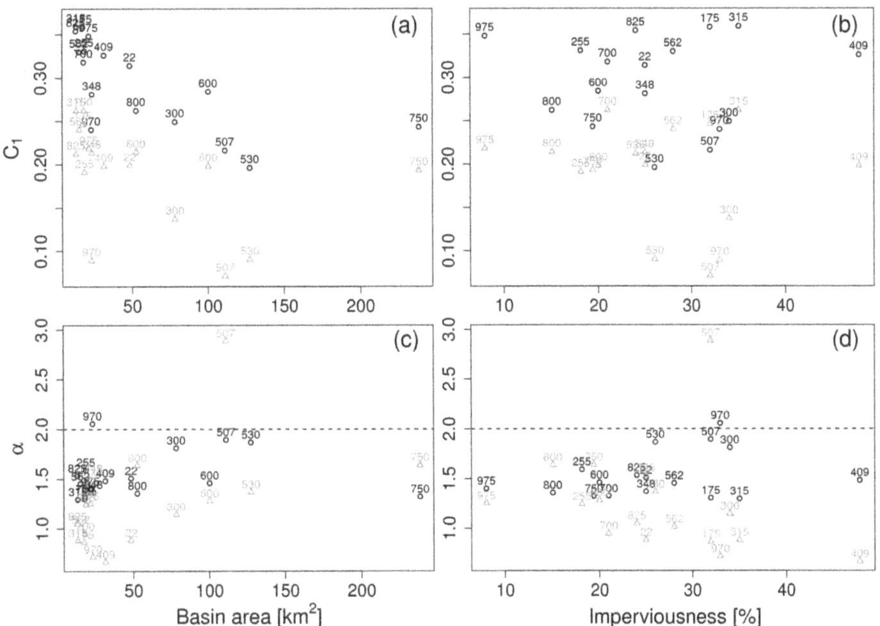

Figure 12. Multifractal parameters C_1 and alpha for scaling analysis of flows and inter-amount times, as a function of drainage area and imperviousness degree. Grey triangle symbols represent inter-amount times, black circles represent flows.

of flows and IATs across a wide range of sampling scales to investigate sensitivity of statistical properties such as distribution quantiles, variance, scaling parameters and flashiness indicators to the sampling scale. We did this based on streamflow time series for 17 (semi-)urbanised basins in North Carolina, USA. The following conclusions were drawn from the analyses:

1. Adaptive sampling of flow time series based on inter-amounts leads to higher sampling frequency during high flow periods compared to conventional sampling based on fixed time windows. This results in a more balanced representation of low flow and peak flow values in the statistical distribution. While conventional sampling gives a lot of weight to low flows, as these are most ubiquitous in flow time series, IAT sampling gives relatively more weight to high flow periods, when given flow amounts are accumulated in shorter time. As a consequence, IAT sampling gives more information about the tail of the distribution associated with high flows, while conventional sampling gives relatively more information about low flow values.

2. Statistical analysis of IATs and flows at the 24 h scale showed that coefficient of variation (CV) and skewness values were much higher for flows than for IATs, while medcouple values were lower for flows, indicating strong asymmetry of the flow distributions and low representation of high flow extremes in the statistical distribution. Larger basins were generally characterised by stronger smoothing of flows, resulting in higher mean flow, lower CV values and lower skewness of the histograms. Flow variability decreased with basin size. Larger imperviousness was associated with higher mean flows and lower variability of IATs, while there was not a clear relation with variability of flows.

3. Comparison of CV across the 17 basins showed that CV values of flows were significantly negatively correlated with basin size. CV values of IAT distributions were not significantly correlated with basin size. This was explained by basin size having a stronger smoothing effect on low flow variability, strongly represented in conventional flow time series, than on peak flows that are more frequently represented in IAT time series. By contrast, CV values of IAT distributions were negatively correlated with imperviousness, while correlation between CV values for flows and imperviousness was not significant. Negative correlation between CV values of IATs and imperviousness probably indicates a stronger influence of flow regulation by detention and capacity constraints of stormwater drains in more urbanised basins, resulting in more uniform runoff during rainy periods. IATs during these periods concentrate relatively more closely to the mean and show fewer extremes. This result is contrary to findings reported in the literature, where urbanisation tends to be associated with higher peak flows. (e.g. Rose and Peters, 2001; Cheng and Wang, 2002; Du et al., 2012; Huang et al., 2008). On the other hand, several studies have found mixed effects of urbanisation on flow peaks asso-

ciated with a combination of imperviousness and flood mitigation measures, especially for basins in the USA where urbanisation has predominantly taken place after implementation of stormwater legislation to lower peak discharges (e.g. Smith et al., 2013; Hopkins et al., 2015; Miller et al., 2014). For the basins in Charlotte watershed, urbanisation has taken place before as well as after stormwater legislation, and a combination of flow regulation by detention facilities and peak flow restrictions induced by capacity constraints results in an overall effect of peak flow reduction associated with urbanisation.

4. Histograms of first-order differences showed negative skewness for IATs and positive skewness for flows, for most of the basins, indicating the prevalence of slow flow recession compared to flow rise. The three basins with low flow regulation could be recognised by their relatively low medcouple values (< 0.4) for IAT differences, showing that hydrographs tend towards being symmetrical in these basins. Significant correlations were found between skewness and medcouple of IAT differences and imperviousness (Spearman correlations 0.75 and -0.55), showing that urbanisation is associated with more regulated flows, thus relatively more symmetrical hydrographs with flow rise and recession at similar rates and lower frequencies of steep flow rise. Here, sub-basin correlation appears to play a role: medcouple values were higher overall in the McAlpine sub-basins than in Little Sugar Creek and Irwin sub-basins. No significant correlations were found for differences in flows.

5. Quantile plots of flows and IATs plotted over a range of sub-daily to seasonal scales showed the influence of the different sampling strategy for IATs compared to conventional flow sampling on median, 25–75, 10–90 and 1–99 percentile ranges of the distributions. The 25–75 and 10–90 percentile ranges for flows remained approximately constant, but the distance between 90 and 99 percentile values rapidly increased towards smaller scales. This reflects the highly skewed nature of flow distributions caused by oversampling of low flows compared to high flows; an effect that increased progressively towards smaller scales. By contrast, 10–90 and 1–99 percentile ranges for IATs increased more or less similarly with scale, for sampling scales ranging from 0.51 mm to approximately 10–16 mm, largely associated with intra-event flow variability. This indicates that the tails of IAT distributions are more or less equally sampled, at least up to the 1 and 99 percentiles.

6. Quantile plots for IATs showed different scaling at small scales (up to inter-amount scale 8–10 mm) and large scales (roughly exceeding 20 mm inter-amounts), with a transition range in between. At smaller scales, IATs are mostly dominated by intra-event variability, while at large-scales IATs span multiple events. Flows sampled over fixed time intervals did not clearly exhibit this transition, probably because peak flow variability is being poorly sampled by fixed time window sampling. Because IATs adapt the sampling rate depending on the level of activity, they still capture a fair amount of peak flow statistics and intra-event properties, even at coarser scales.

7. Comparison of the tails of flows and IAT distributions showed that the distribution tail associated with low flows captures lower flow extremes in conventional sampling than in IAT sampling ($0.02\,\mathrm{mm\,day^{-1}}$ compared to $0.1\,\mathrm{mm\,day^{-1}}$). Conversely, IAT distributions capture more extreme peak flow values than conventional flow sampling, at the same sampling scale: the 99 percentiles for flows are associated with peak flows of 0.38 to $0.78\,\mathrm{mm\,h^{-1}}$ (sampling scales 12 h to 4 days), while 1 percentiles of IATs are associated with peak flows of about $20\,\mathrm{mm\,h^{-1}}$ (sampling scales 0.5 to 4 mm inter-amounts, associated with IATs of 12 h to 4 days).

8. Analysis of CV values of flow and IAT distribution across scales showed that at smaller scales, CV values for flows increase more rapidly than for IATs, indicating that IAT variance remains more stable at smaller scales, while variance rapidly increases at small scales for flows. This is as a result of growing skewness of the statistical distribution of flows, caused by relative oversampling of low flows, or conversely, undersampling of high flows. This shows that for analysis of peak flows, IAT analysis offers advantages of the fixed-time sampling framework, as it samples peak flows more frequently and results in more stable variance across scales. For analysis of low flows, especially in basins characterised by strongly smoothed flow variability, IAT analysis offers little advantage and conventional flow statistics are more suitable.

9. An IAT flashiness indicator was defined as the inter-amount scale at which 1 % of flow accumulations occur in less than 15 min. Comparison between IAT-based flashiness and the commonly applied R–B flashiness index showed that indices were moderately correlated (Spearman rank correlation 0.55), yet there were some striking differences. R–B flashiness was shown to be strongly sensitive to low flow variability, while IAT flashiness was more sensitive to occurrence of peak values. Both flashiness indices showed strong correlation with basin area. R–B flashiness showed no clear relationship with imperviousness. IAT flashiness tends to decrease for a combination of higher imperviousness and larger basin size, basin size playing a stronger role than urbanisation. The effect of urbanisation on flow patterns for the basins in the study area is a mixture of faster run-off flows due to imperviousness and stronger

flow regulation by dams and detention basins. This leads to a mixed effect of basin size, imperviousness and flow regulation on IAT flashiness and peak flows.

10. A minimum observable inter-amount scale was defined as the smallest scale at which flow variations can be studied given a fixed temporal observational resolution. At higher sampling scales, a growing percentage of flow accumulations occurs in less than the given observational resolution, 15 min in this study. This typically coincides with peak flows and implies that during peak events, the observational resolution is too low to measure flow variability. IAT analysis can thus be used to identify a critical resolution for flow observations, if a given peak flow accumulation is of interest. If correct observation of peak flows of a given magnitude is important, flow data need to be collected at a higher than 15 min resolution during times of peak flows. This is typically the case in urban basins, where stormwater drainage systems are often designed for peak flows associated with 10 to 50-year return periods.

11. Multifractal analysis of IATs and flows was applied over a range of sub-daily to seasonal scales. Both approaches exhibited relatively good scaling, as indicated by R^2 values above 0.99. IATs systematically scaled better than flows and showed departures from multifractality for only three basins, subject to low flow regulation, while flows exhibited departures from multifractality for most basins. This showed that IATs can help to better predict peak flow characteristics at small unobservable scales based on coarse-resolution data. Additionally, they provide new interesting alternatives for the stochastic modelling and downscaling of flow data.

This study showed that properties of statistical distributions of flow time series are very sensitive to the scale at which the statistics have been derived. This influences values of summary statistics that are used to characterise flow patterns of hydrological basins, like peak flows at given recurrence intervals and flashiness indices. Adaptive sampling based on inter-amount times helped to achieve more stable variance across scales, yet the behaviour of other statistical properties such as skewness or medcouple is less clear. Further investigations are needed to interpret changes of statistics across scales. Future work will focus on multiscale analysis, how to compare results at different scales and what can be learnt from behaviour at different scales about flow variability in hydrological basins in relation to basin characteristics.

Analyses in this study identified minimum observable scales below which flow variability cannot be captured at a given measurement resolution. The combination of being able to identify these minimum observable scales and to downscale flow data based on IATs is an interesting area for future investigation. Results showed that scaling parameters for IAT time series were more reliable than those based on fixed-time sampling because of smaller departures from linearity in log–log plots. Future work will focus on possible ways to use IATs to downscale coarse-resolution flow data with the help of multifractals and multiplicative random cascades, to see if this leads to more robust and reliable results than downscaling based on conventional flow time series.

Another aspect that remains to be investigated is how IATs computed on flow data compare to IATs computed on associated rainfall time series. Because flow is linked to rainfall, the comparison of the two could help better distinguish which aspects of flow variability are due to rainfall and which relate to basin characteristics and stormwater management.

Competing interests. The authors declare that they have no conflict of interest.

Acknowledgements. The authors would like to acknowledge USGS for making available the datasets of flow gauges in Charlotte. The first author would like to thank NWO Aspasia and Delft University of Technology for the grant that supported this research collaboration. The second author acknowledges the funding provided by the Swiss National Science Foundation, grant P300P2_158499 (project STORMS).

Edited by: K. Arnbjerg-Nielsen

References

Baker, D., Richards, R., Loftus, T., and Kramer, J.: A new flashiness index: Characteristics and applications to Midwestern rivers and streams, J. Am. Water Resour. As., 40, 503–522, 2004.

Berne, A., Delrieu, G., Creutin, J.-D., and Obled, C.: Temporal and spatial resolution of rainfall measurements required for urban hydrology, J. Hydrol., 299, 166–179, doi:10.1016/j.jhydrol.2004.08.002, 2004.

Bloeschl, G. and Sivapalan, M.: Process controls on regional flood frequency: Coefficient of variation and basin scale, Water Resour. Res., 33, 2967–2980, 1997.

Brys, G., Hubert, M., and Struyf, A.: A robust measure of skewness, J. Comput. Graph. Stat., 13, 996–1017, doi:10.1198/106186004X12632, 2004.

Cheng, S.-J. and Wang, R.-Y.: An approach for evaluating the hydrological effects of urbanization and its application, Hydrol. Process., 16, 1403–1418, doi:10.1002/hyp.350, 2002.

Davies, R.: Hypothesis testing when a nuisance parameter is present only under the alternative: Linear model case, Biometrika, 89, 484–489, doi:10.1093/biomet/89.2.484, 2002.

Dippe, M. A. and Wold, E. H.: ANTIALIASING THROUGH STOCHASTIC SAMPLING, Comp. Graph. (ACM), 19, 69–78, 1985.

Du, J., Qian, L., Rui, H., Zuo, T., Zheng, D., Xu, Y., and Xu, C.-Y.: Assessing the effects of urbanization on annual runoff and flood events using an integrated hydrological modeling system for Qinhuai River basin, China, J. Hydrol., 464–465, 127–139, doi:10.1016/j.jhydrol.2012.06.057, 2012.

Emmanuel, I., Andrieu, H., Leblois, E., and Flahaut, B.: Temporal and spatial variability of rainfall at the urban hydrological scale, J. Hydrol., 430–431, 162–172, doi:10.1016/j.jhydrol.2012.02.013, 2012.

Favre, A.-C., Adlouni, S., Perreault, L., Thémonge, N., and Bobée, B.: Multivariate hydrological frequency analysis using copulas, Water Resour. Res., 40, 1–12, W01101, 2004.

Feizi, S., Angelopoulos, G., Goyal, V., and Medard, M.: Energy-efficient time-stampless adaptive nonuniform sampling, SENSORS IEEE, 912–915, doi:10.1109/ICSENS.2011.6127202, 2011.

Fletcher, T., Andrieu, H., and Hamel, P.: Understanding, management and modelling of urban hydrology and its consequences for receiving waters: A state of the art, Adv. Water Resour., 51, 261–279, doi:10.1016/j.advwatres.2012.09.001, 2013.

Goodrich, D., Lane, L., Shillito, R., Miller, S., Syed, K., and Woolhiser, D.: Linearity of basin response as a function of scale in a semiarid watershed, Water Resour. Res., 33, 2951–2965, 1997.

Grimaldi, S. and Serinaldi, F.: Asymmetric copula in multivariate flood frequency analysis, Adv. Water Resour., 29, 1155–1167, doi:10.1016/j.advwatres.2005.09.005, 2006.

Gustafson, D., Carr, K., Green, T., Gustin, C., Jones, R., and Richards, R.: Fractal-based scaling and scale-invariant dispersion of peak concentrations of crop protection chemicals in rivers, Envir. Sci. Tech. Lib., 38, 2995–3003, doi:10.1021/es030522p, 2004.

Hopkins, K., Morse, N., Bain, D., Bettez, N., Grimm, N., Morse, J., Palta, M., Shuster, W., Bratt, A., and Suchy, A.: Assessment of regional variation in streamflow responses to urbanization and the persistence of physiography, Envir. Sci. Tech. Lib., 49, 2724–2732, doi:10.1021/es505389y, 2015.

Huang, H.-J., Cheng, S.-J., Wen, J.-C., and Lee, J.-H.: Effect of growing watershed imperviousness on hydrograph parameters and peak discharge, Hydrol. Process., 22, 2075–2085, doi:10.1002/hyp.6807, 2008.

Kjeldsen, T.: Modelling the impact of urbanization on flood frequency relationships in the UK, Hydrol. Res., 41, 391–405, doi:10.2166/nh.2010.056, 2010.

Labat, D., Mangin, A., and Ababou, R.: Rainfall-runoff relations for karstic springs: Multifractal analyses, J. Hydrol., 256, 176–195, doi:10.1016/S0022-1694(01)00535-2, 2002.

Labat, D., Hoang, C., Masbou, J., Mangin, A., Tchiguirinskaia, I., Lovejoy, S., and Schertzer, D.: Multifractal behaviour of long-term karstic discharge fluctuations, Hydrol. Process., 27, 3708–3717, doi:10.1002/hyp.9495, 2013.

Lang, M., Ouarda, T., and Bobée, B.: Towards operational guidelines for over-threshold modeling, J. Hydrol., 225, 103–117, doi:10.1016/S0022-1694(99)00167-5, 1999.

Lombardo, F., Volpi, E., Koutsoyiannis, D., and Papalexiou, S. M.: Just two moments! A cautionary note against use of high-order moments in multifractal models in hydrology, Hydrol. Earth Syst. Sci., 18, 243–255, doi:10.5194/hess-18-243-2014, 2014.

Miller, J., Kim, H., Kjeldsen, T., Packman, J., Grebby, S., and Dearden, R.: Assessing the impact of urbanization on storm runoff in a peri-urban catchment using historical change in impervious cover, J. Hydrol., 515, 59–70, doi:10.1016/j.jhydrol.2014.04.011, 2014.

Pandey, G., Lovejoy, S., and Schertzer, D.: Multifractal analysis of daily river flows including extremes for basins of five to two million square kilometres, one day to 75 years, J. Hydrol., 208, 62–81, doi:10.1016/S0022-1694(98)00148-6, 1998.

Poff, N.: Ecological response to and management of increased flooding caused by climate change, Philos. T. R. Soc. A, 360, 1497–1510, doi:10.1098/rsta.2002.1012, 2002.

Praskievicz, S. and Chang, H.: A review of hydrological modelling of basin-scale climate change and urban development impacts, Prog. Phys. Geog., 33, 650–671, doi:10.1177/0309133309348098, 2009.

Richter, B.: A method for assessing hydrologic alteration within ecosystems [Un metro para evaluar alteraciones hidrologicas dentro de ecosistemas], Conserv. Biol., 10, 1163–1174, doi:10.1046/j.1523-1739.1996.10041163.x, 1996.

Rose, S. and Peters, N.: Effects of urbanization on streamflow in the Atlanta area (Georgia, USA): A comparative hydrological approach, Hydrol. Process., 15, 1441–1457, doi:10.1002/hyp.218, 2001.

Salvadori, G. and De Michele, C.: Frequency analysis via copulas: Theoretical aspects and applications to hydrological events, Water Resour. Res., 40, 1–17, doi:10.1029/2004WR003133, 2004.

Sauquet, E., Ramos, M.-H., Chapel, L., and Bernardara, P.: Streamflow scaling properties: Investigating characteristic scales from different statistical approaches, Hydrol. Process., 22, 3462–3475, doi:10.1002/hyp.6952, 2008.

Schertzer, D. and Lovejoy, S.: Physical modeling and analysis of rain and clouds by anisotropic scaling mutiplicative processes, J. Geophys. Res., 92, 9693–9714, 1987.

Schertzer, D. and Lovejoy, S.: Multifractals, generalized scale invariance and complexity in geophysics, Int. J. Bifurcat. Chaos, 21, 3417–3456, doi:10.1142/S0218127411030647, 2011.

Schleiss, M. and Smith, J.: Two simple metrics for quantifying rainfall intermittency: The burstiness and memory of interamount times, J. Hydrometeorol., 17, 421–436, doi:10.1175/JHM-D-15-0078.1, 2016.

Scott, D.: On optimal and data-based histograms, Biometrika, 66, 605–610, 1979.

Smakhtin, V.: Low flow hydrology: A review, J. Hydrol., 240, 147–186, doi:10.1016/S0022-1694(00)00340-1, 2001.

Smith, B. and Smith, J.: The flashiest watersheds in the contiguous United States, J. Hydrometeorol., 16, 2365–2381, doi:10.1175/JHM-D-14-0217.1, 2015.

Smith, B., Smith, J., Baeck, M., Villarini, G., and Wright, D.: Spectrum of storm event hydrologic response in urban watersheds, Water Resour. Res., 49, 2649–2663, doi:10.1002/wrcr.20223, 2013.

Smith, J.: Representation of basin scale in flood peak distributions, Water Resour. Res., 28, 2993–2999, doi:10.1029/92WR01718, 1992.

Smith, J., Baeck, M., Meierdiercks, K., Nelson, P., Miller, A., and Holland, E.: Field studies of the storm event hydrologic response in an urbanizing watershed, Water Resour. Res., 41, W10413, doi:10.1029/2004WR003712, 2005.

Stedinger, J.: Estimating a regional flood frequency distribution, Water Resour. Res., 19, 503–510, doi:10.1029/WR019i002p00503, 1983.

Tessier, Y., Lovejoy, S., Hubert, P., Schertzer, D., and Pecknold, S.: Multifractal analysis and modeling of rainfall and river flows and scaling, causal transfer functions, J. Geophys. Res.-Atmos., 101, 26427–26440, 1996.

USGS (United States Geological Survey): National Water Information System, USGS Surface-Water Historical Instantaneous Data for North Carolina, available at: http://waterdata.usgs.gov/nc/nwis, last access: 1 September 2016.

Villarini, G.: On the seasonality of flooding across the continental United States, Adv. Water Resour., 87, 80–91, doi:10.1016/j.advwatres.2015.11.009, 2016.

Villarini, G., Serinaldi, F., Smith, J., and Krajewski, W.: On the stationarity of annual flood peaks in the continental United States during the 20th century, Water Resour. Res., 45, W08417, doi:10.1029/2008WR007645, 2009.

Vittal, H., Singh, J., Kumar, P., and Karmakar, S.: A framework for multivariate data-based at-site flood frequency analysis: Essentiality of the conjugal application of parametric and nonparametric approaches, J. Hydrol., 525, 658–675, doi:10.1016/j.jhydrol.2015.04.024, 2015.

6

Horizontal soil water potential heterogeneity: simplifying approaches for crop water dynamics models

V. Couvreur[1,2], **J. Vanderborght**[3], **L. Beff**[1], **and M. Javaux**[1,3]

[1]Earth and Life Institute, Université catholique de Louvain, Croix du Sud, 2, bte L7.05.02, 1348 Louvain-la-Neuve, Belgium
[2]Department of Land, Air and Water Resources, University of California, 1 Shields Ave., Davis, CA 95616, USA
[3]Institute of Bio- und Geosciences, IBG-3: Agrosphere, Forschungszentrum Juelich GmbH, 52425 Juelich, Germany

Correspondence to: V. Couvreur (vcouvreur@ucdavis.edu)

Abstract. Soil water potential (SWP) is known to affect plant water status, and even though observations demonstrate that SWP distribution around roots may limit plant water availability, its horizontal heterogeneity within the root zone is often neglected in hydrological models. As motive, using a horizontal discretisation significantly larger than one centimetre is often essential for computing time considerations, especially for large-scale hydrodynamics models. In this paper, we simulate soil and root system hydrodynamics at the centimetre scale and evaluate approaches to upscale variables and parameters related to root water uptake (RWU) for two crop systems: a densely seeded crop with an average uniform distribution of roots in the horizontal direction (winter wheat) and a wide-row crop with lateral variations in root density (maize). In a first approach, the upscaled water potential at soil–root interfaces was assumed to equal the bulk SWP of the upscaled soil element. Using this assumption, the 3-D high-resolution model could be accurately upscaled to a 2-D model for maize and a 1-D model for wheat. The accuracy of the upscaled models generally increased with soil hydraulic conductivity, lateral homogeneity of root distribution, and low transpiration rate. The link between horizontal upscaling and an implicit assumption on soil water redistribution was demonstrated in quantitative terms, and explained upscaling accuracy. In a second approach, the soil–root interface water potential was estimated by using a constant rate analytical solution of the axisymmetric soil water flow towards individual roots. In addition to the theoretical model properties, effective properties were tested in order to account for unfulfilled assumptions of the analytical solution: non-uniform lateral root distributions and transient RWU rates. Significant improvements were however only noticed for winter wheat, for which the first approach was already satisfying. This study confirms that the use of 1-D spatial discretisation to represent soil–plant water dynamics is a worthy choice for densely seeded crops. For wide-row crops, e.g. maize, further theoretical developments that better account for horizontal SWP heterogeneity might be needed in order to properly predict soil–plant hydrodynamics in 1-D.

1 Introduction

Even though soil water potential (SWP) is known to affect plant water status, and more specifically plant actual transpiration rate (T_{act}), its horizontal variability within the root zone is neglected in many hydrological models, because of computational efficiency considerations and limitations in the actual monitoring of SWP with high spatial resolution (Beff et al., 2013).

In first-generation land surface schemes, the soil compartment was considered as a spatially homogeneous bucket, filled by precipitation and emptied by evapotranspiration (Manabe, 1969). This approach to plant water availability is considered as a "bulk approach", since the total amount of water in the soil bucket defines its water potential, independently of how water is distributed in the compartment. Later, a vertical discretisation of soil in multiple layers was considered. Root water uptake rates were proportional to relative root length densities and were affected by the water potential in each soil layer (Feddes et al., 1976). This approach allowed explicitly considering vertical capillary water fluxes in the soil and root distribution to evaluate plant water availability. However, the relation between the uptake and local water availability that is used in these models does either not consider the connectivity of the root system or uses rather ad hoc approaches to account for compensation of uptake from regions with a higher water availability (Javaux et al., 2013). Recent developments of models explicitly accounting for three-dimensional (3-D) SWP heterogeneity and water flow in the root system's hydraulic architecture (HA) (Doussan et al., 2006; Javaux et al., 2008) allowed investigating how plant water availability could be inferred from root system hydraulic properties and SWP distribution.

Based on the HA approach, a physically based macroscopic root water uptake (RWU) model, whose three plant-scale parameters can be derived from root segment-scale hydraulic parameters distributed along root system architectures of any complexity, was developed by Couvreur et al. (2012). Since this model provides a 3-D solution of water flow from soil–root interfaces to plant collar, it needs to operate coupled to a 3-D "centimetre-scale" soil water flow model, which drastically increases the computational effort for soil–plant water flow simulations.

In the literature, one can find two contrasting conjectures that allow reducing the computing time by upscaling small-scale 3-D water flow models: (i) neglecting horizontal variations of SWP at the microscopic scale and using a coarser horizontal-scale discretisation to account for lateral fluxes that may be relevant at a larger scale, or (ii) using analytical approaches to account for microscopic gradients of SWP between the bulk soil and the soil–root interface.

By using a coarse discretisation of the soil domain, the first approach assumes that SWP is horizontally homogeneous in zones possibly ranging from the centimetre scale to the plant scale. This configuration most probably occurs under low climatic demand for water, in homogeneously rooted soils with high hydraulic conductivity (Schroeder et al., 2009b).

The second approach relies on a radial axisymmetric expression of the Richards equation around a single root. Approximate analytical solutions of water flow can be obtained by assuming a constant soil hydraulic conductivity or diffusivity (Gardner, 1960), or constant-rate water uptake by roots (Van Noordwijk and De Willigen, 1987; De Jong Van Lier et al., 2006; Schroeder et al., 2007, 2009a). When considering a regular distribution of roots in each soil layer, this approach can be used to create a 1-D RWU model, implicitly accounting for horizontal soil water flow (Raats, 2007; De Jong Van Lier et al., 2008; Jarvis, 2011). Yet, the simplifying assumptions of this approach may be constraining. In reality, local uptake is not at constant rate, but highly variable on a daily basis, notably due to variations of plant transpiration (Jolliet and Bailey, 1992; Sperling et al., 2012). In addition, differences in root hydraulic properties between different root types and horizontal heterogeneity of root density may lead to biased predictions of RWU when homogeneously distributed roots with similar hydraulic properties are assumed (Schneider et al., 2010; Durigon et al., 2012).

The objective of this paper is to provide a theoretical framework and an exploratory analysis of methods aiming at simplifying horizontal soil water flow calculation within the root zone, for soil–plant water flow models. Therefore, an approach to upscale the macroscopic RWU model that was derived based on the fully discretised hydraulic root architecture by Couvreur et al. (2012) will be presented. The upscaling approach corresponding to the first conjecture will be tested under different conditions regarding atmospheric demand, soil type and rooting heterogeneity, so as to discuss its applicability field. The opportunities and obstacles tied to the second conjecture will be analysed in the last part.

2 Theory

When a soil system at hydrostatic equilibrium is impacted by external processes, like evaporation, transpiration or aquifer level rise, the uniform SWP distribution is perturbed. Internal fluxes like soil capillary fluxes, drainage and hydraulic lift, driven by SWP heterogeneity then come into play to dissipate this heterogeneity and stabilise the system to another equilibrium state, unless other external perturbations arise in the meantime. The resulting system state heterogeneity may hinder the accuracy of its upscaled representation. Such accuracy thus highly relies on system properties influencing the rates of processes generating and dissipating heterogeneity.

In this section, we present soil- and plant-water flow equations that generate and dissipate SWP heterogeneity.

2.1 Equations for three-dimensional explicit water flow simulation

Soil water capillary flow is driven by local gradients of SWP and tends to dissipate SWP heterogeneity. In this study, we assume 3-D soil water flow to be well described by the Richards equation:

$$\frac{\partial \theta}{\partial t} = \nabla \cdot [K \nabla \psi_s] - S, \quad (1)$$

where θ is the volumetric water content ($L^3\ L^{-3}$), t is time (T), K is the unsaturated soil hydraulic conductivity ($L^2\ P^{-1}\ T^{-1}$) here considered as isotropic, ψ_s is the SWP (P) including matric and gravimetric components of water potential, and S is the sink term ($L^3\ L^{-3}\ T^{-1}$), which accounts for RWU. Note that the units of K and ψ_s differ from standards of soil physics (in which $L T^{-1}$ and L are more commonly used for K and ψ_s, respectively) but were chosen for consistency with those used in plant physiology.

In fine soil elements, the macroscopic RWU model based on the HA approach proposed by Couvreur et al. (2012) provides an expression for sink terms of the Richards equation:

$$S_k.V_k = T_{\text{act}}.\text{SSF}_k + K_{\text{comp}}.\left(\psi_{\text{s},k} - \psi_{\text{s eq}}\right).\text{SSF}_k, \quad (2)$$

where S_k (T^{-1}) is the sink term in the kth soil element, V_k (L^3) is the volume of the kth soil element, T_{act} ($\text{L}^3\ \text{T}^{-1}$) is the plant's actual transpiration rate, SSF_k (-) is the standard sink fraction in the kth soil element (the sum of these individual fractions being one by definition), K_{comp} ($\text{L}^3\ \text{P}^{-1}\ \text{T}^{-1}$) is the compensatory RWU conductance of the plant, $\psi_{\text{s},k}$ (P) is the SWP of the kth soil element, and $\psi_{\text{s eq}}$ (P) is the equivalent SWP sensed by the plant, which is a function of local SWPs and of the standard sink fraction distribution:

$$\psi_{\text{s eq}} = \sum_{j=1}^{M} \psi_{\text{s},j}.\text{SSF}_j, \quad (3)$$

where the j index ranges from the first to the last of the M soil elements (SSF_j being zero for soil elements that do not contain any root segment).

Equations (2) and (3) rely on the assumption that the water potentials at soil–root interfaces located inside a soil element equal the element bulk SWP $\psi_{\text{s},k}$. If sufficiently small soil elements are used, this assumption may be satisfied (Schroeder et al., 2009a, b). Another simplifying assumption that needs to be fulfilled for Eq. (2) to be valid is that root radial conductances should be much lower than root axial conductances.

Equation (2) provides a conceptual split of the RWU variable into a "standard RWU" ($T_{\text{act}}.\text{SSF}_k$) and a "compensatory RWU" ($K_{\text{comp}}.\left(\psi_{\text{s},k} - \psi_{\text{s eq}}\right).\text{SSF}_k$). While the former creates SWP heterogeneity as long as the plant transpires, the latter is driven by, and tends to dissipate, SWP heterogeneity as long as SWP heterogeneity exists in the rooting zone.

With the HA approach, a link between water potential in the soil, at the plant collar, and actual transpiration rate is also provided by Couvreur et al. (2012):

$$\psi_{\text{collar}} = \psi_{\text{s eq}} - \frac{T_{\text{act}}}{K_{\text{rs}}}, \quad (4)$$

where K_{rs} ($\text{L}^3\ \text{P}^{-1}\ \text{T}^{-1}$) is the equivalent conductance of the root system, and ψ_{collar} (P) is the water potential in xylem vessels at the plant collar, which will be referred to as the "plant collar water potential".

It is worth noting that, through Eq. (4), plant collar water potential can be interpreted as being the sum of the equivalent SWP sensed by the plant and of the water potential loss due to water flow in the root system.

The pathway of water from plant collar xylem vessels to leaves is considered as one of the least resistive from a hydraulic perspective, the main resistances being located in soil (Draye et al., 2010), between soil and root xylem (Frensch and Steudle, 1989), and between the inner leaf and atmosphere. For the purpose of simplification, we considered the hydraulic resistance from plant collar to leaves to be negligible as compared to the root system's hydraulic resistance. This is equivalent to assuming leaf water potential as equal to ψ_{collar}. By using Eq. (4), one can then estimate the plant transpiration rate from leaf water potential under water stress, $\psi_{\text{leaf stress}}$ (P):

$$T_{\text{water stress}} = K_{\text{rs}}.\left(\psi_{\text{s eq}} - \psi_{\text{leaf stress}}\right), \quad (5)$$

where $T_{\text{water stress}}$ ($\text{L}^3\ \text{T}^{-1}$) is the plant transpiration rate under water stress, and $\psi_{\text{leaf stress}}$ is a constant for isohydric plants such as maize (Tardieu and Simonneau, 1998).

The assumption on collar to leaf hydraulic resistance may however be inappropriate for certain types of plants (Domec and Pruyn, 2008), in which case the whole plant conductance should be used instead of K_{rs}. Also, processes such as cavitation or aquaporin gating were not accounted for in this study, but may affect the plant conductance. Future prospects may concentrate on these aspects.

Considering that T_{act} neither exceeds the plant's potential transpiration rate nor $T_{\text{water stress}}$, we obtain the following simplistic water stress function:

$$T_{\text{act}} = \min\left(T_{\text{pot}}, T_{\text{water stress}}\right), \quad (6)$$

where T_{pot} ($\text{L}^3\ \text{T}^{-1}$) is the plant's potential transpiration rate, which depends on both atmospheric conditions and plant leave properties.

It is worth noting that the variables and parameters presented in this section are representative for a single plant. They could also be used to obtain the average transpiration rate of several plants under water stress having the same K_{rs} (average $\psi_{\text{leaf stress}}$ and $\psi_{\text{s eq}}$ then apply). However, as soon as the considered plants have significantly different K_{rs}, such averaging method might not provide accurate estimates of average transpiration rate, and plants should be considered individually.

2.2 Upscaling of water flow parameters and state variables

2.2.1 Plant water flow

Equation (2) was set up for 3-D soil–plant water dynamics modelling on small soil elements (centimetre scale). Understanding the implications of its application to larger elements requires the definition of upscaled variables in terms of the original "fine-scale" variables and parameters (S_k, V_k, SSF_k and $\psi_{\text{s},k}$). Here, we consider that upscaled soil elements are groups of smaller soil elements.

Since soil element volumes and standard sink fractions are extensive entities (i.e. additive for independent subsystems), their value for a group of soil elements is the sum of the soil elements values:

$$V_{\text{Up},g} = \sum_{k=1}^{M} \varepsilon_{k,g}.V_k, \quad (7)$$

$$\mathrm{SSF}_{\mathrm{Up},g} = \sum_{k=1}^{M} \varepsilon_{k,g}.\mathrm{SSF}_k, \tag{8}$$

where $V_{\mathrm{Up},g}$ (L^3) is the "upscaled" volume of the gth group, $\mathrm{SSF}_{\mathrm{Up},g}$ (-) is the standard sink fraction of the gth group, $\varepsilon_{k,g}$ (-) is one when the kth element belongs to the gth group and zero otherwise, and the k index ranges from the first to the last of the M soil elements. Note that groups are non-overlapping, so that the summation of $\mathrm{SSF}_{\mathrm{Up},g}$ on the whole soil domain is 1, like for SSF_k.

The sink term only becomes an extensive variable when multiplied by the associated soil element volume (then it becomes an additive flux). We can thus write

$$S_{\mathrm{Up},g}.V_{\mathrm{Up},g} = \sum_{k=1}^{M} \varepsilon_{k,g}.S_k.V_k, \tag{9}$$

where $S_{\mathrm{Up},g}$ (T^{-1}) is the sink term in the gth group.

Upscaling the left and right hand sides of Eq. (2) leads to

$$S_{\mathrm{Up},g}.V_{\mathrm{Up},g} = T_{\mathrm{act}}.\mathrm{SSF}_{\mathrm{Up},g} + K_{\mathrm{comp}}.\left(\psi_{\mathrm{sr\ Up},g} - \psi_{\mathrm{s\ eq}}\right) .\mathrm{SSF}_{\mathrm{Up},g}. \tag{10}$$

From Eqs. (2) and (8)–(10), the upscaled soil–root interface water potential, $\psi_{\mathrm{sr\ Up},g}$ (P), is defined as

$$\psi_{\mathrm{sr\ Up},g} = \frac{\sum_{k=1}^{M} \varepsilon_{k,g}.\psi_{\mathrm{s},k}.\mathrm{SSF}_k}{\sum_{k=1}^{M} \varepsilon_{k,g}.\mathrm{SSF}_k}. \tag{11}$$

According to Eq. (11), the upscaled soil–root interface water potential represents the SSF-weighted mean SWP of the individual soil elements that constitute the upscaled soil element.

It is worth noting that the upscaled soil-root interface water potential $\psi_{\mathrm{sr\ Up}}$ represents the SWP sensed by the plant in a part of the root zone. When it comprises the entire root zone, $\psi_{\mathrm{sr\ Up}}$ is the plant sensed SWP $\psi_{\mathrm{s\ eq}}$ (Eq. 3).

So as to illustrate this concept, three simple examples are shown in Fig. 1. In the first example, only soil element # 3 contains a root segment. Following Eq. (11), $\psi_{\mathrm{sr\ Up},1}$ should equal the SWP of element # 3. In other words, in group # 1, the root segment only senses the SWP of element # 3, which is its direct environment. In the second example, each soil element contains a root segment. Considering all non-null SSF_k as equal to each other, $\psi_{\mathrm{sr\ Up},2}$ would be the arithmetic mean of the three individual SWPs. In the third example, no soil element contains a root segment so that $\mathrm{SSF}_{\mathrm{Up},3}$ is zero and no water potential sensed by root segments needs to be calculated for this element.

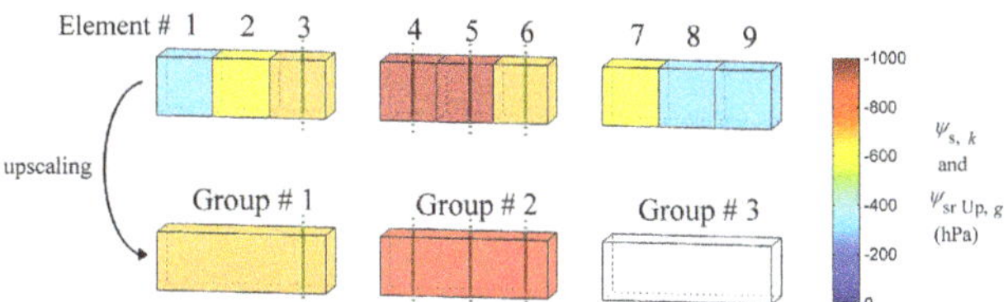

Fig. 1. Examples of the relation between $\psi_{\mathrm{s},k}$ and $\psi_{\mathrm{sr\ Up},g}$. Cubes are soil elements whose SWP, $\psi_{\mathrm{s},k}$, is represented by the colour scale. Parallelepipeds are groups of three, upscaled, soil elements, whose upscaled soil–root interface water potential $\psi_{\mathrm{sr\ Up},g}$ is represented by the same colour scale. Green vertical lines, in elements 3–6 and groups 1 and 2, are root segments.

Eventually, by using Eqs. (3), (8) and (11), it can be demonstrated that the equivalent SWP sensed by the plant can be calculated from $\mathrm{SSF}_{\mathrm{Up}}$ (vector size: $[G \times 1]$) using

$$\psi_{\mathrm{s\ eq}} = \sum_{f=1}^{G} \psi_{\mathrm{sr\ Up},f}.\mathrm{SSF}_{\mathrm{Up},f}, \tag{12}$$

where the f index ranges from the first to the last of the G groups of soil elements. The equations that are used to determine the plant-sensed soil water content (Eqs. 3, 12) and the local water uptake (Eqs. 2, 10) are scale invariant, which follows directly from the fact that these relations are linear at the small scale. Similarly, the water stress equations (Eqs. 5, 6) are scale invariant and do not depend on the scale at which SSF and ψ_{sr} are defined. A problem though is that for the calculation of the upscaled soil–root interface water potentials, $\psi_{\mathrm{sr\ Up}}$, using Eq. (11) the distribution of the SWPs and SSF at the smaller scale must be known. In the following, we will make two assumptions to derive $\psi_{\mathrm{sr\ Up}}$ directly from simulated upscaled SWPs and upscaled SSF.

2.2.2 Soil water flow

In this study, soil water flow state variables of upscaled elements were estimated with a simple "bulk" approach (i.e. the distribution of water inside upscaled soil elements was not accounted for). Their SWP, $\psi_{\mathrm{s\ Up}}$ (P), and hydraulic conductivity, K_{Up} ($L^2\ P^{-1}\ T^{-1}$), were directly deduced from their bulk water content θ_{Up} ($L^3\ L^{-3}$) and, respectively, water retention curve and hydraulic conductivity curve (these properties being uniform in space and time).

In consequence, the following upscaled expression of the Richards equation was used:

$$\frac{\partial \theta_{\mathrm{Up}}}{\partial t} = \nabla \cdot \left[K_{\mathrm{Up}} \nabla \psi_{\mathrm{s\ Up}}\right] - S_{\mathrm{Up}}, \tag{13}$$

where S_{Up} is provided by Eq. (10).

2.3 Simplifying assumptions for horizontal soil water flow

2.3.1 First conjecture: homogeneous soil water potential in upscaled soil elements

In simulations with upscaled soil elements (for instance in a 1-D soil domain), detailed SWPs around individual root segments are not available. In the first proposed approach, upscaled soil–root interface water potentials were approximated by the corresponding element bulk SWP:

$$\psi_{\mathrm{sr\ Up},g} = \psi_{\mathrm{m}}\left(\theta_{\mathrm{Up},g}\right) + z_g, \tag{14}$$

where $\psi_{\mathrm{m}}(\theta)$ (P) is the function providing soil matric potential from soil water content, $\theta_{\mathrm{Up},g}$ ($L^3\ L^{-3}$) is the bulk water content of the gth upscaled soil element, and z_g (P) is the gravitational potential of water at the centre of the gth upscaled soil element. Note that z_g is defined zero at the soil surface and positive upwards.

This assumption is generally considered as consistent either on short distances (as in fine elements of reference scenarios), or in conditions of high soil hydraulic conductivity (when lateral redistribution of water occurs almost instantaneously).

When water is redistributed by soil capillary flow (or by compensatory RWU), a positive divergence of water flow is generated at points where water is removed, while a negative divergence occurs where water is added. Considering water mass conservation, the volumetric integration of positive water divergences related to the process of redistribution must equal the volumetric integration of negative water divergences. Both integrated terms represent a volume of water moved from a place to another one per time unit, and equal a rate of water redistribution.

By assuming SWP as permanently homogeneous in an environment where water uptake is actually local, it is implicitly hypothesised that the divergence of soil water flow is high enough to instantly compensate for the removal of water by roots. For a given uptake rate in an upscaled element, and knowing the fine distribution of the standard sink fractions inside the element, it can be demonstrated (see Appendix A) that the soil water redistribution rate required to maintain SWP homogeneous inside the element should be the following:

$$R_{\mathrm{soil}\leftrightarrow\mathrm{hyp},g} = \left|S_{\mathrm{Up},g}\right| . V_{\mathrm{Up},g} . \frac{\sum_{k=1}^{M} \varepsilon_{k,g} . \left| \frac{\mathrm{SSF}_k}{\mathrm{SSF}_{\mathrm{Up},g}} - \frac{V_k}{V_{\mathrm{Up},g}} \right|}{2}, \tag{15}$$

where $R_{\mathrm{soil}\leftrightarrow\mathrm{hyp},g}$ ($L^3\ T^{-1}$) is the soil water redistribution rate required in order to keep the SWP horizontally homogeneous in the gth group of soil elements.

Note that soil water flow divergence at scales lower than the fine scale of the reference scenarios is not considered in the latter equation.

2.3.2 Second conjecture: solution for implicit SWP horizontal heterogeneity in soil layers

In the second proposed approach, the De Jong Van Lier et al. (2008) model provides a solution for differences between bulk soil and soil–root interface water potentials within 1-D soil elements, which does not require explicitly solving horizontal soil water flow. The latter is coupled to the upscaled macroscopic RWU model (Eq. 10), which simulates the consequent vertical water flow in root system HA.

The solution for horizontal soil water flow around roots relies on the concept of matric flux potential (MFP), which is the integral of soil hydraulic conductivity curve $K(\psi_{\mathrm{m}})$, over soil matric potential ψ_{m} (P), and, equivalently, the integral of the soil diffusivity curve $D(\theta)$ ($L^2\ T^{-1}$), over soil water content θ ($L^3\ L^{-3}$):

$$M(\psi_{\mathrm{m}},\theta) = \int_{\psi_{\mathrm{w}}}^{\psi_{\mathrm{m}}} K(\psi_{\mathrm{m}}) . \mathrm{d}\psi_{\mathrm{m}} = \int_{\theta_{\mathrm{w}}}^{\theta} D(\theta) . \mathrm{d}\theta, \tag{16}$$

where $M(\psi_{\mathrm{m}},\theta)$ ($L^2\ T^{-1}$) is the soil MFP at soil matric potential ψ_{m} or soil water content θ, ψ_{w} (P) is the soil matric potential at permanent wilting point, and θ_{w} ($L^3\ L^{-3}$) the soil water content at permanent wilting point.

By assuming root distribution as horizontally regular and the rate of uptake as constant, De Jong Van Lier et al. (2008) provide a simple relation between RWU rate in a soil layer ($S_{\mathrm{Up},g}$), bulk soil layer MFP $M_{\mathrm{s\ Up},g}$ ($L^2\ T^{-1}$), and MFP at soil–root interfaces in that soil layer $M_{\mathrm{sr\ Up},g}$ ($L^2\ T^{-1}$), which implicitly accounts for SWP horizontal heterogeneity:

$$M_{\mathrm{sr\ Up},g} = M_{\mathrm{s\ Up},g} - \frac{S_{\mathrm{Up},g}}{\rho_g}, \tag{17}$$

where $S_{\mathrm{Up},g}$ is given by Eq. (10), and ρ_g (L^{-2}) is a geometrical factor depending on rooting density and root radius at the gth depth (see Eq. B1). The factor ρ decreases with decreasing rooting density (and thus typically with depth). Decreasing ρ or increasing sink terms induce larger differences between $M_{\mathrm{s\ Up},g}$ and predicted $M_{\mathrm{sr\ Up},g}$.

By using the MFP curve, which links a soil matric potential to its MFP, one can derive $\psi_{\mathrm{sr\ Up},g}$ from $M_{\mathrm{sr\ Up},g}$:

$$\psi_{\mathrm{sr\ Up},g} = \psi_{\mathrm{m}}\left(M_{\mathrm{sr\ Up},g}\right) + z_g, \tag{18}$$

where $\psi_{\mathrm{m}}(M)$ (P) is the function providing soil matric potential from soil MFP.

As compared to Eq. (14), Eq. (18) is an alternative way to estimate soil–root interface's water potential in relatively large soil elements.

Knowing $\psi_{\mathrm{sr\ Up},g}$ in every soil layer, the equivalent SWP sensed by the plant can be calculated (Eq. 12), which allows further calculations of the plant's actual transpiration (Eqs. 5, 6) and RWU distribution (Eq. 10).

3 Methodology

So as to discuss up to what point the first soil water flow simplification leads to worthy compromises between accuracy and computing time, the conjecture of homogeneous SWP in upscaled soil elements was tested in different scenarios. These scenarios further described in Sect. 3.1 varied in (i) rooting heterogeneity, (ii) soil type, and (iii) atmospheric demand for water. Section 3.2 explains in detail the methods used to evaluate both conjectures implemented as options in R-SWMS (Root-Soil Water Movement and Solute transport; Javaux et al., 2008).

3.1 Scenarios description

3.1.1 Root systems architecture and hydraulic properties

Two crops with typically contrasting root distributions in the field were chosen for this study.

The first one is maize, whose horizontal rooting density varies more in the perpendicular direction than in the parallel direction to the row, due to its "wide row" sowing pattern (here corresponding to 75 cm × 15 cm). The generation and parameterisation of the 80 days-old virtual maize root system used in this study is fully described by Couvreur et al. (2012).

The second crop is winter wheat, whose horizontal rooting density is more homogeneous than that of maize, due to a dense seeding pattern. A density of 140 plants m^{-2} with a distance between plants of 10 cm in the x direction and 7 cm in the y direction was considered.

A winter wheat root system at early spring of 17 000 segments was generated with RootTyp (Pages et al., 2004). This model generates root systems based on plant-specific genetic properties like insertion angles of the different root types, their trajectories, average growth speed and distances between lateral roots, which were characterised for a winter wheat during early spring, in Nebraska (USA), by Weaver et al. (1924). They were also used to adapt RootTyp environmental parameters so as to reproduce measured root-length density profiles. The optimised wheat root system architecture is shown in Fig. 2a.

Wheat root's hydraulic properties were dependent on root segment age and type (shown in Fig. 2b and c) and were obtained from the literature. Root segments radial conductivities were measured by Tazawa et al. (1997) and Bramley et al. (2007, 2009). Root segment's axial conductance were measured by Sanderson et al. (1988) and Bramley et al. (2007) for primary roots, while Watt et al. (2008) estimated this property for lateral roots by using the Poiseuille–Hagen law.

So as to represent winter wheat root distribution in the field and accounting for the effect of overlapping root zones from neighbouring plants, while limiting the computational needs, the virtual root system was located in a horizontally periodic soil domain of 10 cm × 7 cm, which corresponds to the spacing between plants. Periodicity was applied for root system architecture at the vertical boundaries of the domain. Viewed from a larger scale than the individual plant scale, this case would correspond to a field containing identical root system architectures regularly spaced. In consequence, SWP variability is only accounted for at scales lower or equal to the plant scale.

3.1.2 Soil hydraulic properties

Two soil types with typically contrasting hydraulic properties were chosen for this study. The first one is a silt loam, whose water capacity and hydraulic conductivity are relatively high for a wide range of soil matric potentials (properties represented in blue, respectively in Fig. 3a and b).

The second soil type is a sandy loam, whose hydraulic conductivity is quite high close to water saturation, but soon becomes resistive to water flow when SWP decreases (properties represented in red, respectively in Fig. 3a and b).

Note that Mualem–van Genuchten equations (Van Genuchten, 1980) were used to define the soil hydraulic property curves, and that Carsel and Parrish (1988) parameterisations were chosen for both soil types.

In the scenarios, SWP was initially uniform (hydrostatic equilibrium) and set to field capacity (−300 hPa) for the silt loam. Sandy loam initial water potential was set to −130 hPa, so that water availability would not be limiting the uptake during the first days of the scenarios.

The soil domain was 123 cm deep, which means that for an initially uniform SWP, and neglecting the effect of osmotic potential, there was a difference of approximately 123 hPa between top and bottom matric potentials. This implied that soil water content and hydraulic conductivity were changing along the soil profile, already at initial conditions, as illustrated by the coloured bands in Fig. 3.

3.1.3 Boundary conditions

In order to focus on RWU and soil capillary flow as processes generating or reducing SWP heterogeneity, no other processes were considered in the scenarios. Therefore, no-flux boundary conditions were set at the top and bottom boundaries of the soil domain, while plant transpiration was the only process removing water from the system. In addition to being periodic for the root system architecture, vertical boundaries of the domain were periodic for soil- and root-water flow.

High- and low-transpiration-rate cases were selected in order to investigate whether these rates impact the validity of simplifying assumptions about lateral SWP distributions in the root zone. Atmospheric demand for water reflected the geographical position and period of the year for which the root system architectures were determined. The FAO approach (Allen et al., 1998) was used to determine the daily

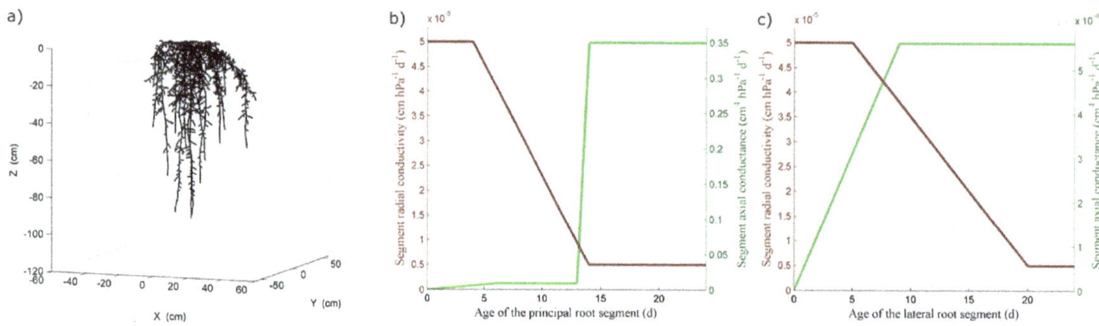

Fig. 2. Virtual winter wheat root system **(a)** architecture at early spring, and **(b)** principal and **(c)** lateral root segments' hydraulic properties.

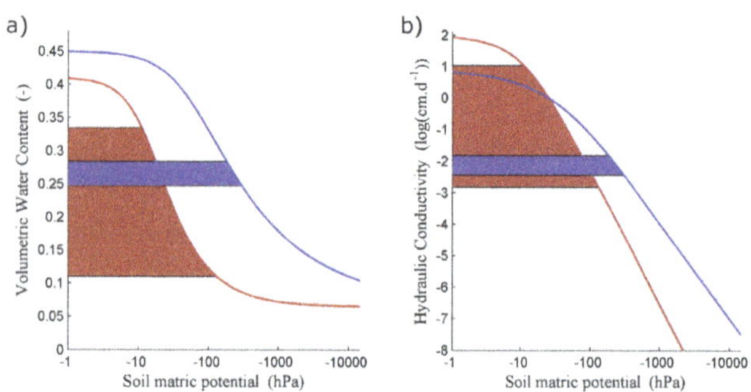

Fig. 3. Silt loam (blue) and sandy loam (red) hydraulic properties: **(a)** water retention curves and **(b)** hydraulic conductivity curves. The coloured bands show the ranges of **(a)** water content and **(b)** hydraulic conductivities initially met in the soil profile.

potential transpiration rate of single plants, T_{daily} ($L^3\ T^{-1}$), from selected reference evapotranspiration rates:

$$T_{\text{daily}} = \text{ET}_{\text{ref}}.K_{\text{c}}.\text{Surf}, \tag{19}$$

where ET_{ref} ($L\,T^{-1}$) is the reference evapotranspiration, K_{c} (-) is the crop coefficient, and Surf (L^2) is the horizontal surface occupied by a single plant in a field. Note that the part of evaporation in ET_{ref} was considered as negligible. Accounting for it would have led to slightly lower transpiration rates.

For the French maize crop in July, K_{c} was 1.2, Surf was 1125 cm^2, and the high ET_{ref} was 4.5 mm d^{-1} while the low ET_{ref} was 2.25 mm d^{-1}. For the Nebraskan winter wheat crop at early spring, K_{c} was 1, Surf was 70 cm^2, and the high ET_{ref} was 3.9 mm d^{-1} while the low ET_{ref} was 1.95 mm d^{-1}.

Sinusoidal daily variations of T_{pot} were expressed as a function of T_{daily} with the following expression:

$$T_{\text{pot}} = T_{\text{daily}}.\left(\sin\left(\frac{2\pi.t}{\tau} - \frac{\pi}{2}\right) + 1\right), \tag{20}$$

where t (T) is the time after midnight, and τ (T) is the number of time units in a day–night cycle (e.g. τ is 24 h if t is given in hours, and 1 day if t is given in days).

$\psi_{\text{leaf stress}}$, which triggers stomata partial closure due to water stress (see Eqs. 5 and 6), was −15 000 hPa for both crops.

The duration of scenarios is 14 days, except for high ET_{ref} on sandy loam (10 days).

3.2 Testing the simplifying approaches

The simplifying approaches described in Sect. 2.3 were tested by comparing their results with simulated reference results. In the reference simulations, Richards equation was solved for a fine 3-D soil grid, and the model of Doussan et al. (1998) was used to predict RWU by the root system HA in R-SWMS. Due to computing power considerations, the reference maize crop scenarios could not be run with soil elements smaller than cubes of 1.5 cm length. Since the winter wheat domain dimensions were smaller, its reference scenarios could be run with cubic soil elements of 0.5 cm length. Consequently, reference scenarios do not account for additional SWP gradients around roots at scales smaller than, respectively, 1.5 and 0.5 cm. Accounting for this feature may increase differences between reference results and results obtained from upscaled soil grids (Schroeder et al., 2009b).

3.2.1 Simplifying approaches features

In order to test the first conjecture (homogeneous SWP in upscaled soil elements), each of the eight scenarios defined in Sect. 3.1 (combinations of the following properties: maize or winter wheat; silt loam or sandy loam; high or low T_{daily}) were run with soil elements of increasing horizontal surface, as summarised in Table 1 and illustrated in Fig. 4.

For maize, the assumption on SWP homogeneity was firstly applied to the direction parallel to the row. Subsequently, the discretisation was coarsened in the direction perpendicular to the rows. Therefore, all intermediate soil discretisations, between the finest one and 1-D, are 2-D (see Table 1). This is not the case for winter wheat, for which no preferential direction was considered to group soil elements.

In opposition, the second conjecture (soil–root interface water potential predicted from the approximate analytical solution of water flow towards a root) was directly tested for 1-D soil layers (75 cm × 15 cm × 1.5 cm and

Table 1. Sizes of upscaled soil elements and domain properties for both maize and winter wheat crops in the runs testing the first conjecture.

Plant type	Element properties	Case 1	Case 2	Case 3	Case 4	Case 5	Case 6
Maize	Horizontal area (cm^2)	2.25	22.5	45	112.5	225	1125
	x and y lengths (cm)	1.5×1.5	1.5×15	4×15	7.5×15	15×15	75×15
	Elements per layer (-)	500	50	25	10	5	1
	Domain dimensionality	3-D	2-D	2-D	2-D	2-D	1-D
Winter wheat	Horizontal area (cm^2)	0.25	1	7	70		
	x and y lengths (cm)	0.5×0.5	1×1	2×3.5	10×7		
	Elements per layer (-)	280	70	10	1		
	Domain dimensionality	3-D	3-D	3-D	1-D		

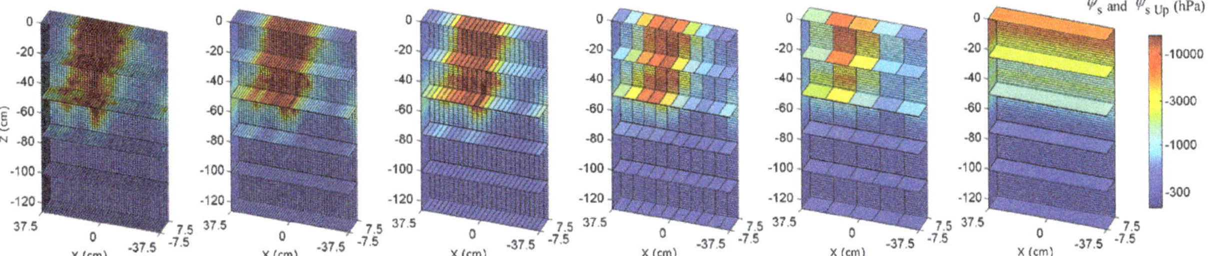

Fig. 4. Discretisations of the maize crop soil domain used for the first simplifying approach. The colour scale gives the soil water potential distribution at the end of the high-transpiration-rate scenario on silt loam.

10 cm × 7 cm × 0.5 cm, respectively for maize and winter wheat).

3.2.2 Comparison with reference scenarios

In order to evaluate the accuracy of the first simplified approach, differences between the reference and different upscaling scenarios were estimated for ψ_{collar} and horizontally averaged sink term and water content profiles. The mean of the absolute differences for all times and depths was divided by the mean value for the reference case, which provided one relative mean absolute difference for each scenario. The relative computation time of the simplified to reference simulations was also determined.

Eventually, horizontal and vertical redistribution of water by both soil and roots from 1-D and reference results were compared, in order to understand which process dissipating SWP heterogeneity would be responsible of possibly wrong representations of 1-D soil–plant water dynamics. For simulations directly run in 1-D, the total horizontal redistribution of water by soil was estimated as the integration of the redistribution necessary to keep each layer's inner water potential homogeneous (i.e. vertical integration of Eq. 15). Other equations quantifying vertical and horizontal water redistribution by soil and roots from reference and 1-D simulation results are detailed in Appendix C.

With the second conjecture, simple effective methods that allow overcoming basic assumptions of the De Jong Van Lier et al. (2006) model were discussed. These concern (i) horizontal heterogeneity of root distribution, and (ii) transient rate of water uptake. For reasons discussed in Sect. 4.4.2, a proper coupling with the Richards equation could not be achieved with this conjecture. However, using bulk SWP data from the reference simulation and keeping past uptake rates in memory, we could evaluate the accuracy of the second conjecture at each individual time step.

Effective values of the geometrical parameter ρ were first estimated from reference simulations and compared to theoretical values (calculated from each layer's root length density and assuming a regular distribution of roots), in order to understand how this parameter may be affected by horizontal rooting heterogeneity. Then, $\psi_{\text{sr Up}}$ was predicted from either the current sink term, or a weighted mean of sink terms on time windows of chosen length (weights linearly decreasing to zero with passed time), in order to understand if the history of past sink terms should be accounted for when RWU is transient.

Considering that the simplifying approaches presented in this paper introduce structural errors in the model, differences as compared to reference scenarios were considered as "errors". However, also the reference model is subject to structural errors (supposed relatively small). These basic errors were not accounted for in the next pages.

Table 2. Relative absolute differences on ψ_{collar}, 1-D sink terms, 1-D water contents and computing times in the maize scenarios, for increasing soil element sizes. Refer to Table 1 for the detailed geometry of cases 1–6.

	Maize scenario	Case # 1	2	3	4	5	6
Relative difference on ψ_{collar} (%)	Low T_{daily} – silt loam	0.5	0.3	0.9	5.1	10.6	14.8
	High T_{daily} – silt loam	0.9	1.5	4.5	15.5	26.8	30.3
	Low T_{daily} – sandy loam	1.9	2.9	8.8	25.9	30.6	32.7
	High T_{daily} – sandy loam	3.7	4.6	8.1	13.2	15.0	18.7
Relative difference 1-D sink (%)	Low T_{daily} – silt loam	1.0	1.2	1.2	5.3	12.1	17.1
	High T_{daily} – silt loam	1.9	3.2	4.2	11.1	19.7	24.2
	Low T_{daily} – sandy loam	3.4	5.0	6.8	21.3	35.4	38.5
	High T_{daily} – sandy loam	6.3	8.0	10.9	24.3	44.9	47.4
Relative difference 1-D water cont. (%)	Low T_{daily} – silt loam	0.6	0.6	0.5	2.6	5.4	10.0
	High T_{daily} – silt loam	1.3	1.9	2.1	5.1	9.5	17.0
	Low T_{daily} – sandy loam	2.0	2.8	3.4	8.6	13.6	22.4
	High T_{daily} – sandy loam	2.4	4.4	5.1	9.3	14.2	22.8
Relative comput. time (%)	Low T_{daily} – silt loam	3.0	1.5	1.2	1.1	0.81	0.37
	High T_{daily} – silt loam	1.9	0.21	0.13	0.10	0.10	0.09
	Low T_{daily} – sandy loam	3.9	0.46	0.36	0.33	0.29	0.26
	High T_{daily} – sandy loam	0.98	0.04	0.03	0.02	0.02	0.02

4 Results and discussion

4.1 First conjecture: homogeneous soil water potential in upscaled soil elements

Tables 2 and 3 show the relative errors of predicted state variables and relative computing time for each scenario, with increasing element size inside which SWP is assumed homogeneous. Errors that occur at the finest spatial discretisation (i.e. horizontal surfaces of respectively 2.25 and 0.25 cm^2 for maize and winter wheat) are due to the replacement of the Doussan RWU model by Eq. (2) to calculate the sink terms.

It is notable that 1-D sink terms and ψ_{collar} were generally more sensitive to errors than 1-D water contents, even though water content differences are a consequence of sink term differences. This can be explained by the fact that SWP heterogeneity is the driver of soil water flow. Thus, for instance, locally overestimating RWU leads to higher SWP heterogeneity, which leads to higher "compensation" by soil water flow. Consequently, errors of RWU tend to be larger than errors of soil water content, especially in cases of high soil hydraulic conductivity.

In the next sections, we study the impact of element size, daily transpiration rate and soil type on the reported relative absolute differences, and further analyse where these differences take place in space and time. Illustrations are mostly given for the scenario "high T_{daily} on silt loam", but complementary explanations are given for other scenarios in case their trends differ from the illustrations.

4.1.1 Impact of element size and crop type

For maize, the simplification from 3-D to 2-D soil discretisation results in a relatively small increase of model errors (see Fig. 5a) since SWP is quite homogeneous in the direction of maize rows (see left subplot in Fig. 4). Conversely, further increases of element size in the direction perpendicular to maize rows (in which a big part of SWP variability is observed in reference scenarios) result in significant increase of model errors, particularly beyond case #3 (elements of 3 cm in the direction perpendicular to the row). This result encourages the use of 2-D soil discretisation for simulating water dynamics in a maize crop, whereas considering a 1-D approach with homogeneous SWP in horizontal soil layers leads to strong errors in predicted state variables (approaching 50 % of relative error on 1-D sink terms and ψ_{collar} over a period of 10 days on sandy loam).

For winter wheat, while changing the element dimension from 3-D to 1-D (see Fig. 5b), model errors stayed remarkably low (below 1 % for scenarios on silt loam over a period of 14 days). This feature can be related to the dense sowing pattern of the winter wheat crop (140 plants m^{-2}, against 9 for maize), which naturally induces rather homogeneous horizontal rooting, uptake and SWP patterns.

One of the main interests of simplifying approaches is model computing time reduction. As shown in Table 2 (and illustrated in Fig. 5c and d), for maize, if computing time was already reduced by a factor of 25–100 due to the replacement of the Doussan model by Eq. (2), another factor of 3–30 was gained by using a 2-D soil discretisation. For winter wheat,

Table 3. Relative absolute differences on ψ_{collar}, 1-D sink terms, 1-D water contents and computing times in the winter wheat scenarios, for increasing soil element sizes. Refer to Table 1 for the detailed geometry of cases 1–4.

		Case #			
	Winter wheat scenario	1	2	3	4
Relative difference on H_{collar} (%)	Low T_{daily} – silt loam	0.0	0.0	0.1	0.1
	High T_{daily} – silt loam	0.1	0.2	0.5	0.6
	Low T_{daily} – sandy loam	0.3	1.8	2.5	2.5
	High T_{daily} – sandy loam	4.0	4.8	5.2	5.6
Relative difference 1-D sink (%)	Low T_{daily} – silt loam	0.2	0.2	0.2	0.2
	High T_{daily} – silt loam	0.4	0.5	0.7	0.8
	Low T_{daily} – sandy loam	0.9	2.9	4.6	4.9
	High T_{daily} – sandy loam	5.9	10.9	14.1	15.8
Relative difference 1-D water cont. (%)	Low T_{daily} – silt loam	0.1	0.1	0.1	0.1
	High T_{daily} – silt loam	0.1	0.2	0.3	0.3
	Low T_{daily} – sandy loam	0.19	1.1	2.0	2.4
	High T_{daily} – sandy loam	2.8	3.6	4.9	5.9
Relative comput. time (%)	Low T_{daily} – silt loam	6.9	4.8	1.9	1.5
	High T_{daily} – silt loam	9.0	3.3	1.5	0.98
	Low T_{daily} – sandy loam	17	3.6	1.3	0.79
	High T_{daily} – sandy loam	11	1.1	0.27	0.10

using Eq. (2) only reduced computing time by a factor of 6–14 because its root system has half of the segments than maize (using Doussan model is computationally cheaper for small root systems, while computing time of Eq. (2) does not discriminate between big and small root systems). Computation time was reduced by another factor of 5–100 as compared to the high resolution 3-D winter wheat scenarios, by using 1-D soil elements.

Such results suggest that using the first conjecture in, respectively, 2-D (maize) and 1-D (winter wheat) soil elements as simplifying hypothesis for SWP distribution, is a worthy compromise maintaining accuracy while reducing computation time.

4.1.2 Impact of daily transpiration and soil type

Even though crop type and soil elements size had major impact on the simplifying approach accuracy, two other features also clearly impacted this accuracy: T_{daily} and soil type.

Almost systematically, the simplified model accuracy was higher when decreasing T_{daily}, and in the silt loam than in the sandy loam. Since accuracy under the first conjecture is highly related to the absence of SWP horizontal heterogeneity, the previous statement can be explained through processes involving creation and dissipation of SWP heterogeneity.

Firstly, standard RWU is a process creating SWP heterogeneity in a soil with an initial hydrostatic equilibrium state; increasing T_{daily} (and obviously standard RWU) will thus lead to increased SWP heterogeneity and decreased accuracy under the first conjecture. Note that as defined in the theory, RWU is conceptually the superimposing of two processes: standard RWU, which creates SWP heterogeneity, and compensatory RWU, which dissipates (and is driven by) SWP heterogeneity but is independent of the plant's instantaneous transpiration rate.

Secondly, soil water flow is a process dissipating SWP heterogeneity; a high soil hydraulic conductivity thus favours SWP heterogeneity dissipation and leads to better predictions by approximations that use the first conjecture. Note that, even though silt loam hydraulic conductivity is mostly lower than that of sandy loam at the beginning of the simulations (see conductivity ranges in Fig. 3b), it stays relatively high at low soil matric potentials, which explains the higher accuracy of the silt loam than the sandy loam scenarios.

It is also worth noting that, in general, structural and parameterisation errors in a RWU model may have a limited impact on SWP distributions when soil water flow is a dominating process, as previously discussed by Hupet et al. (2002).

4.1.3 Spatio-temporal distribution of processes: comparison with 1-D results

This section clarifies the underlying assumption on soil water horizontal redistribution when using 1-D soil discretisation, and provides further insight on how it may impact model errors in space and time.

As shown in Fig. 6 for scenario "high T_{daily} on silt loam", the intensity of each process redistributing water can be rated

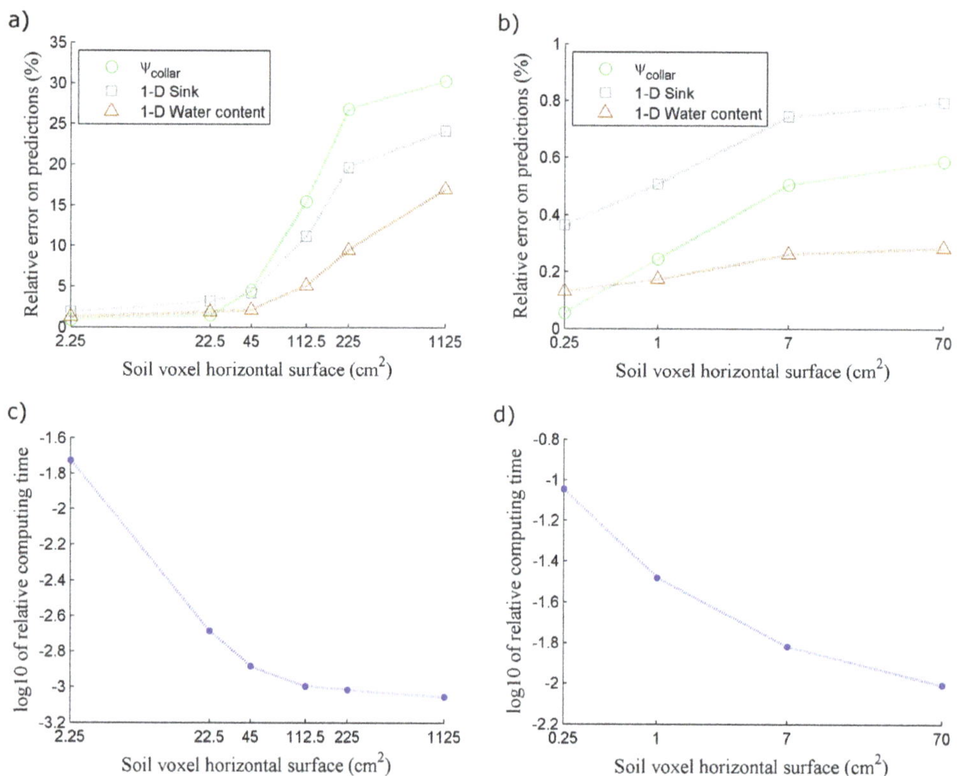

Fig. 5. Relative errors on three state variable predictions (ψ_{collar}, 1-D sink and 1-D water content) when using upscaled soil elements whose inner SWP is considered as homogeneous, for maize **(a)** and winter wheat **(b)**. Relative computing time for maize **(c)** and wheat **(d)**. x axes on logarithmic scale. Scenario: high T_{daily} on silt loam.

in terms of its total positive volumetric divergence of water flow (total negative volumetric divergence being equivalent to the positive one, by definition, since these processes only redistribute water in the system). Blue lines correspond to processes as they occurred in the reference scenarios while the red ones are for 1-D scenarios. Solid and dotted lines correspond, respectively, to horizontal and vertical spatial components of the processes. Figure 6a and c shows water redistribution rates by soil, evolving with time, while Fig. 6b and d shows water redistribution rates by roots. Eventually, Fig. 6a and b corresponds to maize, while Fig. 6c and d corresponds to winter wheat.

In Fig. 6a (maize), one can see that the assumed horizontal redistribution rate of water by soil in 1-D is overestimated during daytime; reference horizontal soil water flow is, thus, far from sustaining the necessary flow rate to keep SWP homogeneous. Also, during nighttime, even though decreased, reference horizontal soil water flow continues, due to the persistence of SWP horizontal heterogeneities, while in 1-D, the assumed horizontal water flow stops as soon as the plant stops transpiring (except once compensatory RWU significantly compensates vertical SWP heterogeneities at night). Conversely, in Fig. 6c (wheat), similar peaks of divergence of horizontal soil water flow can be noticed in both reference and 1-D scenarios. This can be attributed to the fact that water needs to flow on much shorter horizontal distances to compensate wheat SWP heterogeneities, and thus is much more effective in dissipating these heterogeneities (which almost disappear at night). For both maize and wheat, the vertical component of divergence of soil water flow is slightly underestimated in 1-D, which suggests that this process is affected by the hypothesis of horizontally homogeneous SWP, and may actually participate in dissipating SWP horizontal heterogeneities in reference scenarios.

For maize, both components of compensatory RWU are largely underestimated in 1-D (especially the horizontal one, which is null in 1-D, since SWP is considered as horizontally uniform), which is not the case for wheat, whose dominant vertical component of compensatory RWU is well represented in 1-D (see Fig. 6d).

During the second week of simulation, compensatory RWU rates reach increasingly high values (approximately 10 and 250 $\text{cm}^3\,\text{day}^{-1}$ redistributed in the profile, respectively for wheat and maize). For maize, compensatory RWU rates are similar or even higher than water redistribution rates by soil. Such integrated values of redistribution of water uptake are also non-negligible as compared to each plant's daily transpiration rate (respectively 27 and 600 $\text{cm}^3\,\text{d}^{-1}$). This confirms that the process of compensatory RWU might have a major impact on plant water availability (Feddes et al.,

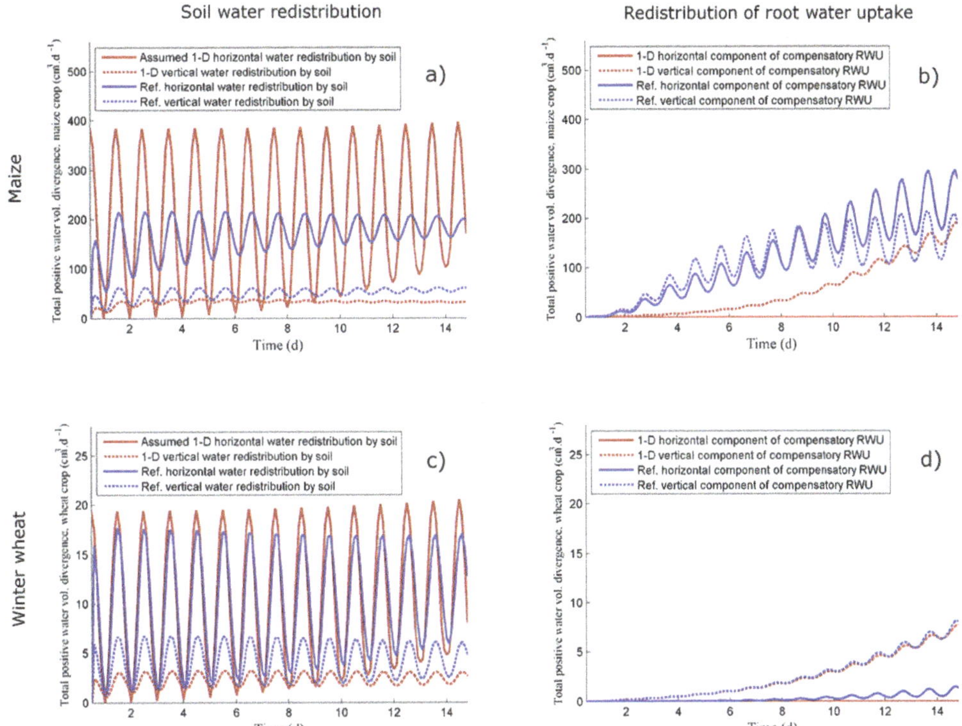

Fig. 6. Rating of processes dissipating soil water potential heterogeneity by **(a, c)** soil and **(b, d)** roots, in scenario "high T_{daily} on silt loam", for **(a, b)** maize and **(c, d)** wheat.

2001; Teuling et al., 2006). However, compensatory RWU takes some time to become significant, as compared to horizontal and vertical water redistribution by soil. This can be explained by the fact that, while SWP heterogeneity increases with time, root system hydraulic conductances do not change; redistribution of water by the root system thus increases. At the same time, soil hydraulic conductivities tend to decrease (due to soil water content reduction); redistribution of water by soil capillary flow thus becomes of lesser importance as compared to compensatory RWU. That sort of reflection was previously raised by Gardner and Ehlig (1963), who stated that, with soil drying, "while processes such as capillary rise see their rate reduced, due to a decreased soil hydraulic diffusivity, an increasing proportion of water moves upward through roots, which somehow short-circuits the path of water movement through soil."

As illustrated in Fig. 6 (left subplots), vertical soil water redistribution was generally the least important process, in terms of rates; which can be explained by the fact that, on long vertical distances, equivalent soil hydraulic resistances are high enough to limit redistribution (water has to flow through a larger number of hydraulic resistances in series), and thus prevent SWP heterogeneities from being dissipated.

In case horizontal soil water flow would actually not be fast enough to equilibrate a layer's SWP, the assumed water potential at soil–root interfaces would be overestimated in 1-D. This is exactly the observed response in scenarios of RWU by maize, where local SWP sensed by the plant decreases slower than in reference scenarios (Fig. 7a vs. 7d). This overestimation of local SWP sensed by the plant has two main consequences: (i) an underestimation of compensatory RWU (Fig. 7b vs. 7e, and dotted lines in Fig. 7f), and (ii) an overestimation of total SWP sensed by the plant (Fig. 7c) inducing underestimation of plant water stress (Fig. 7f).

It is notable that, for the same T_{act}, errors on ψ_{collar} equal errors on $\psi_{\mathrm{s\ eq}}$ since the difference between these variables is $\frac{T_{\mathrm{act}}}{K_{\mathrm{rs}}}$, which has no spatial dimension, and thus, is not affected by a spatial dimension reduction. Also, values of compensatory RWU in Fig. 7b and e are given as fluxes per plant in soil layers of 1.5 cm height. As a matter of comparison, the spatial integration of positive terms is given in Fig. 7f, while the integration of all terms would be zero by definition.

Figure 8 is the equivalent of Fig. 7 for winter wheat on sandy loam instead of maize on silt loam. The 1-D system state appears to be very close to the reference one for all variables. Even though ψ_{collar} and $\psi_{\mathrm{s\ eq}}$ are slightly overestimated at night and underestimated during daytime, T_{act} follows the same trend in both simulations. Conversely to results shown in Fig. 7, compensatory RWU is slightly overestimated in 1-D (see Fig. 8f), possibly due to water depletion around deep roots of wheat in the reference scenario, which limited the compensation rate. Proportionally to the total uptake rate, the compensation rate was always more intense on

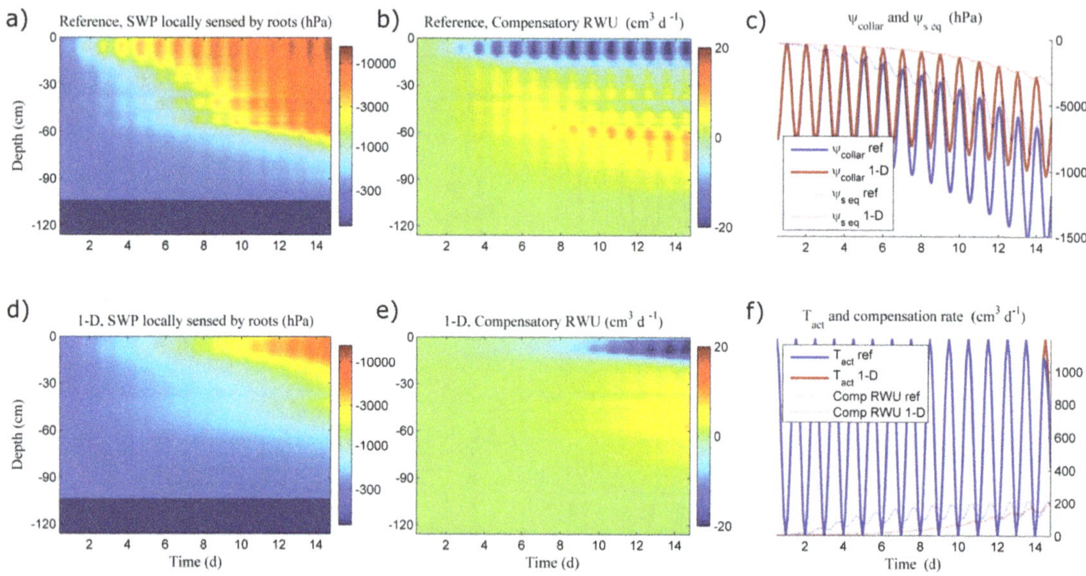

Fig. 7. Spatio-temporal distribution of **(a, d)** SWP locally sensed by roots and **(b, e)** compensatory RWU rates (spatial integration of positive terms), respectively in reference and 1-D scenarios. Temporal evolution of **(c)** plant collar water potential and SWP sensed by the plant, and **(f)** actual transpiration and compensation rates (scenario: maize, high T_{daily} on silt loam).

sandy loam than on silt loam, seemingly because water is not as efficiently redistributed by the sandy soil.

A conclusion of the detailed comparison between 1-D and reference maize scenarios is that, when horizontal redistribution of water by soil is a limiting process, there is a clear need to account for differences between bulk SWP and water potential sensed by roots in soil layers, in order to avoid biased predictions of compensatory RWU and plant water stress, in dimensionally simplified soil–plant systems. A physical approach presented in Sect. 2.3 was developed by De Jong Van Lier et al. (2006) for that purpose, of which opportunities and limitations are discussed in the next section.

4.2 Second conjecture: solution for water potential differences between bulk soil and root surface in 1-D soil layers

In this section, limitations of the second conjecture and tested adaptations aiming at better accounting for unfulfilled assumptions are discussed.

4.2.1 Horizontally heterogeneous rooting pattern

Like macroscopic RWU models using a "microscopic approach" (Raats, 2007; De Jong Van Lier et al., 2008; Jarvis, 2011), the second conjecture allows predicting SWP variations between the bulk soil ($\psi_{\text{s Up}}$) and soil–root interfaces ($\psi_{\text{sr Up}}$) by assuming a horizontally homogeneous root distribution, which implies that the water dynamics around roots is the same (their properties being considered as identical).

Yet, for maize crops, due to the wide-row sowing pattern, two features are in contradiction with the second conjecture's assumptions: (i) water potentials at soil–root interfaces are not horizontally homogeneous (see for instance left subplot of Fig. 4), and (ii) the horizontal rooting pattern is not uniform. As demonstrated in Eq. (11), in each soil layer, a unique value of $\psi_{\text{sr Up},g}$ may lead to the right average sink term for the layer. The microscopic approach might help finding this layer's "equivalent soil–root interface water potential", which makes it unnecessary to search for the full range of soil–root interface water potentials in each soil layer. The second contradiction is more of an issue since no definition of the geometrical factor ρ (see Eq. B1 for its theoretical formulation) accounts for horizontal rooting pattern heterogeneity. However, knowing values of $S_{\text{Up},g}$, $M_{\text{sr Up},g}$ and $M_{\text{s Up},g}$ (from the reference scenarios), an effective value of ρ_g was calculated at each depth for each time step of the scenarios, by using Eq. (17). As shown in Fig. 9a for scenario "maize high T_{daily} on silt loam", the effective values grouped by depth are significantly lower than theoretical values of ρ (blue dotted line), which means that the system behaves as if there were much fewer roots, or maybe, one "big root". This necessity to use smaller values of ρ was already noticed in comparisons with experimental data, by Faria et al. (2010), who interpreted that feature as a consequence of rooting heterogeneity, poor contact at soil–root interfaces and inactivity of a significant percentage of roots (approximately 95 %), which thus should not be taken into account when calculating ρ. Through this modelling study, we investigated and confirmed the expected impact of horizontal rooting heterogeneity on ρ.

Note that since root geometry does not change during scenarios, effective ρ values at a certain depth should

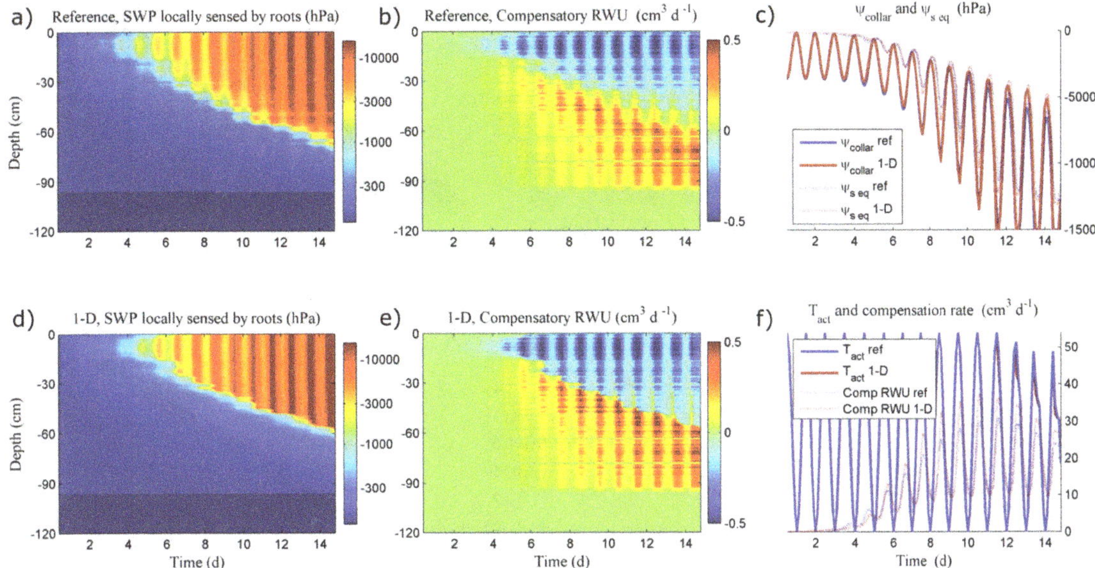

Fig. 8. Spatio-temporal distribution of **(a, d)** SWP locally sensed by roots and **(b, e)** compensatory RWU rates (spatial integration of positive terms), respectively in reference and 1-D scenarios. Temporal evolution of **(c)** plant collar water potential and SWP sensed by the plant, and **(f)** actual transpiration and compensation rates (scenario: winter wheat, high T_{daily} on sandy loam).

theoretically remain constant with time. As shown in Fig. 9a, they actually cover a certain range of effective values, which are also strongly sensitive to soil type (not shown). One should thus be careful when using the theoretical parameterisation of ρ for root systems with heterogeneous horizontal distribution.

Figure 9c shows the same comparison for wheat, whose theoretical ρ values are much closer to the effective ones. This confirms that the theoretical parameterisation is more reliable for wheat, whose horizontal root distribution is indeed rather uniform.

Note that negative effective values of ρ are not displayed in Fig. 9a and c. These however occur in reference simulations when roots exude water while $\psi_{\text{sr Up}}$ is still lower than the corresponding layer's bulk SWP. This transient situation cannot be predicted by the default model of water depletion around roots, since the geometrical factor ρ is defined positive (see Eq. B1).

4.2.2 Transient rate of root water uptake

Another assumption of macroscopic RWU models using a "microscopic approach" to predict SWP depletion at soil–root interfaces is that rates of water uptake are constant with time. Water uptake rates at a soil–root interface change over time due to temporal changes in plant transpiration but also due to compensation mechanisms in the connected root system. Since the soil system has a memory due to its buffer capacity, the water potential profile around a root at a certain time does not depend only on the extraction rate at that time but also on previous extraction rates. Thus, using a weighted-mean of past sink terms in Eq. (17) rather than the sink term at a given moment might be better to predict the difference between soil–root interface $\psi_{\text{sr Up}}$ and bulk soil $\psi_{\text{s Up}}$.

In this section, we tested if reference values of $M_{\text{sr Up},g}$ (from which $\psi_{\text{sr Up},g}$ can directly be deduced) could be predicted from Eq. (17), either by using the theoretical values of ρ_g and instantaneous $S_{\text{Up},g}$ ("default method"), or by using the mean values of effective ρ_g (red vertical lines in the box plots in Fig. 9a and c) and instantaneous $S_{\text{Up},g}$ ("average ρ" method), or eventually by using time-averaged values of $S_{\text{Up},g}$, in addition of the mean effective ρ_g ("average ρ & S" method).

Figure 9b shows the results obtained for maize at all time-steps of the "high T_{daily} on silt loam" scenario. The "1:1 line" illustrates the position of the reference $M_{\text{sr Up},g}$, while black circles correspond to the layer's bulk MFP (and to the $M_{\text{sr Up},g}$ predicted under the first conjecture). Mostly, even though more accurate than the first conjecture, using the "default method" (red crosses) still resulted in an overestimation of $M_{\text{sr Up},g}$, mainly due to the theoretical overestimation of ρ. Effective methods "average ρ" and "average ρ & S" allowed increasing the accuracy of the predictions around the 1:1 line, however significant differences persist, mainly in dry conditions (where small errors on $M_{\text{sr Up},g}$ moreover have a high impact on $\psi_{\text{sr Up},g}$). The prediction of negative values of $M_{\text{sr Up},g}$ is also problematic since the function providing MFP values from soil matric potentials is positive by definition. Consequently, no $\psi_{\text{sr Up},g}$ value can be deduced from a negative $M_{\text{sr Up},g}$. Even though both effective methods were sensitive to the chosen averaging function, none of the tested functions allowed reaching satisfying results for

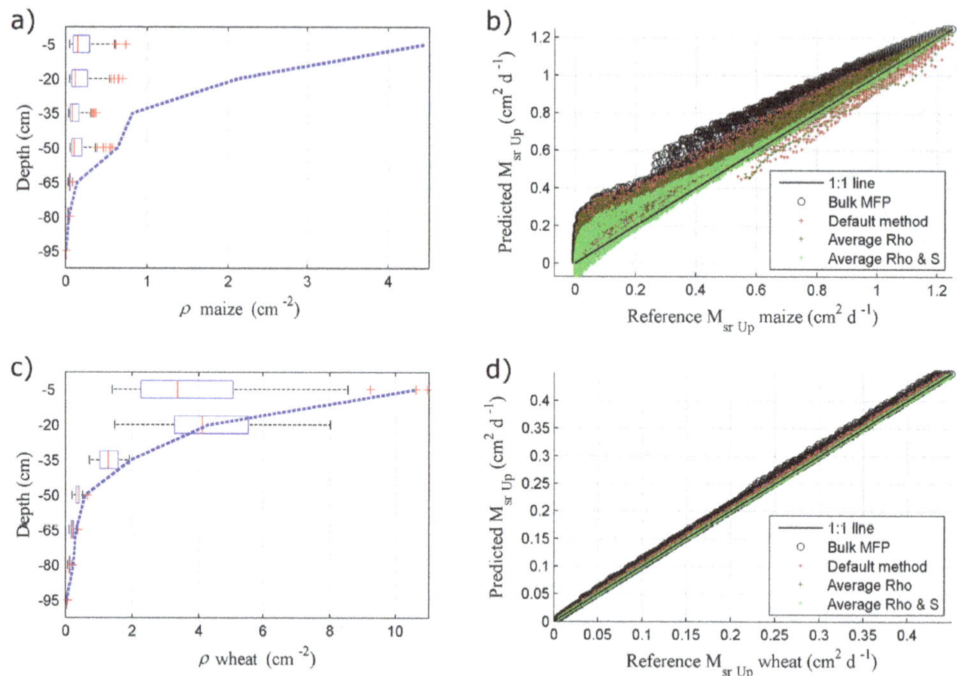

Fig. 9. System effective properties and state from the scenario "high T_{daily} on silt loam". Theoretical (blue dotted lines) and effective (box plots) values of ρ for maize **(a)** and wheat **(c)**. The layer's matric flux potential at soil–root interfaces predicted using Eq. (17) with default and effective methods, compared with reference values, for maize **(b)** and wheat **(d)**.

maize (averaging functions used for results shown in Fig. 9b and d: mean function for ρ, and 36 h average for the sink term). For wheat (Fig. 9d), results were already satisfying under the first conjecture, but could be improved by using the second conjecture, as shown by Fig. 9d.

When the RWU model using the second conjecture was further coupled to the Richards equation, the frequent prediction of negative values of $M_{\mathrm{sr\ Up}}$ (happens when $M_{\mathrm{s\ Up},g} < \frac{S_{\mathrm{Up},g}}{\rho_g}$, typically when the soil becomes dry) and oscillating $\psi_{\mathrm{sr\ Up},g}$ caused non-convergence issues (mainly for simulations on sandy loam). These could not be solved in this study.

4.2.3 Closing remarks on the second conjecture

Ideally, exact physical expressions would allow accounting for transient RWU rates and heterogeneous rooting distribution, with a resulting model shape that would possibly have to be adapted as compared to Eqs. (17) and (B1). However, such opportunity does not exist today, and a simple alternative is to use effective parameters and variables such as described in previous paragraphs and suggested by Faria et al. (2010), even though they entail a loss of physical meaning of the model.

The proposed effective methods, accounting for unfulfilled assumptions of the De Jong Van Lier et al. (2006) model, did not allow significantly improving predictions of differences between bulk SWP and SWP sensed by roots for 1-D spatial discretisation, except in conditions in which the first conjecture was already satisfying (winter wheat crop on silt loam). There is however a clear need for accurate functions predicting soil–root interface water potential, in order to correctly predict compensatory RWU and plant water stress. In the future, that problem might be solved through the development of specific analytical solutions for each type of system properties.

5 Conclusions and outlook

The objective of this paper was to provide a theoretical framework and exploratory analysis regarding the use of "upscaled" RWU models, partly or fully neglecting SWP horizontal heterogeneity within the root zone. We demonstrated how to derive upscaled RWU parameters and state variables (among which the upscaled soil–root interface water potential) from small scale information. Two simplified approaches aiming at estimating such upscaled water potential (when small-scale information is not available) were then tested in soil–plant hydrodynamics scenarios, for two crops with rather heterogeneous (maize) or homogeneous (winter wheat) horizontal rooting distributions.

With the first approach, SWP was considered as homogeneous in upscaled soil elements. For maize, neglecting SWP heterogeneities in the direction of the row was shown to be a good compromise between accuracy (relative errors mostly below 5 %) and computing time (reduced by 67–96 %).

However, in 1-D, the assumed horizontal water redistribution rate by soil was far above reference 3-D values during daytime and far below them at night. Consequently, the intensity of compensatory RWU was underestimated while plant collar water potential was overestimated. For winter wheat, the rather uniform rooting distribution tended to generate short-distance SWP heterogeneities, and favoured a fast horizontal redistribution of water by soil. Therefore, 1-D processes of water redistribution were in agreement with reference values (relative errors mostly below 5 %), and computation time could be reduced by 80–99 %. More generally, the accuracy of the first approach was improved when processes creating SWP heterogeneity were reduced (e.g. low plant transpiration rate) and processes dissipating SWP heterogeneity were dominant (e.g. high soil hydraulic conductivity). A conclusion of the first conjecture is that a 1-D soil geometry is enough to represent soil–plant water dynamics for winter wheat, but not for maize. Representing the latter case in 1-D would require accounting for water depletion around roots, which is the aim of the second conjecture.

With the second conjecture, the difference between bulk SWP and SWP sensed by roots in 1-D soil layers was estimated with an approximate analytical solution of soil water flow towards roots. The validity of the latter model, when two of its assumptions are not met (regular rooting distribution and constant RWU rate) was questioned. First, horizontal rooting heterogeneity was shown to impact effective values of the geometrical parameter ρ for maize, while a better agreement between theoretical and effective values of ρ were noticed for the rather regular rooting distribution of winter wheat. Second, accounting for past uptake rates over a time window of 36 h improved the agreement with reference results, whose local RWU rates were transient. However, for maize, the layers' soil–root interface water potentials could not be accurately predicted, especially in dry conditions.

This study confirmed that the use of 1-D spatial discretisation to represent soil–plant water dynamics is a worthy choice for densely seeded crops. It also highlighted that, for wide-row crops, further theoretical developments, better accounting for actual system properties, might be needed to properly predict plant collar water potential and compensatory RWU, as compared to fine-scale simulations.

Future prospects in line with this study could also focus on the analysis of implications of using even coarser grids when modelling soil–plant hydrodynamics at the plot or larger scales.

Acknowledgements. During the preparation of this manuscript, V. Couvreur was supported by the Fonds National de la Recherche Scientifique (FNRS) of Belgium as a research fellow, by the Belgian American Educational Foundation (BAEF), as UCLouvain fellow, and by the Wallonie-Bruxelles International (WBI) with a WBI.WORLD excellence grant. The authors thank these funding agencies for their financial support as well as Quirijn de Jong van Lier and an anonymous referee for their constructive comments. This work is a contribution of the Transregio Collaborative Research Center 32, Patterns in Soil-Vegetation-Atmosphere Systems: Monitoring, Modelling and Data Assimilation, which is funded by the German research association, DFG.

Edited by: I. Neuweiler

Appendix A

Definition of soil water flow divergence necessary to keep soil water potential homogeneous during root water uptake in upscaled soil elements

From an initially uniform distribution of SWP inside a horizontally upscaled soil element, taking up a flux "$S_{\mathrm{Up},g}.V_{\mathrm{Up},g}$" of water would generate SWP heterogeneity around roots if water was not redistributed. Leading SWP to a new homogeneous state inside the upscaled soil element instantly requires a horizontal divergence of soil water flow (mostly negative in regions where RWU occurs), which depends on the characteristic distribution of RWU inside the upscaled soil element.

When using upscaled soil elements, one indirectly assumes that the element is an entity keeping its inner water potential homogeneous, independently of other upscaled elements. In other words, the equilibration of inner SWP requires soil water redistribution, which is assumed to come from the inside of the upscaled element only. The divergence of soil capillary flow over the upscaled soil element is thus zero regarding the equilibration step, while divergences may locally be different from zero in its constituting elements. Note that when calculating soil water flow between different upscaled soil elements, their divergence of water flow may of course be different from zero.

The following forms of the Richards equation thus apply, respectively for upscaled and fine soil elements, regarding the instantaneous equilibration of upscaled elements inner SWP:

$$\frac{\partial \theta_{\mathrm{Up},g}}{\partial t} = -S_{\mathrm{Up},g}, \tag{A1}$$

$$\frac{\partial \theta_k}{\partial t} = -\mathrm{Div}_k - S_k, \tag{A2}$$

where Div_k ($\mathrm{L^3\,L^{-3}\,T^{-1}}$) is the divergence of soil water flow in the kth fine element, more commonly expressed as "$-\nabla \cdot \left[K\left(\psi_{\mathrm{m},k}\right)\nabla\psi_{\mathrm{s},k}\right]$".

In order to keep SWP horizontally homogeneous inside an upscaled element (and considering soil hydraulic properties as uniform), all local $\frac{\partial\theta_k}{\partial t}$ need to equal $\frac{\partial\theta_{\mathrm{Up},g}}{\partial t}$. From Eqs. (A1) and (A2) we thus obtain

$$\mathrm{Div}_k = S_{\mathrm{Up},g} - S_k. \tag{A3}$$

Considering initial SWP as homogeneous inside the upscaled soil element, local uptake rates can be defined as standard fractions of the total uptake rate of the upscaled element:

$$S_k.V_k = S_{\mathrm{Up},g}.V_{\mathrm{Up},g}.\frac{\mathrm{SSF}_k}{\mathrm{SSF}_{\mathrm{Up},g}}, \tag{A4}$$

where $\sum_{k=1}^{M} \varepsilon_{k,g}.\mathrm{SSF}_k = \mathrm{SSF}_{\mathrm{Up},g}$.

From Eqs. (A3) and (A4), the local divergence of soil water flow can be defined as follows:

$$\mathrm{Div}_k = S_{\mathrm{Up},g}.\frac{V_{\mathrm{Up},g}}{V_k}.\left(\frac{V_k}{V_{\mathrm{Up},g}} - \frac{\mathrm{SSF}_k}{\mathrm{SSF}_{\mathrm{Up},g}}\right). \tag{A5}$$

Since in our case, soil water flow divergence is simply a redistribution of water inside the upscaled element, the volumetric integration of positive terms equals that of negative terms, and half the volumetric integration of all absolute terms. We thus obtain the following definition of the volumetric integration of positive water flow divergence necessary to keep SWP uniform inside an upscaled soil element, $R_{\mathrm{soil}\leftrightarrow\mathrm{hyp},g}$ ($\mathrm{L}^3\,\mathrm{T}^{-1}$):

$$R_{\mathrm{soil}\leftrightarrow\mathrm{hyp},g} = \frac{\sum_{k=1}^{M} \varepsilon_{k,g}.V_k.\left|\mathrm{Div}_k\right|}{2}$$

$$= \left|S_{\mathrm{Up},g}\right|.V_{\mathrm{Up},g}.\frac{\sum_{k=1}^{M} \varepsilon_{k,g}.\left|\frac{\mathrm{SSF}_k}{\mathrm{SSF}_{\mathrm{Up},g}} - \frac{V_k}{V_{\mathrm{Up},g}}\right|}{2}. \tag{A6}$$

Note that soil water flow divergence at scales lower than the scale of fine elements is not considered in the latter equation.

The coefficient $\frac{\sum_{k=1}^{M} \varepsilon_{k,g}.\left|\frac{\mathrm{SSF}_k}{\mathrm{SSF}_{\mathrm{Up},g}} - \frac{V_k}{V_{\mathrm{Up},g}}\right|}{2}$ appears to be an indicator of how "generator of SWP heterogeneity" a HA is, inside an upscaled soil element (which could be enlarged up to the whole soil domain). Its value tends to zero for uniform standard sink distributions inside the upscaled element, which do not create SWP heterogeneities, and tends to one for a single root inside an infinitesimal part of the upscaled element, which corresponds to the case generating the biggest amount of heterogeneity for a given water uptake or exudation rate.

Appendix B

Theoretical equation for the geometrical parameter ρ_g for regular root distribution in a soil layer

De Jong Van Lier et al. (2006) provides the following theoretical equation for the geometrical parameterρ_g for regular root distribution in a soil layer:

$$\rho_g = \frac{4}{r_{0,g}^2 - \frac{a^2}{\pi.\mathrm{RLD}_g} + 2.\left(\frac{1}{\pi.\mathrm{RLD}_g} + r_{0,g}^2\right).\ln\left(\frac{a}{r_{0,g}.\sqrt{\pi.\mathrm{RLD}_g}}\right)} \tag{B1}$$

where $r_{0,g}$ (L) is the mean roots radius at the gth depth, a (-) is a parameter considered as equal to 0.53 (De Jong Van Lier et al., 2006), and RLD_g (L^{-2}) is the root length density at the gth depth.

Appendix C

Equations for vertical and horizontal water redistribution rates by soil and roots

Vertical and horizontal water redistribution rates by soil were calculated as the volumetric integration of the corresponding absolute components of water flow divergence between soil elements:

$$R_{\mathrm{soil}\leftrightarrow} = \frac{1}{2}.\sum_{k=1}^{M}\left|\left(J_{\mathrm{x}2,k} - J_{\mathrm{x}1,k}\right).d_{\mathrm{y}}.d_{\mathrm{z}} + \left(J_{\mathrm{y}2,k} - J_{\mathrm{y}1,k}\right).d_{\mathrm{x}}.d_{\mathrm{z}}\right| \tag{C1}$$

$$R_{\mathrm{soil}\updownarrow} = \frac{1}{2}.\sum_{k=1}^{M}\left|\left(J_{\mathrm{z}2,k} - J_{\mathrm{z}1,k}\right).d_{\mathrm{x}}.d_{\mathrm{y}}\right| \tag{C2}$$

where $R_{\mathrm{soil}\leftrightarrow}$ and $R_{\mathrm{soil}\updownarrow}$ ($\mathrm{L}^3\,\mathrm{T}^{-1}$) are, respectively, the horizontal and vertical components of water redistribution rates by soil, $J_{\mathrm{x}1,k}$ and $J_{\mathrm{x}2,k}$ ($\mathrm{L\,T}^{-1}$) are soil water flow densities in the x direction, respectively on the first and second side of soil element # k, and d_{x} (L) is the length of soil elements in the x direction (same logic for y and z directions).

Even though RWU rates have no direction per se, water redistribution between layers was considered vertical while redistribution resulting from horizontal heterogeneities was considered horizontal.

Vertical water redistribution rates by roots were calculated as the integration of absolute net compensatory RWU of each soil layer:

$$R_{\text{root}\updownarrow} = \frac{1}{2} \cdot \sum_{l=1}^{L} \left| \sum_{k=1}^{M} \varepsilon_{k,l} . \beta_k \right| \tag{C3}$$

where $R_{\text{root}\updownarrow}$ (L^3 T^{-1}) is the vertical water redistribution rate by roots, $\beta_k = S_k.V_k - \text{SSF}_k.T_{\text{act}}$ (L^3 T^{-1}) is the compensatory RWU in the kth soil element, l (-) is the soil layer index, L is the total number of soil layers, and $\varepsilon_{k,l}$ (-) equals 1 when the kth soil element is included in the lth soil layer and equals 0 otherwise.

Horizontal water redistribution rates by roots were calculated as the integration of absolute deviations of compensatory RWU as compared to the expected distribution of layers net compensatory RWU for horizontally uniform SWP:

$$R_{\text{root}\leftrightarrow} = \sum_{l=1}^{L} \frac{1}{2} \cdot \left(\sum_{k=1}^{M} \varepsilon_{k,l} . \left| \beta_k - \frac{SSF_k}{\sum_{k=1}^{M} \varepsilon_{k,l}.\text{SSF}_k} . \left(\sum_{k=1}^{M} \varepsilon_{k,l} . \beta_k \right) \right| \right) \tag{C4}$$

where $R_{\text{root}\leftrightarrow}$ (L^3 T^{-1}) is the horizontal water redistribution rate by roots, $\sum_{k=1}^{M} \varepsilon_{k,l}.\beta_k$ (L^3 T^{-1}) is the net compensatory RWU in the lth soil layer, $\frac{\text{SSF}_k}{\sum_{k=1}^{M} \varepsilon_{k,l}.\text{SSF}_k}$ (-) is the fraction of net compensatory RWU expected in the kth soil element in case SWP would be horizontally uniform in the lth soil layer.

References

Allen, R. G., Pereira, L. S., Raes, D., and Smith, M.: Crop evapotranspiration – Guidelines for computing crop water requirements, FAO Irrigation and drainage, Paper 56, Rome, 1998.

Beff, L., Günther, T., Vandoorne, B., Couvreur, V., and Javaux, M.: Three-dimensional monitoring of soil water content in a maize field using Electrical Resistivity Tomography, Hydrol. Earth Syst. Sci., 17, 595–609, doi:10.5194/hess-17-595-2013, 2013.

Bramley, H., Turner, N. C., Turner, D. W., and Tyerman, S. D.: Comparison between gradient-dependent hydraulic conductivities of roots using the root pressure probe: the role of pressure propagations and implications for the relative roles of parallel radial pathways, Plant Cell Environ., 30, 861–874, 2007.

Bramley, H., Turner, N. C., Turner, D. W., and Tyerman, S. D.: Roles of morphology, anatomy, and aquaporins in determining contrasting hydraulic behavior of roots, Plant Physiol., 150, 348–364, 2009.

Carsel, R. F. and Parrish, R. S.: Developping joint probability-distributions of soil-water retention characteristics, Water Resour. Res., 24, 195–200, 1988.

Couvreur, V., Vanderborght, J., and Javaux, M.: A simple three-dimensional macroscopic root water uptake model based on the hydraulic architecture approach, Hydrol. Earth Syst. Sci., 16, 2957–2971, doi:10.5194/hess-16-2957-2012, 2012.

De Jong Van Lier, Q., Metselaar, K., and Van Dam, J. C.: Root water extraction and limiting soil hydraulic conditions estimated by numerical simulation, Vadose Zone J., 5, 1264–1277, 2006.

De Jong Van Lier, Q., Van Dam, J. C., Metselaar, K., De Jong, R., and Duijnisveld, W. H. M.: Macroscopic root water uptake distribution using a matric flux potential approach, Vadose Zone J., 7, 1065–1078, 2008.

Domec, J. C. and Pruyn, M. L.: Bole girdling affects metabolic properties and root, trunk and branch hydraulics of young ponderosa pine trees, Tree Physiol., 28, 1493–1504, 2008.

Doussan, C., Pages, L., and Vercambre, G.: Modelling of the hydraulic architecture of root systems: An integrated approach to water absorption – Model description, Ann. Bot.-London, 81, 213–223, 1998.

Doussan, C., Pierret, A., Garrigues, E., and Pages, L.: Water uptake by plant roots: II – Modelling of water transfer in the soil root-system with explicit account of flow within the root system – Comparison with experiments, Plant. Soil, 283, 99–117, 2006.

Draye, X., Kim, Y., Lobet, G., and Javaux, M.: Model-assisted integration of physiological and environmental constraints affecting the dynamic and spatial patterns of root water uptake from soils, J. Exp. Bot., 61, 2145–2155, 2010.

Durigon, A., dos Santos, M. A., van Lier, Q. D., and Metselaar, K.: Pressure Heads and Simulated Water Uptake Patterns for a Severely Stressed Bean Crop, Vadose Zone J., 11, 14 pp., doi:10.2136/vzj2011.0187, 2012.

Faria, L. N., Da Rocha, M. G., Van Lier, Q. D., and Casaroli, D.: A split-pot experiment with sorghum to test a root water uptake partitioning model, Plant. Soil, 331, 299–311, 2010.

Feddes, R. A., Kowalik, P., Kolinska-Malinka, K., and Zaradny, H.: Simulation of field water uptake by plants using a soil water dependent root extraction function, J. Hydrol., 31, 13–26, 1976.

Feddes, R. A., Hoff, H., Bruen, M., Dawson, T., De Rosnay, P., Dirmeyer, P., Jackson, R. B., Kabat, P., Kleidon, A., Lilly, A., and Pitman, A. J.: Modeling root water uptake in hydrological and climate models, B. Am. Meteorol. Soc., 82, 2797–2809, 2001.

Frensch, J. and Steudle, E.: Axial and radial hydraulic resistance to roots of maize (Zea-mays-L), Plant Physiol., 91, 719–726, 1989.

Gardner, W. R.: Dynamic aspects of water availability to plants, Soil. Sci., 89, 63–73, 1960.

Gardner, W. R. and Ehlig, C. F.: The influence of soil water on transpiration by plants, J. Geophys. Res., 68, 5719–5724, 1963.

Hupet, F., Lambot, S., Javaux, M., and Vanclooster, M.: On the identification of macroscopic root water uptake parameters from soil water content observations, Water Resour. Res., 38, 1–14, 2002.

Jarvis, N. J.: Simple physics-based models of compensatory plant water uptake: concepts and eco-hydrological consequences, Hydrol. Earth Syst. Sci., 15, 3431–3446, doi:10.5194/hess-15-3431-2011, 2011.

Javaux, M., Schroder, T., Vanderborght, J., and Vereecken, H.: Use of a three-dimensional detailed modeling approach for predicting root water uptake, Vadose Zone J., 7, 1079–1088, 2008.

Javaux, M., Couvreur, V., Vanderborght, J., and Vereecken, H.: Root Water Uptake: From 3D Biophysical Processes to Macroscopic Modeling Approaches, Vadose Zone J., 16 pp., doi:10.2136/vzj2013.02.0042, 2013.

Jolliet, O. and Bailey, B. J.: The effect of climate on tomato transpiration in greenhouses: measurements and models comparison, Agr. Forest Meteorol., 58, 43–62, 1992.

Manabe, S.: Climate and ocean circulation .I. Atmospheric circulation and hydrology of earth surface, Mon. Weather Rev., 97, 739–774, 1969.

Pages, L., Vercambre, G., Drouet, J. L., Lecompte, F., Collet, C., and Le Bot, J.: Root Typ: a generic model to depict and analyse the root system architecture, Plant Soil, 258, 103–119, 2004.

Raats, P. A. C.: Uptake of water from soils by plant roots, Transport Porous Med., 68, 5–28, 2007.

Sanderson, J., Whitbread, F. C., and Clarkson, D. T.: Persistent Xylem Cross-Walls Reduce The Axial Hydraulic Conductivity In The Apical 20 Cm Of Barley Seminal Root Axes – Implications For The Driving Force For Water-Movement, Plant Cell Environ., 11, 247–256, 1988.

Schneider, C. L., Attinger, S., Delfs, J.-O., and Hildebrandt, A.: Implementing small scale processes at the soil-plant interface – the role of root architectures for calculating root water uptake profiles, Hydrol. Earth Syst. Sci., 14, 279–289, doi:10.5194/hess-14-279-2010, 2010.

Schroeder, T., Javaux, M., Vanderborght, J., and Vereecken, H.: Comment on "Root water extraction and limiting soil hydraulic conditions estimated by numerical simulation", Vadose Zone J., 6, 524–526, 2007.

Schroeder, T., Javaux, M., Vanderborght, J., Korfgen, B., and Vereecken, H.: Implementation of a Microscopic Soil-Root Hydraulic Conductivity Drop Function in a Three-Dimensional Soil-Root Architecture Water Transfer Model, Vadose Zone J., 8, 783–792, 2009a.

Schroeder, T., Tang, L., Javaux, M., Vanderborght, J., Körfgen, B., and Vereecken, H.: A grid refinement approach for a three-dimensional soil-root water transfer model, Water Resour. Res., 45, W10412, doi:10.1029/2009WR007873, 2009b.

Sperling, O., Shapira, O., Cohen, S., Tripler, E., Schwartz, A., and Lazarovitch, N.: Estimating sap flux densities in date palm trees using the heat dissipation method and weighing lysimeters, Tree Physiol., 32, 1171–1178, 2012.

Tardieu, F. and Simonneau, T.: Variability among species of stomatal control under fluctuating soil water status and evaporative demand: modelling isohydric and anisohydric behaviours, J. Exp. Bot., 49, 419-432, doi:10.1093/jexbot/49.suppl_1.419, 1998.

Tazawa, M., Ohkuma, E., Shibasaka, M., and Nakashima, S.: Mercurial-sensitive water transport in barley roots, J. Plant Res., 110, 435–442, 1997.

Teuling, A. J., Uijlenhoet, R., Hupet, F., and Troch, P. A.: Impact of plant water uptake strategy on soil moisture and evapotranspiration dynamics during drydown, Geophys. Res. Lett., 33, L03401, doi:10.1029/2005GL025019, 2006.

Van Genuchten, M. T.: A closed form equation for predicting the hydraulic conductivity of unsaturated soils, Soil Sci. Soc. Am. J., 44, 892–898, 1980.

Van Noordwijk, M. and De Willigen, P.: Agricultural concepts of roots: From morphogenetic to functional equilibrium between root and shoot growth, Neth. J. Agric. Sci., 35, 487–496, 1987.

Watt, M., Magee, L. J., and McCully, M. E.: Types, structure and potential for axial water flow in the deepest roots of field-grown cereals, New Phytol., 178, 135–146, 2008.

Weaver, J. E., Kramer, J., and Reed, M.: Development of root and shoot of winter wheat under field environment, Ecology, 5, 26–50, 1924.

7

Evaluation of statistical methods for quantifying fractal scaling in water-quality time series with irregular sampling

Qian Zhang[1], **Ciaran J. Harman**[2], **and James W. Kirchner**[3,4,5]

[1]University of Maryland Center for Environmental Science, US Environmental Protection Agency Chesapeake Bay Program Office, 410 Severn Avenue, Suite 112, Annapolis, Maryland 21403, USA

[2]Department of Environmental Health and Engineering, Johns Hopkins University, 3400 North Charles Street, Baltimore, Maryland 21218, USA

[3]Department of Environmental System Sciences, ETH Zurich, Universitätstrasse 16, 8092 Zurich, Switzerland

[4]Swiss Federal Research Institute WSL, Zürcherstrasse 111, 8903 Birmensdorf, Switzerland

[5]Department of Earth and Planetary Science, University of California, Berkeley, Berkeley, California 94720, USA

Correspondence: Qian Zhang (qzhang@chesapeakebay.net)

Abstract. River water-quality time series often exhibit fractal scaling, which here refers to autocorrelation that decays as a power law over some range of scales. Fractal scaling presents challenges to the identification of deterministic trends because (1) fractal scaling has the potential to lead to false inference about the statistical significance of trends and (2) the abundance of irregularly spaced data in water-quality monitoring networks complicates efforts to quantify fractal scaling. Traditional methods for estimating fractal scaling – in the form of spectral slope (β) or other equivalent scaling parameters (e.g., Hurst exponent) – are generally inapplicable to irregularly sampled data. Here we consider two types of estimation approaches for irregularly sampled data and evaluate their performance using synthetic time series. These time series were generated such that (1) they exhibit a wide range of prescribed fractal scaling behaviors, ranging from white noise ($\beta = 0$) to Brown noise ($\beta = 2$) and (2) their sampling gap intervals mimic the sampling irregularity (as quantified by both the skewness and mean of gap-interval lengths) in real water-quality data. The results suggest that none of the existing methods fully account for the effects of sampling irregularity on β estimation. First, the results illustrate the danger of using interpolation for gap filling when examining autocorrelation, as the interpolation methods consistently underestimate or overestimate β under a wide range of prescribed β values and gap distributions. Second, the widely used Lomb–Scargle spectral method also consistently underestimates β. A previously published modified form, using only the lowest 5 % of the frequencies for spectral slope estimation, has very poor precision, although the overall bias is small. Third, a recent wavelet-based method, coupled with an aliasing filter, generally has the smallest bias and root-mean-squared error among all methods for a wide range of prescribed β values and gap distributions. The aliasing method, however, does not itself account for sampling irregularity, and this introduces some bias in the result. Nonetheless, the wavelet method is recommended for estimating β in irregular time series until improved methods are developed. Finally, all methods' performances depend strongly on the sampling irregularity, highlighting that the accuracy and precision of each method are data specific. Accurately quantifying the strength of fractal scaling in irregular water-quality time series remains an unresolved challenge for the hydrologic community and for other disciplines that must grapple with irregular sampling.

1 Introduction

1.1 Autocorrelations in time series

It is well known that time series from natural systems often exhibit autocorrelation; that is, observations at each time step are correlated with observations one or more time steps in the past. This property is usually characterized by the autocorrelation function (ACF), which is defined as follows for a process X_t at lag k:

$$\gamma(k) = \mathrm{cov}(X_t, X_{t+k}). \quad (1)$$

In practice, autocorrelation has been frequently modeled with classical techniques such as autoregressive (AR) or autoregressive moving-average (ARMA) models (Darken et al., 2002; Yue et al., 2002; Box et al., 2008). These models assume that the underlying process has short-term memory; i.e., the ACF decays exponentially with lag k (Box et al., 2008).

Although the short-term memory assumption holds sometimes, it cannot adequately describe many time series whose ACFs decay as a power law (thus much slower than exponentially) and may not reach zero even for large lags, which implies that the ACF is non-summable. This property is commonly referred to as long-term memory or fractal scaling, as opposed to short-term memory (Beran, 2010).

Fractal scaling has been increasingly recognized in studies of hydrological time series, particularly for the common task of trend identification. Such hydrological series include river flows (Montanari et al., 2000; Khaliq et al., 2008, 2009; Ehsanzadeh and Adamowski, 2010), air and sea temperatures (Fatichi et al., 2009; Lennartz and Bunde, 2009; Franzke, 2012a, b), conservative tracers (Kirchner et al., 2000, 2001; Godsey et al., 2010), and non-conservative chemical constituents (Kirchner and Neal, 2013; Aubert et al., 2014). Because for fractal scaling processes the variance of the sample mean converges to zero much slower than the rate of n^{-1} (n: sample size), the fractal scaling property must be taken into account to avoid false positives (Type I errors) when inferring the statistical significance of trends (Cohn and Lins, 2005; Fatichi et al., 2009; Ehsanzadeh and Adamowski, 2010; Franzke, 2012a). Unfortunately, as stressed by Cohn and Lins (2005), it is "surprising that nearly every assessment of trend significance in geophysical variables published during the past few decades has failed [to do so]", and a similar tendency is evident in the decade following that statement as well.

1.2 Overview of approaches for quantification of fractal scaling

Several equivalent metrics can be used to quantify fractal scaling. Here we provide a review of the definitions of such processes and several typical modeling approaches, including both time-domain and frequency-domain techniques, with special attention to their reconciliation. For a more comprehensive review, readers are referred to Beran et al. (2013), Boutahar et al. (2007), and Witt and Malamud (2013).

Strictly speaking, X_t is called a stationary long-memory process if the condition

$$\lim_{k\to\infty} k^{\alpha}\gamma(k) = C_1 > 0, \quad (2)$$

where C_1 is a constant and is satisfied by some $\alpha \in (0, 1)$ (Boutahar et al., 2007; Beran et al., 2013). Equivalently, X_t is a long-memory process if, in the spectral domain, the condition

$$\lim_{\omega\to 0} |\omega|^{\beta} f(\omega) = C_2 > 0 \quad (3)$$

is satisfied by some $\beta \in (0, 1)$, where C_2 is a constant and $f(\omega)$ is the spectral density function of X_t, which is related to ACF as follows (which is also known as the Wiener–Khinchin theorem):

$$f(\omega) = \frac{1}{2\pi}\sum_{k=-\infty}^{\infty} \gamma(k) e^{-ik\omega}, \quad (4)$$

where ω is angular frequency (Boutahar et al., 2007).

One popular model for describing long-memory processes is the so-called fractional autoregressive integrated moving-average model, or ARFIMA (p, q, d), which is an extension of ARMA models and is defined as follows:

$$(1 - B)^d \varphi(B) X_t = \psi(B)\varepsilon_t, \quad (5)$$

where ε_t is a series of independent, identically distributed Gaussian random numbers ($0, \sigma_\varepsilon^2$), B is the backshift operator (i.e., $BX_t = X_{t-1}$), and functions $\varphi(\bullet)$ and $\psi(\bullet)$ are polynomials of order p and q, respectively. The fractional differencing parameter d is related to the parameter α in Eq. (2) as follows:

$$d = \frac{1-\alpha}{2} \in (-0.5, 0.5) \quad (6)$$

(Beran et al., 2013; Witt and Malamud, 2013).

In addition to a slowly decaying ACF, a long-memory process manifests itself in two other equivalent fashions. One is the so-called Hurst effect, which states that, on a log–log scale, the range of variability of a process changes linearly with the length of the time period under consideration. This power-law slope is often referred to as the Hurst exponent or Hurst coefficient H (Hurst, 1951), which is related to d as follows:

$$H = d + 0.5 \quad (7)$$

(Beran et al., 2013; Witt and Malamud, 2013).

The second equivalent description of long-memory processes, this time from a frequency-domain perspective, is fractal scaling, which describes a power-law decrease in spectral power with increasing frequency, yielding power spectra that are linear on log–log axes (Lomb, 1976; Scargle, 1982; Kirchner, 2005). Mathematically, this inverse proportionality can be expressed as

$$f(\omega) = C_3|\omega|^{-\beta}, \quad (8)$$

where C_3 is a constant and the scaling exponent β is termed the spectral slope. In particular, for spectral slopes of zero,

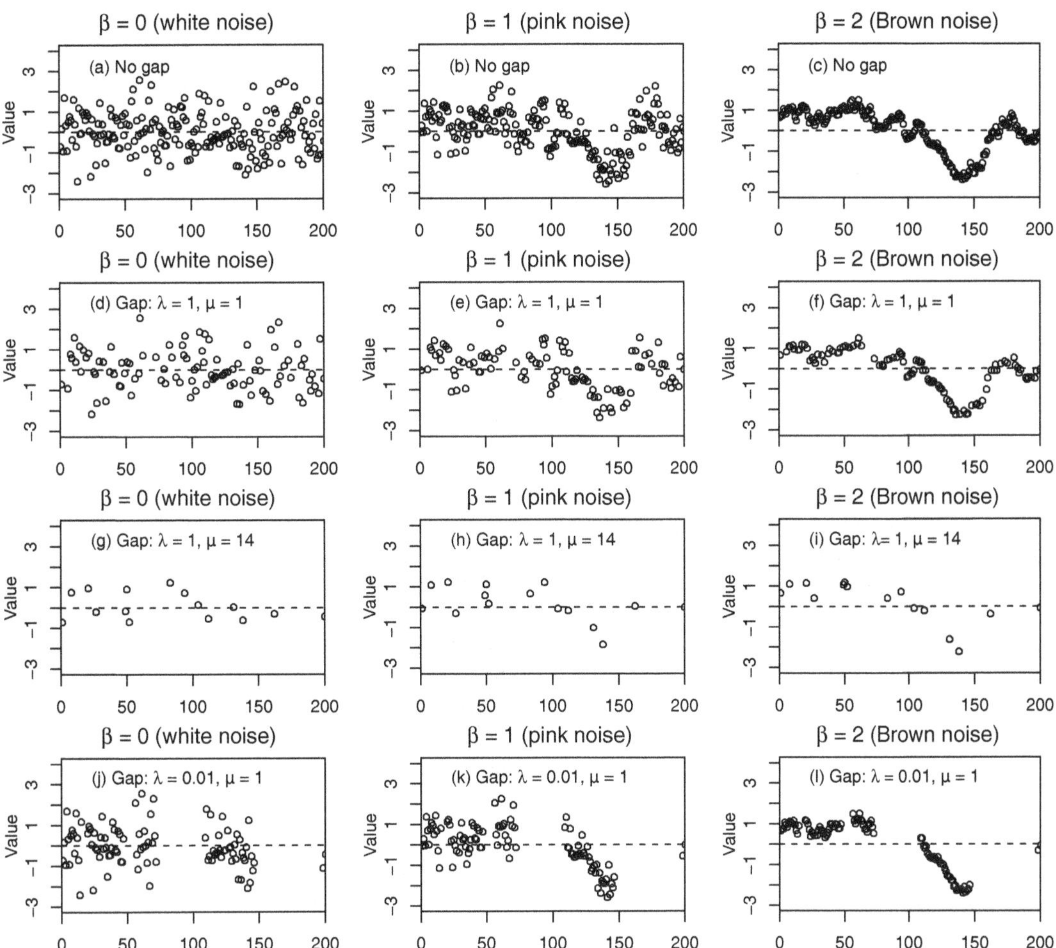

Figure 1. Synthetic time series with 200 time steps for three representative fractal scaling processes that correspond to white noise ($\beta = 0$), pink noise ($\beta = 1$), and Brown noise ($\beta = 2$). **(a–c)** show the simulated time series without any gap. **(d–l)** show the same time series as in **(a–c)** but with data gaps that were simulated using three different negative binomial (NB) distributions – that is, **(d–f)**: NB($\lambda = 1$, $\mu = 1$); **(g–i)**: NB($\lambda = 1$, $\mu = 14$); **(j–l)**: NB($\lambda = 0.01$, $\mu = 1$).

one, and two, the underlying processes are termed as "white", "pink" (or "flicker"), and "Brown" (or "red") noises, respectively (Witt and Malamud, 2013). Illustrative examples of these three noises are shown in Fig. 1a–c.

In addition, it can be shown that the spectral density function for ARFIMA (p, d, q) is

$$f(\omega) = \frac{\sigma_\varepsilon^2}{2\pi} \frac{\left|\psi(e^{-i\omega})\right|^2}{\left|\varphi(e^{-i\omega})\right|^2} \left|1 - e^{-i\omega}\right|^{-2d} \tag{9}$$

for $-\pi < \omega < \pi$ (Boutahar et al., 2007; Beran et al., 2013). For $|\omega| \ll 1$, Eq. (9) can be approximated by

$$f(\omega) = C_4 |\omega|^{-2d} \tag{10}$$

with

$$C_4 = \frac{\sigma_\varepsilon^2}{2\pi} \frac{|\psi(1)|^2}{|\varphi(1)|^2}. \tag{11}$$

Equation (10) thus exhibits the asymptotic behavior required for a long-memory process given by Eq. (3). In addition, a comparison of Eqs. (10) and (8) reveals that

$$\beta = 2d. \tag{12}$$

Overall, these derivations indicate that these different types of scaling parameters (i.e., α, d, and H and β) can be used equivalently to describe the strength of fractal scaling. Specifically, their equivalency can be summarized as follows:

$$\beta = 2d = 1 - \alpha = 2H - 1. \tag{13}$$

It should be noted, however, that the parameters d, α, and H are only applicable over a fixed range of fractal scaling, which is equivalent to $(-1, 1)$ in terms of β.

1.3 Motivation and objective of this work

To account for fractal scaling in trend analysis, one must be able to first quantify the strength of fractal scaling for a given time series. Numerous estimation methods have been developed for this purpose, including the Hurst rescaled range analysis, Higuchi's method, Geweke and Porter-Hudak's method, Whittle's maximum likelihood estimator, detrended fluctuation analysis, and others (Taqqu et al., 1995; Montanari et al., 1997, 1999; Rea et al., 2009; Stroe-Kunold et al., 2009). For brevity, these methods are not elaborated here; readers are referred to Beran (2010) and Witt and Malamud (2013) for details. While these estimation methods have been extensively adopted, they are unfortunately only applicable to regular (i.e., evenly spaced) data, e.g., daily streamflow discharge, monthly temperature. In practice, many types of hydrological data, including river water-quality data, are often sampled irregularly or have missing values, and hence their strengths of fractal scaling cannot be readily estimated with the above traditional estimation methods.

Thus, estimation of fractal scaling in irregularly sampled data is an important challenge for hydrologists and practitioners. Many data analysts may be tempted to interpolate the time series to make it regular and hence analyzable (Graham, 2009). Although technically convenient, interpolation can be problematic if it distorts the series' autocorrelation structure (Kirchner and Weil, 1998). In this regard, it is important to evaluate various types of interpolation methods using carefully designed benchmark tests and to identify the scenarios under which the interpolated data can yield reliable (or, alternatively, biased) estimates of spectral slope.

Moreover, quantification of fractal scaling in real-world water-quality data is subject to several common complexities. First, water-quality data are rarely normally distributed; instead, they are typically characterized by log-normal or other skewed distributions (Hirsch et al., 1991; Helsel and Hirsch, 2002), with potential consequences for β estimation. Moreover, water-quality data also tend to exhibit long-term trends, seasonality, and flow dependence (Hirsch et al., 1991; Helsel and Hirsch, 2002), which can also affect the accuracy of β estimates. Thus, it may be more plausible to quantify β in transformed time series after accounting for the seasonal patterns and discharge-driven variations in the original time series, which is the approach taken in this paper. For the trend aspect, however, it remains a puzzle whether the data set should be detrended before conducting β estimation. Such detrending treatment can certainly affect the estimated value of β and hence the validity of (or confidence in) any inference made regarding the statistical significance of temporal trends in the time series. This somewhat circular issue is beyond the scope of our current work – it has been previously discussed in the context of short-term memory (Zetterqvist, 1991; Darken et al., 2002; Yue et al., 2002; Noguchi et al., 2011; Clarke, 2013; Sang et al., 2014), but it is not well understood in the context of fractal scaling (or long-term memory) and hence presents an important area for future research.

In the above context, the main objective of this work was to use Monte Carlo simulation to systematically evaluate and compare two broad types of approaches for estimating the strength of fractal scaling (i.e., spectral slope β) in irregularly sampled river water-quality time series. Specific aims of this work include the following:

1. to examine the sampling irregularity of typical river water-quality monitoring data and to simulate time series that contain such irregularity, and
2. to evaluate two broad types of approaches for estimating β in simulated irregularly sampled time series.

The first type of approach includes several forms of interpolation techniques for gap filling, thus making the data regular and analyzable by traditional estimation methods. The second type of approach includes the well-known Lomb–Scargle periodogram (Lomb, 1976; Scargle, 1982) and a recently developed wavelet method combined with a spectral aliasing filter (Kirchner and Neal, 2013). The latter two methods can be directly applied to irregularly spaced data; here we aim to compare them with the interpolation techniques. Details of these various approaches are provided in Sect. 3.1.

This work was designed to make several specific contributions. First, it uses benchmark tests to quantify the performance of a wide range of methods for estimating fractal scaling in irregularly sampled water-quality data. Second, it proposes an innovative and general approach for modeling sampling irregularity in water-quality records. Third, while this work was not intended to compare all published estimation methods for fractal scaling, it does provide and demonstrate a generalizable framework for data simulation (with gaps) and β estimation, which can be readily applied toward the evaluation of other methods that are not covered here. Last but not least, while this work was intended to help hydrologists and practitioners understand the performance of various approaches for water-quality time series, the findings and approaches may be broadly applicable to irregularly sampled data in other scientific disciplines.

The rest of the paper is organized as follows. We propose a general approach for modeling sampling irregularity in typical river water-quality data and discuss our approach for simulating irregularly sampled data (Sect. 2). We then introduce various methods for estimating fractal scaling in irregular time series and compare their estimation performance (Sect. 3). We close with a discussion of the results and implications (Sect. 4).

2 Quantification of sampling irregularity in river water-quality data

2.1 Modeling of sampling irregularity

River water-quality data are often sampled irregularly. In some cases, samples are taken more frequently during particular periods of interest, such as high flows or drought periods; here we will address the implications of the irregularity, but not the (intentional) bias, inherent in such a sampling strategy. In other cases, the sampling is planned with a fixed sampling interval (e.g., 1 day) but samples are missed (or lost, or fail quality-control checks) at some time steps during implementation. In still other cases, the sampling is intrinsically irregular because, for example, one cannot measure the chemistry of rainfall on rainless days or the chemistry of a stream that has dried up. Theoretically, any deviation from fixed-interval sampling can affect the subsequent analysis of the time series.

To quantify sampling irregularity, we propose a simple and general approach that can be applied to any time series of monitoring data. Specifically, for a given time series with N points, the time intervals between adjacent samples are calculated; these intervals themselves make up a time series of N-1 points that we call Δt. For this time series, the following parameters are calculated to quantify its sampling irregularity:

- $L =$ the length of the period of record;
- $N =$ the number of samples in the record;
- $\Delta t_{\text{nominal}} =$ the nominal sampling interval under regular sampling (e.g., $\Delta t_{\text{nominal}} = 1$ day for daily samples);
- $\Delta t^* = \Delta t \,/\, \Delta t_{\text{nominal}}$, the sample intervals non-dimensionalized by the nominal sampling interval;
- $\Delta t_{\text{average}} = L \,/\, (N - 1)$ the average of all the entries in Δt.

The quantification is illustrated with two simple examples. The first example contains data sampled every hour from 01:00 to 11:00 UTC on 1 day. In this case, $L = 10$ h, $N = 11$ samples, $\Delta t = \{1, 1, 1, 1, 1, 1, 1, 1, 1, 1\}$ h, and $\Delta t_{\text{nominal}} = \Delta t_{\text{average}} = 1$ h. The second example contains data sampled at 01:00, 03:00, 04:00, 08:00, and 11:00. In this case, $L = 10$ h, $N = 5$ samples, $\Delta t = \{2, 1, 4, 3\}$ h, $\Delta t_{\text{nominal}} = 1$ h, and $\Delta t_{\text{average}} = 2.5$ h. It is readily evident that the first case corresponds to fixed-interval (regular) sampling that has the property of $\Delta t_{\text{average}} \,/\, \Delta t_{\text{nominal}} = 1$ (dimensionless), whereas the second case corresponds to irregular sampling for which $\Delta t_{\text{average}} \,/\, \Delta t_{\text{nominal}} > 1$.

The dimensionless set Δt^* contains essential information for determining sampling irregularity. This set is modeled as independent, identically distributed values drawn from a negative binomial (NB) distribution. This distribution has two dimensionless parameters, the shape parameter (λ) and the mean parameter (μ), which collectively represent the irregularity of the samples. The NB distribution is a flexible distribution that provides a discrete analogue of a gamma distribution. The geometric distribution, itself the discrete analogue of the exponential distribution, is a special case of the NB distribution when $\lambda = 1$.

The parameters μ and λ represent different aspects of sampling irregularity, as illustrated by the examples shown in Fig. 2. The mean parameter μ represents the fractional increase in the average interval between samples due to gaps: $\mu = \text{mean}(\Delta t^*) - 1 = (\Delta t_{\text{average}} - \Delta t_{\text{nominal}}) \,/\, \Delta t_{\text{nominal}}$. Thus, the special case of $\mu = 0$ corresponds to regular sampling (i.e., $\Delta t_{\text{average}} = \Delta t_{\text{nominal}}$), whereas any larger value of μ corresponds to irregular sampling (i.e., $\Delta t_{\text{average}} > \Delta t_{\text{nominal}}$) (Fig. 2c). The shape parameter λ characterizes the similarity of gaps to each other; that is, a small λ indicates that the samples contain gaps of widely varying lengths, whereas a large λ indicates that the samples contain many gaps of similar lengths (Fig. 2a, b).

To visually illustrate these gap distributions, representative samples of irregular time series are presented in Fig. 1 for the three special processes described above (Sect. 1.2), i.e., white noise, pink noise, and Brown noise. Specifically, three different gap distributions, namely, NB($\lambda = 1$, $\mu = 1$), NB($\lambda = 1$, $\mu = 14$), and NB($\lambda = 0.01$, $\mu = 1$), were simulated and each was applied to convert the three original (regular) time series (Fig. 1a–c) to irregular time series (Fig. 1d–l1). These simulations clearly illustrate the effects of the two parameters λ and μ. In particular, compared with NB($\lambda = 1$, $\mu = 1$), NB($\lambda = 1$, $\mu = 14$) shows a similar level of sampling irregularity (same λ) but a much longer average gap interval (larger μ). Again compared with NB($\lambda = 1$, $\mu = 1$), NB($\lambda = 0.01$, $\mu = 1$) shows the same average interval (same μ) but a much more irregular (skewed) gap distribution that contains a few very large gaps (smaller λ).

2.2 Examination of sampling irregularity in real river water-quality data

The above modeling approach was applied to real water-quality data from two large river monitoring networks in the United States to examine sampling irregularity. One such network is the Chesapeake Bay River Input Monitoring Program, which typically samples streams roughly once or twice monthly, accompanied with additional sampling during storm flows (Langland et al., 2012; Zhang et al., 2015). These data were obtained from the US Geological Survey National Water Information System (http://doi.org/10.5066/F7P55KJN). The other network is the Lake Erie and Ohio Tributary Monitoring Program, which typically samples streams at a daily resolution (National Center for Water Quality Research, 2015). For each site, we determined the NB parameters to quantify sampling irregularity. The mean parameter μ can be estimated as described above, and the shape parameter λ can be calculated directly from the mean and variance of Δt^* as follows: $\lambda = \mu^2 \,/\, [\text{var}(\Delta t^*) - \mu] = (\text{mean}(\Delta t^*) - 1)^2 \,/\, [\text{var}(\Delta t^*) - \text{mean}(\Delta t^*) + 1]$. Alternatively, a maximum likelihood approach can be used, which employs the *fitdist* function in the *fitdistrplus* R package (Delignette-Muller and Dutang, 2015). In general, the two approaches produce similar results, which are summarized in

Table 1. Quantification of sampling irregularity for selected water-quality constituents at nine sites of the Chesapeake Bay River Input Monitoring Program and six sites of the Lake Erie and Ohio Tributary Monitoring Program. (λ: shape parameter estimated using maximum likelihood; λ': shape parameter estimated using the direct approach (see Sect. 2.2); μ: mean parameter; $\Delta t_{average}$: average gap interval; N: total number of samples.)

I. Chesapeake Bay River Input Monitoring program

Site ID	River and station name	Drainage area (km^2)	Total nitrogen (TN)					Total phosphorus (TP)				
			λ	λ'	μ	$\Delta t_{average}$ (days)	N	λ	λ'	μ	$\Delta t_{average}$ (days)	N
01578310	Susquehanna River at Conowingo, MD	70 189	0.8	1.1	13.5	14.5	876	0.8	1.0	13.4	14.4	881
01646580	Potomac River at Chain Bridge, Washington D.C.	30 044	0.9	0.6	9.5	10.5	1385	1.1	1.0	24.4	25.4	579
02035000	James River at Cartersville, VA	16 213	0.8	1.0	13.9	14.9	960	0.8	1.1	13.7	14.7	974
01668000	Rappahannock River near Fredericksburg, VA	4144	0.8	0.6	15.6	16.6	776	0.8	0.6	15.2	16.2	796
02041650	Appomattox River at Matoaca, VA	3471	0.8	0.8	15.1	16.1	798	0.8	0.8	14.9	15.9	810
01673000	Pamunkey River near Hanover, VA	2774	0.8	0.9	15.1	16.1	873	0.8	1.0	14.7	15.7	894
01674500	Mattaponi River near Beulahville, VA	1557	0.7	0.9	14.3	15.3	810	0.8	0.9	14.2	15.2	820
01594440	Patuxent River at Bowie, MD	901	0.9	1.1	15.3	16.3	787	0.8	0.8	14.0	15.0	861
01491000	Choptank River near Greensboro, MD	293	1.2	1.5	19.6	20.6	680	1.1	1.0	20.5	21.5	690

II. Lake Erie and Ohio tributary monitoring program

Site ID	River and station name	Drainage area (km^2)	Nitrate-plus-nitrite (NO_x)					Total phosphorus (TP)				
			λ	λ'	μ	$\Delta t_{average}$ (days)	N	λ	λ'	μ	$\Delta t_{average}$ (days)	N
04193500	Maumee River at Waterville, OH	16 395	0.005	0.0003	0.19	1.19	9101	0.005	0.0003	0.19	1.19	9101
04198000	Sandusky River near Fremont, OH	3245	0.01	0.003	0.22	1.22	9641	0.01	0.003	0.22	1.22	9655
04208000	Cuyahoga River at Independence, OH	1834	0.007	0.006	0.13	1.13	7421	0.007	0.006	0.13	1.13	7426
04212100	Grand River near Painesville, OH	1777	0.01	0.005	0.21	1.21	5023	0.01	0.005	0.22	1.22	4994
04197100	Honey Creek at Melmore, OH	386	0.007	0.005	0.06	1.06	9914	0.007	0.005	0.06	1.06	9914
04197170	Rock Creek at Tiffin, OH	90	0.007	0.008	0.06	1.06	8422	0.007	0.008	0.06	1.06	8440

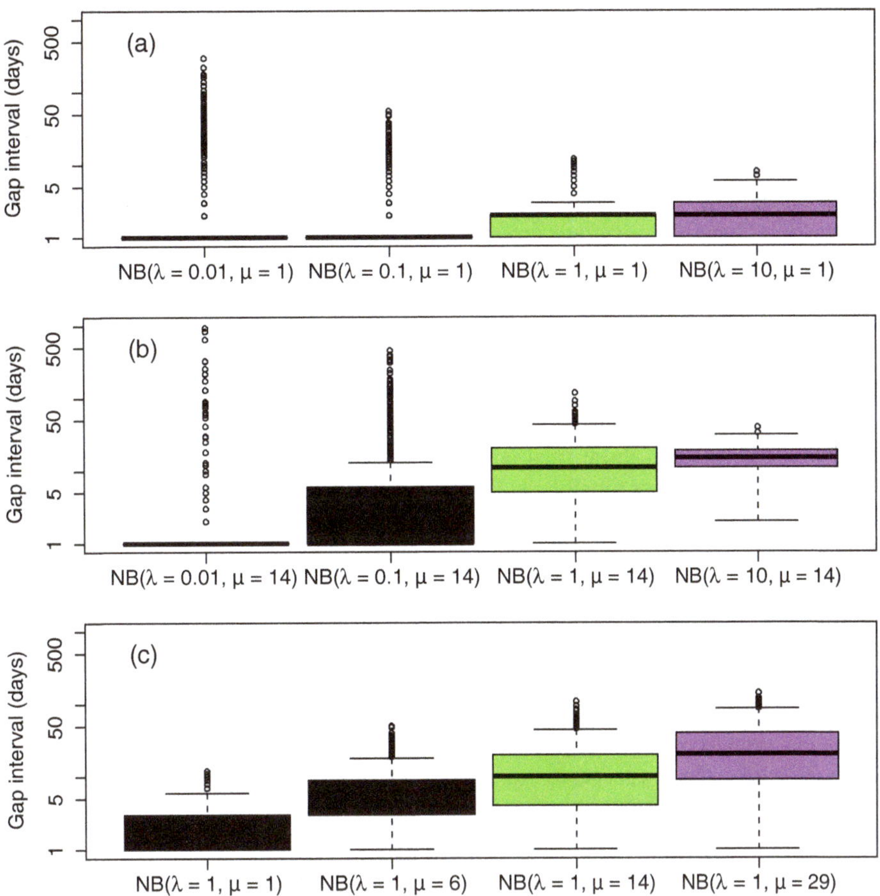

Figure 2. Examples of gap-interval simulation using negative binomial distributions, NB (shape λ, mean μ). Simulation parameters: $L = 9125$ days, $\Delta t_{\text{nominal}} = 1$ day. The three panels show simulation with fixed **(a)** $\mu = 1$, **(b)** $\mu = 14$, and **(c)** $\lambda = 1$. Note that $\Delta t_{\text{average}}/\Delta t_{\text{nominal}} = \mu + 1$.

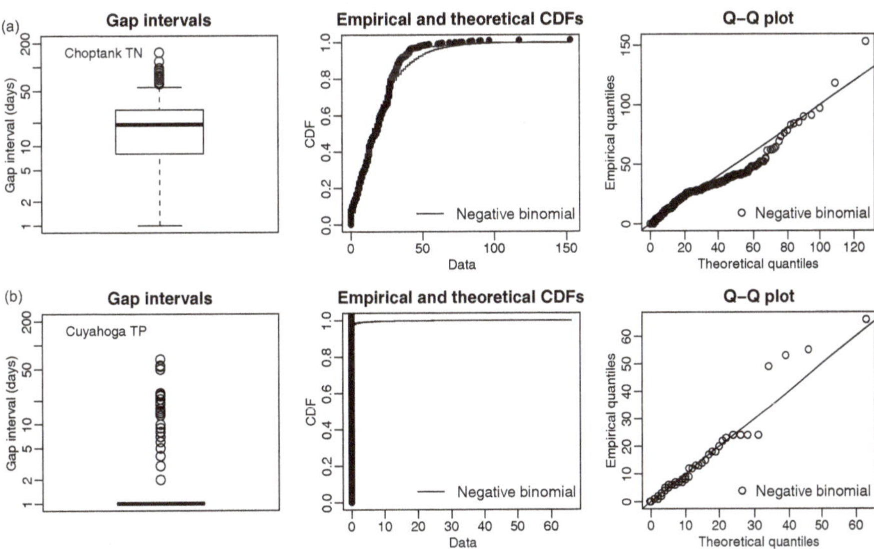

Figure 3. Examples of quantified sampling irregularity with negative binomial (NB) distributions: total nitrogen in Choptank River **(a)** and total phosphorus in Cuyahoga River **(b)**. Theoretical CDF (cumulative distribution function) and quantiles are based on the fitted NB distributions. See Table 1 for estimated mean and shape parameters.

Table 1, with two examples of fitted NB distributions shown in Fig. 3.

For the Chesapeake Bay River Input Monitoring Program (nine sites), total nitrogen (TN) and total phosphorus (TP) are taken as representatives of water-quality constituents. According to the maximum likelihood approach, the shape parameter λ varies between 0.7 and 1.2 for TN and between 0.8 and 1.1 for TP (Table 1). These λ values are around 1.0, reflecting the fact that these sites have relatively even gap distributions (i.e., relatively balanced counts of large and small gaps). The mean parameter μ varies between 9.5 and 19.6 for TN and between 13.4 and 24.4 for TP in the Chesapeake monitoring network, corresponding to $\Delta t_{\text{average}}$ of 10.5–20.6 days for TN and 14.4–25.4 days for TP, respectively. This is consistent with the fact that these sites have typically been sampled roughly once or twice monthly, along with additional sampling during storm flows (Langland et al., 2012; Zhang et al., 2015).

For the Lake Erie and Ohio Tributary Monitoring Program (six sites), records of nitrate plus nitrite (NO_x) and TP were examined. According to the maximum likelihood approach, the shape parameter λ is approximately 0.01 for both constituents (Table 1). These very low λ values occur because these time series contain a few very large gaps, ranging from 35 days to 1109 days ($\sim$3 years). The mean parameter μ varies between 0.06 and 0.22, corresponding to $\Delta t_{\text{average}}$ of 1.06 and 1.22 days, respectively. This is consistent with the fact that these sites have been sampled at a daily resolution with occasional missing values on some days (Zhang and Ball, 2017).

2.3 Simulation of time series with irregular sampling

To evaluate the various β estimation methods, our first step was to use the Monte Carlo simulation to produce time series that mimic the sampling irregularity observed in real water-quality monitoring data. We began by simulating regular (gap-free) time series using the fractional noise simulation method of Witt and Malamud (2013), which is based on inverse Fourier filtering of white noises. Our analysis showed this method performed reasonably well compared to other simulation methods for β values between 0 and 1 (see the Supplement). In addition, this method can also simulate β values beyond this range. The noises simulated by the Witt and Malamud method, however, are band limited to the Nyquist frequency (half of the sampling frequency) of the underlying white noise time series, whereas true fractional noises would contain spectral power at all frequencies, extending well above the Nyquist frequency for any sampling. Thus, these band-limited noises will be less susceptible to spectral aliasing than true fractional noises would be (see Kirchner, 2005, for detailed discussions of the aliasing issue).

A total of 100 replicates of regular (gap-free) time series were produced for nine prescribed spectral slopes, which vary from $\beta = 0$ (white noise) to $\beta = 2$ (Brownian motion or "random walk") with an increment of 0.25 (i.e., 0, 0.25, 0.5, 0.75, 1.0, 1.25, 1.5, 1.75, and 2). These regular time series each have a length (N) of 9125, which can be interpreted as 25 years of regular daily samples (that is, $\Delta t_{\text{nominal}} = 1$ day).

The simulated regular time series were converted to irregular time series using gap intervals that were simulated with NB distributions. To make these gap intervals mimic those in typical river water-quality time series, representative NB parameters were chosen based on results from Sect. 2.2. Specifically, μ was set at 1 and 14, corresponding to $\Delta t_{\text{average}}$ of 2 and 15 days, respectively. For λ, we chose four values that span 3 orders of magnitude, i.e., 0.01, 0.1, 1, and 10. Note that when $\lambda = 1$ the generated time series corresponds to a Bernoulli process. With the chosen values of μ and λ, a total of eight scenarios were generated, which were implemented using the *rnbinom* function in the *stats* R package (R Development Core Team, 2014):

1. $\mu = 1$ (i.e., $\Delta t_{\text{average}} / \Delta t_{\text{nominal}} = 2$), $\lambda = 0.01$,
2. $\mu = 1$, $\lambda = 0.1$,
3. $\mu = 1$, $\lambda = 1$,
4. $\mu = 1$, $\lambda = 10$,
5. $\mu = 14$ (i.e., $\Delta t_{\text{average}} / \Delta t_{\text{nominal}} = 15$), $\lambda = 0.01$,
6. $\mu = 14$, $\lambda = 0.1$,
7. $\mu = 14$, $\lambda = 1$,
8. $\mu = 14$, $\lambda = 10$.

Examples of these simulations are shown with box plots in Fig. 2.

3 Evaluation of proposed estimation methods for irregular time series

3.1 Summary of estimation methods

For the simulated irregular time series, β was estimated using the aforementioned two types of approaches. The first type includes 11 different interpolation methods (designated as B1–B11 below) to fill the data gaps, thus making the data regular and analyzable by traditional methods.

- B1 Global mean: all missing values replaced with the mean of all observations.
- B2 Global median: all missing values replaced with the median of all observations.
- B3 Random replacement: all missing values replaced with observations randomly drawn (with replacement) from the time series.

B4 Next observation carried backward (NOCB): each missing value replaced with the next available observation.

B5 Last observation carried forward (LOCF): each missing value replaced with the preceding available observation.

B6 Average of the two nearest samples: each missing value replaced with the mean of its next and preceding available observations.

B7 LOWESS (locally weighted scatterplot smoothing) with a smoothing span of 1: missing values replaced using fitted values from a LOWESS model determined using all available observations (Cleveland, 1981).

B8 LOWESS with a smoothing span of 0.75: same as B7 except that the smoothing span is 75 % of the available data (similar distinction follows for B9–B11).

B9 LOWESS with a smoothing span of 50 %.

B10 LOWESS with a smoothing span of 30 %.

B11 LOWESS with a smoothing span of 10 %.

B4 and B5 were implemented using the *na.locf* function in the *zoo* R package (Zeileis and Grothendieck, 2005). B7–B11 were implemented using the *loess* function in the *stats* R package (R Development Core Team, 2014). An illustration of these interpolation methods is provided in Fig. 4. The interpolated data, along with the original regular data (designated as A1) were analyzed using Whittle's maximum likelihood method for β estimation, which was implemented using the *FDWhittle* function in the *fractal* R package (Constantine and Percival, 2014).

The second type of approaches estimates β directly from the irregularly sampled data, using several variants of the Lomb–Scargle periodogram (designated as C1a–C1c below), and a recently developed wavelet-based method (designated as C2 below). Specifically, these approaches are as follows.

C1a Lomb–Scargle periodogram: the spectral density of the time series (with gaps) is estimated and the spectral slope is fit using all frequencies (Lomb, 1976; Scargle, 1982). This is a classic method for examining periodicity in irregularly sampled data, which is analogous to the more familiar fast Fourier transform method often used for regularly sampled data.

C1b Lomb–Scargle periodogram with 5 % data: same as C1a except that the fitting of the spectral slope considers only the lowest 5 % of the frequencies (Montanari et al., 1999).

C1c Lomb–Scargle periodogram with "binned" data: same as C1a except that the fitting of the spectral slope is performed on binned data in three steps as follows.

(a) The entire range of frequency is divided into 100 equal-interval bins on logarithmic scale.

(b) The respective medians of frequency and power spectral density are calculated for each of the 100 bins.

(c) The 100 pairs of median frequency and median spectral density are used to estimate the spectral slope on a log–log scale.

C2 Kirchner and Neal (2013)'s wavelet method: uses a modified version of Foster's weighted wavelet spectrum (Foster, 1996) to suppress spectral leakage from low frequencies and applies an aliasing filter (Kirchner, 2005) to remove spectral aliasing artifacts at high frequencies.

C1a was implemented using the *spec.ls* function in the *cts* R package (Wang, 2013). C2 was run in *C*, using codes modified from those in Kirchner and Neal (2013).

3.2 Evaluation of methods' performance

Each estimation method listed above was applied to the simulated data (Sect. 2.3) to estimate β, which were then compared with the prescribed ("true") β to quantify the performance of each method. Plots of method evaluation for all simulations are provided as Figs. S3–S12 (Supplement S2). Close inspections of these plots reveal some general patterns of the methods' performance. For brevity, these patterns are presented with a subset of the plots, which correspond to the cases where true $\beta = 1$ and shape parameter $\lambda = 0.01$, 0.1, 1, and 10 (Fig. 5). In general, β values estimated using the regular data (A1) are very close to 1.0, which indicates that the adopted fractional noise generation method and Whittle's maximum likelihood estimator have small combined simulation and estimation bias. This is perhaps unsurprising, since the estimator is based on the Fourier transform and the noise generator is based on an inverse Fourier transform; thus, one method is essentially just the inverse of the other. One should also note that when fractional noises are not arbitrarily band limited at the Nyquist frequency (as they inherently are with the noise generator that is used here), spectral aliasing should lead to spectral slopes that are flatter than expected (Kirchner, 2005) and thus to underestimates of β.

For the simulated irregular data, the estimation methods differ widely in their performance. Specifically, three interpolation methods (i.e., B4–B6) consistently overestimate β, indicating that they introduce additional correlations into the time series, reducing its short-timescale variability. In contrast, the other eight interpolation methods (i.e., B1–B3 and B7–B11) generally underestimate β, indicating that the interpolated points are less correlated than the original time series, thus introducing additional variability on short timescales. As expected, results from the LOWESS methods (B7–B11) depend strongly on the size of the smoothing window; that is, β is more severely underestimated as the smoothing window becomes wider. In fact, when the smoothing window is 1.0 (i.e., method B7), LOWESS performs the interpolation using all data available and thus behaves similarly to interpolations

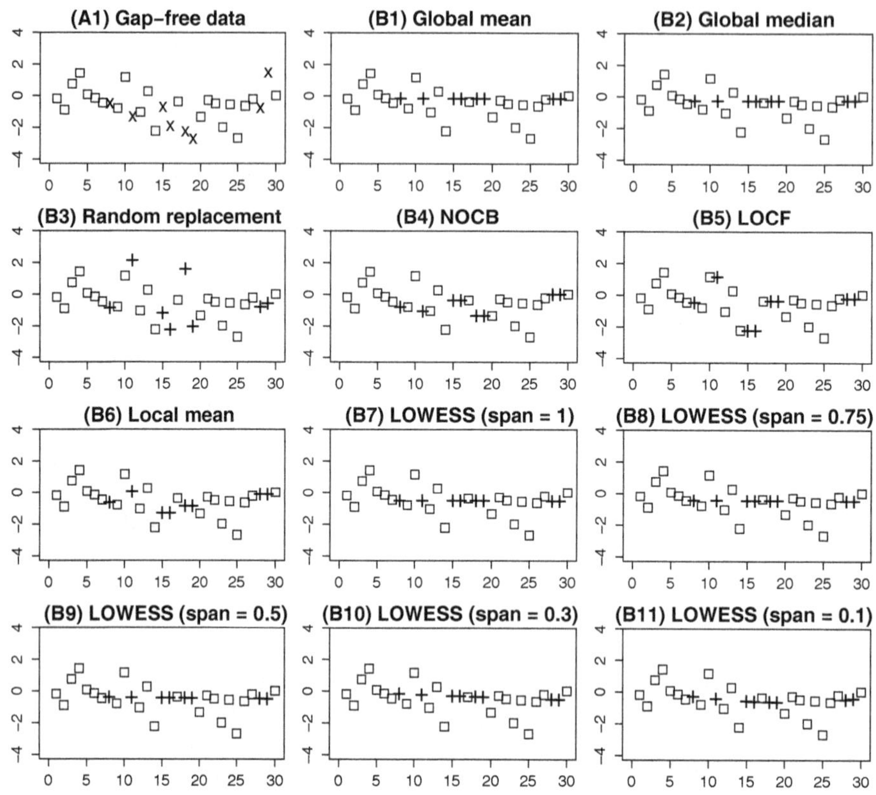

Figure 4. Illustration of the interpolation methods for gap filling. The gap-free data (A1) was simulated with a series length of 500, with the first 30 data shown. (×: omitted data for gap filling; +: interpolated data; NOCB: next observation carried backward; LOCF: last observation carried forward; LOWESS: locally weighted scatterplot smoothing.)

based on global means (B1) or global medians (B2), except that LOWESS fits a polynomial curve instead of constant values. However, whenever a sampling gap is much shorter than the smoothing window, the infilled LOWESS value will be close to the local mean or median, and the abrupt jumps produced by these infilled values will artificially increase the variance in the time series at high frequencies, leading to an artificially reduced spectral slope β and, correspondingly, an underestimate of β. This mechanism explains why LOWESS interpolation distorts β more when there are many small gaps (large λ) and therefore more jumps to, and away from, the infilled values than when there are only a few large gaps (small λ).

Among the direct methods (i.e., C1a, C1b, C1c, and C2), the Lomb–Scargle method, with original data (C1a) or binned data (C1c) tends to underestimate β, though the underestimation by C1c is generally less severe. The modified Lomb–Scargle method (C1b), using only the lowest 5 % of frequencies, yields estimates that are centered around 1.0 for large λ. However, C1b has the highest variability (i.e., least precision) in β estimates among all methods. Compared with all the above methods, the wavelet method (C2) has much better performance in terms of both accuracy and precision when λ is 1 or 10, a slightly better or similar performance when λ is 0.1, but worse performance when λ is 0.01.

The shape parameter λ greatly affects the performance of the estimation methods. All the interpolation methods that underestimate β (i.e., B1–B3 and B7–B11) perform worse as λ increases from 0.01 to 10. This effect can be interpreted as follows: when the time series contains a large number of relatively small gaps (e.g., $\lambda = 1$ or 10), there are many jumps (which, as noted above, contain mostly high-frequency variance) between the original data and the infilled values, resulting in more severe underestimation. In contrast, when the data contain only a small number of very large gaps (e.g., $\lambda = 0.01$ or 0.1), there are fewer of these jumps, resulting in minimal underestimation. Similar effects of λ are also observed with the interpolation methods that show overestimation (i.e., B4–B6) – that is, overestimation is more severe when λ is larger. Similarly, the Lomb–Scargle method (C1a and C1c) performs worse (more serious underestimation) as λ increases. Finally, method C2 seems to perform the best when λ is large (1 or 10), but not well when λ is very small (0.01), as noted above. This result highlights the sensitivity of the wavelet method to the presence of a few large gaps in the time series. For such cases, a potentially more feasible approach is to break the whole time series into several seg-

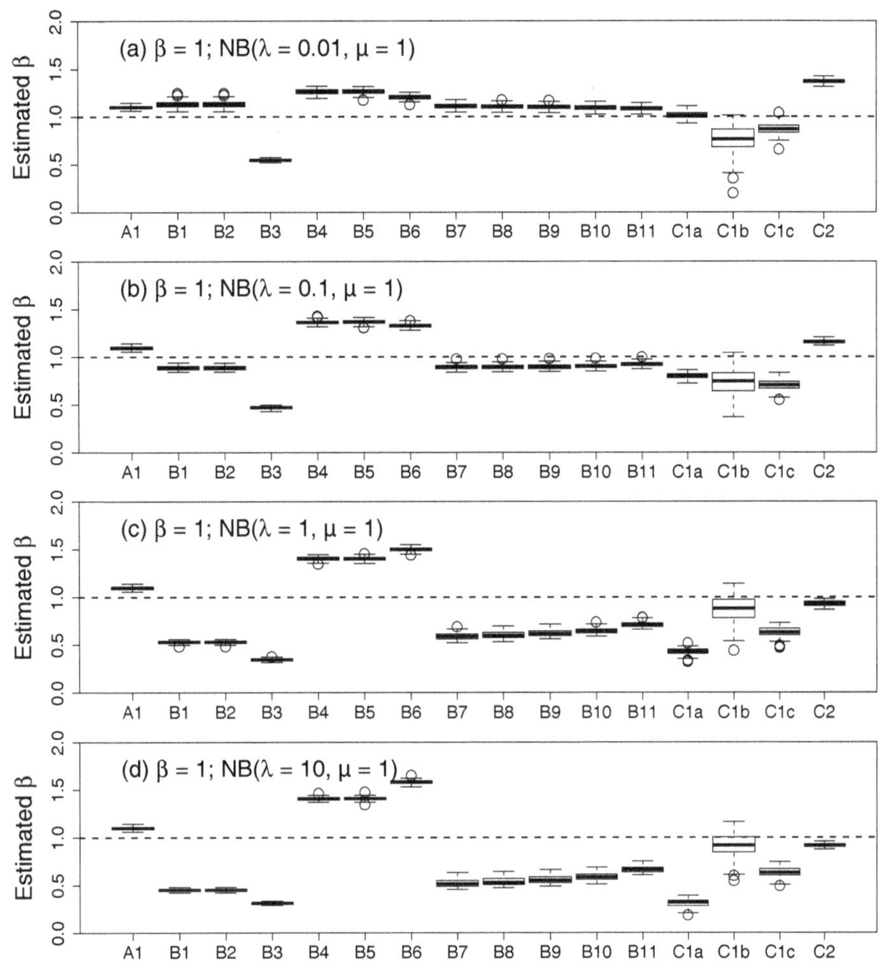

Figure 5. Comparison of bias in estimated spectral slope in irregular data that are simulated with prescribed $\beta = 1$ (100 replicates), a series length of 9125, and gap intervals simulated with **(a)** NB ($\lambda = 0.01$, $\mu = 1$), **(b)** NB ($\lambda = 0.1$, $\mu = 1$), **(c)** NB ($\lambda = 1$, $\mu = 1$), and **(d)** NB ($\lambda = 10$, $\mu = 1$). The blue dashed lines indicate the true β value.

ments (each without long gaps) and then apply the wavelet method (C2) to analyze each segment separately. If this can yield more accurate estimates, then further simulation experiments should be designed to systematically determine how long the gap needs to be to invoke such an approach.

Next, the method evaluation is extended to all the simulated spectral slopes, that is, $\beta = 0$, 0.25, 0.5, 0.75, 1.0, 1.25, 1.5, 1.75, and 2. For ease of discussion, three quantitative criteria were proposed for evaluating performance, namely, bias (B), standard deviation (SD), and root-mean-squared error (RMSE), as defined below:

$$B_i = \overline{\beta_i} - \beta_{\text{true}}, \tag{14}$$

$$\text{SD}_i = \sqrt{\frac{1}{99}\sum_{j=1}^{100}(\beta_{i,j} - \overline{\beta_i})^2}, \tag{15}$$

$$\text{RMSE}_i = \sqrt{B_i^2 + SD_i^2}, \tag{16}$$

where $\overline{\beta_i}$ is the mean of 100 β values estimated by method i, and β_{true} is the prescribed β value for simulation of the initial regular time series. In general, B and SD can be considered as the models' systematic error and random error, respectively, and RMSE serves as an integrated measure of both errors. For all evaluations, plots of bias and RMSE are provided in the main text. (Plots of SD are provided as Figs. S7 and S12 in the Supplement for simulations with $\mu = 1$ and $\mu = 14$, respectively.)

For simulations with $\mu = 1$, results of estimation bias and RMSE are summarized in Figs. 6 and 7, respectively. (More details are provided in Figs. S3–S6 in the Supplement.) For brevity, we focus on three direct methods (C1a, C1b, and C2) and three representative interpolation methods. (Specifically, B1 represents B1–B3 and B7, B6 represents B4–B6, and B8 represents B8–B11.) Overall, these six methods show mixed performances. In terms of bias (Fig. 6), B1 (global mean) and B8 (LOWESS with a smoothing span of 0.75) tend to have negative bias, particularly for time series with

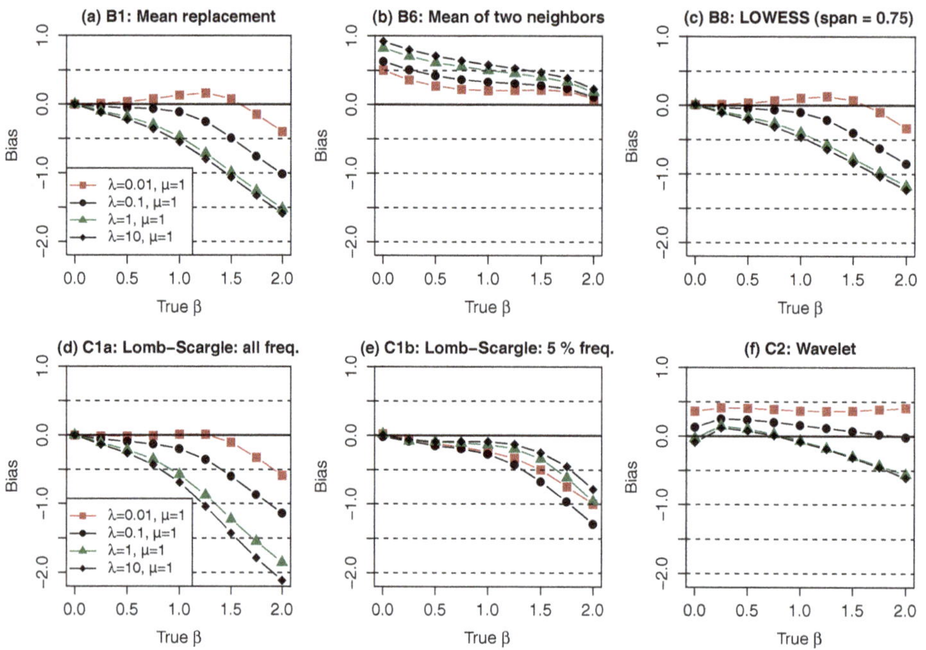

Figure 6. Comparison of bias in estimated spectral slope in irregular data that are simulated with varying prescribed β values (100 replicates), a series length of 9125, and a mean gap interval of 2 (i.e., $\mu = 1$).

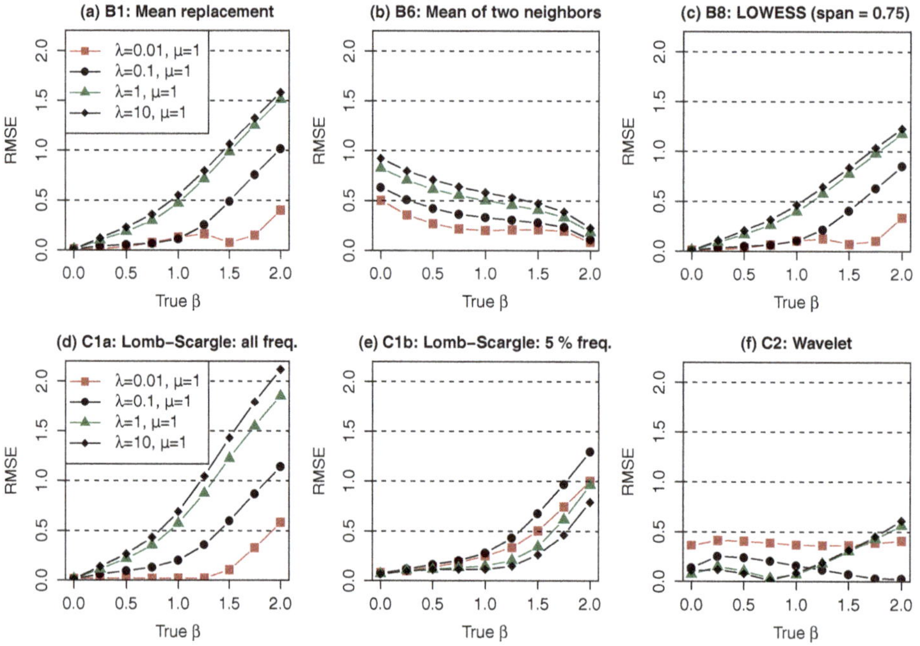

Figure 7. Comparison of root-mean-squared error (RMSE) in estimated spectral slope in irregular data that are simulated with varying prescribed β values (100 replicates), a series length of 9125, and a mean gap interval of 2 (i.e., $\mu = 1$).

(1) moderate-to-large β_{true} values and (2) large λ values (i.e., less skewed gap intervals). By contrast, B1 and B8 generally have minimal bias when (1) β_{true} is close to zero (i.e., when the simulated time series is close to white noise) and (2) λ is small (e.g., 0.01), since interpolating a few large gaps cannot significantly affect the overall correlation structure. In addition, LOWESS interpolation with a larger smoothing window tends to yield more negatively biased estimates (data not shown). The other interpolation method, B6 (mean of the two nearest neighbors) tends to overestimate β, particularly for

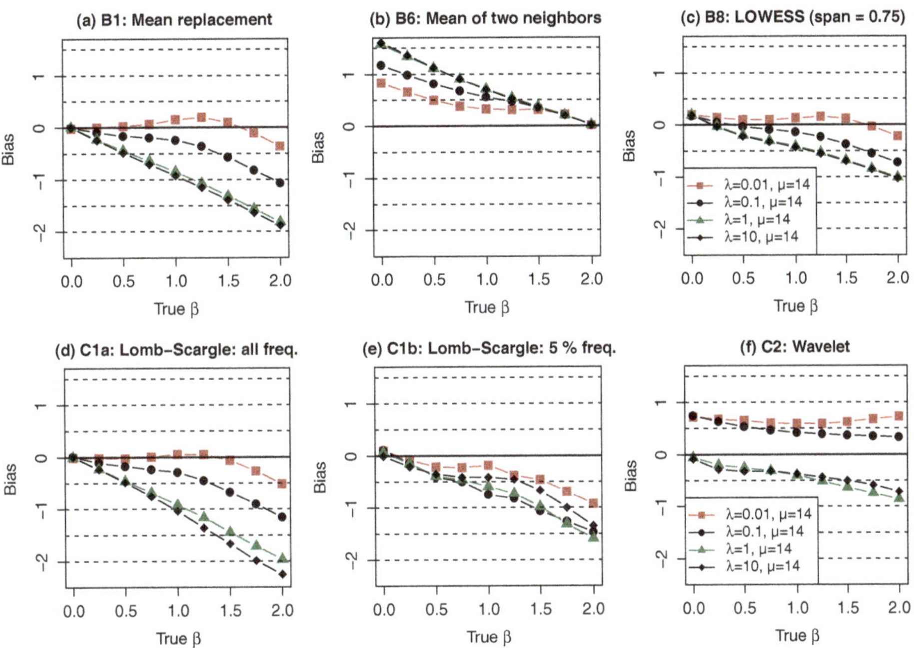

Figure 8. Comparison of bias in estimated spectral slope in irregular data that are simulated with varying prescribed β values (100 replicates), a series length of 9125, and a mean gap interval of 15 (i.e., $\mu = 14$).

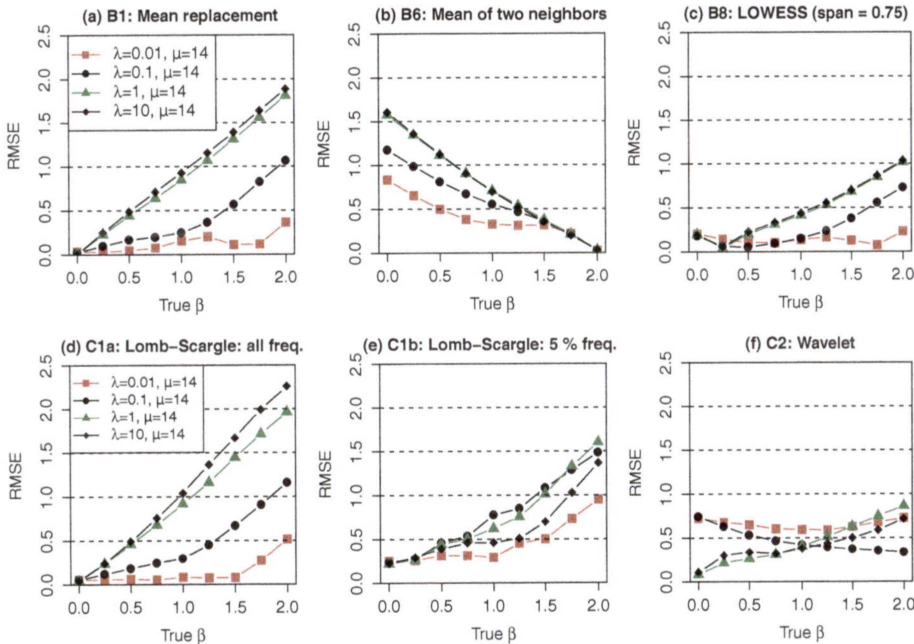

Figure 9. Comparison of root-mean-squared error (RMSE) in estimated spectral slope in irregular data that are simulated with varying prescribed β values (100 replicates), a series length of 9125, and a mean gap interval of 15 (i.e., $\mu = 14$).

time series with (1) small β_{true} values and (2) large λ values. At large β_{true} values (e.g., 2.0), the autocorrelation is already very strong such that taking the mean of two neighbors for gap filling does not introduce much additional correlation, as opposed to the case of small β_{true} values. The Lomb–Scargle methods (C1a and C1b) generally have negative bias, particularly for time series with (1) moderate-to-large β_{true} values (for both methods) and (2) large λ values (for C1a), which is similar to B1 and B8. However, C1b overall shows less severe bias than C1a. Finally, the wavelet method (C2) shows generally the smallest bias among all methods. However, its performance advantage is not as great when the time series has small λ values (i.e., very skewed gap intervals), as noted above, which may be due to the fact that the aliasing filter was designed for regular time series. In terms of SD (Fig. S7 in the Supplement), method C1b performs the worst among all methods (as noted above), method B6 and B8 perform poorly for large β_{true} values, and method C2 performs poorly for $\beta_{\text{true}} = 0$. In terms of RMSE (Fig. 7), methods B1, B8, C1a, and C1b perform well for small β_{true} values and small λ values, whereas method B6 performs well for large β_{true} values and small λ values. In comparison, method C2 generally has the smallest RMSEs among all methods, and its RMSEs are similarly small for the wide range of β_{true} and λ values. In general, the wavelet method can be considered the best among all the tested methods.

For simulations with $\mu = 14$, results of estimation bias and RMSE are summarized in Figs. 8 and 9, respectively. (More details are provided in Figs. S8–S11 in the Supplement.) Overall, these methods show mixed performances that are generally similar to the cases when $\mu = 1$, as discussed above. These results highlight the generality of these methods' performances, which applies at least to the range of $\mu = [1, 14]$. In addition, all methods show generally larger RMSE for $\mu = 14$ than $\mu = 1$, indicating their dependence on the mean gap interval (Fig. 9). Perhaps the most notable difference is observed with method C2, which in this case shows positive bias for small λ values (0.01 and 0.1) and negative bias for large λ values (1 and 10) (Fig. 8f). It nonetheless generally shows the smallest RMSEs among all the tested methods as in the cases of $\mu = 1$ above.

3.3 Quantification of spectral slopes in real water-quality data

In this section, the proposed estimation approaches were applied to quantify β in real water-quality data from the two monitoring programs presented in Sect. 2.2 (Table 1). As noted in Sect. 1.3, such real data are typically much more complex than our simulated time series, because of (1) strong deviations from normal distributions and (2) effects of flow dependence, seasonality, and temporal trends (Hirsch et al., 1991; Helsel and Hirsch, 2002). In this regard, future research may simulate time series with these important characteristics and evaluate the performance of various estimation approaches, perhaps following the modeling framework described here. Alternatively, one may quantify β in transformed time series after accounting for the above aspects. In this work, we have taken the latter approach for a preliminary investigation. Specifically, we have used the published weighted regressions on time, discharge, and season (WRTDS) method (Hirsch et al., 2010) to transform the original time series. This widely accepted method estimates daily concentrations based on discretely collected concentration samples using time, season, and discharge as explanatory variables, i.e.,

$$\begin{aligned}\ln(C) &= \beta_0 + \beta_1 t + \beta_2 \ln(Q) + \beta_3 \sin(2\pi t) \\ &+ \beta_4 \cos(2\pi t) + \varepsilon, \end{aligned} \tag{17}$$

where C is concentration, Q is daily discharge, t is time in decimal years, β_i are fitted coefficients, and ε is the error term. The second and third terms on the right represent time and discharge effects, respectively, whereas the fourth and fifth terms collectively represent cyclical seasonal effects. For a full description of this method, see Hirsch et al. (2010). In this work, WRTDS was applied to obtain time series of estimated daily concentrations for each constituent at each site. The difference between observed concentration (C_{obs}) and estimated concentration (C_{est}) was calculated in logarithmic space to obtain the concentration residuals,

$$\text{residuals} = \ln(C_{\text{obs}}) - \ln(C_{\text{est}}). \tag{18}$$

For our data sets, histograms of concentration residuals (expressed in natural log concentration units) are shown in Figs. S13–S16 in the Supplement. Compared with the original concentration data, these model residuals are much more nearly normal and homoscedastic. Moreover, the model residuals are less susceptible to the issues of temporal, seasonal, and discharge-driven variations than the original concentrations. Therefore, the model residuals are more appropriate than the original concentrations for β estimation using the simulation framework adopted in this work.

The estimated β values for the concentration residuals are summarized in Fig. 10. Clearly, the estimated β varies considerably with the estimation method. In addition, the estimated β varies with site and constituent (i.e., TP, TN, or NO_x). Our discussion below focuses on the wavelet method (C2), because it is established above that this method performs better than the other estimation methods under a wide range of gap conditions. We emphasize that it is beyond our current scope to precisely quantify β in these water-quality data sets, but our simulation results presented above (Sect. 3.2) can be used as references to qualitatively evaluate the reliability of C2 and/or other methods for these data sets.

For TN and TP concentration data at the Chesapeake River input monitoring sites (Table 1), μ varies between 9.5 and 24.4, whereas λ is ~ 1.0. Thus, the simulated gap scenario of NB($\mu = 14$, $\lambda = 1$) can be used as a reasonable reference to assess methods' reliability (Fig. 8). Based on method C2, the estimated β ranges between $\beta = 0.36$ and $\beta = 0.61$ for TN and between $\beta = 0.30$ and $\beta = 0.58$ for TP at these sites (Fig. 10). For such ranges, the simulation results indicate that method C2 tends to moderately underestimate β under this gap scenario (Fig. 8), and hence spectral slopes for TN and TP at these Chesapeake sites are probably slightly higher than those presented here (Fig. 10).

For NO_x and TP concentration data at the Lake Erie and Ohio sites (Table 1), μ varies between 0.06 and 0.22, whereas λ is ~ 0.01. Thus, the simulated gap scenario of NB($\mu = 1$,

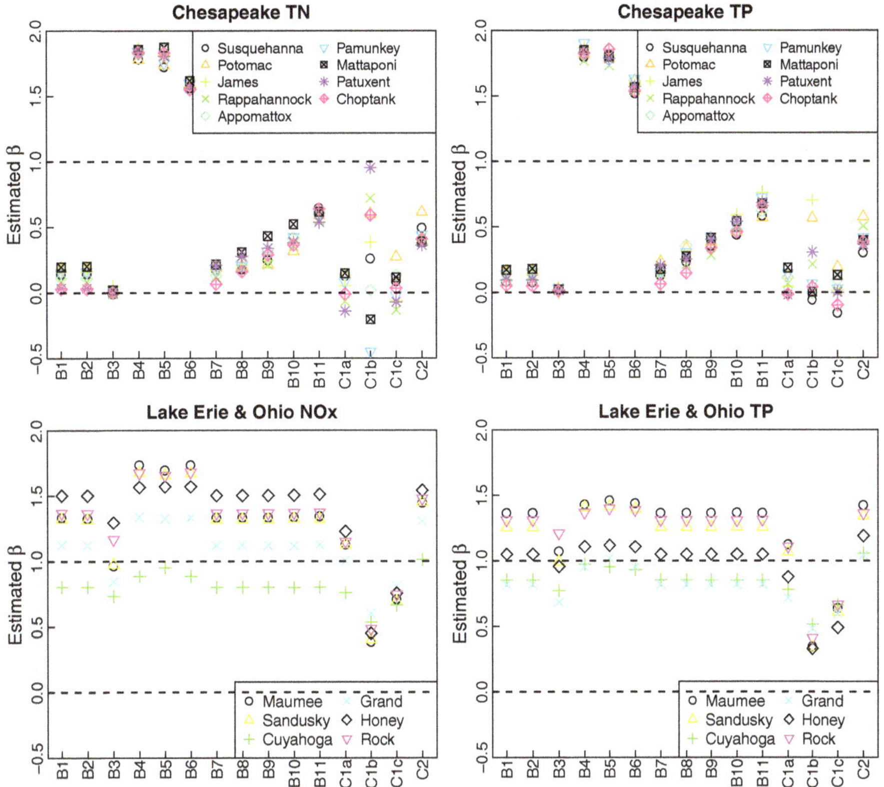

Figure 10. Quantification of spectral slope in real water-quality data from the two regional monitoring networks, as estimated using the set of examined methods. All estimations were performed on concentration residuals (in natural log concentration units) after accounting for effects of time, discharge, and season. The two dashed lines in each panel indicate white noise ($\beta = 0$) and pink (flicker) noise ($\beta = 1$), respectively. See Table 1 for site and data details.

$\lambda = 0.01$) can be used as a reasonable reference to assess the methods' reliability (Fig. 6). For such small λ (i.e., a few gaps that are very dissimilar from others), C2 is not reliable for β estimation, as reflected by the generally positive bias in the simulation results. By contrast, methods B1 (interpolation with global mean) and B8 (LOWESS with span 0.75) both perform quite well under this gap scenario (Fig. 6). These two methods provide almost identical β estimates for each site–constituent combination, ranging from $\beta = 0.8$ to $\beta = 1.5$ for NO_x and TP (Fig. 10).

Overall, the above analysis of real water-quality data has illustrated the wide variability in β estimates, with different choices of estimation methods yielding very different results. To our knowledge, these water-quality data have not previously been analyzed in this context. As illustrated above, our simulation experiments (Sect. 3.2) can be used as references to coarsely evaluate the reliability of each method under specific gap scenarios, thereby considerably narrowing the likely range of the estimated spectral slopes. Nonetheless, our results demonstrate that the analyzed water-quality time series can exhibit strong fractal scaling, particularly at the Lake Erie and Ohio tributary sites. Thus, an important implication is that researchers and analysts should be cautious when applying standard statistical methods to identify temporal trends in such water-quality data sets (Kirchner and Neal, 2013). In future work, one may consider applying Bayesian statistical analysis or other approaches to more accurately quantify the spectral slope and associated uncertainty for real water-quality data analysis. In addition, the modeling framework presented here (including both gap simulation and β estimation) may be extended to simulations of irregular time series that have prescribed spectral slopes and also superimposed temporal trends, which can then be used to evaluate the validity of various statistical methods for identifying trends and their associated statistical significance.

4 Conclusions

River water-quality time series often exhibit fractal scaling behavior, which presents challenges to the identification of deterministic trends. Because traditional spectral estimation methods are generally not applicable to irregularly sampled time series, we have examined two broad types of estimation approaches and evaluated their performances against synthetic data with a wide range of prescribed β values and gap intervals that are representative of the sampling irregularity of real water-quality data.

The results of this work suggest several important messages. First, the results remind us of the risks in using interpolation for gap filling when examining autocorrelation, as the interpolation methods consistently underestimate or overestimate β under a wide range of prescribed β values and gap distributions. Second, the widely used Lomb–Scargle spectral method also consistently underestimates β. Its modified form, using the 5 % lowest frequencies for spectral slope estimation, has very poor precision, although the overall bias is small. Third, the wavelet method, coupled with an aliasing filter, has the smallest bias and root-mean-squared error among all methods for a wide range of prescribed β values and gap distributions, except for cases with small prescribed β values (i.e., close to white noise) or small λ values (i.e., very skewed gap distributions). Thus, the wavelet method is recommended for estimating spectral slopes in irregular time series until improved methods are developed. In this regard, future research should aim to develop an aliasing filter that is more applicable to irregular time series with very skewed gap intervals. Finally, all methods' performances depend strongly on the sampling irregularity in terms of both the skewness and mean of gap-interval lengths, highlighting that the accuracy and precision of each method are data specific.

Overall, these results provide new contributions in terms of better understanding and quantification of the proposed methods' performances for estimating the strength of fractal scaling in irregularly sampled water-quality data. In addition, the work has provided an innovative and general approach for modeling sampling irregularity in water-quality records. Moreover, this work has proposed and demonstrated a generalizable framework for data simulation (with gaps) and estimation, which can be readily applied to evaluate other methods that are not covered in this work. More generally, the findings and approaches may also be broadly applicable to irregularly sampled data in other scientific disciplines. Last but not least, we note that accurate quantification of fractal scaling in irregular water-quality time series remains an unresolved challenge for the hydrologic community and for many other disciplines that must grapple with irregular sampling.

Competing interests. The authors declare that they have no conflict of interest.

Acknowledgements. Zhang was supported by the Maryland Sea Grant through awards NA10OAR4170072 and NA14OAR1470090 and by the Maryland Water Resources Research Center through a graduate fellowship while he was a doctoral student at Johns Hopkins University. Subsequent support to Zhang was provided by the US EPA under grant "EPA/CBP Technical Support 2017" (no. 07-5-230480). Harman's contribution to this work was supported by the National Science Foundation through grants CBET-1360415 and EAR-1344664. We thank Bill Ball (Johns Hopkins University) and Bob Hirsch (US Geological Survey) for many useful discussions. We are very grateful to the Editor and two anonymous reviewers for their comments and suggestions. This is contribution no. 5449 of the University of Maryland Center for Environmental Science.

Edited by: Erwin Zehe

References

Aubert, A. H., Kirchner, J. W., Gascuel-Odoux, C., Faucheux, M., Gruau, G., and Mérot, P.: Fractal water quality fluctuations spanning the periodic table in an intensively farmed watershed, Environ. Sci. Technol., 48, 930–937, https://doi.org/10.1021/es403723r, 2014.

Beran, J.: Long-range dependence, Wiley Interdiscip. Rev. Comput. Stat., 2, 26–35, https://doi.org/10.1002/wics.52, 2010.

Beran, J., Feng, Y., Ghosh, S., and Kulik, R.: Long-Memory Processes: Probabilistic Properties and Statistical Methods, Berlin, Heidelberg, Springer Berlin Heidelberg, 884 pp., 2013.

Boutahar, M., Marimoutou, V., and Nouira, L.: Estimation Methods of the Long Memory Parameter: Monte Carlo Analysis and Application, J. Appl. Stat., 34, 261–301, https://doi.org/10.1080/02664760601004874, 2007.

Box, G. E. P., Jenkins, G. M., and Reinsel, G. C.: Time Series Analysis, Fourth Edition. Hoboken, NJ, John Wiley & Sons, Inc., 47–92, 2008.

Clarke, R. T.: Calculating uncertainty in regional estimates of trend in streamflow with both serial and spatial correlations, Water Resour. Res., 49, 7120–7125, https://doi.org/10.1002/wrcr.20465, 2013.

Cleveland, W. S.: LOWESS: A program for smoothing scatterplots by robust locally weighted regression, Am. Stat., 35, 54, https://doi.org/10.2307/2683591, 1981.

Cohn, T. A. and Lins, H. F.: Nature's style: Naturally trendy, Geophys. Res. Lett., 32, L23402, https://doi.org/10.1029/2005GL024476, 2005.

Constantine, W. and Percival, D.: fractal: Fractal Time Series Modeling and Analysis, available at: https://cran.r-project.org/web/packages/fractal (last access: 6 April 2015.), 2014.

Darken, P. F., Zipper, C. E., Holtzman, G. I., and Smith, E. P.: Serial correlation in water quality variables: Estimation and implications for trend analysis, Water Resour. Res., 38, 1117, https://doi.org/10.1029/2001WR001065, 2002.

Delignette-Muller, M. L. and Dutang, C.: fitdistrplus: An R Package for Fitting Distributions, J. Stat. Softw., 64, 1–34, 2015.

Ehsanzadeh, E. and Adamowski, K.: Trends in timing of low stream flows in Canada: impact of autocorrelation and long-term persistence, Hydrol. Process., 24, 970–980, https://doi.org/10.1002/hyp.7533, 2010.

Fatichi, S., Barbosa, S. M., Caporali, E., and Silva, M. E.: Deterministic versus stochastic trends: Detection and challenges, J. Geophys. Res., 114, D18121, https://doi.org/10.1029/2009JD011960, 2009.

Foster, G.: Wavelets for period analysis of unevenly sampled time series, Astron. J., 112, 1709–1729, 1996.

Franzke, C.: Nonlinear Trends, Long-Range Dependence, and Climate Noise Properties of Surface Temperature, J. Clim., 25, 4172–4183, https://doi.org/10.1175/JCLI-D-11-00293.1, 2012a.

Franzke, C.: On the statistical significance of surface air temperature trends in the Eurasian Arctic region, Geophys. Res. Lett., 39, L23705, https://doi.org/10.1029/2012GL054244, 2012b.

Godsey, S. E., Aas, W., Clair, T. A., de Wit, H. A., Fernandez, I. J., Kahl, J. S., Malcolm, I. A., Neal, C., Neal, M., Nelson, S. J., Norton, S. A., Palucis, M. C., Skjelkvåle, B. L., Soulsby, C., Tetzlaff, D., and Kirchner, J. W.: Generality of fractal 1/f scaling in catchment tracer time series, and its implications for catchment travel time distributions, Hydrol. Process., 24, 1660–1671, https://doi.org/10.1002/hyp.7677, 2010.

Graham, J.: Missing Data Analysis: Making It Work in the Real World, Annu. Rev. Psychol., 60, 549–576, https://doi.org/10.1146/annurev.psych.58.110405.085530, 2009.

Helsel, D. R. and Hirsch, R. M.: Statistical Methods in Water Resources, US Geological Survey Techniques of Water-Resources Investigations Book 4, Chapter A3, US Geological Survey, Reston, VA, p. 522, http://pubs.usgs.gov/twri/twri4a3/ (last access: 11 June 2016.), 2002.

Hirsch, R. M., Alexander, R. B., and Smith, R. A.: Selection of methods for the detection and estimation of trends in water quality, Water Resour. Res., 27, 803–813, https://doi.org/10.1029/91WR00259, 1991.

Hirsch, R. M., Moyer, D. L., and Archfield, S. A.: Weighted regressions on time, discharge, and season (WRTDS), with an application to Chesapeake Bay river inputs, J. Am. Water Resour. Assoc., 46, 857–880, https://doi.org/10.1111/j.1752-1688.2010.00482.x, 2010.

Hurst, H. E.: Long-term storage capacity of reservoirs, Trans. Amer. Soc. Civil Eng., 116, 770–808, 1951.

Kasmarek, M. C. and Ramage, J. K.: Water-Level Measurement Data Collected during 2015-2016 and Approximate Long-term Water-Level Altitude Changes of Wells Screened in the Chicot, Evangeline, and Jasper Aquifers, Houston-Galveston Region, Texas: US Geological Survey data release, https://doi.org/10.5066/F77H1GP3, 2016.

Khaliq, M. N., Ouarda, T. B. M. J., and Gachon, P.: Identification of temporal trends in annual and seasonal low flows occurring in Canadian rivers: The effect of short- and long-term persistence, J. Hydrol., 369, 183–197, https://doi.org/10.1016/j.jhydrol.2009.02.045, 2009.

Khaliq, M. N., Ouarda, T. B. M. J., Gachon, P., and Sushama, L.: Temporal evolution of low-flow regimes in Canadian rivers, Water Resour. Res., 44, W08436, https://doi.org/10.1029/2007WR006132, 2008.

Kirchner, J.: Aliasing in 1?f\{^α} noise spectra: Origins, consequences, and remedies, Phys. Rev. E, 71, 066110, https://doi.org/10.1103/PhysRevE.71.066110, 2005.

Kirchner, J. W. and Neal, C.: Universal fractal scaling in stream chemistry and its implications for solute transport and water quality trend detection, P. Natl. Acad. Sci. USA, 110, 12213–12218, https://doi.org/10.1073/pnas.1304328110, 2013.

Kirchner, J. W. and Weil, A.: No fractals in fossil extinction statistics, Nature, 395, 337–338, https://doi.org/10.1038/26384, 1998.

Kirchner, J. W., Feng, X., and Neal, C.: Fractal stream chemistry and its implications for contaminant transport in catchments, Nature, 403, 524–527, https://doi.org/10.1038/35000537, 2000.

Kirchner, J. W., Feng, X., and Neal, C.: Catchment-scale advection and dispersion as a mechanism for fractal scaling in stream tracer concentrations, J. Hydrol., 254, 82–101, https://doi.org/10.1016/s0022-1694(01)00487-5, 2001.

Langland, M. J., Blomquist, J. D., Moyer, D. L., and Hyer, K. E.: Nutrient and suspended-sediment trends, loads, and yields and development of an indicator of streamwater quality at nontidal sites in the Chesapeake Bay watershed, 1985–2010, US Geological Survey Scientific Investigations Report 2012-5093, Reston, VA, p. 26., available at: http://pubs.usgs.gov/sir/2012/5093/pdf/sir2012-5093.pdf (last access: 6 April 2015), 2012.

Lennartz, S. and Bunde, A.: Trend evaluation in records with long-term memory: Application to global warming, Geophys. Res. Lett., 36, L16706, https://doi.org/10.1029/2009GL039516, 2009.

Lomb, N. R.: Least-squares frequency analysis of unequally spaced data, Astrophys. Space Sci., 39, 447–462, https://doi.org/10.1007/BF00648343, 1976.

Montanari, A., Rosso, R., and Taqqu, M. S.: Fractionally differenced ARIMA models applied to hydrologic time series: Identification, estimation, and simulation, Water Resour. Res., 33, 1035–1044, https://doi.org/10.1029/97WR00043, 1997.

Montanari, A., Taqqu, M. S., and Teverovsky, V.: Estimating long-range dependence in the presence of periodicity: An empirical study, Math. Comput. Model., 29, 217–228, https://doi.org/10.1016/S0895-7177(99)00104-1, 1999.

Montanari, A., Rosso, R., and Taqqu, M. S.: A seasonal fractional ARIMA Model applied to the Nile River monthly flows at Aswan, Water Resour. Res., 36, 1249–1259, https://doi.org/10.1029/2000WR900012, 2000.

National Center for Water Quality Research: Tributary Data Download, https://ncwqr.org/monitoring/data/ (last access: 23 July 2015), 2015.

Noguchi, K., Gel, Y. R., and Duguay, C. R.: Bootstrap-based tests for trends in hydrological time series, with application to ice phenology data, J. Hydro., 410, 150–161, https://doi.org/10.1016/j.jhydrol.2011.09.008, 2011.

R Development Core Team: R: A language and environment for statistical computing, R Foundation for Statistical Computing, Vienna, Austria, http://www.r-project.org (last access: 6 April 2015), , 2014.

Rea, W., Oxley, L., Reale, M., and Brown, J.: Estimators for Long Range Dependence: An Empirical Study, Electron. J. Stat., http://arxiv.org/abs/0901.0762 (last access: 6 April 2015), 2009.

Sang, Y.-F., Wang, Z., and Liu, C.: Comparison of the MK test and EMD method for trend identification in hydrological time series, J. Hydrol., 510, 293–298, https://doi.org/10.1016/j.jhydrol.2013.12.039, 2014.

Scargle, J. D.: Studies in Astronomical Time-Series Analysis. II. Statistical Aspects of Spectral-Analysis of Unevenly Spaced Data, Astrophys. J., 263, 835–853, https://doi.org/10.1086/160554, 1982.

Stroe-Kunold, E., Stadnytska, T., Werner, J., and Braun, S.: Estimating long-range dependence in time series: an evaluation of estimators implemented in R, Behav. Res. Meth., 41, 909–923, https://doi.org/10.3758/BRM.41.3.909, 2009.

Taqqu, M. S., Teverovsky, V., and Willinger, W.: Estimators for long-range dependence: an empirical study, Fractals, 3, 785–798, https://doi.org/10.1142/S0218348X95000692, 1995.

Wang, Z.: cts: An R Package for Continuous Time Autoregressive Models via Kalman Filter, J. Stat. Softw., 53, 1–19, 2013.

Witt, A. and Malamud, B. D.: Quantification of Long-Range Persistence in Geophysical Time Series: Conventional and Benchmark-Based Improvement Techniques, Surv. Geophys., 34, 541–651, https://doi.org/10.1007/s10712-012-9217-8, 2013.

Yue, S., Pilon, P., Phinney, B., and Cavadias, G.: The influence of autocorrelation on the ability to detect trend in hydrological series, Hydrol. Process., 16, 1807–1829, https://doi.org/10.1002/hyp.1095, 2002.

Zeileis, A. and Grothendieck, G.: zoo: S3 Infrastructure for Regular and Irregular Time Series, J. Stat. Softw., 14, 1–27, 2005.

Zetterqvist, L.: Statistical Estimation and Interpretation of Trends in Water Quality Time Series, Water Resour. Res., 27, 1637–1648, https://doi.org/10.1029/91wr00478, 1991.

Zhang, Q. and Ball, W. P.: Improving Riverine Constituent Concentration and Flux Estimation by Accounting for Antecedent Discharge Conditions, J. Hydrol., 547, 387–402, https://doi.org/10.1016/j.jhydrol.2016.12.052, 2017.

Zhang, Q., Brady, D. C., Boynton, W. R., and Ball, W. P.: Long-Term Trends of Nutrients and Sediment from the Nontidal Chesapeake Watershed: An Assessment of Progress by River and Season, J. Am. Water Resour. Assoc., 51, 1534–1555, https://doi.org/10.1111/1752-1688.12327, 2015.

8

Monitoring the variations of evapotranspiration due to land use/cover change in a semiarid shrubland

Tingting Gong, Huimin Lei, Dawen Yang, Yang Jiao, and Hanbo Yang

State Key Laboratory of Hydroscience and Engineering, Department of Hydraulic Engineering, Tsinghua University, Beijing, 100084, China

Correspondence to: Huimin Lei (leihm@tsinghua.edu.cn)

Abstract. Evapotranspiration (E_T) is an important process in the hydrological cycle, and vegetation change is a primary factor that affects E_T. In this study, we analyzed the annual and inter-annual characteristics of E_T using continuous observation data from eddy covariance (EC) measurement over 4 years (1 July 2011 to 30 June 2015) in a semiarid shrubland of Mu Us Sandy Land, China. The Normalized Difference Vegetation Index (NDVI) was demonstrated as the predominant factor that influences the seasonal variations in E_T. Additionally, during the land degradation and vegetation rehabilitation processes, E_T and normalized E_T both increased due to the integrated effects of the changes in vegetation type, topography, and soil surface characteristics. This study could improve our understanding of the effects of land use/cover change on E_T in the fragile ecosystem of semiarid regions and provide a scientific reference for the sustainable management of regional land and water resources.

1 Introduction

Arid and semiarid biomes cover approximately 40 % of the Earth's terrestrial surface (Fernández, 2002). Previous studies have shown that more than 50 % of precipitation (P) is consumed by evapotranspiration (E_T) (Yang et al., 2007; Liu et al., 2002). Moreover, a slight change in E_T could have significant influences on water cycle and the ratio of E_T/P could increase to even 90 % or more in these regions (Mo et al., 2004; Glenn et al., 2007). In terms of physical processes, E_T is affected by net radiation (Valipour et al., 2015), water vapor pressure deficit (Zhang et al., 2014), wind speed (Falamarzi et al., 2014), and soil water stress (Allen et al., 1998). Moreover, vegetation condition is also a crucial factor influencing E_T (Tian et al., 2015; Wang et al., 2011; Piao et al., 2006; Mackay et al., 2007).

Vegetation change mainly includes phenological change (temporal) and land use/cover change (spatial). Phenological change reflects the response of plants to climate change (vegetation greening and browning processes) (Ge et al., 2015), which actively controls E_T through internal physiologies such as stomatal conductance (Pearcy et al., 1989), as well as the number and sizes of stomata (Turrell, 1947). In general, transpiration is directly proportional to stomatal conductance at the leaf scale (Leuning et al., 1995). At the canopy scale, E_T is positively proportional to surface conductance, which is an integration of stomatal conductance and leaf area (Ding et al., 2014). Thus, as a good indicator of vegetation phenological change, many studies have found that E_T is positively related to vegetation indexes such as the Normalized Difference Vegetation Index (NDVI) (Gu et al., 2007). Land use/cover change influences E_T by modifying vegetation species with different transpiration rates, radiation transfers within the canopy (Martens et al., 2000; Panferov et al., 2001), topography (Lv et al., 2006), albedos (Zeng and Yoon, 2009), soil texture (Maayar and Chen, 2006), litter coverage (Wang, 1992), and biological soil crusts (BSCs) (Yang et al., 2015; Fu et al., 2010; Liu, 2012). These complex processes result in no consensus on the effects of land use/cover change on E_T. For example, during the land degradation process, some researchers found that warming air temperature was the main cause of making E_T increase (Zeng and Yang, 2008; Li et al., 2013; Feddema and Freire, 2001). By contrast, a decline in E_T was found along with the deforestation process because of less transpiration (Snyman,

2001; Souza and Oyama, 2011) or higher albedo (Zeng et al., 2002). Moreover, no changes in E_T during the land degradation process were reported either (Hoshino et al., 2009). Thus, there has been an important push to better understand how E_T responds to vegetation change, especially to the land use/cover change.

Three methods were usually employed to assess the effects of vegetation change on E_T: numerical models, paired comparative approaches, and in situ field observations. In these methods, numerical models are widely used (Twine et al., 2004; Kim et al., 2005; Li et al., 2009; Cornelissen et al., 2013; Mo et al., 2004). However, model parameterization of vegetation conditions is a big challenge, as the aforementioned complex underlying mechanisms may not be completely considered in the models. Therefore, the simulated effects of vegetation change on E_T are highly dependent on model parameterizations, which may induce uncertainty (Cornelissen et al., 2013; Li et al., 2009). The paired comparative approach is often considered the best method; nonetheless, it is difficult to find two sites with similar meteorological conditions but different vegetation conditions (Li et al., 2009; Lorup et al., 1998). Moreover, the method of in situ field observations is widely used to investigate long-term land–atmosphere exchanges. However, the land use/cover conditions at sites are generally stable, and only the response of E_T to vegetation phenological change can be observed, such as the E_T variations in grassland (Y. Zhang et al., 2005), mixed plantation (cork oak, black locust, and arborvitae) (Tong et al., 2017), vineyard (Li et al., 2015), and grazed steppe (Chen et al., 2009; Vetter et al., 2012). Continuous field observations under both land degradation and vegetation rehabilitation processes have rarely been documented, especially in the semiarid shrubland.

The Mu Us Sandy Land is a semiarid shrubland ecosystem on the northern margin of the Loess Plateau in China. The area covers only 40 000 km^2 (Dong and Zhang, 2001) and is ecologically fragile (Yang et al., 2007). In such an ecosystem, sand dunes and BSCs are commonly observed (Gao et al., 2014; Yang et al., 2015; Li and Li, 2000; Liu, 2012). Due to the existence of BSCs and dry sand layers (Z. Wang et al., 2006; Feng, 1994; Liu et al., 2006; Yuan et al., 2008), soil evaporation has been effectively retained; therefore, the Mu Us Sandy Land contains abundant groundwater (Li and Li, 2000). During the past decades, rapid land use/cover changes have occurred in this region due to agricultural reclamation (Wu and Ci, 2002; Ostwald and Chen, 2006; Zhang et al., 2006), leading to dramatic changes in vegetation conditions. With respect to the specific question of whether land use/cover change will lead to increases in E_T or not, a continuous measurement of E_T under different land use/cover conditions is required in this region. Coincidentally, two processes of land use/cover changes (land degradation and vegetation rehabilitation) have occurred at the edge of the Mu Us Sandy Land, providing us with a unique opportunity to study the effects of land use/cover change on E_T.

Hence, based on the 4-year measurement of E_T by eddy covariance techniques, this study analyzed the seasonal and inter-annual variations in E_T, and discussed the possible reasons for the responses of E_T to land use/cover change. Our results were expected to provide a scientific reference for the sustainable management of regional land and water resources in the context of intensive agricultural reclamation.

2 Case study and data

2.1 Site description

The study was conducted at the Yulin flux site (38°26′ N, 109°28′ E, 1233 m), which was established in June 2011. This site is located in a landform transition zone that changes from the Mu Us Sandy Land to the north Shaanxi Loess Plateau (Fig. 1). This site is a semiarid area with temperate continental monsoon climate. According to long-term climate data (1951–2012) from a meteorological station in Yulin (Fig. 1), the annual precipitation varied from 235 to 685 mm, with a mean of 402 mm, and more than 50 % of annual precipitation fell in the monsoon season (July–September). The mean annual air temperature was 8.4 °C over the past 61 years. The dominant soil type is sand (98 % sand) (saturated soil water content of 0.43 m^3 m^{-3}, field capacity of 0.16 m^3 m^{-3}, residual moisture content of 0.045 m^3 m^{-3}). There are widely distributed fixed sand dunes and semi-fixed sand dunes around the site, and the depth of the dry sand layer is 10 cm (Z. Wang et al., 2006). The mean groundwater depth at our study site from 1 July 2011 to 30 June 2015 was 3.5 m.

Shortage of water is the critical limiting factor for vegetation growth in this site, and drought-enduring vegetation (e.g., shrubs) prevails as a result of droughts (Wang et al., 2002; Wu, 2006). The study site is mainly covered with mixed vegetation: the native drought-enduring shrubs with low water demand (e.g., *Artemisia ordosica* and *Salix psammophila*) (Fig. 2a) and the sparse grass (mainly distributed at the bottom of sand dunes because of the better soil moisture condition) (Lv et al., 2006). The maximum root depth of the shrubs was approximately 160 cm. Xiao et al. (2005) reported that the growing season of *Artemisia ordosica* and *Salix psammophila* spanned from late April to late September. Therefore, we defined the period from 1 May to 30 September as the vegetation growing season for data analysis in this study. On 15 August and 7 September 2011, we did surveys of the vegetation coverage by randomly selecting seven samples around the flux tower (5 × 500 cm × 500 cm and 2 × 1000 cm × 1000 cm). We found that the vegetation coverage was 28.2 % in August and 27.9 % in September.

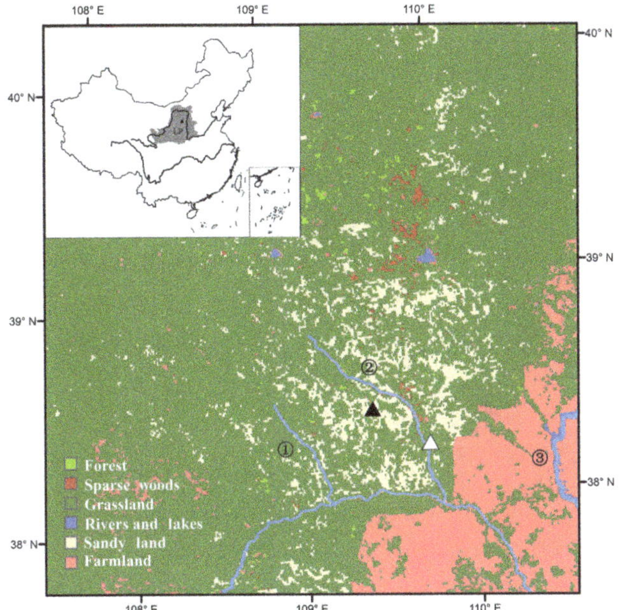

Figure 1. Location of the Loess Plateau and map of the study site (LP: the Loess Plateau; black triangle: flux tower; white triangle: Yulin meteorological station; (1): Tu River; (2): Yuxi River; (3): Yellow River).

At the end of June 2012, the land use/cover condition around the eastern portion of the flux tower began to be changed by farmers (leaves and branches were cut, and the sand dunes were bulldozed) (Fig. 2c), converting part of the natural vegetated land to bare land, with the planning of planting potatoes in the future. As time went on, natural grass gradually grew out in the area of bare land before potatoes were planted. Thus, our study period (1 July 2011 to 30 June 2015) was divided into four periods according to the land use/cover conditions: (a) Period I (1 July 2011 to 30 June 2012), the period with the natural land use/cover condition (i.e., mixed sparsely distributed shrubs and grass) (Fig. 2a and b); (b) Period II (1 July 2012 to 30 June 2013), the transitional period when the land use/cover condition started to change (some natural vegetation removed and sand dunes bulldozed); (c) Period III (1 July 2013 to 30 June 2014), the period when the land use/cover condition constituted two parts: the natural vegetation zone and the bare soil zone (Fig. 2c); and (d) Period IV (1 July 2014 to 30 June 2015), the period when the bare soil zone was gradually covered by regrowing grass (Fig. 2d).

2.2 Field measurements

2.2.1 Eddy covariance system measurements

Net exchange of water vapor between atmosphere and canopy at this site is measured by the eddy covariance (EC) flux measurement, which assesses the fluxes of land–atmosphere (such as water and energy) (Baldocchi et al., 2001). The data are essential for the estimation of the water and energy balance (Franssen et al., 2010). At our site, the EC system is installed at a height of 7.53 m above the ground surface, using CSAT3 three-dimensional sonic anemometers (Campbell Scientific Inc., Logan, UT, USA) for wind and temperature fluctuation measurements and a LI-7500A open-path infrared gas analyzer (LI-COR, Inc., Lincoln, NE, USA) for water vapor content measurement.

2.2.2 Other measurements

Net radiation (R_n) is measured by a net radiometer (CNR-4; KIPP&ZONEN, Delft, the Netherlands), including four radiometers measuring the incoming and reflected short-wave radiation (R_S), and incoming and outgoing long-wave radiation (R_L). Sunshine duration (D_S) is measured by a sunshine recorder (CSD3; KIPP&ZONEN, Delft, the Netherlands). Wind speed and direction (05103, Young Co. Traverse City, MI, USA) are measured at 10 m above the ground surface. Precipitation (P, mm) is recorded with a tipping bucket rain gauge (TE525MM; Campbell Scientific Inc., Logan, UT, USA) installed at a height of 0.7 m above the ground surface. Air temperature (T_a) and relative humidity (R_H) are measured by a temperature and relative humidity probe (HMP45C; Campbell Scientific Inc., Logan, UT, USA) at a height of 2.6 m above the ground surface. Soil water content (θ) is measured by time domain reflectometry (TDR) sensors (CS616; Campbell Scientific Inc., Logan, UT, USA), soil temperature (T_s) is measured by thermocouples (109; Campbell Scientific Inc., Logan, UT, USA), and soil heat flux (G) is measured by heat flux plates (HFP01SC; Campbell Scientific Inc., Logan, UT, USA) at a depth of 0.03 m below the ground surface. These ground variables (G, θ, T_s) are measured beneath the surface at two profiles: a plant canopy profile and a bare soil profile. θ and T_s are measured at depths of 5, 10, 20, 40, 60, 80, 120, and 160 cm below the ground surface. The groundwater table is measured by an automatic sensor (CS450-L; Campbell Scientific Inc., Logan, UT, USA), which is installed in a groundwater well close to the tower.

2.3 Flux data processing

The 10 Hz three-dimensional wind speed and water vapor concentrations that were collected by the EC technique were processed to half-hourly latent heat flux (λE_T) using Eddypro processing software (v5.2.0, LI-COR, Lincoln, NE, USA). The main principle is that λE_T can be expressed as $\rho_a \overline{w'q'}$ (where w' is the fluctuation of vertical wind speed, q' is the fluctuation of specific humidity, and ρ_a is the air density). The software also applies the quality control of data, including spike removal, tilt correction, time lag compensation, turbulent fluctuation blocking, and spectral corrections. The percentages of half-hourly λE_T values removed (includ-

Figure 2. Land use/cover conditions at the study site: **(a)** the natural land use/cover condition of shrubland (photo was taken on 6 August 2011); **(b)** the natural land use/cover condition of grassland (photo was taken on 7 September 2011); **(c)** the undisturbed zone (natural vegetation) and the disturbed zone (bare soil) in the land degradation process (photo was taken on 26 April 2013); **(d)** the undisturbed zone (natural vegetation) and the disturbed zone (grassland) during the vegetation rehabilitation process (photo was taken on 16 August 2014).

ing missing and rejected) through the quality control procedure were 17.3 % in Period I, 20.2 % in Period II, 16.5 % in Period III, and 18.6 % in Period IV. Almost all the removed λE_T values occurred during the nighttime (89.1 % in Period I, 91.3 % in Period II, 92.6 % in Period III, and 88.7 % in Period IV). During the nighttime, the change in λE_T was small, and E_T values were close to zero. Therefore, after removal of the nighttime λE_T values, the errors of the gap-filled nighttime values based on the neighboring good data were small. Moreover, nighttime λE_T values accounted for only a small proportion of the daily E_T. Furthermore, the percentages of rejected and missing data in our study are similar to those reported by other scholars, and these percentages are in a range of 15–31 % (Falge et al., 2001; Wever et al., 2002; Mauder et al., 2006). Therefore, the λE_T data set was considered reliable after a quality control procedure.

After quality control, missing and rejected data were gap-filled in order to create continuous data sets. Three methods were applied in the gap-filling procedure: (1) linear interpolation was used to fill gaps of less than 1 h by calculating an average of the values before and after the data gap; (2) for gaps that are larger than 1 h but smaller than 7 days, the mean diurnal variation (MDV) method (Falge et al., 2001) was used; (3) for gaps that are larger than 7 days but smaller than 15 days in daily λE_T values, we fitted the relationship between daily λE_T ($\mathrm{W\,m^{-2}}$) and the daily available energy flux ($R_n - G$) ($\mathrm{W\,m^{-2}}$) in each period. We chose the function f with the highest coefficient of correlation (R) in each period (Yan et al., 2013), and the function was expressed as $f = a \cdot (R_n - G)^2 + b \cdot (R_n - G) + c$ (Period I: $a = 0.0014\,\mathrm{m^2\,W^{-1}}$, $b = 0.075$, $c = 10.69\,\mathrm{W\,m^{-2}}$, $R = 0.77$; Period II: $a = 0.0012\,\mathrm{m^2\,W^{-1}}$, $b = 0.056$, $c = 17.69\,\mathrm{W\,m^{-2}}$, $R = 0.67$; Period III: $a = 0.0014\,\mathrm{m^2\,W^{-1}}$, $b = 0.16$, $c = 13.24\,\mathrm{W\,m^{-2}}$, $R = 0.75$; and Period IV: $a = 0.0015\,\mathrm{m^2\,W^{-1}}$, $b = -0.083$, $c = 25.87\,\mathrm{W\,m^{-2}}$, $R = 0.69$). Then, we used the fitted function f in each period to estimate the daily λE_T values of large gaps. In addition, gaps that are larger than 7 days but smaller than 15 days mostly appeared in the winter, which accounted for a small proportion of annual λE_T.

3 Methodology

3.1 Footprint model

In order to determine the contributing source area of flux at our site, the scalar flux footprint model proposed by Hsieh et al. (2000) was used. The analytic model accurately describes the relationship between the footprint, observation height, surface roughness, and atmospheric stability. The fetch F_f was calculated as follows:

$$F_f / Z_m = D / (0.105 \times k^2) Z_m^{-1} |L|^{1-Q} Z_u^Q, \tag{1}$$

where k is the von Karman constant (= 0.40), D and Q are similarity constants (for stable conditions, $D = 0.28$ and $Q = 0.59$; for near neutral and neutral conditions, $D = 0.97$ and $Q = 1$; for unstable conditions, $D = 2.44$ and $Q = 1.33$),

L is the Obukhov length, Z_m is the height of the wind instrument (= 10.0 m), and Z_u is defined as (Hsieh et al., 2000)

$$Z_u = Z_m(\ln(Z_m/Z_0) - 1 + Z_m/Z_0), \quad (2)$$

where Z_0 is the height of the momentum roughness (0.05 m).

3.2 Method of analyzing controlling factors on E_T

It is generally recognized that potential evapotranspiration (E_{TP}), vegetation condition, and soil water stress are the three main factors that control E_T (Lettenmaier and Famiglietti, 2006; Chen et al., 2014). In order to decouple the effect of vegetation change from the integrated effects of these three factors on E_T, we used a simple equation which was similar to the FAO single crop coefficient method (Irrigation and Drainage Paper No. 56 (FAO-56)). This equation can be expressed as follows:

$$E_T = E_{TP} \times f_v\,(\text{vegetation}) \times f_s\,(\text{soil water}), \quad (3)$$

where f_v (vegetation) represents the effect of vegetation change on E_T and f_s (soil water) represents the effect of soil water stress on E_T.

Moreover, f_v (vegetation) can be regarded as the normalized E_T, which eliminates the effects of atmospheric and soil water stress on E_T and can be expressed by rearranging Eq. (3):

$$f_v\,(\text{vegetation}) = E_T/[E_{TP} \times f_s\,(\text{soil water})]. \quad (4)$$

3.2.1 Potential evapotranspiration

E_{TP} (mm day^{-1}) was estimated by the following equation (Maidment, 1993), which is a modification of the Penman equation:

$$E_{TP} = \frac{\Delta}{\Delta+\gamma}(R_n - G) + \frac{\rho_a c_p/r_a}{\Delta+\gamma}\frac{\text{VPD}}{\lambda}, \quad (5)$$

where the units of R_n and G are mm day^{-1}; ρ_a is the air density ($= 3.486\frac{P_a}{275+T_a}$, kg m^{-3}, where P_a is the atmospheric pressure in kPa and T_a is in degrees Celsius); c_p is the specific heat of moist air (= 1.013 kJ kg^{-1} °C^{-1}); Δ is the slope of the saturation vapor–pressure–temperature curve (kPa °C^{-1}); VPD is the difference between the mean saturation vapor pressure (e_s, kPa) and actual vapor pressure (e_a, kPa); and λ is the latent heat of vaporization of water (= 2.51 MJ kg^{-1}). γ is the psychrometric constant (kPa °C^{-1}), which is calculated by the following equation:

$$\gamma = \frac{c_p P_a}{\varepsilon\lambda}, \quad (6)$$

where ε is the ratio of the molecular weight of water vapor to that of dry air (= 0.622).

r_a is the aerodynamic resistance, which can be calculated as follows (Penman, 1948):

$$r_a = \frac{4.72\left[\ln\left(\frac{Z_h}{Z_{a0}}\right)\right]\left[\ln\left(\frac{Z_h}{Z_{ao}}\right)\right]}{1+0.536U_2}, \quad (7)$$

where Z_h is the height at which meteorological variables are measured (2 m), and Z_{ao} is the aerodynamic roughness of the surface (0.00137 m) (Penman, 1963); U_2 is the daily wind speed at a height of 2.0 m (m s^{-1}), and it was calculated by the wind speed at a height of 10.0 m (U_{10}, m s^{-1}):

$$U_2 = U_{10}\frac{4.87}{\ln(67.8\cdot 10 - 5.42)}. \quad (8)$$

3.2.2 Vegetation parameters

In this study, vegetation phenology was represented by Moderate Resolution Imaging Spectroradiometer (MODIS) NDVI data when the land use/cover conditions were fixed. NDVI is sufficiently stable to reflect the seasonal changes of any vegetation (Huete et al., 2002). Higher NDVI generally reflects the greater photosynthetic capacity (greenness) of the vegetation canopy (Gu et al., 2007; Tucker, 1979). The daily NDVI was calculated by daily surface reflectance data:

$$\text{NDVI} = \frac{\text{NIR}-\text{VIS}}{\text{NIR}+\text{VIS}}, \quad (9)$$

where NIR is the spectral response in the near-infrared band (857 nm) and VIS is the visible red radiation band (645 nm). In this study, NDVI was calculated by using MODIS/Terra data (MOD09GQ) ($\text{NDVI}_{\text{Terra}}$) and MODIS/Aqua data (MYD09GQ) ($\text{NDVI}_{\text{Aqua}}$) (http://reverb.echo.nasa.gov), respectively. As we found that there were slight differences ($|\text{NDVI}_{\text{Terra}} - \text{NDVI}_{\text{Aqua}}| = 0.01 \pm 0.0075$) between $\text{NDVI}_{\text{Terra}}$ and $\text{NDVI}_{\text{Aqua}}$, we calculated NDVI by averaging $\text{NDVI}_{\text{Terra}}$ and $\text{NDVI}_{\text{Aqua}}$ in order to eliminate the impacts of such differences. The calculated NDVI values were then filtered to remove anomalous hikes and drops (Lunetta et al., 2006), and the smoothing spline method was used to produce a smoother profile.

Theoretically, land use/cover change can be evaluated by comparing the land use/cover maps in two different periods. However, transient land use/cover maps were unavailable at our site. Therefore, we separated the study area within the footprint into two zones: the undisturbed zone without any land use/cover change was deemed zone A and the disturbed zone with land use/cover change was deemed zone B. In zone A, vegetation change included only vegetation phenological change; however, in zone B, there were not only vegetation phenological changes, but also land use/cover changes. Based on the assumption that the phenological changes caused by climate in the two zones were the same, we defined an indicator (D_{lu}) as a measure of land use/cover change:

$$D_{lu} = M_A - M_B, \quad (10)$$

where M_A and M_B are the monthly vegetation coverages of zone A and zone B, respectively. The monthly vegetation coverage was calculated by monthly NDVI values (Gutman and Ignatov, 1998):

$$M = (NDVI - NDVI_{min})/(NDVI_{max} - NDVI_{min}), \quad (11)$$

where $NDVI_{max}$ is the maximum value (0.8 in this study) and $NDVI_{min}$ is the minimum value (0.05 in this study) (Gutman and Ignatov, 1998). The calculated monthly M values (27.6 and 24.2 %) were consistent with the measured vegetation coverages in August 2011 (28.2 %) and September 2011 (27.9 %) at our study site.

3.2.3 Soil water stress

The effects of the soil water stress on E_T can be described in three stages (Idso et al., 1974). Stage 1: the soil water is enough to satisfy the potential evaporation rate ($f_s = 1$); stage 2: the soil is drying and water availability limits E_T ($0 < f_s < 1$); and stage 3: the soil is dry and evaporation can be considered negligible ($f_s = 0$). We used daily soil water content in the root depth (θ_r) to estimate f_s by the following expression (Morillas et al., 2013):

$$f_s = \begin{cases} = 1 & \theta_r > \theta_k, \\ = 0 & \theta_r < \theta_w, \\ = \dfrac{\theta_r - \theta_w}{\theta_k - \theta_w} & \theta_w \le \theta_r \le \theta_k. \end{cases} \quad (12)$$

where θ_w is the wilting value and θ_k is the stable field capacity which is considered to be equivalent to 60 % of the field capacity (Lei et al., 1988; Wang et al., 2008). θ_r was calculated by measured soil water contents at different depths (θ_i where $i = 5$, 10, 20, 40, 60, 80, 120, and 160 cm). From land surface to the depth of 5 cm, the soil water profile was assumed triangular, while at other depths, the soil water profiles were assumed trapezoidal. Therefore, the soil moisture of root zone was calculated as

$$\theta_r = \frac{0.5\left[\begin{array}{l} 5\theta_5 + (\theta_5 + \theta_{10}) \cdot (10 - 5) + (\theta_{10} + \theta_{20}) \cdot (20 - 10) \\ + (\theta_{20} + \theta_{40}) \cdot (40 - 20) + (\theta_{40} + \theta_{60}) \cdot (60 - 40) \\ + (\theta_{60} + \theta_{80}) \cdot (80 - 60) + (\theta_{80} + \theta_{120}) \cdot (120 - 80) \\ + (\theta_{120} + \theta_{160}) \cdot (160 - 120) \end{array}\right]}{160}, \quad (13)$$

where $\theta_i (i = 5$, 10, 20, 40, 60, 80, 120, and 160 cm) was calculated by taking a weighted average of the measured values in the canopy and bare surface patches,

$$\theta_i = M_A \times \theta_{i,c} + (1 - M_A) \times \theta_{i,b}, \quad (14)$$

where $\theta_{i,c}$ and $\theta_{i,b}$ refer to the measured soil water contents of the canopy patch and bare soil patch at the depth of i cm, respectively.

3.3 Statistical analysis

In this study, we chose daily data in Period I to analyze the correlations between E_T and the three controlling factors (E_{TP}, NDVI, and f_s). We used several common functions (e.g., an exponential function, a linear function, a logarithmic function, and a quadratic function) to fit these correlations. We found that the determination coefficient (R^2) of the linear function was generally the highest. Therefore, in this study, we chose the linear function to fit the correlations between E_T and the three controlling factors. Additionally, a significant t test was performed to evaluate the degrees of these correlations. Moreover, data on rainy days were removed because E_T values were gap-filled rather than measured.

4 Results

4.1 Footprint and energy balance closure

Based on the footprint model, we got the half-hourly scatter data (Eq. 2), and, according to the wind rose diagram (Fig. 3a), the prevailing wind directions at this site were northwesterly and southeasterly. Therefore, we chose an ellipse to enclose the scatters and simulate the footprint (Fig. 3b). Under unstable conditions, 93 % of the half-hourly flux data are plotted within the ellipse.

Additionally, we measured the boundary of zone B in October 2013 when the land use/cover condition in zone B had stopped changing (Fig. 3b). There were 11 pixels (250 m × 250 m) in zone A and 19 pixels (250 m × 250 m) in zone B, and thus, when calculating the weight-averaged NDVI ($NDVI_w$) within the footprint, we chose the weighted coefficient as $\beta = 11/(11 + 19)$.

EC system performance was assessed by the energy balance closure which was calculated by conducting the linear regression between available energy ($R_n - G$) and the sum of surface fluxes ($\lambda E_T + H$), which is also used to examine the quality of flux data (Wilson et al., 2002). The linear regression yielded a slope of 0.87, an intercept of $-1.42\,W\,m^{-2}$, and an R^2 of 0.82. These indicators suggested that the measurements at our experimental site provided reliable flux data and that the EC measurements underestimated the sum of the surface fluxes to the extent of 13 %. Many researchers have investigated energy imbalance (Barr et al., 2006; Wilson et al., 2002; Franssen et al., 2010), and there is a consensus that it is difficult to examine the exact reasons for the imbalance.

4.2 Characteristics of environmental variables

A brief summary of key environmental variables is presented in this section. Four-year and long-term (1954–2014) average monthly values of D_s, T_a, R_H, and P are shown in Fig. 4. Monthly D_s was much higher than the long-term average monthly values, except in July and September. The highest value of D_s was observed in May (299.5 h) and the low-

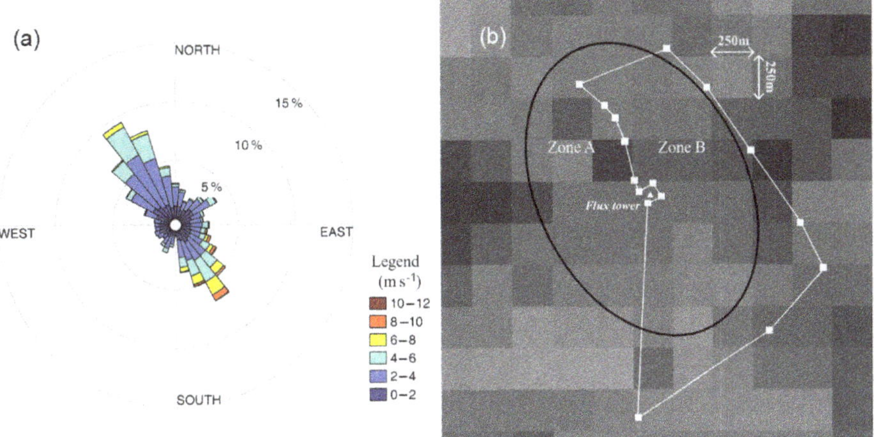

Figure 3. Diagrams of the wind rose and footprint: **(a)** wind rose of the study site by using half-hourly wind speed and wind direction data and **(b)** simulated footprint by ellipse (the long axis is 1682 m, and the short axis is 1263 m; zone A is the source area in which the land use/cover condition did not change, while zone B is the source area in which the land use/cover condition did change due to human activities; the white triangle is the flux tower).

Table 1. Daily air temperature (T_a, °C), relatively humidity (R_H, %), total sunshine duration (D_S, h), soil water content of the root zone (θ_r, m^3 m^{-3}), the groundwater level (GWL, m), and total precipitation (P, mm) in 1954–2014 and in the growing season of each period (because there were some missing data in Period IV (from 12 September to 23 November 2014 and from 13 March to 22 April 2015), we excluded data in these two time ranges of Periods I–III and 1954–2014).

Variable	1954–2014	I	II	III	IV
T_a (°C)	19.8	19.6	20.4	19.9	19.3
R_H (%)	57.7	57.3	54.9	53.4	52
D_S (h)	213.3	220.7	215.8	218.2	220.7
P (mm)	329.8	357.1	384.1	330.2	199.8
θ_r (m^3 m^{-3})	–	0.077	0.077	0.076	0.064
GWL (m)	–	−3.8	−3.6	−3.0	−3.5

est was observed in February (206.6 h). The seasonal characteristics of T_a showed a highly similar pattern to that of long-term average monthly values, and the differences were less than 1 °C, except in July, January, and March. The highest value of T_a was observed in July (22.1 °C) and the lowest was observed in December (−8.1 °C). The values of R_H were almost lower than the long-term average monthly values, especially in March and April. The highest R_H was observed in September (65.4 %) and the lowest was observed in March (35.1 %). The seasonal distributions of P were consistent with the long-term average monthly values, and 89.7 % of P occurred in the growing season. P was highest in July (120.5 mm) and lowest in January (0.3 mm).

The inter-annual characteristics of daily T_a, D_s, R_H, θ_r, groundwater level (GWL), and total P in the growing season of each period are listed in Table 1.

The values of T_a, R_H, P, and θ_r in the growing season of Period IV were the lowest compared to those in the other three periods. Periods I–III were all wet years, while Period IV was a dry year. The values of θ_r in Periods I–III were similar; however, θ_r decreased by 0.0113 m^3 m^{-3} in Period IV. The mean GWL in Period III was the shallowest.

4.3 Seasonal variations in E_T due to climate variability and vegetation phenology

The seasonal curve of E_T in each year had a single peak value (Fig. 5a), with higher E_T appearing mostly in the growing season, while lower E_T appeared in the non-growing season. The daily E_T ranged from 0.0 mm day^{-1} to 6.8 mm day^{-1} during the four periods; the highest E_T was observed on 22 June 2013, which was the day after a continuous rainfall event that extended from 19 to 21 June 2013 (90.3 mm). The lowest E_T appeared on 28 November 2012, which was in the frozen period (late November to early March at our study site). On rainy days, E_{TP} (Fig. 5b) was low due to low net radiation and air temperature. E_{TP} ranged from 0.2 mm day^{-1} in December 2011 to 17.9 mm day^{-1} in September 2013.

The seasonal NDVI curve for natural land use/cover conditions (in zone A during Periods I–IV and in zone B during Period I) represented the process of natural vegetation phenology, and it had a single peak value in each year (Fig. 5c). In early May, the seasonal NDVI curve began to increase as the native vegetation entered the growing season, and a maximum value (0.27 ± 0.01) was reached in July or August. In the winter, the daily NDVI remained relatively constant (0.13 ± 0.01). f_s (Fig. 5d) increased rapidly in response

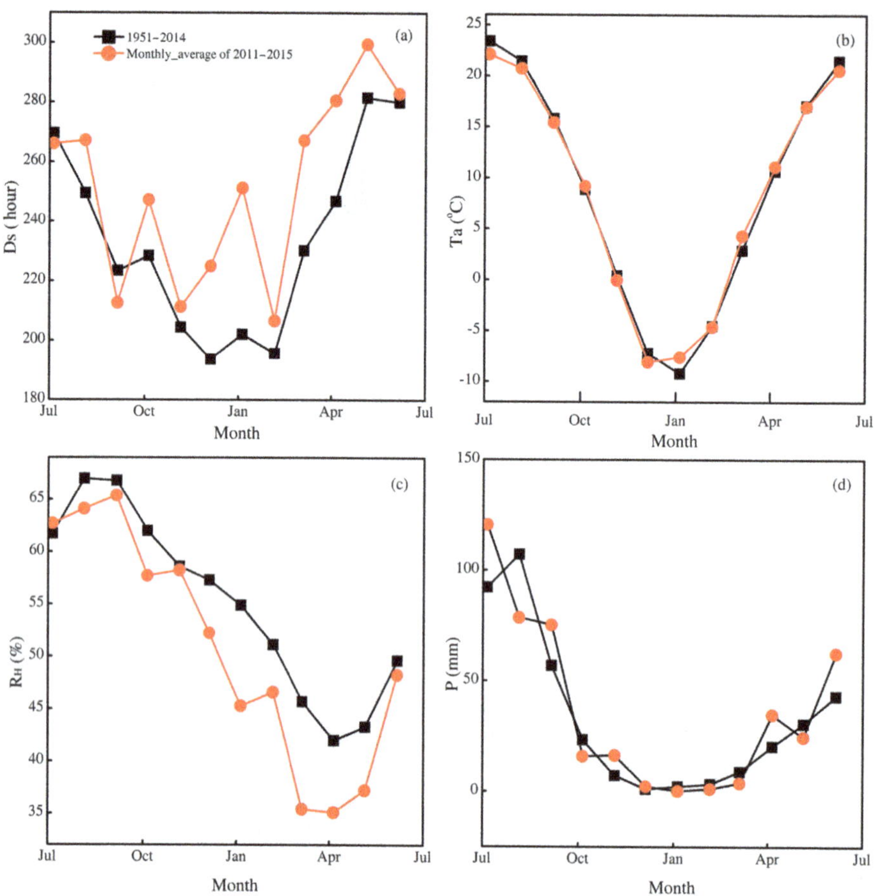

Figure 4. Seasonal characteristics of 4-year and long-term (1954–2014, from Yulin meteorological station) average monthly values of **(a)** sunshine duration (D_S); **(b)** air temperature (T_a); **(c)** relative humidity (R_H); and **(d)** total precipitation (P).

to rainfall events of more than 5 mm a day and decreased rapidly 1 or 2 days after rainfall events. From late November to early March, there was a frozen period when the soil water content was below the wilting point. The groundwater level changed obviously in the monsoon season (July to September) and mildly in the winter (December to February).

The linear correlations between E_T and the three controlling factors all passed the t test at a 95 % confidence level. The R^2 value of the correlation between E_T and $NDVI_w$ ($NDVI_w = NDVI_A \times \beta + NDVI_B \times (1 - \beta)$) was the largest, indicating that NDVI was highly correlated with the daily variations in E_T. To better quantify the effects of the phenological process on E_T, the correlation between daily f_v and $NDVI_w$ in Period I was analyzed (Fig. 7a).

A positive linear regression was found between f_v and $NDVI_w$ (Fig. 7a). The slope of the linear regression was used to evaluate the degree of the correlation between f_v and the vegetation phenological process. We found that when $NDVI_w$ increased by 1 unit, f_v increased by approximately 1.86 units.

4.4 Inter-annual variations in E_T due to land use/cover change

During the four periods, in zone A, the NDVI values of each period were similar because the land use/cover condition did not change, while in zone B, the peak values of NDVI first declined from 0.28 to 0.15 (Period I to Period III) due to the land use/cover condition changing from mixed vegetation to bare soil. The peak NDVI values then increased to 0.22 (Period IV) due to grass recovery (Fig. 5c). An interesting phenomenon was observed accompanied by the changing process of land use/cover conditions: E_T in the growing season gradually increased from Period I to Period III (Table 2), while it increased greatly in Period IV even with less precipitation, because a mass of soil water and groundwater was consumed to satisfy the E_T demand (Fig. 5e).

Compared with Period I, D_{lu} values in Period II and Period III gradually increased, while D_{lu} in Period IV decreased. Taking August in each period as an example, in Period I, D_{lu} was 0.2 %, while in Periods II–IV, D_{lu} were 2.9, 12.6, and 8.6 %, respectively. In order to eliminate the influence of vegetation phenological change on E_T, we chose the

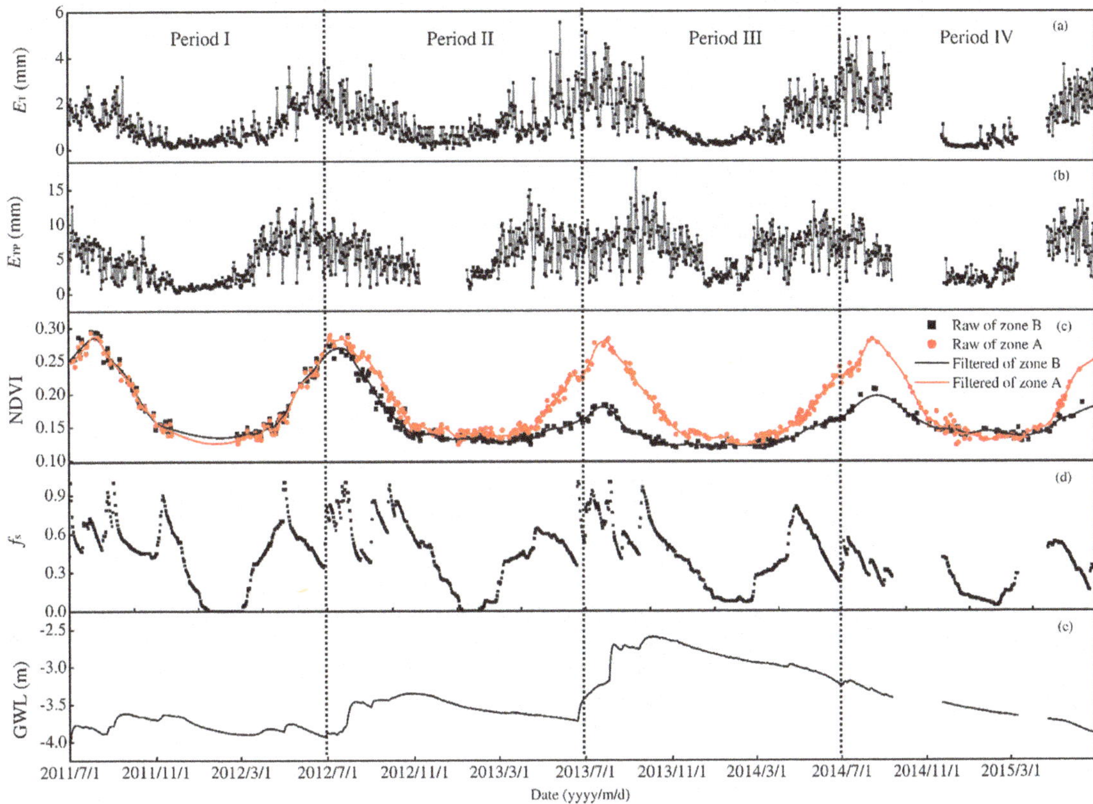

Figure 5. Seasonal and inter-annual characteristics of daily **(a)** evapotranspiration (E_T, mm); **(b)** potential evapotranspiration (E_{TP}, mm); **(c)** NDVI in zone A and zone B within the footprint; **(d)** the soil water stress of the root zone (f_S); and **(e)** the groundwater level (GWL, m) from 1 July 2011 to 30 June 2015.

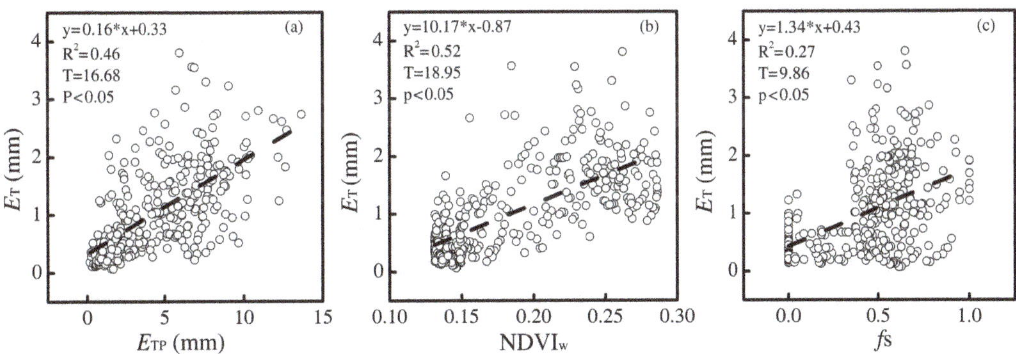

Figure 6. The correlations between daily evapotranspiration (E_T, mm) and its controlling factors: **(a)** daily potential evapotranspiration (E_{TP}, mm); **(b)** daily weight-averaged NDVI ($NDVI_w$) within the footprint; and **(c)** daily soil water stress of the root zone (f_S) in Period I by excluding the data on rainy days (r: Pearson's correlation coefficient; T: t test significance).

growing season of each period to analyze the correlation between f_v and D_{lu}.

The quantitative results of the correlation between D_{lu} and f_v are shown in Fig. 7b. From Period I to Period III, as land surface characteristics changed (the natural vegetation in zone B was cleared, the fixed and semi-fixed sand dunes were bulldozed, and the BSCs and dry sand layers disappeared), f_v increased, and this increase was more evident in Period III (from 78.5 to 88.1). When the land use/cover conditions in zone B gradually changed from bare soil to sparse grassland due to the self-restoring capacity of nature, f_v increased significantly (from 88.1 to 111.3).

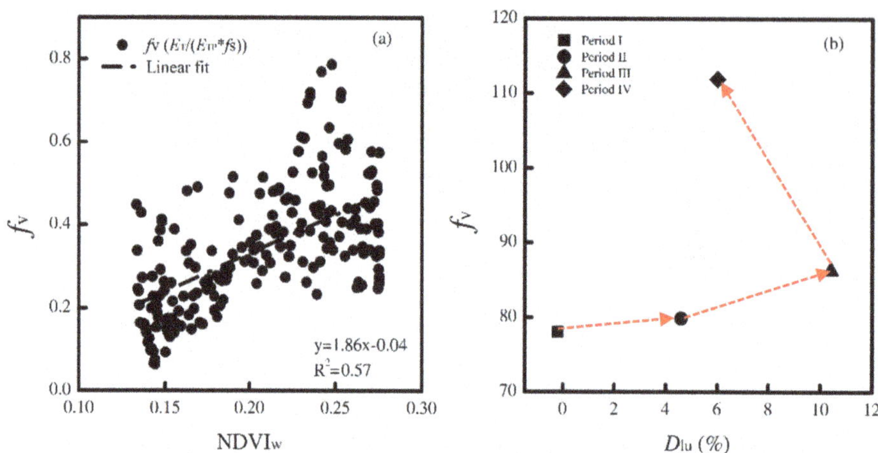

Figure 7. Quantitative analysis of the correlations between **(a)** vegetation phenological change (NDVI_w) and daily normalized E_T ($f_\mathrm{v} = E_\mathrm{T}/(E_\mathrm{TP} \times f_\mathrm{s})$) in Period I (excluding the data on rainy days and frozen days) and **(b)** the indicator of land use/cover change (D_lu) and total normalized E_T ($f_\mathrm{v} = E_\mathrm{T}/(E_\mathrm{TP} \times f_\mathrm{s})$) in the growing season of each period.

Table 2. Typical values of total evapotranspiration (E_T, mm), total potential evapotranspiration (E_TP, mm), the indicator of land use/cover change (D_lu, %), the soil water stress of the root zone (f_s), and normalized E_T (f_v ($= E_\mathrm{T}/(E_\mathrm{TP} \times f_\mathrm{s})$)) in the growing season of each period (because there were some missing data in Period IV (from 12 September to 23 November 2014 and from 13 March to 22 April 2015), we removed the values of E_T, E_TP, and f_s in these two time ranges of Periods I–III).

	Period	E_T (mm)	E_TP (mm)	D_lu (%)	f_s (dimensionless)	f_v (dimensionless)
Growing season	I	238.4	876.1	−0.2	0.62	78.1
	II	236.5	870.7	4.6	0.63	79.9
	III	292.1	956	10.4	0.59	86.3
	IV	332.2	937	6	0.37	111.9

5 Discussion

5.1 Implications of the effects of phenological change on E_T

The correlations between E_T and its controlling factors suggest that at our experimental site, NDVI is the predominant factor that influences the seasonal variations in E_T. The positive linear relationship between f_v and NDVI suggests that transpiration is likely controlled by the stomatal conductance and the numbers of stomata, which are proportional to the leaf area (Pearcy et al., 1989; Turrell, 1947), rather than the atmospheric water demand represented by E_TP.

Various studies have assessed the correlation between vegetation phenological change and E_T, and these results generally reflected consistent and positive linear relationships (Nouri et al., 2014; Rossato et al., 2005; Duchemin et al., 2006; Glenn et al., 2007). However, for different vegetation species, phenological change has effects on E_T to different degrees. Relatively strong regressions between NDVI and E_T have been reported at forested sites (Loukas et al., 2005; Nouri et al., 2014; Lo Seen Chong et al., 1993) and grass-covered sites (Kondoh and Higuchi, 2001; Nouri et al., 2014), with determination coefficients higher than 0.7. These results reflect the strong control between phenological changes and E_T. Thus, we speculate that for high vegetated ecosystems, phenological change may have a significant control on E_T. However, in low vegetated ecosystems such as the sparse shrubland in this study, the relationship between E_T and phenological change is thus positive but relatively weak.

5.2 Possible reasons for the effects of land use/cover changes

During Periods I–IV, the land use/cover conditions at our experimental site underwent changes associated with two processes: land degradation process (Periods II–III) and vegetation rehabilitation process (Period IV). Notable results were observed during these two processes: (1) E_T and normalized E_T values both increased and (2) normalized E_T increased much faster during the vegetation rehabilitation process than it did during the land degradation process.

The effect of phenological change on E_T demonstrates that E_T decreases with leaf browning. Thus, we expect that E_T will also decrease if leaves are cleared by human activities. However, during Periods II–III, not only were leaves cleared, but other land surface properties were also changed (all branches were cut, sand dunes (fixed and semi-fixed) were bulldozed, and the dry sand layers and BSCs were destroyed), resulting in complex land use/cover conditions. These altered land surface properties might contribute to the increase in E_T. Previous studies demonstrated that dry sand layers and BSCs could effectively restrict the soil evaporation rate (Z. Wang et al., 2006; Lv et al., 2006; Liu et al., 2006; Dong et al., 1999; Yang et al., 2015; Fu et al., 2010; Liu, 2012). However, the bulldozing of sand dunes at our experimental site made the elevation of the flat soil surface lower than the average elevation of the undisturbed soil surface (approximately 1.5 m lower, Fig. 2d), making the groundwater depth much shallower than the pre-disturbance depth. Thus, the formation of dry sand layers was restricted due to the shallow groundwater level. In this situation with the destroyed BSCs and the disappeared dry sand layers, the sufficient groundwater supply (Li and Li, 2000) accelerated the loss of water that was stored in shallow soil through evaporation. The enhanced soil evaporation offset the inhibiting effect of transpiration due to leaves clearing, which made E_T increase.

A secondary reason for the increase in soil evaporation was that the soil layer absorbed more solar radiation during the land degradation process. In Period I, the radiation absorbed by the shadowed soil was the solar radiation transmitted into the canopy of shrubs and grass. However, when the natural vegetation was cleared, the leaves and the branches were also removed, which made the shadowed soil exposed and enhanced the radiation absorbed by the soil, thereby increasing soil evaporation (Martens et al., 2000; Panferov et al., 2001). Moreover, the removal of leaves and branches and the disappearance of sand dunes both altered the land surface albedo, which could have directly altered the solar radiation absorbed by the land surface (Dirmeyer and Shukla, 1994; Greene et al., 1999), subsequently leading to the change in E_T.

Some inconsistent results regarding the E_T dynamics during land degradation process were reported. A portion of studies reported that E_T decreased during the land degradation process for different reasons. For example, Souza and Oyama (2011) and Snyman (2001) demonstrated that E_T decreased during the land degradation process due to decreased transpiration in semiarid regions. Lu et al. (2011) considered that the low soil water content was the main reason for the decrease in E_T during the land degradation process. Mao and Cherkauer (2009) also reported a decrease in E_T when the land use/cover condition was converted from forest to grass or cropland in the Great Lakes region. However, contrasting results were also reported regarding the effects of land degradation on E_T. Hoshino et al. (2009) found that there was no difference in E_T during the land degradation process associated with overgrazing in a semiarid Mongolian grassland, and they hypothesized that the reason for this lack of change might be the short grazing time (2 years). Li et al. (2013) demonstrated that the warming air temperature was the main cause of increased E_T during the land degradation process on the Qinghai–Tibet Plateau. Throughout the above studies of E_T during land degradation processes, we found it difficult to accurately describe the trends in E_T, even when the land degradation was only manifested by less vegetation coverage. Therefore, at our study site with complex land surface properties (sand dunes, dry sand layers, and BSCs), the effect of land degradation on E_T was much more complicated.

During the vegetation rehabilitation process (Period IV), f_v increased significantly due to the rehabilitation of grass in zone B, even though less precipitation was observed compared with other periods (Periods I, II, and III). The rehabilitation of grass, rather than shrubs, was due to the sufficient groundwater supply, which resulted from bulldozing the sand dunes. Previous researchers reported that sparse shrubs more commonly grew at the top of sand dunes and that grass grew at the bottom of sand dunes because the difference between groundwater level and the top of sand dunes was larger than that between groundwater level and the bottom of the sand dunes (Lv et al., 2006; Dong et al., 1999). Because transpiration increases with vegetation greening (as demonstrated in Sect. 4.3), the regrowing grass would enhance plant transpiration supplied by the sufficient groundwater. More importantly, the transpiration rate of grass is higher than that of shrubs because shrubs are easier to survive in water-limited conditions (Yang et al., 2014; Wang et al., 2002; Wu, 2006). Therefore, in the vegetation rehabilitation process, the enhancement of the transpiration rate in Period IV was much higher than that in Periods I–III. Similar conclusions regarding increased E_T due to the enhanced transpiration during the vegetation rehabilitation process were reported (Qiu et al., 2011; Yang et al., 2014; Sun et al., 2006; Li et al., 2009). Meanwhile, the regrowing grass could reduce the radiation absorbed by the soil and hence reduce soil evaporation. However, the interception of radiation by the grass canopy was expected to be smaller than that by the mixed shrub and grass canopy in Periods I–III because the leaf area index of grass was smaller than the sum of leaf area and stem area indexes of the mix of shrubs and grass. Therefore, the reduction in soil evaporation in Period IV might be small compared with the increase in soil evaporation in Periods I–III.

We noticed that the GWL decreased continuously from Period III to Period IV due to the enhanced E_T by the regrowth of grass and relative low precipitation, and the regrowing grass has a higher transpiration rate than that of the native mixed shrub and grass. Therefore, we hypothesize that if the land use/cover condition of zone B continues to be grassland over the next several years, the groundwater level will decrease due to the larger consumption, making the soil water condition gradually become poorer for the growth of grass. Then, in this situation, the grassland is expected to degrade to shrubland in zone B because shrubs are easier to survive in water-limited ecosystems. Furthermore, in the next few years, potatoes will be planted in zone B. However, the water requirement of potato is more than 320 mm in the growing season (Qin et al., 2013; Liu et al., 2010) and the water consumption is more than that of natural grass (Qin et al., 2013, 2014; Hou et al., 2010). Thus, irrigation is necessary for planting potatoes during the growing season in water-limited ecosystems (Liu et al., 2010; Fabeiro et al., 2001). Our results imply that the groundwater level might continue to decrease faster with the growth of potatoes in the future, which may lead to a more fragile ecosystem.

6 Conclusion

In this study, seasonal and inter-annual features of E_T were analyzed. Daily E_T was in a range from 0.0 to 6.8 mm day^{-1} during the four periods. NDVI was the predominant factor that influences the seasonal variations in E_T, and vegetation greening had a positive effect on E_T. During the land degradation process (Periods II–III), when natural vegetation (including leaves and branches), sand dunes, dry sand layers, and BSCs were all bulldozed, E_T increased at a mild rate. During the vegetation rehabilitation process (Period IV) with less precipitation, E_T increased at a faster rate than that in the degradation process. Our study demonstrated that when the land use/cover condition was changed by human activities, the underlying mechanisms that influence E_T were complex, and vegetation type, topography, and soil surface characteristics may all contribute to the changes in E_T. Furthermore, our results suggest that when we simulate the effects of land use/cover change on hydrological processes, the vegetation factor might not be the unique factor to parameterize; instead, the integrated effects of land surface and vegetation conditions should be considered. Our study also provides a scientific reference to the regional sustainable management of water resources in the context of intensive agricultural reclamation.

Competing interests. The authors declare that they have no conflict of interest.

Acknowledgements. This research was supported by the National Natural Science Foundation of China (project no. 91225302), the National Key Research and Development Program of China (2016YFC0402404 and 2016YFC0402406), the Basic Research Fund Program of State key Laboratory of Hydroscience and Engineering (grant no. 2014-KY-04) and the Basic Research Plan of Natural Science of Shaanxi Province (2016JQ5105). We thank A. W. Jayawardena for language suggestions and constructive comments of the manuscript.

Edited by: F. Fenicia

References

Allen, R. G., Pereira, L. S., Raes, D., and Smith, M.: Crop evapotranspiration-Guidelines for computing crop water requirements-FAO Irrigation and drainage paper 56, FAO, Rome, 300, D05109, 1998.

Baldocchi, D. D. and Wilson, K. B.: Modeling CO_2 and water vapor exchange of a temperate broadleaved forest across hourly to decadal time scales, Ecol. Model., 142, 155–184, 2001.

Barr, A. G., Morgenstern, K., Black, T. A., McCaughey, J. H., and Nesic, Z.: Surface energy balance closure by the eddy-covariance method above three boreal forest stands and implications for the measurement of the CO_2 flux, Agr. Forest Meteorol., 140, 322–337, 2006.

Chen, S., Chen, J., Lin, G., Zhang, W., Miao, H., Wei, L., Huang, J., and Han, X.: Energy balance and partition in Inner Mongolia steppe ecosystems with different land use types, Agr. Forest Meteorol., 149, 1800–1809, 2009.

Chen, Y., Xia, J. Z., Liang, S. L., Feng, J. M., Fisher, J. B., Li, X., Li, X. L., Liu, S. G., Ma, Z. G., Miyata, A., Mu, Q. Z., Sun, L., Tang, J. W., Wang, K. C., Wen, J., Xue, Y. J., Yu, G. R., Zha, T. G., Zhang, L., Zhang, Q., Zhao, T. B., Zhao, L., and Yuan, W. P.: Comparison of satellite-based evapotranspiration models over terrestrial ecosystems in China, Remote Sens. Environ., 140, 279–293, 2014.

Cornelissen, T., Diekkrüger, B., and Giertz, S.: A comparison of hydrological models for assessing the impact of land use and climate change on discharge in a tropical catchment, J. Hydrol., 498, 221–236, 2013.

Ding, R. S., Kang, S. Z., Zhang, Y. Q., and Du, T. S.: Multi-layer model of water vapor and heat fluxes over maize field in an arid inland region, ShuiLiXueBao, 45, 27–35, 2014 (in Chinese).

Dirmeyer, P. A. and Shukla, J.: Albedo as a modulator of climate response to tropical deforestation, J. Geophys. Res.-Atmos., 99, 20863–20877, 1994.

Dong, X. J. and Zhang, X. S.: Some observations of the adaptions of sandy shrubs to the arid environment in the Mu Us Sandland: leaf water relations and anatomic features, J. Arid Environ., 48, 41–48, 2001.

Dong, X. J., Chen, Z. X., Ala, T. B., Liu, Z. M., and Sideng D. B.: A preliminary study on the water regimes of Sabina Vulgaris in Maowusu sandland, China, Acta Phytoecologica Sinica, 23, 311–319, 1999.

Duchemin, B., Hadria, R., Erraki, S., Boulet, G., Maisongrande, P., Chehbouni, A., Escadafal, R., Ezzahar, J., Hoedjes, J. C. B., Kharrou, M. H., Khabba, S., Mougenot, B., Olioso, A., Rodriguez, J. C., and Simonneaux, V.: Monitoring wheat phenology and irrigation in Central Morocco: On the use of relationships between evapotranspiration, crops coefficients, leaf area index and remotely-sensed vegetation indices, Agr. Water Manage., 79, 1–27, 2006.

Fabeiro, C., de Santa Olalla, F. M., and De Juan, J. A.: Yield and size of deficit irrigated potatoes, Agr. Water Manage., 48, 255–266, 2001.

Falamarzi, Y., Palizdan, N., Huang, Y. F., and Lee, T. S.: Estimating evapotranspiration from temperature and wind speed data using artificial and wavelet neural networks (WNNs), Agr. Water Manage., 140, 26–36, 2014.

Falge, E., Baldocchi, D., Olson, R., Anthoni, P., Aubinet, M., Bernhofer, C., Burba, G., Ceulemans, R., Clement, R., Dolman, H.,

Granier, A., Gross, P., Grunwald, T., Hollinger, D., Jensen, N. O., Katul, G., Keronen, P., Kowalski, A., Lai, C. T., Law, B. E., Meyers, T., Moncrieff, H., Moors, E., Munger, J. W., Pilegaard, K., Rannik, U., Rebmann, C., Suyker, A., Tenhunen, J., Tu, K., Verma, S., Vesala, T., Wilson, K., and Wofsy, S.: Gap filling strategies for long term energy flux data sets, Agr. Forest Meteorol., 107, 71–77, 2001.

Fernández, R. J.: Do humans create deserts?, Trends Ecol. Evol., 17, 6–7, 2002.

Feddema, J. J. and Freire, S. C.: Soil degradation, global warming and climate impacts, Clim. Res., 17, 209–216, 2001.

Feng, Q.: Preliminary study on the dry sand layer of sandy land in semi-humid region, Arid zone research, 11, 24–26, 1994 (in Chinese).

Franssen, H. J., Stoeckli, R., Lehner, I., Rotenberg, E., and Seneviratne, S. I.: Energy balance closure of eddy-covariance data: A multisite analysis for European FLUXNET stations, Agr. Forest Meteorol., 150, 1553–1567, 2010.

Fu, G. J., Liao, C. Y., and Sun, C. Z.: The effect of soil crust to water movement in maowusu sandland, Journal of northwest forestry university, 25, 7–10, 2010 (in Chinese).

Fulton, J. M.: Relationship of root extension to the soil moisture level required for maximum yield of potatoes, tomatoes and corn, Can. J. Soil Sci., 50, 92–94, 1970.

Gao, S., Pan, X., Cui, Q., Hu, Y., Ye, X., and Dong, M.: Plant Interactions with Changes in Coverage of Biological Soil Crusts and Water Regime in Mu Us Sandland, China, PLoS one, 9, e87713, doi:10.1371/journal.pone.0087713, 2014.

Ge, Q., Wang, H., and Dai, J.: Phenological response to climate change in China: a meta-analysis, Glob. Change Biol., 21, 265–274, 2015.

Glenn, E. P., Huete, A. R., Nagler, P. L., Hirschboeck, K. K., and Brown, P.: Integrating remote sensing and ground methods to estimate evapotranspiration, Cr. Rev. Plant Sci., 26, 139–168, 2007.

Greene, E. M., Liston, G. E., and Pielke, R. A. S.: Relationships between landscape, snowcover depletion, and regional weather and climate, Hydrol. Process., 13, 2453–2466, 1999.

Gu, Y., Brown, J. F., Verdin, J. P., and Wardlow, B.: A five-year analysis of MODIS NDVI and NDWI for grassland drought assessment over the central Great Plains of the United States, Geophys. Res. Lett., 34, L06407, 2007.

Gutman, G. and Ignatov, A.: The derivation of the green vegetation fraction from NOAA/AVHRR data for use in numerical weather prediction models, Int. J. Remote Sens., 19, 1533–1543, 1998.

Hoshino, A., Tamura, K., Fujimaki, H., Asano, M., Ose, K., and Higashi, T.: Effects of crop abandonment and grazing exclusion on available soil water and other soil properties in a semi-arid Mongolian grassland, Soil Till. Res., 105, 228–235, 2009.

Hou, X. Y., Wang, F. X., Han, J. J., Kang, S. Z., and Feng, S. Y.: Duration of plastic mulch for potato growth under drip irrigation in an arid region of Northwest China, Agr. Forest Meteorol., 150, 115–121, 2010.

Hsieh, C. I., Katul, G., and Chi, T. W.: An approximate analytical model for footprint estimation of scalar fluxes in thermally stratified atmospheric flows, Adv. Water Resour., 23, 765–772, 2000.

Huete, A., Didan, K., Miura, T., Rodriguez, E. P., Gao, X., and Ferreira, L. G.: Overview of the radiometric and biophysical performance of the MODIS vegetation indices, Remote Sens. Environ., 83, 195–213, 2002.

Idso, S. B., Reginato, R. J., Jackson, R. D., Kimball, B. A., and Nakayama, F. S.: The three stages of drying of a field soil, Soil Sci. Soc. Am. J., 38, 831–837, 1974.

Kim, W., Kanae, S., Agata, Y., and Oki, T.: Simulation of potential impacts of land use/cover changes on surface water fluxes in the Chaophraya river basin, Thailand, J. Geophys. Res.-Atmos., 110, D08110, 2005.

Kondoh, A. and Higuchi, A.: Relationship between satellite-derived spectral brightness and evapotranspiration from a grassland, Hydrol. Process., 15, 1761–1770, 2001.

Lei, Z. D., Yang, S. X., and Xie, S. C.: Soil water dynamics, Tsing-Hua University Press, Beijing, 1988 (in Chinese).

Lettermaier, D. P. and Famiglietti, J. S.: Water from on high, Nature, 444, 562–563, 2006.

Leuning, R., Kelliher, F. M., Pury, D. D., and Schulze, E. D.: Leaf nitrogen, photosynthesis, conductance and transpiration: scaling from leaves to canopies, Plant Cell Environ., 18, 1183–1200, 1995.

Li, P. F. and Li, B. G.: Study on some characteristics of evaporation of sand dune and evapotranspiration of grassland in Mu Us desert, Shuili Xuebao, 3, 23–28, 2000 (in Chinese).

Li, S., Kang, S., Zhang, L., Du, T., Tong, L., Ding, R., Guo, W., Zhao, P., Chen, X., and Xiao, H.: Ecosystem water use efficiency for a sparse vineyard in arid northwest China, Agr. Water Manage., 148, 24–33, 2015.

Li, X. L., Gao, J., Brierley, G., Qiao, Y. M., Zhang, J., and Yang, Y. W.: Rangeland degradation on the Qinghai-Tibet plateau: Implications for rehabilitation, Land Degrad. Dev., 24, 72–80, 2013.

Li, Z., Liu, W. Z., Zhang, X. C., and Zheng, F. L.: Impacts of land use change and climate variability on hydrology in an agricultural catchment on the Loess Plateau of China, J. Hydrol., 377, 35–42, 2009.

Liu, C., Zhang, X., and Zhang, Y.: Determination of daily evaporation and evapotranspiration of winter wheat and maize by large-scale weighing lysimeter and micro-lysimeter, Agr. Forest Meteorol., 111, 109–120, 2002.

Liu, F.: Point pattern of Artemisia ordosica and the impact to soil crust thickness in Mu Us Sandland, Mater Thesis of Beijing Forestry University, 2012 (in Chinese).

Liu, X. P., Zhang, T. H., Zhao, H. L., He, Y. H., Yun, J. Y., and Li, Y. Q.: Influences of dry sand bed thickness on soil moisture evaporation in mobile dune, Arid land geography, 29, 523–526, 2006 (in Chinese).

Liu, Z. D., Xiao, J. F., and Yu, X. Q.: Effects of different soil moisture treatments on morphological index, water consumption and yield of potatoes, China Rural Water and Hydropower, 8, 1–7, 2010.

Lo Seen Chong, D., Mougin, E., and Gastellu-Etchegorry, J. P.: Relating the global vegetation index to net primary productivity and actual evapotranspiration over Africa, Int. J. Remote Sens., 14, 1517–1546, 1993.

Loukas, A., Vasiliades, L., Domenikiotis, C., and Dalezios, N. R.: Basin-wide actual evapotranspiration estimation using NOAA/AVHRR satellite data, Phys. Chem. Earth, 30, 69–79, 2005.

Lorup, J. K., Refsgaard, J. C., and Mazvimavi, D.: Assessing the effect of land use change on catchment runoff by combined use of statistical tests and hydrological modelling: case studies from Zimbabwe, J. Hydrol., 205, 147–163, 1998.

Lu, N., Chen, S., Wilske, B., Sun, G., and Chen, J.: Evapotranspiration and soil water relationships in a range of disturbed and undisturbed ecosystems in the semi-arid Inner Mongolia, China, J. Plant Ecol., 4, 49–60, 2011.

Lunetta, R. S., Knight, J. F., Ediriwickrema, J., Lyon, J. G., and Worthy, L. D.: Land-cover change detection using multi-temporal MODIS NDVI data, Remote Sens. Environ., 105, 142–154, 2006.

Lv, Y. Z., Hu, K. L., and Li, B. G.: The spatio-temporal variability of soil water in sand dunes in maowusu desert, Acta Pedologica Sinica, 43, 152–154, 2006 (in Chinese).

Maayar, M. and Chen, J. M.: Spatial scaling of evapotranspiration as affected by heterogeneities in vegetation, topography, and soil texture, Remote Sens. Environ., 102, 33–51, 2006.

Mackay, D. S., Ewers, B. E., Cook, B. D., and Davis, K. J.: Environmental drivers of evapotranspiration in s shrub wetland and an upland forest in northern Wisconsin, Water Resour. Res., 43, W03442, doi:10.1029/2006WR005149, 2007.

Maidment, D. R.: Handbook of hydrology, New York, McGraw-Hill, 1993.

Mao, D. and Cherkauer, K. A.: Impacts of land-use change on hydrologic responses in the Great Lakes region, J. Hydrol., 374, 71–82, 2009.

Martens, S. N., Breshears, D. D., and Meyer, C. W.: Spatial distributions of understory light along the grassland/forest continuum: effects of cover, height, and spatialpattern of tree canopies, Ecol. Model., 126, 79–93, 2000.

Mauder, M., Liebethal, C., Göckede, M., Leps, J. P., Beyrich, F., and Foken, T.: Processing and quality control of flux data during LITFASS-2003, Bound.-Lay. Meteorol., 121, 67–88, 2006.

Mo, X. G., Liu, S. X., Lin, Z. H., and Chen, D.: Simulating the water balance of the wuding river basin in the Loess Plateau with a distributed eco-hydrological model, Acta Geographica Sinica, 59, 341–347, 2004 (in Chinese).

Morillas, L., Leuning, R., Villagarcía, L., García, M., Serrano-Ortiz, P., and Domingo, F.: Improving evapotranspiration estimates in Mediterranean drylands: The role of soil evaporation, Water Resour. Res., 49, 6572–6586, 2013.

Nouri, H., Beecham, S., Anderson, S., and Nagler, P.: High spatial resolution WorldView-2 imagery for mapping NDVI and its relationship to temporal urban landscape evapotranspiration factors, Remote Sens., 6, 580–602, 2014.

Ostwald, M. and Chen, D.: Land-use change: Impacts of climate variations and policies among small-scale farmers in the Loess Plateau, China, Land Use Policy, 23, 361–371, 2006.

Panferov, O., Knyazikhin, Y., Myneni, R. B., Szarzynski, J., Engwald, S., Schnitzler, K. G., and Gravenhorst, G.: The role of canopy structure in the spectral variation of transmission and absorption of solar radiation in vegetation canopies, IEEE T. Geosci. Remote, 39, 241–253, 2001.

Pearcy, R. W., Schulze, E. D., and Zimmermann, R.: Measurement of transpiration and leaf conductance, Plant physiological ecology, Springer Netherlands, 137–160, 1989.

Penman, H. L.: National evaporation from open water, bare soil and grass, P. R. Soc. Lond. A, 193, 120–145, 1948.

Penman, H. L.: Vegetation and hydrology, Tech. Comm. 53, Commenwealth Bureau of Soils, Harpenden, England, 1963.

Piao, S., Fang, J., Zhou, L., Ciais, P., and Zhu, B.: Variations in satellite-derived phenology in China's temperate vegetation, Glob. Change Biol., 12, 672–685, 2006.

Qin, S. H., Li, L. L., Wang, D., Zhang, J. L., and Pu, Y. L.: Effects of limited supplemental irrigation with catchment rainfall on rain-fed potato in semi-arid areas on the Western Loess Plateau, China, Am. J. Potato Res., 90, 33–42, 2013.

Qin, S., Zhang, J., Dai, H., Wang, D., and Li, D.: Effect of ridge–furrow and plastic-mulching planting patterns on yield formation and water movement of potato in a semi-arid area, Agr. Water Manage., 131, 87–94, 2014.

Qiu, G. Y., Xie, F., Feng, Y. C., and Tian, F.: Experimental studies on the effects of the "Conversion of Cropland to Grassland Program" on the water budget and evapotranspiration in a semi-arid steppe in Inner Mongolia, China, J. Hydrol., 411, 120–129, 2011.

Rossato, L., Alvala, R. C. S., Ferreira, N. J., and Tomasella, J.: Evapotranspiration estimation in the Brazil using NDVI data, Proc. SPIE, 5976, 377–385, 2005.

Snyman, H. A.: Water-use efficiency and infiltration under different rangeland conditions and cultivation in a semi-arid climate of South Africa, Proceedings of the XIX International Grassland Congress, Sao Paulo, Brazil, 965–966, 11–20 February 2001.

Souza, D. C. and Oyama, M. D.: Climatic consequences of gradual desertification in the semi-arid area of Northeast Brazil, Theor. Appl. Climatol., 103, 345–357, 2011.

Sun, G., Zhou, G., Zhang, Z., Wei, X., McNulty, S. G., and Vose, J. M.: Potential water yield reduction due to forestation across China, J. Hydrol., 328, 548–558, 2006.

Tian, H., Cao, C., Chen, W., Bao, S., Yang, B., and Myneni, R. B.: Response of vegetation activity dynamic to climatic change and ecological restoration programs in Inner Mongolia from 2000 to 2012, Ecol. Eng., 82, 276–289, 2015.

Tong, X., Zhang, J., Meng, P., Li, J., and Zheng, N.: Environmental controls of evapotranspiration in a mixed plantation in North China, Int. J. Biometeorol., 61, 227–238, doi:10.1007/s00484-016-1205-0, 2017.

Tucker, C. J.: Red and photographic infrared linear combinations for monitoring vegetation, Remote Sens. Environ.,8, 127–150, 1979.

Turrell, F. M.: Citrus leaf stomata: structure, composition, and pore size in relation to penetration of liquids, The University of Chicago Press, 108, 476–483, 1947.

Twine, T. E., Kucharik, C. J., and Foley, J. A.: Effects of land cover change on the energy and water balance of the Mississippi River basin, J. Hydrometeorol., 5, 640–655, 2004.

Valipour, M.: Study of different climatic conditions to assess the role of solar radiation in reference crop evapotranspiration equations, Arch. Acker Pfl. Boden, 61, 679–694, 2015.

Vetter, S. H., Schaffrath, D., and Bernhofer, C.: Spatial simulation of evapotranspiration of semi-arid Inner Mongolian grassland based on MODIS and eddy covariance data, Environmental Earth Sciences, 65, 1567–1574, 2012.

Wang, Y. H.: The hydrological influence of black locust plantations in the loess area of northwest China, Hydrol. Process., 6, 241–251, 1992.

Wang, M. Y., Guan, S. H., and Wang, Y.: Soil moisture regime and application for plants in Maowusu Transition Zone from sandland to desert, 16, 37–44, 2002.

Wang, L., Wang, Q. J., Wei, S. P., Shao, M. A., and Yi, L.: Soil desiccation for Loess soils on natural and regrown areas, Forest Ecol. Manag., 255, 2467–2477, 2008.

Wang, S., Wilkes, A., Zhang, Z., Chang, X., Lang, R., Wang, Y., and Niu, H.: Management and land use change effects on soil carbon in northern China's grasslands: a synthesis, Agr. Ecosyst. Environ., 142, 329–340, 2011.

Wang, Y. Q., Shao, M. A., Zhu, Y. J., and Liu, Z. P.: Impacts of land use and plant characteristics on dried soil layers in different climatic regions on the Loess Plateau of China, Agr. Forest Meteorol., 151, 437–448, 2011.

Wang, Z., Wang, L., Liu, L. Y., and Zheng, Q. H.: Preliminary study on soil moisture content in dried layer of sand dunes in the Mu Us sandland, 26, 2006 (in Chinese).

Wever, L. A., Flanagan, L. B., and Carlson, P. J.: Seasonal and interannual variation in evapotranspiration, energy balance and surface conductance in a northern temperate grassland, Agr. Forest Meteorol., 112, 31–49, 2002.

Wilson, K., Goldstein, A., Falge, E., Aubinet, M., Baldocchi, D., Berbigier, P., Bernhofer, C., Ceulemans, R., Dolman, H., Field, C., Grelle, A., Ibrom, A., Law, B. E., Kowalski, A., Meyers, T., Moncrieff, J., Monson, R., Oechel, W., Tenhunen, J., Valentini, R., and Verma, S.: Energy balance closure at FLUXNET sites, Agr. Forest Meteorol., 113, 223–243, 2002.

Wu, B. and Ci, L. J.: Landscape change and desertification development in the Mu Us Sandland, Northern China, J. Arid Environ., 50, 429–444, 2002.

Wu, G. X.: Roots' distribution characteristics and fine root dynamics of Sabina vulgaris and Artemisia ordosica in Mu Us Sandland, Master thesis of Inner Mongolia Agricultural University, 2006.

Xiao, C. W., Zhou, G. S., Zhang, X. S., Zhao, J. Z., and Wu, G.: Responses of dominant desert species Artemisia ordosica and Salix psammophila to water stress, Photosynthetica, 43, 467–471, 2005.

Yan, K., Li, S. Y., Lei, J. Q., Wang, H. F., Sun C., Yan, F. S., and Li, C.: Characteristics of surface energy exchange in the artificial shelter forest land of the hinterland of Taklimakan Desert, Arid Land Geography, 36, 433–440, 2013.

Yang, L., Wei, W., Chen, L., Chen, W., and Wang, J.: Response of temporal variation of soil moisture to vegetation restoration in semi-arid Loess Plateau, China, Catena, 115, 123–133, 2014.

Yang, Y. M., Yang, G. H., and Feng, Y. Z.: Climatic variation and its effect on desertification in 45 recent years in Mu Us sandland, Journal of Northwest A&F University (Nat. Sci. Ed.), 35, 87–92, 2007 (in Chinese).

Yang, Y., Bu, C., Mu, X., and Zhang, K.: Effects of differing coverage of moss-dominated soil crusts on hydrological processes and implications for disturbance in the Mu Us Sandland, China, Hydrol. Process., 29, 3112–3123, 2015.

Yuan, P. F., Ding, G. D., Wang, W. W., Wang, X. Y., and Shi, H. S.: Characteristics of rainwater infiltration and evaporation in Mu Us Sandland, Science of Soil and Water Conservation, 4, 23–27, 2008.

Zeng, B. and Yang, T.-B.: Impacts of climate warming on vegetation in Qaidam area from 1990 to 2003, Environ. Monit. Assess., 144, 403–417, 2008.

Zeng, N. and Yoon, J.: Expansion of the world's deserts due to vegetation-albedo feedback under global warming, Geophys. Res. Lett., 36, L17401, 2009.

Zeng, X., Shaikh, M., Dai, Y., Dickinson, R. E., and Myneni, R.: Coupling of the common land model to the NCAR community climate model, J. Climate, 15, 1832–1854, 2002.

Zhang, Q., Manzoni, S., Katul, G., Porporato, A., and Yang, D.: The hysteretic evapotranspiration – Vapor pressure deficit relation, J. Geophys. Res.-Biogeo., 119, 125–140, 2014.

Zhang, Y., Munkhtsetesg, E., Kadota, T., and Ohata, T.: An observational study of ecohydrology of a sparse grassland at the edge of the Eurasian cryosphere in Mongolia, J. Geophys. Res.-Atmos., 110, D14103, doi:10.1029/2004JD005474, 2005.

Zhang, Z. P.: Vegetation pattern changes in Mu Us desert and the analysis of water income and expenses: a case study in wushen county, Mater thesis of Inner Mongolia University, 2006 (in Chinese).

9

A nonparametric statistical technique for combining global precipitation datasets: development and hydrological evaluation over the Iberian Peninsula

Md Abul Ehsan Bhuiyan[1], **Efthymios I. Nikolopoulos**[1,2], **Emmanouil N. Anagnostou**[1], **Pere Quintana-Seguí**[3], and **Anaïs Barella-Ortiz**[3,4]

[1]Department of Civil and Environmental Engineering, University of Connecticut, Storrs, CT, USA
[2]Innovative Technologies Center S.A., Athens, Greece
[3]Ebro Observatory, Ramon Llull University – CSIC, Roquetes (Tarragona), Spain
[4]Castilla-La Mancha University, Toledo, Spain

Correspondence: Emmanouil N. Anagnostou (manos@uconn.edu)

Abstract. This study investigates the use of a nonparametric, tree-based model, quantile regression forests (QRF), for combining multiple global precipitation datasets and characterizing the uncertainty of the combined product. We used the Iberian Peninsula as the study area, with a study period spanning 11 years (2000–2010). Inputs to the QRF model included three satellite precipitation products, CMORPH, PERSIANN, and 3B42 (V7); an atmospheric reanalysis precipitation and air temperature dataset; satellite-derived near-surface daily soil moisture data; and a terrain elevation dataset. We calibrated the QRF model for two seasons and two terrain elevation categories and used it to generate ensemble for these conditions. Evaluation of the combined product was based on a high-resolution, ground-reference precipitation dataset (SAFRAN) available at $5\,\mathrm{km}\,1\,\mathrm{h}^{-1}$ resolution. Furthermore, to evaluate relative improvements and the overall impact of the combined product in hydrological response, we used the generated ensemble to force a distributed hydrological model (the SURFEX land surface model and the RAPID river routing scheme) and compared its streamflow simulation results with the corresponding simulations from the individual global precipitation and reference datasets. We concluded that the proposed technique could generate realizations that successfully encapsulate the reference precipitation and provide significant improvement in streamflow simulations, with reduction in systematic and random error on the order of 20–99 and 44–88 %, respectively, when considering the ensemble mean.

1 Introduction

Accurate estimates of precipitation on a global scale, which are essential to hydrometeorological applications (Stephens and Kummerow, 2007), rely primarily on satellite-based observations and atmospheric reanalysis simulations. Although advancement in both satellite retrievals and reanalysis-based precipitation datasets has been continuous (Seyyedi et al., 2014; Dee et al., 2011; Huffman et al., 2007; Mo et al., 2012), they are still associated with several sources of error (Derin et al., 2016; Mei et al., 2014; Seyyedi et al., 2014; Gottschalck et al., 2005; Peña-Arancibia et al., 2013) that limit their use in water resource applications. Quantifying and correcting the sources of the error and characterizing its propagation are important for improving and promoting the use of satellite and reanalysis precipitation estimates in hydrological applications on a global scale.

During the past two decades, research investigations have focused on characterizing the error in satellite precipitation products and its propagation in streamflow simulations (Hossain and Anagnostou, 2004; Li et al., 2009; Bitew and Gebremichael, 2011; Nikolopoulos et al., 2013; Mei et al., 2016). These studies have highlighted the dependence of the error on a multitude of factors, including seasonality, topography, soil wetness, and vegetation cover (Derin et al., 2016; Mei et al., 2014; Seyyedi et al., 2014; Hou et al., 2014). Other studies have used stochastic satellite rainfall error models to investigate uncertainty characteristics and their dependencies (Hossain and Anagnostou, 2006; Teo and Grimes, 2007; Maggioni et al., 2014; Adler et al., 2001; AghaKouchak et al., 2009) and have used stochastically generated ensemble rainfall fields as input in hydrological models to study the satellite precipitation uncertainty propagation in the simulation of various hydrological variables. The two-dimensional satellite rainfall error model, SREM2-D (Hossain et al., 2006), has been used to evaluate the significance of surface soil moisture (Seyyedi et al., 2014) and seasonality (Maggioni et al., 2017) in modeling the error structure of satellite rainfall products.

Given the multidimensionality of error dependence and the lack of a clear winner among the various precipitation datasets established by these studies, we argue that, to mitigate the errors and uncertainties, one should combine the different precipitation datasets, taking into account the different climatological and land surface factors. A promising approach to modeling appears to be the application of statistical nonparametric techniques, which efficiently combine information on several factors (Ciach et al., 2007; Gebremichael et al., 2011). In fact, although nonparametric statistical techniques are not widely used in rainfall estimation, some notable examples exist in the literature, with encouraging results. Ciach et al. (2007) established a nonparametric estimation technique based on weather radar data to characterize the uncertainties in radar precipitation estimates as a function of range, temporal scale, and season. Lakhankar et al. (2009) introduced a nonparametric technique to retrieve soil moisture from satellite remote sensing products in reliable ways with sufficient accuracy. Moreover, Gebremichael et al. (2011) developed a nonparametric technique for satellite rainfall error modeling using rain-gauge-adjusted, ground-based radar rainfall and reported improved satellite precipitation performance with relatively large variation at low and high rainfall rates.

The use of nonparametric statistical techniques in error modeling has also gained popularity in weather forecasting, climate change prediction, and the modeling of hydrological processes (Croley, 2003; Brown et al., 2010; Mujumdar et al., 2008; Yenigun et al., 2013). Recently, a wide variety of nonparametric techniques have been developed for error analytics (Taillardat et al., 2016; He et al., 2017). Nonparametric statistical techniques require fewer assumptions for the form of the relationship and data. The advantages over parametric techniques for prediction are explained in detail in Guikema et al. (2010). Specifically, the authors exhibited better results (lower prediction error) with nonparametric techniques than with parametric analysis models. The techniques they used were classification and regression trees (CART) and Bayesian additive regression trees (BART) (Chipman et al., 2010). Another nonparametric technique, random forest (RF) regression, which provides information about the full conditional distribution of the response variable, was used by Breiman (2001) and found to yield more robust predictions by stretching the use of the training data partition.

This paper investigates the use of a nonparametric statistical technique for optimally combining globally available precipitation sources from satellite and reanalysis products. Specifically, we use the quantile regression forests (QRF) tree-based regression model (Meinshausen, 2006) to combine dynamic (for example, temperature and soil moisture) and static (for example, elevation) land surface variables with multiple global precipitation sources to stochastically generate improved precipitation ensemble. The proposed framework provides a consistent formalism for optimally combining several rainfall products by using information from these datasets. It is, furthermore, able to characterize uncertainty through the ensemble representation of the combined precipitation product. We present the development of the proposed framework and evaluate relative improvements in the combined rainfall product in detail. We also evaluate the new combined product in terms of hydrological simulations to assess the importance of precipitation improvement for streamflow simulations, thus highlighting the usefulness of this approach for global hydrological applications.

The paper is structured as follows: Sect. 2 briefly explains the study area and the datasets used. Section 3 describes the QRF model, the rainfall error analysis, and the hydrological model setup. Performance evaluation of the combined product in precipitation and corresponding hydrological simulations is presented in Sect. 4. Conclusions and recommendations are discussed in Sect. 5.

2 Study area and data

The study area we selected for this investigation is the Iberian Peninsula, which has three main climatic zones: Mediterranean, oceanic, and semiarid. The peninsula's climate is primarily Mediterranean, except in its northern and southern parts, which are characterized mostly as oceanic and semiarid, respectively. The topography varies from almost zero elevation to altitudes of 3500 m in the Pyrenees. For the hydrological analysis, we focused the study over the Ebro River basin and, specifically, on five subbasins of different spatial scale: (1) the Ebro River at Tortosa (84 230 km^2); (2) the Ebro River at Zaragoza (40 434 km^2); (3) the Cinca River at Fraga (9612 km^2); (4) the Segre River at Lleida (11 369 km^2); and (5) the Jalon River at Grisen (9694 km^2) (Fig. 1). The datasets we used are described below.

2.1 Reference precipitation (SAFRAN)

The default reference dataset was recently created by Quintana-Seguí et al. (2016, 2017) using the SAFRAN meteorological analysis system (Durand et al., 1993), which is the same as the one used in earlier studies over France (Quintana-Seguí et al., 2008; Vidal et al., 2010). SAFRAN uses optimal interpolation to combine the outputs of a meteorological model and all available observations, which in

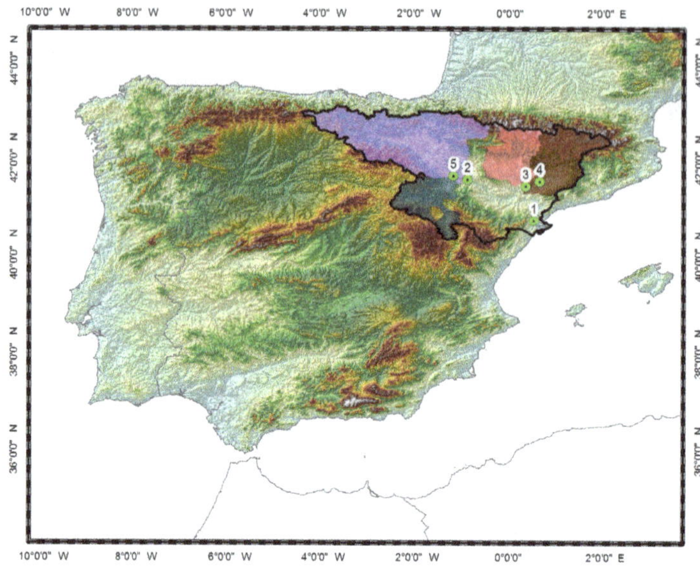

Figure 1. Map of the Iberian Peninsula case study area.

this case were provided by the Spanish State Meteorological Agency (AEMET). The variables analyzed were precipitation, temperature, relative humidity, wind speed, and cloudiness. In the case of precipitation, the first guess was deduced from the observations themselves instead of coming from a numerical model, like the other variables. The observations were analyzed daily (as opposed to every 6 h for the other variables), but the resulting product had a time resolution of 1 h. This was achieved by an interpolation method that used relative humidity to distribute precipitation throughout the day. Spatially, the outputs are presented on a regular grid with 5 km resolution. The dataset (Quintana-Seguí et al., 2016), which spans 35 years, covers mainland Spain and the Balearic Islands.

2.2 Satellite-based precipitation

We used three gauge-adjusted quasi-global satellite precipitation products – CMORPH, PERSIANN, and 3B42 (V7) – in this study. CMORPH (Climate Prediction Center Morphing technique of the National Oceanic and Atmospheric Administration, or NOAA) is a global precipitation product based on passive microwave (PMW) satellite precipitation fields spatially propagated by motion vectors calculated from infrared (IR) data (Joyce et al., 2004). PERSIANN (Precipitation Estimation from Remotely Sensed Information using Artificial Neural Networks) is IR-based and uses a neutral network technique to connect IR observations to PMW rainfall estimates (Sorooshian et al., 2000). TMPA (Tropical Rainfall Measuring Mission Multisatellite Precipitation Analysis), or 3B42 (V7), is a merged IR and passive microwave precipitation product from NASA that is gauge-adjusted and available in both near-real time and post-real time (Huffman et al., 2010). Spatial and temporal resolutions of the satellite precipitation products are 0.25° and 3-hourly time intervals, respectively.

2.3 Atmospheric reanalysis

For meteorological forcing, we selected the WATCH1 (Water and Global Change FP7 project) Forcing Dataset ERA-Interim (hereafter WFDEI) (Weedon et al., 2014), a contemporary state-of-the-art database. WFDEI, a dataset that follows up on the European Union's WATCH project (Harding et al., 2011), is built on the ECMWF ERA-Interim reanalysis (Dee et al., 2011) with a geographical resolution of $0.5° \times 0.5°$ and a sequential frequency of 3 h for the time span 1979–2012, with particular bias corrections using gridded monitoring. Finally, we chose two atmospheric products (atmospheric precipitation and air temperature) among WFDEI variables as predictors for the nonparametric statistical technique.

2.4 Soil moisture

The soil moisture information used in this study was obtained from the satellite-based soil moisture estimates produced by the European Space Agency (ESA) Climate Change Initiative (CCI) project under the ESA Programme on Global Monitoring of Essential Climate Variables (ECV) (Liu et al., 2011; Owe et al., 2008; De Jeu, 2003; http://www.esa-soilmoisture-cci.org/node/145). The ESA CCI (v02.0) soil moisture product is derived from passive and active microwave satellite-based sensors (Liu et al., 2011, 2012; Wagner et al., 2012) and provides information on daily surface soil moisture at 0.25° spatial resolution and quasi-global scale.

2.5 Terrain elevation

The Shuttle Radar Topography Mission (SRTM) dataset included in this study has, in recent years, been one of the most extensively used publicly accessible terrain elevation datasets. Available at ~90 m spatial resolution, it was obtained using 1° digital elevation model (DEM) tiles from the US Geological Survey and interpolated to the 0.25° grid resolution to match the resolution of precipitation and soil moisture products.

3 Methodology

3.1 Blending technique

In this study, we applied a nonparametric, tree-based regression model, quantile regression forests (QRF) (Meinshausen, 2006), to produce a rainfall ensemble with respect to the reference precipitation. The model input includes the three global satellite precipitation datasets (CMORPH, PERSIANN, and 3B42-V7), the global reanalysis rainfall and air temperature datasets, and the satellite near-surface soil moisture and terrain elevation datasets, described in the previous section. The atmospheric products were interpolated in space using the nearest neighbor interpolation technique to match the resolution of precipitation satellite precipitation datasets. The spatial and temporal resolutions of the atmospheric products are 0.25° and 3-hourly time intervals, correspondingly. The high-resolution SAFRAN data were matched to the satellite precipitation datasets on a pixel-by-pixel basis by averaging all high-resolution pixels within a 0.25° pixel. Finally, all 3-hourly data were mapped to the grid resolution of 0.25° chosen to be the final spatial resolution for the combined product.

QRF is derived from random forest regression (Meinshausen, 2006), which is capable of handling data from large samples; it has desirable built-in features, such as variable selection, interaction detection, incorporation of missing data, and the ability to save the trained model for future prediction (Nateghi et al., 2014). QRF uses a bagged version (bootstrapped aggregating) of decision trees by randomly sampling from the bootstrapped sample, which reduces variance and helps to avoid overfitting that improves the stability and accuracy of the algorithm (Meinshausen, 2006). QRF provides a nonparametric way to evaluate conditional quantiles for high-dimensional predictors of variables. The conditional distribution function of Y is defined by

$$\hat{F}\left(y|X=x\right)=P\left(Y\leq y|X=x\right)=E\left(1_{\{Y\leq y\}}|X=x\right), \quad (1)$$

where Y refers to observations of the response variable, X is a covariate or predictor variable, and E is the conditional mean, $E(1_{\{Y\leq y\}}|X=x)$, which is approximated by the weighted mean over the observation of $1_{\{Y\leq y\}}$ (Meinshausen, 2006). Then Eq. (1) can be expressed as

$$\hat{F}\left(y|X=x\right)=\sum_{i=1}^{n}\omega_i\left(x\right)1_{\{Y_i\leq y\}}, \quad (2)$$

where weight vector $\omega_i(x)=k^{-1}\sum_{i=1}^{k}\omega_i(x,\theta_t)$ using random forest regression; k indicates the number of single trees ($t=1,\ldots,k$); and each tree is built with an independent and identically distributed vector θ_t (Meinshausen, 2006).

This nonparametric technique utilizes the weighted average of all trees to compute the empirical distribution function. It keeps not only the mean but also all observation values in nodes and, building on this information, it calculates the conditional distribution. In this method, consistency of the empirical quantities is induced based on a large number of instances in terminal nodes. The overall framework of the QRF scheme is shown in Fig. 2. Higher number of trees reduces the variance of the model. So, increasing the number of trees in the ensemble will not have any impact on the bias of the model. Furthermore, a higher variance reduction can be achieved by decreasing the correlation between trees in the ensemble. Therefore, QRF utilizes the optimal number “mtry” (size of the random subset of predictors) for split point selection at each node. This approach introduces randomness in the ensemble to reduce the correlation between trees, which helps to avoid overfitting (Meinshausen, 2006). In this study we used the default value ($k=1000$) (Meinshausen, 2006) throughout all simulations to create the empirical distribution at each grid cell and used the cross-validation experiments (see next section) to demonstrate stability of the method.

Specifically, we initialized a random forest of 1000 trees for each terminal node of each of the classified dataset and calculated the 95 % prediction intervals at each grid cell. QRF utilizes the same weights to calculate the empirical distribution function and a weighted average of all trees for the predicted expected response values to calculate the empirical distribution. To conduct the hydrological simulations in this study, we resampled from the empirical distribution function 20 times per grid cell to obtain “reference”-like rainfall ensemble members.

To build the rainfall error model, we grouped available rainfall estimates from all input datasets (three satellite and reanalysis) into three subsets: (1) all rainfall products that report rainfall greater than zero; (2) all rainfall products that report zero rainfall; and (3) at least one product that reports nonzero rainfall. We categorized each case into two seasons: the “warm season”, which included data from May through October, and the “cold season”, which included data from November through April. We then classified each season category into two levels based on two terrain elevation ranges: above (high) and below (low) 1000 m a.s.l. Finally, for each subset, we prepared four groups (warm-high, warm-low, cold-high, cold-low) for the error model. For each group, leave-one-pixel-out cross-validation is applied where each point in the statistics is not included in the calibra-

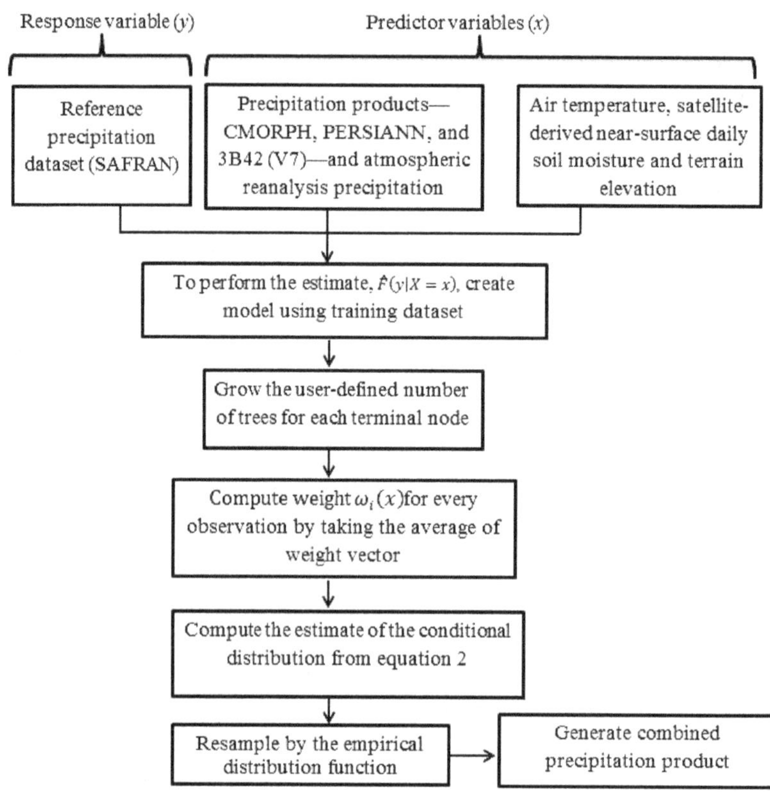

Figure 2. A schematic representation of the quantile regression forests (QRF) framework used.

tion of the technique. In general, prominent "mtry" is obtained by this method by extending the sample size and therefore preventing overfitting. Applying this validation method, the model exhibits great skill on both the training dataset and the unseen test data. Similarly, to strengthen the validation results, we also performed a test using leave-one-year-out cross-validation. Namely, for each year of the database hold out for validation, we calibrated on the rest of the years (10 years) and repeated the experiment for all 11 years of the study period. The model validation results based on leave-one-pixel-out and leave-one-year-out are described in Sect. 4.2.

3.2 Hydrological modeling

To perform the streamflow simulations, we used the SASER (SAfran-Surfex-Eaudyssée-Rapid) hydrological modeling suite. SASER is a physically based and distributed hydrological model for Spain based on SURFEX (Surface Externalisée), a land surface modeling platform developed by Météo-France (Masson et al., 2013) that integrates several schemes for different kinds of surfaces (natural, urban, lakes, and so on). The scheme for natural surfaces, ISBA (Noilhan and Planton, 1989; Noilhan and Mahfouf, 1996), has different versions, with differing degrees of complexity. Within SASER we used the explicit multilayer version (Boone, 2000; Decharme et al., 2011) with prescribed vegetation. The physiography was provided by the ECOCLIMAP dataset (Champeaux et al., 2005).

Since SURFEX has no river routing scheme, we chose the RAPID river routing scheme (David et al., 2011a, b) within the Eau-dyssée framework (http://www.geosciences.mines-paristech.fr/fr/equipes/systemes-hydrologiques-et-reservoirs/projets/eau-dyssee). Eau-dyssée transfers SURFEX runoff (surface and subsurface or drainage) from the SURFEX grid cells to the river cells using its own isochrony algorithm. Then, RAPID uses a matrix-based version of the Muskingum method to calculate flow and volume of water for each reach of a river network. The current application of SASER uses HYDROSHEDS (Lehner et al., 2008) to describe the river network. As the current setup cannot simulate dams, canals, or irrigation, the resulting river flows are estimations of the natural system (that is, the system without direct human intervention in the form of irrigation or hydraulic infrastructure, such as dams or canals).

It is important to note that, since the current version of SASER uses the default parameters for its different schemes, it has not been specifically calibrated for the target basin. This has some implications. The benefit is that the model is not overfitted, which makes it directly comparable to global applications of SURFEX, which are not calibrated either. The downside is that the model might not perform optimally. In the future, we plan to improve the options used in the land surface model to adapt its structure better to the necessary physical processes that take place in the basin, while limiting the need for parameter calibration. For the purpose of this study, which involved the relative comparison of multiple rainfall forcing-based simulations, the current model setup was considered adequate.

3.3 Metrics of model performance evaluation

We based quantification of the systematic and random error of model-generated ensemble on different error metrics. We evaluated the random error component based on the normalized centered root mean square error (NCRMSE), which is defined as

$$\mathrm{NCRMSE} = \frac{\sqrt{\frac{1}{n}\sum_{i=1}^{n}\left[\hat{y}_i - y_i - \frac{1}{n}\sum_{i=1}^{n}\left(\hat{y}_i - y_i\right)\right]^2}}{\frac{1}{n}\sum_{i=1}^{n} y_i}. \quad (3)$$

Note that y_i is reference rainfall, $\hat{y}_i$ is estimated rainfall for times i from the blended technique, and n is the total number of data points used in the calculations. An NCRMSE value of 0 indicates no random error, while 1 indicates that the random error is equal to 100 % of the mean reference rainfall.

To measure the systematic error, we used the bias ratio (BR) metric, which indicates the mean of the ratio of estimated rainfall to reference rainfall and is defined as

$$\mathrm{BR} = \frac{1}{n}\sum_{i=1}^{n}\left(\frac{\hat{y}_i}{y_i}\right). \quad (4)$$

For an unbiased model, the BR would be 1.

To assess the ability of the QRF-generated ensemble to encapsulate the reference rainfall, we used the exceedance probability (EP) metric, which indicates the probability that the reference value will exceed the prediction interval:

$$\mathrm{EP} = 1 - \frac{1}{n}\sum_{i=1}^{n} 1_{\left\{Q_{\mathrm{lower}_i} < y_i < Q_{\mathrm{upper}_i}\right\}}. \quad (5)$$

Here, Q_{lower} and Q_{upper} denote lower and upper boundaries of prediction interval, respectively. The EP would be 0 for an ideal model; that means a perfect encapsulation of the reference within the prediction interval.

To evaluate the accuracy of the QRF-generated ensemble, we used the uncertainty ratio (UR), which measures uncertainty from the prediction interval (Q_{lower}, Q_{upper}), as used in Eq. (3):

$$\mathrm{UR} = \frac{\sum_{i=1}^{n}\left(Q_{\mathrm{upper}} - Q_{\mathrm{lower}}\right)}{\sum_{i=1}^{n} y_i}. \quad (6)$$

To achieve accurate and successful prediction, comparatively small prediction intervals are expected. A UR value of 1 means the best estimate of the actual uncertainty which indicates the maximum possible uncertainty of the prediction interval. The UR quantifies the prediction interval width relative to the magnitude of observations. A UR value close to 1 indicates confidence intervals being in the order of magnitude of the predicted values.

For the evaluation of the accuracy of the ensemble, we also calculated the rank histogram, which is computed by counting the rank of observations and comparing this with values from a compiled ensemble, in ascending order. The rank of the actual value is denoted by $r_j(r_1, r_2, \ldots, r_{m+1})$ and r_j is expressed as follows:

$$r_j = \hat{P}\{\hat{y}_{i,j-1} < y_i < \hat{y}_{i,j}\}. \quad (7)$$

A flat rank histogram diagram means precise prediction of error distribution (Hamil, 2001; Hamil and Colucci, 1997). A U-shaped rank histogram (convex) represents conditional biases, and a concave shape means an over-spread. Skewing to the right denotes negative bias and vice versa.

The Nash–Sutcliffe efficiency (NSE) is widely used in hydrology to assess model performance (Nash and Sutcliffe, 1970) and is defined as

$$\mathrm{NSE} = 1 - \frac{\sum_{i=1}^{n}\left(\hat{y}_i - y_i\right)^2}{\sum_{i=1}^{n}\left(y_i - \overline{y}\right)^2}. \quad (8)$$

The NSE bounds from negative infinity to positive 1, where positive 1 means ideal consistency. A negative value of NSE denotes the performance of the estimator being worse than the mean of reference.

4 Results and discussion

4.1 Sensitivity analysis

In this paper we present a blending technique that leads to an improved characterization of precipitation estimation uncertainty through an optimal combination of precipitation and other datasets. The technique is designed to evaluate the sensitivity of the blending technique to the different forcing variables. The selection of variables was based upon recent research works (Seyyedi et al., 2014; Bhuiyan et al., 2017; Mei et al., 2016) that have examined the factors related to precipitation error characteristics and have shown that soil moisture, temperature, precipitation products, and elevation are important predictors in the error modeling of rainfall estimates. Bhuiyan et al. (2017) have recently used a nonparametric statistical technique (QRF) to evaluate the significance of surface soil moisture in modeling the error structure of satellite products and successfully assessed the impact of surface soil moisture information on the model's performance. Therefore, soil moisture is identified as a potential factor for the proposed blending technique instead of the vegetation indicator. In addition, we have daily quality controlled surface soil moisture data with a global coverage and 0.25° spatial resolution while the available vegetation indices are provided at much coarser (multi-day) temporal scales. Also the precipitation products, despite using similar observations to some extent, exhibit different error characteristics and could provide complementary information. Therefore, dynamic (temperature and soil moisture) and static (elevation) land surface variables were used as input along with the multiple global precipitation sources (CMORPH, PERSIANN, and 3B42 (V7); and atmospheric reanalysis precipitation).

After choosing these predictor variables, for any nonparametric statistical technique, it is essential to know the sensitivity of the result to the different input predictor variables and to quantify the impact of change from one variable to another. The variable importance methodology (Breiman, 2001) is a measure of sensitivity in that case, which helps to recognize crucial input variables that demonstrate the relative contribution of each variable. The importance of the predictor variables depends on the magnitude of the percentage increase in mean square error (% IncMSE) of the model (Breiman, 2001). Higher values of % IncMSE indicate higher importance of the predictor variables. Briefly, the mean squared error (MSE) computed from the original model (i.e., considering all variables) is compared against MSE from a new model that holds all variables the same as the original model except one, the one of which we want to determine its relative importance.

Results from the variable importance test show that all variables used in this technique contribute valuable information in the modeling process. Results from the variable importance test for two groups included in our methodology (for warm and cold periods, high elevation and rainfall greater than zero for all products) are presented in Fig. 3, which shows the variable importance of the seven predictor variables. According to these results all variables are important but the level of significance varies considerably among the different variables. Soil moisture, reanalysis, and the three satellite precipitation datasets were ranked as the most important predictor variables (Fig. 3), showing their strong impact in model prediction. Similar results (not shown) were obtained from examination of the other scenarios (e.g., for low elevation cases). From the variable importance test, it is argued that all variables selected have a significant impact on the model prediction, justifying our choice for including them.

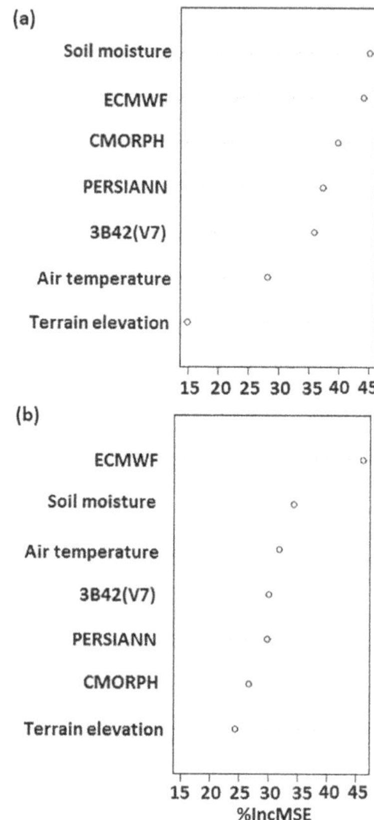

Figure 3. Variable importance plot, where % IncMSE is the percentage increase in mean square error. **(a)** Warm period–high elevation. **(b)** Cold period–high elevation.

4.2 Evaluation of the blending technique

We first evaluated the method by applying a leave-one-pixel-out cross validation, where each pixel was treated as an independent dataset and was predicted based on QRF parameters determined based on the remaining pixels in the database. The time series of 20 ensemble members' cumulative rainfall simulated by QRF for high and low elevations in warm and cold seasons are shown in Fig. 4. The ensemble envelope encapsulates the actual rainfall time series, with better convergence of QRF ensemble members for the warm season but with overall satisfying results for the cold season as well. As is shown, the model was capable of generating stochastic realizations that successfully encapsulated the reference precipitation dynamics. Apart from the ensemble performance, one can note from the results in Fig. 4 the variability in the performance of different precipitation products and their inconsistencies relative to the reference precipitation. For example, PERSIANN overestimated high elevation in the warm season, while CMORPH, 3B42 (V7), and atmospheric reanalysis precipitation underestimated it. Overall, CMORPH underestimated the most for both seasons, which indicated poor performance of the QRF ensemble.

Figure 5 compares the combined rainfall product precipitation accumulation maps to the corresponding reference rainfall accumulation maps for the warm and cold seasons. In general, the spatial distribution of rainfall was consistent for both seasons, with the northwestern part of the study area,

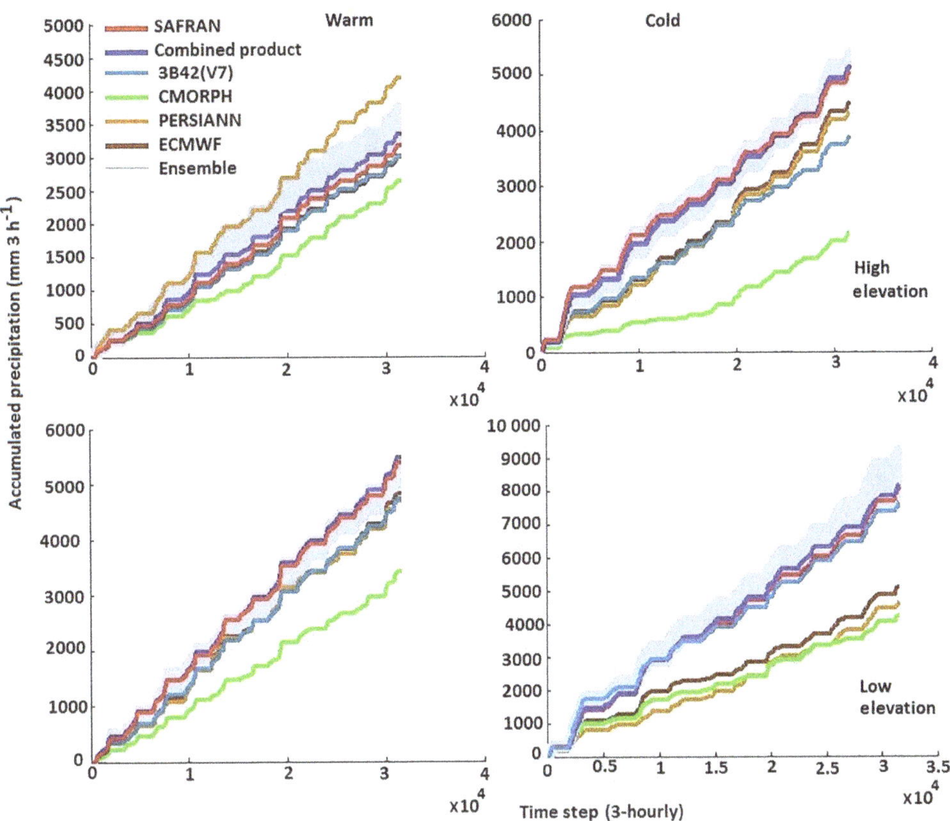

Figure 4. Time series of cumulative rainfall of CMORPH, PERSIANN, 3B42 (V7), reanalysis rainfall, and QRF ensemble (blue envelope) for warm and cold seasons.

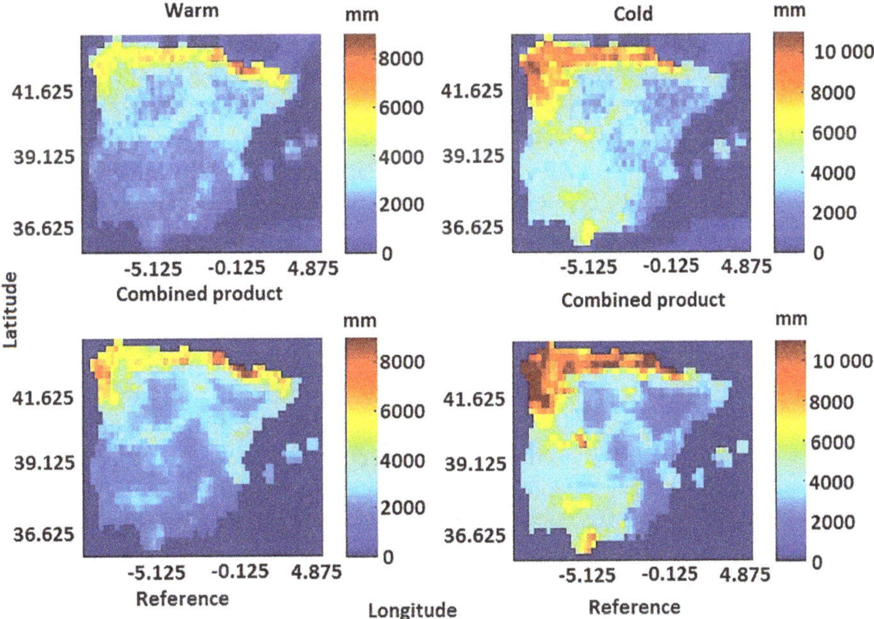

Figure 5. QRF-generated mean ensemble and reference rainfall maps for warm and cold seasons.

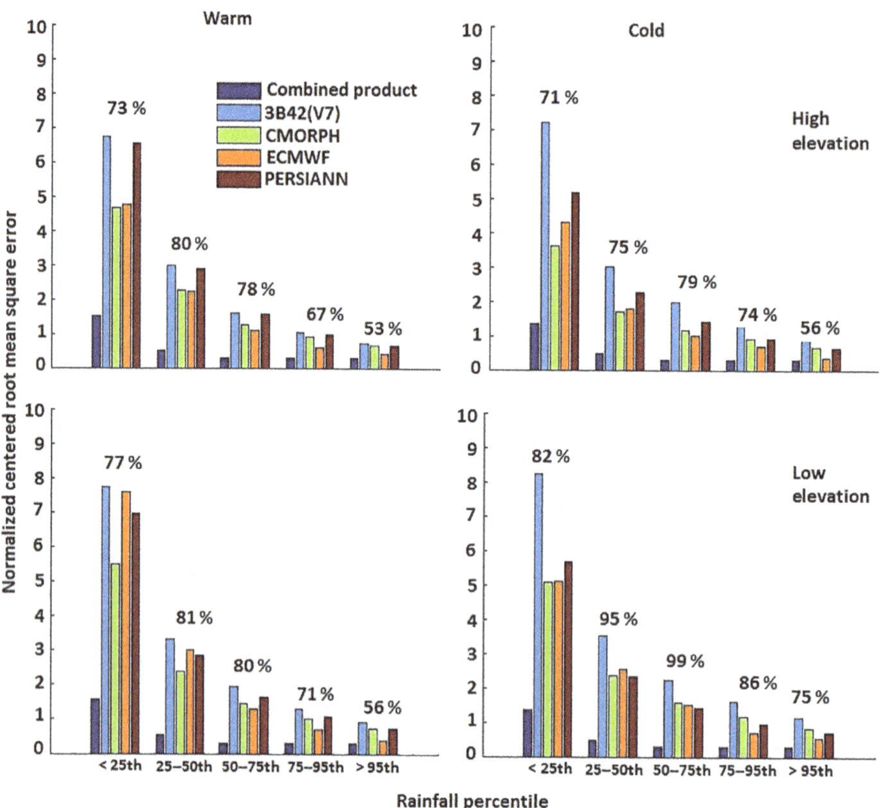

Figure 6. Normalized centered root mean square error for warm and cold seasons. The relative error reductions for the combined product (relative to other products) are shown above the bars.

which is near to the ocean, associated with more precipitation. In the cold season, the combined product gave higher precipitation in the southwestern and northwestern parts of the Iberian Peninsula than in the warm season. These patterns were consistent with those presented in the reference dataset.

We calculated NCRMSE for the various precipitation products (Fig. 6) for five reference precipitation categories, with values in the percentile ranges of < 25th, 25th–50th, 50th–75th, 75th–95th, and > 95th. The results showed the QRF-based combined product could reduce the random error for all rainfall rate categories in both seasons. It is also shown that the random error reduced consistently in all products as the rainfall rate increased.

To quantify the performance of the combined product in contrast to the individual precipitation datasets, we calculated the relative reduction of the NCRMSE for the different precipitation ranges; Fig. 6 presents these performance metric statistics. The relative reduction of the values was defined as the difference between the average of the different datasets and the combined product over the average NCRMSE of the datasets. We noted that relative NCRMSE reduction was greater during the cold season, particularly in regions of low elevation (75 to 99 %). During the warm season, the relative reduction varied between 53 and 81 % for both high and low elevations. Overall, results from all metrics examined showed that the random error of the combined product was significantly lower than those of the individual global precipitation datasets used in the technique.

The combined product's accuracy was further assessed using BR for both seasons (Fig. 7). The results indicated QRF improved accuracy for rain rates beyond the 50th percentile threshold, exhibiting lower BR values. For moderate to high rainfall in both seasons, all individual rainfall datasets exhibited underestimation, which was reduced in the combined product. For the low rainfall, the systematic error reduction for the combined product was not prominent, resulting in a comparatively higher BR value. Generally, the QRF model is expected not to capture very low and extremely high values well due to the weakness of the empirical distribution function to model probabilities close to 0 or 1. The sample size plays an important role in empirical distribution function. Therefore, very large sample sizes are required for low values to quantify the rate of convergence to the underlying cumulative distribution function. Moreover, studies have shown that QRF can perform better in generating one-sided prediction intervals, which is the case in Juban et al. (2007), Francke et al. (2008), and Zimmermann et al. (2012). In terms of el-

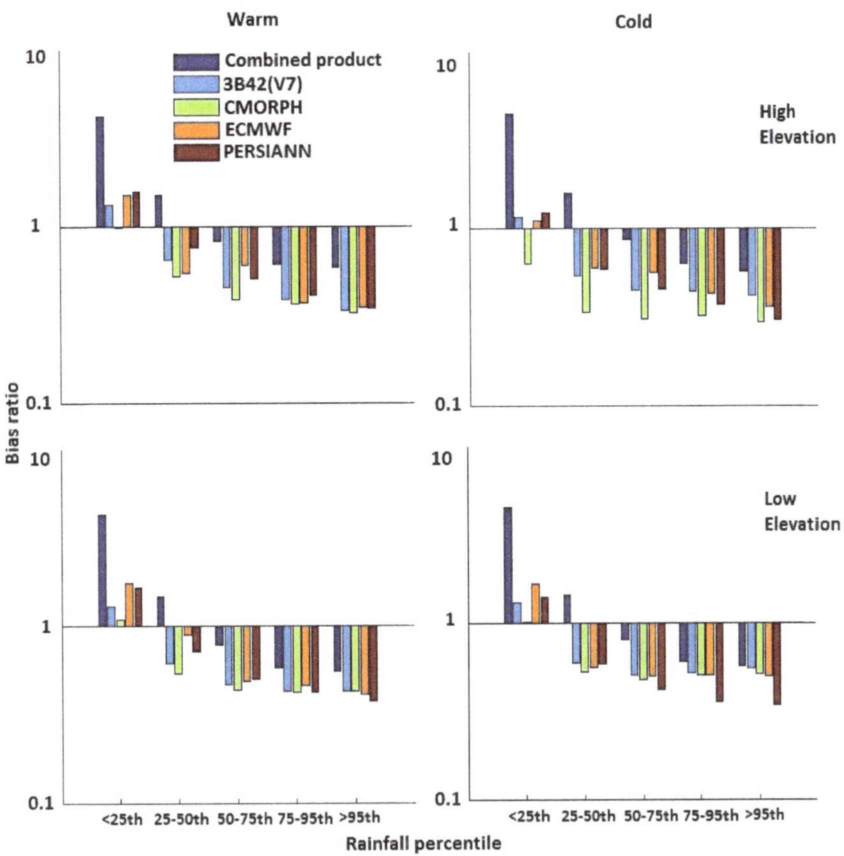

Figure 7. Bias ratio for warm and cold seasons.

evation, the magnitude of BR was considerably less for the high elevations in the warm season. The model used elevation as a control parameter, which reflected its ability to reduce the systematic error at high elevations noticeably. For the higher rain rate category (> 95th percentile), the BR value in the warm season ranged between 0.5 and 0.6, which was close to estimations for the cold season over the study area. The relative reduction was high for the combined product, varying from 17 to 76 % for both seasons. Overall, the combined product exhibited BR values closer to 1 than the individual precipitation datasets did, demonstrating superior performance.

Results for exceedance probability (EP) values (Fig. 8), which are used to assess the ability of the QRF-generated ensemble to encapsulate the reference data, suggest that reference precipitation is captured well within the ensemble envelope. Specifically, considerably reduced exceedance probability values (< 0.26) were reported for rain rates below the 95th percentile threshold for both seasons. A season-based comparison revealed that cold-season EP values were smaller than the corresponding values for the warm season across all rain rate thresholds. Even for the high rain rates (> 95th percentile), EP values were found to be acceptable (~ 0.5).

Analysis of UR showed the QRF ensemble envelope was associated with slightly wider prediction intervals in the warm season than in the cold season, indicating varying degrees of uncertainty throughout the year (Fig. 8). A UR closer to 1 was exhibited for higher rain rates, demonstrating that the ensemble envelope represented the uncertainty for the moderate to high rain rates well, while the variability of the ensemble envelope for rain rates below the 50th percentile threshold overestimated the product uncertainty.

Finally, to evaluate the accuracy of the QRF ensemble, we calculated the rank histogram for the cold and warm seasons. Figure 9 shows the rank histogram of reference values in the posterior sample of QRF ensemble prediction for both seasons. QRF produced a nearly uniformly distributed rank histogram for low to moderate rain rates, which means the rank test was more promising for QRF ensemble predictions for these rain rates. However, the rank histogram exhibited larger values on the right-hand side, indicating underestimation of high rain rates in the ensemble prediction, which is potentially attributed to the inclusion of the entire spectrum of values in the training dataset (i.e., large sample of zeroes and low values) that although allow for reproducing certain

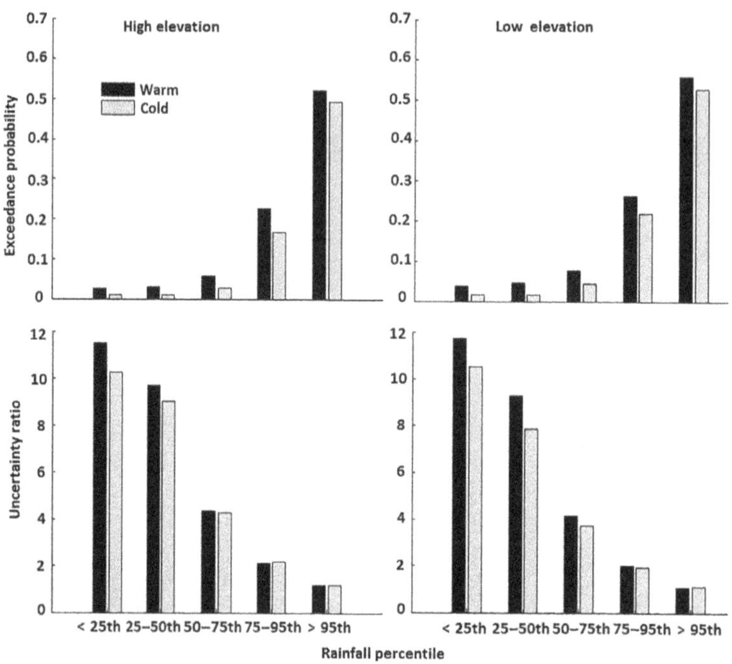

Figure 8. Exceedance probability and uncertainty ratio for warm and cold seasons.

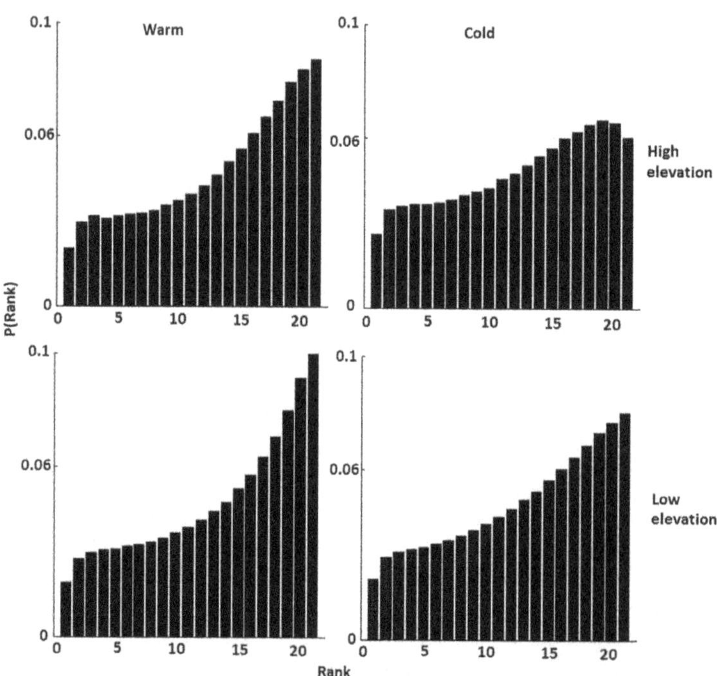

Figure 9. Rank histogram for warm and cold seasons.

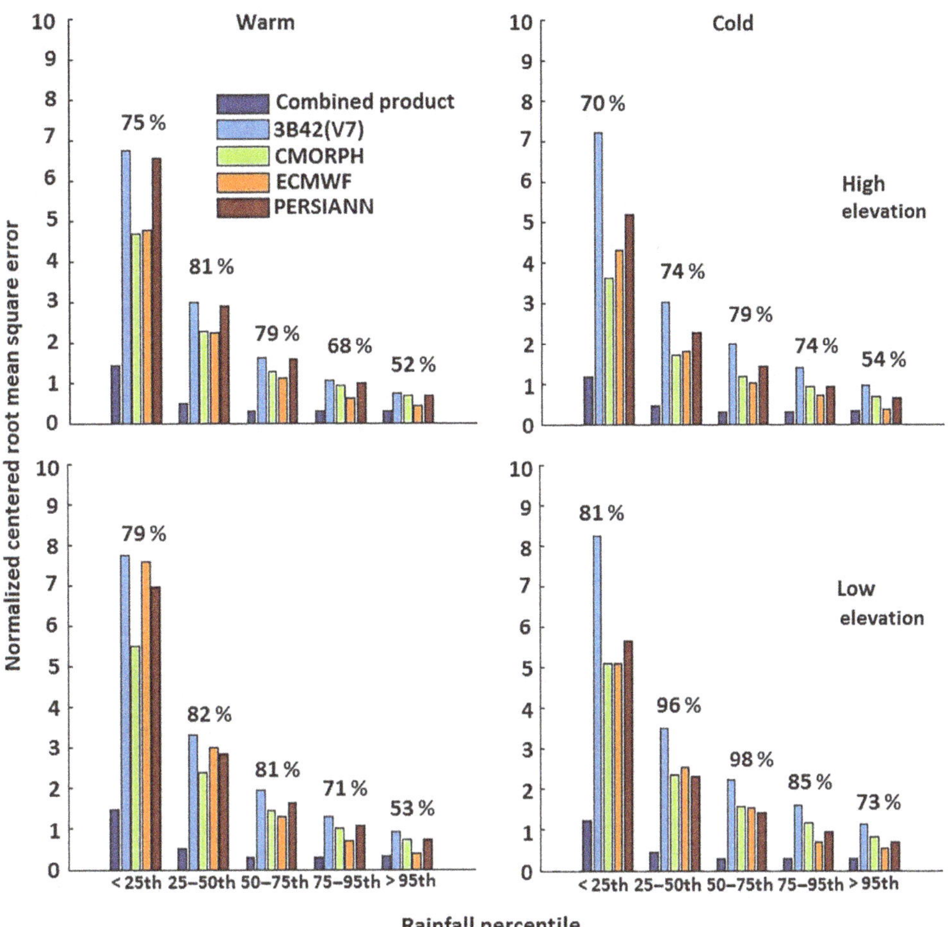

Figure 10. Normalized centered root mean square error for warm and cold season. The relative error reductions for the combined product (relative to other products) are shown above the bars.

important features of precipitation (e.g., intermittency), impact the simulation of high rainfall regime.

As an additional evaluation step, we applied a leave-one-year-out cross-validation, where we kept a whole year as an independent dataset and the model was trained on the remaining years of the study period. Results for NCRMSE are shown in Fig. 10, which are consistent with the leave-one-pixel-out cross-validation analysis in terms of the reduction of the random error for both seasons (warm and cold) and precipitation percentile ranges. The random error decreases with increasing scale, and for all cases, results from the combined product are associated with an error reduction (relative to other products) on the order of 52–98 %. Overall, results indicate that the random error of the combined product was significantly lower than those of the individual precipitation products used in this study.

The performance of the estimates for the model was also evaluated in terms of systematic error, as shown in Fig. 11. Results show that the magnitude of systematic error for the combined product is substantially lower than for the individual precipitation products. Overall, BR values are closer to 1 for moderate to high rain rates in both seasons for the combined product, which indicates that QRF is able to reduce the systematic error in moderate to high rain rates.

In summary, both validation approaches demonstrated that the QRF model is able to reduce the systematic and random error of precipitation estimates exhibited by individual precipitation products significantly, which indicates that QRF was appropriately trained and does not exhibit limitations due to overfitting. To conduct the hydrological simulations in this study, the validation of the results based on the leave-one-pixel-out method is provided in Sect. 4.3.

4.3 Evaluation of ensemble hydrological simulations

This section summarizes the results of the streamflow simulations associated with the different precipitation forcing data (satellite, reanalysis, and combined product). We carried out the evaluation using the SAFRAN-based simulations for reference. Comparisons allowed us to understand the per-

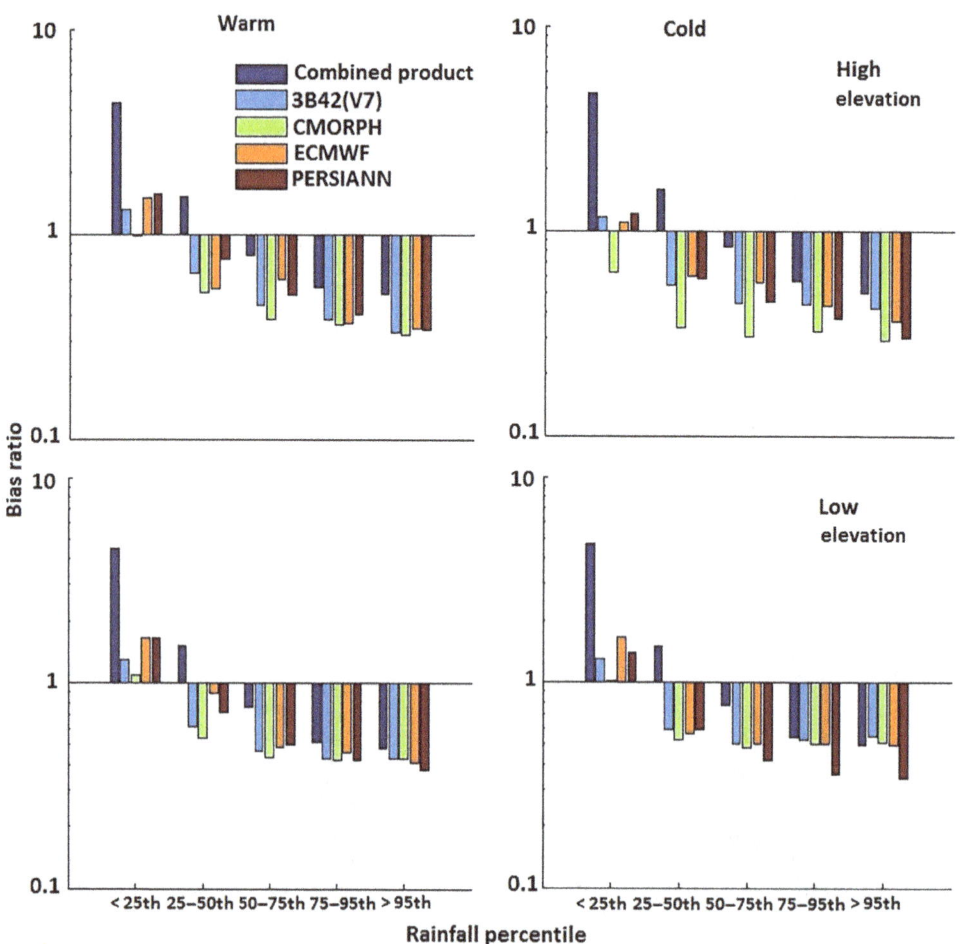

Figure 11. Bias ratio for warm and cold seasons.

formance of each individual precipitation product in terms of streamflow simulations and to evaluate the impact of the QRF-based blending technique in terms of hydrological simulations. The NCRMSE and BR quantitative statistics are used here to assess the performance of basin-scale precipitation forcing data and corresponding generated streamflows.

Since SASER does not simulate dams or canals or irrigation, the simulated streamflow reflects how the flow would be if the water resources of the basin were not managed (that is, naturalized streamflow). The Ebro basin is heavily managed, with hundreds of dams and an important canal network. This raised the problem that model flows could not be compared to the observed flows on those subbasins that are influenced by water management, which is most of them. The Spanish Ministry of Agriculture, Fisheries, Food, and the Environment (MAPAMA) had, however, produced monthly naturalized river flows using the SIMPA rain-runoff model (Ruiz et al., 1998; Álvarez et al., 2004). These are the reference values used by the managers, and they currently offer the only means of reference for validating SASER results.

Table 1 compares the bias and NSE of SASER to those of SIMPA. In terms of monthly accumulation of precipitation, the precipitation data used by SIMPA were similar to SAFRAN's; thus, the difference may have resided in evapotranspiration, which is calculated differently in both models, in terms of both formulation and land-use maps. In terms of NSE, the scores are acceptable at the outlet (between 0.4 and 0.6) and better at most Pyrenean basins (between 0.4 and 0.8), with some exceptions.

Although SASER had room for improvement, mainly with regard to bias, the model was generally able to simulate the main dynamics of the basin. In fact, given that it is essentially evaluated against another model (SIMPA) the reported bias cannot be attributed with certainty to a deficiency of the SASER model, and therefore this evaluation exercise mostly shows that the model we used can represent flows consistently (to a certain degree) with another model that is widely used in the region. As this study aimed to evaluate streamflow simulations in a relative sense (that is, with respect to SAFRAN-based simulations) and not to reproduce the observed river flows with precision, the current version of the

Table 1. The bias and NSE of high-resolution SAFRAN-SURFEX simulation.

River	Station	Area NSE (km^2)	Bias	
Ega	Andosilla	1445	0.735	−13.52
Arga	Funes	2759	0.651	−38.068
Aragón	Caparroso	5462	0.733	−25.655
Jiloca	Daroca	2202	0.069	−23.28
Ebro	Zaragoza	40 434	0.716	−29.324
Gállego	Ardisa	2040	0.477	−26.613
Ésera	Graus	893	0.37	−28.177
Cinca	El Grado	2127	0.629	−21.223
Segre	Lleida	11 369	0.256	−35.016
Ebro	Tortosa	84 230	0.576	−27.2
Najerilla	Mansilla	242	0.114	−62.214
Albercos	Ortigosa	45	0.193	−45.826
Cidacos	Yanguas	223	0.392	−42.551
Salazar	Aspurz	396	0.731	−27.012
Irati	Liedena	1546	0.721	−30.451
Arga	Echauri	1756	0.508	−46.398
Ega	Estella	943	0.622	−36.172
Aragón	Yesa	2191	0.716	−22.334
Noguera Pallaresa	Collegats	1518	−0.768	2.102
Huerva	Mezcalocha	620	0.427	−24.375
Ebro	Mendavia	12 010	0.592	−36.507
Ebro	Flix	82 416	0.59	−27.564
Ésera	Barasona	1511	0.446	−32.846
Ebro	Ribarroja	81 060	0.595	−27.604
Jalón	Calatayud	6841	0.222	−22.072

SASER model was considered adequate for this purpose. Being physically based, it simulated the interplay among different physical processes realistically, and, thus, it could be used to assess their impact on uncertainty.

To provide an overview of the differences in error magnitude between forcing and response variables, we present our analysis of error metrics for simulated streamflow along with corresponding error metrics in basin-average precipitation. Results for NCRMSE are shown in Fig. 12, which notes that for the two larger basins, Ebro at Tortosa (84 230 km^2) and Ebro at Zaragoza (40 434 km^2), the NCRMSE (0.1–0.3) of the combined product was significantly lower than those of the other subbasins for all intervals of streamflow values; this was related to the significant smoothing effect on random error associated with larger basin scales.

Consistent with streamflow, these two basins also exhibited considerably lower NCRMSE values (0.2–0.5) for the combined product in terms of basin-average precipitation. The random component of error was generally slightly higher for low precipitation rates than for the moderate and higher rates for the five subbasins. Similar findings in NCRMSE values for the low streamflow rates were observed. For the basin-average precipitation, the relative NCRMSE reduction was high for the combined product, ranging from 84 to 88 % (below the 25th percentile group). A product-wise comparison showed the combined product had more significant error-dampening effects than reanalysis and satellite precipitation products in streamflow simulation for all the subbasins. Specifically, the combined product (above the 50th percentile) was characterized by a noticeably relative error reduction (44 to 78 %) for streamflow. Moreover, we also observed that relative error reduction for the combined product decreased remarkably (56 to 88 %) for low streamflow. These results indicated random error was reduced through the rainfall–streamflow transformation in all subbasins. Overall, results show combining information from reanalysis and satellite precipitation datasets could decrease random error in streamflow simulations.

The bias ratio (BR) for basin-average precipitation ranges from overestimation to underestimation as a function of precipitation magnitude, with precipitation rates above the 50th percentile strongly underestimated (BR in the range of 0.07–0.25) (Fig. 13). The magnitude of BR for precipitation indicated lower systematic errors in estimates of low to high basin-average precipitation for all subbasins. The corresponding BR values for the simulated streamflows provided a general appreciation of how the magnitude of systematic error in basin-average precipitation translates to systematic error in streamflow simulations. While a one-to-one correspondence between rainfall and streamflow classes was not possible (note that an event from the highest rainfall class might have resulted in moderate flow values depending on antecedent conditions and so on), we could, however, compare the overall range of BR values between basin-average precipitation and streamflow. As Fig. 13 indicates for the combined product, BR values were closer to 1 for the different streamflow classes, indicating that streamflow was relatively stable. For the two larger basins, Ebro at Tortosa and Ebro at Zaragoza, the combined product underestimated actual values slightly. Overall, the relative systematic error reduction for streamflow ranged from 20 to 99 %. These results highlight the usefulness of optimally combining satellite and reanalysis precipitation datasets. Overall, after reducing systematic error, the QRF-generated ensemble corrections brought rainfall products closer to the reference rainfall and simulated runoff.

5 Conclusions

A new framework was presented in this study that uses a nonparametric technique (QRF) to combine multiple globally available data sources, including reanalysis and gauge-adjusted satellite precipitation datasets, for generating an improved ensemble precipitation product. The study investigated the accuracy of the combined product using a high-resolution reference rainfall dataset (SAFRAN) over the Iberian Peninsula. The QRF-generated ensemble members

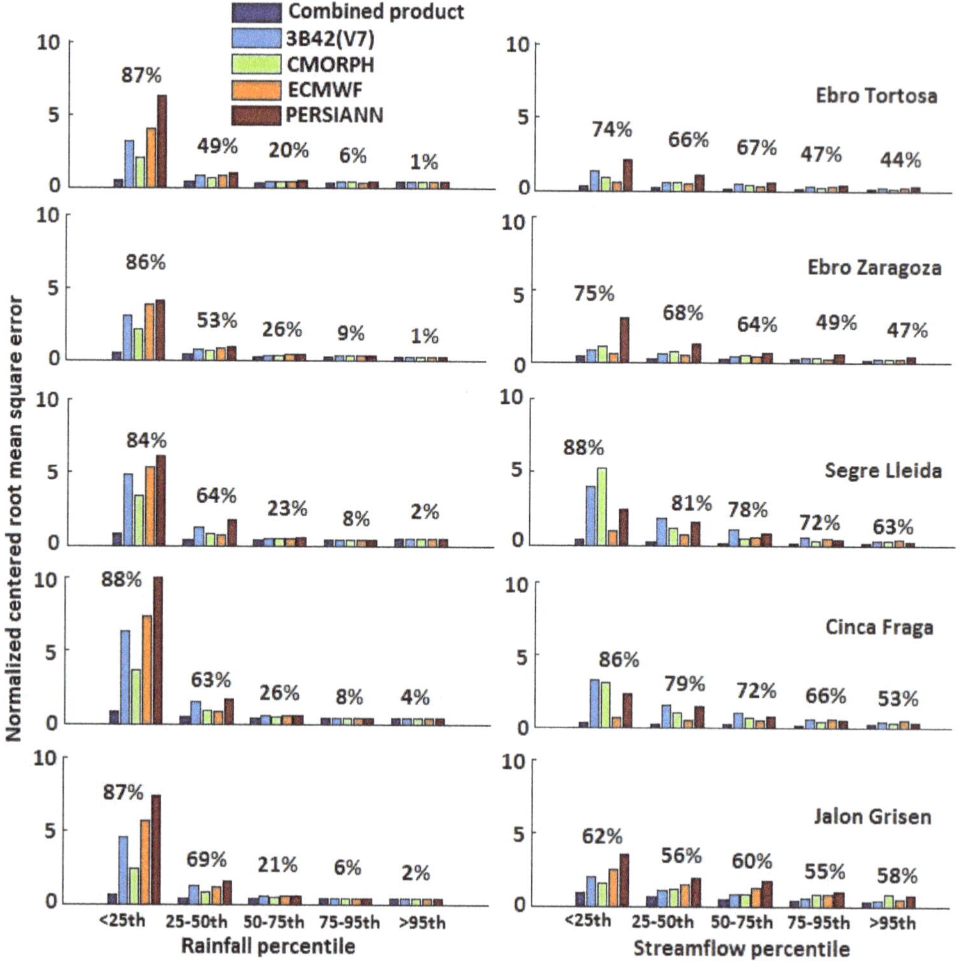

Figure 12. Normalized centered root mean square error for basin-average rainfall and streamflow. The relative error reductions for the combined product (relative to other products) are shown above the bars.

are evaluated in terms of precipitation for both warm- and cold-season weather patterns, representing a wide variety of precipitation events. Furthermore, the QRF-based streamflow simulations from a distributed hydrological model are evaluated against the SAFRAN-based simulations for a range of basin scales of the Ebro River basin.

Results from the analysis carried out demonstrate clearly that the proposed blending technique has the potential to generate a realistic precipitation ensemble that is statistically consistent with the reference precipitation and is associated with considerably reduced errors. In terms of seasonality effects, the random error significantly decreased for the combined product with increasing rainfall magnitude, and this reduction was greater during the cold season. The systematic error of the combined product varied from over- to underestimation as rain rate increased during both seasons. In terms of elevation, among all individual products, the magnitude of systematic error for the combined product was noticeably decreased for the higher elevations, which is a strong indication for using the proposed scheme in retrieving global precipitation in high-elevation regions. Overall, the reduction of the combined product (relative to other products) for the systematic and random error ranged between 17 and 76 and 53 and 99 %, respectively.

Evaluation of the impact on streamflow simulations showed that the magnitude of systematic and random error for simulations corresponding to the combined product was significantly lower than for the individual precipitation products. In addition, for the combined product, the large-scale basins exhibited considerably lower systematic and random error values than the small-size basins, which shows the dependency on basin scale. Specifically, the relative reduction for the combined product in systematic and random error ranged between 20 and 99 and 44 and 88 %, respectively, which highlights the potential of the proposed technique in advancing hydrologic simulations.

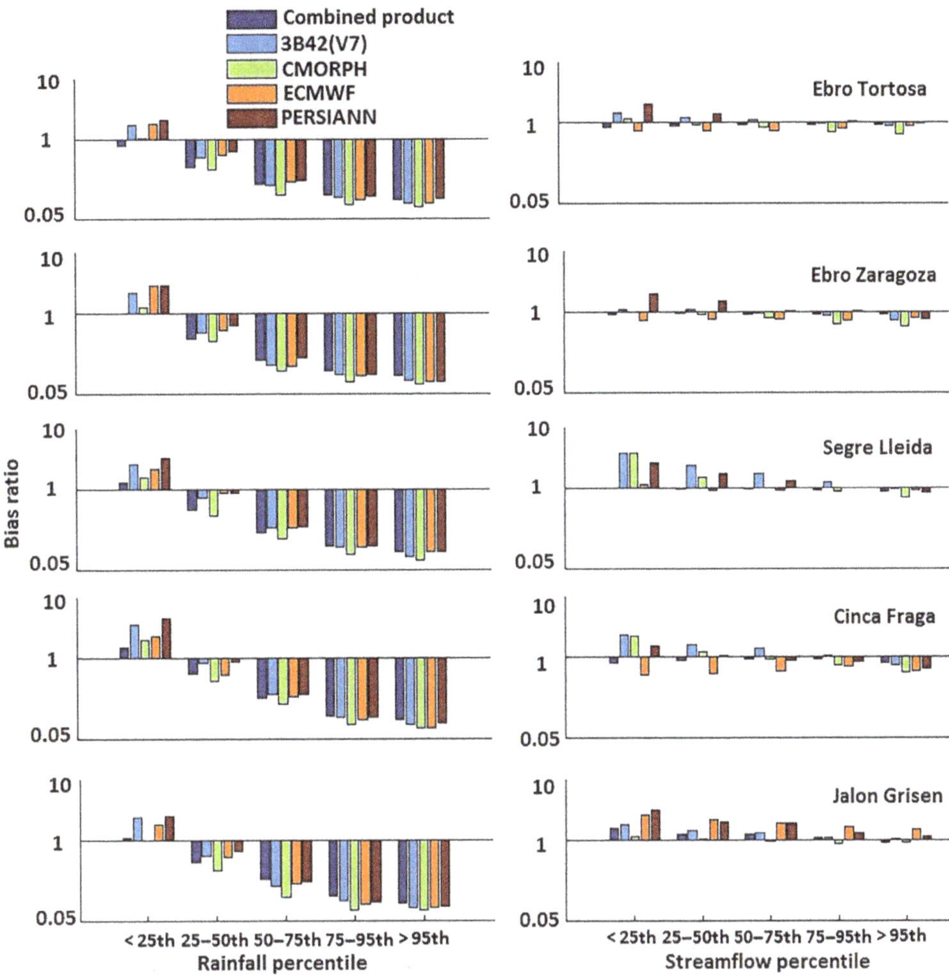

Figure 13. Bias ratio of precipitation and streamflow.

Our overall conclusion is that the proposed framework offers a robust way of blending globally available precipitation datasets, providing at the same time an improved precipitation product and characterization of its uncertainty. This can have important applications in studies dealing with water resources reanalysis and quantification of uncertainty in hydrologic simulations. Future work will include evaluation of the proposed framework in different hydroclimatic regions, also considering the sensitivity of its performance to availability (e.g., record length and spatial coverage) of in situ reference precipitation.

Data availability. Several datasets were used for this paper. The SAFRAN dataset for Spain was obtained from the MISTRALS HyMeX database (https://doi.org/10.14768/MISTRALS-HYMEX.1388). The other datasets are also available online: CMORPH (ftp://ftp.cpc.ncep.noaa.gov/precip/CMORPH_V1.0/RAW/0.25deg-3HLY/), PERSIANN (http://fire.eng.uci.edu/PERSIANN/data/3hrly_adj_cact_tars/), 3B42 (V7) (https://mirador.gsfc.nasa.gov), atmospheric reanalysis dataset (https://wci.earth2observe.eu/portal/), satellite-derived near-surface daily soil moisture data (http://www.esa-soilmoisture-cci.org/node/145/), terrain elevation data (http://srtm.csi.cgiar.org/SELECTION/inputCoord.asp), and the combined product (https://sites.google.com/uconn.edu/ehsanbhuiyan/research).

Competing interests. The authors declare that they have no conflict of interest.

Acknowledgements. This research was supported by the FP7 project eartH2Observe (grant agreement no. 603608).

Edited by: Hannah Cloke

References

Adler, R. F., Kidd, C., Petty, G., Morissey, M., and Goodman, H. M.: Intercomparison of global precipitation products: the third precipitation intercomparison project (PIP–3), B. Am. Meteorol. Soc., 82, 1377–1396, 2001.

AghaKouchak, A., Nasrollahi, N., and Habib, E.: Accounting for uncertainties of the TRMM Satellite Estimates, Remote Sens., 1, 606–619, 2009.

Álvarez, J., Sánchez, A., and Quintas, L.: SIMPA, a GRASS based tool for hydrological studies, in: Proceedings of the FOSS/GRASS Users Conference, Bangkok, Thailand, 2004.

Bhuiyan, M. A. E., Anagnostou, E. N, and Kirstetter, P. E.: A nonparametric statistical technique for modeling overland TMI (2A12) rainfall retrieval error, IEEE Geosci. Remote S., 14, 1898–1902, 2017.

Bitew, M. M. and Gebremichael, M.: Evaluation of satellite rainfall products through hydrologic simulation in a fully distributed hydrologic model, Water Resour. Res., 47, W06526, https://doi.org/10.1029/2010WR009917, 2011.

Boone, A.: Modélisation des processus hydrologiques dans le schéma de surface ISBA: Inclusion d'un réservoir hydrologique, du gel et modélisation de la neige, Université Paul Sabatier (Toulouse III), Toulouse, available at: http://www.cnrm.meteo.fr/IMG/pdf/boone_thesis_2000.pdf, 2000.

Breiman, L.: Random forests, Mach. Learn., 45, 5–32, 2001.

Brown, J. D. and Seo, D. J.: A nonparametric postprocessor for bias correction of hydrometeorological and hydrologic ensemble forecasts, J. Hydrometeorol., 11, 642–665, 2010.

Champeaux, J. L., Masson, V., and Chauvin, F.: ECOCLIMAP: a global database of land surface parameters at 1 km resolution, Meteorol. Appl., 12, 29–32, 2005.

Chipman, H. A., George, E. I., and McCulloch, R. E.: Bart: Bayesian additive regression trees, Ann. Appl. Stat., 4, 266–298, 2010.

Ciach, G. J., Krajewski, W. F., and Villarini, G.: Product-error-driven uncertainty model for probabilistic quantitative precipitation estimation with NEXRAD data, J. Hydrometeorol., 8, 1325–1347, https://doi.org/10.1175/2007JHM814.1, 2007.

Croley, T. E.: Weighted-climate parametric hydrologic forecasting, J. Hydrol. Eng., 8, 8171–8180, 2003.

David, C. H., Maidment, D. R., Niu, G. Y., Yang, Z. L., Habets, F., and Eijkhout, V.: River network routing on the NHDPlus dataset, J. Hydrometeorol., 12, 913–934, 2011a.

David, C. H., Habets, F., Maidment, D. R., and Yang, Z. L.: RAPID applied to the SIM-France model, Hydrol. Process., 25, 3412–3425, 2011b.

De Jeu, R. A.: Retrieval of land surface parameters using passive microwave remote sensing, PhD Dissertation, VU Amsterdam, the Netherlands, 120 pp., 2003.

Decharme, B., Boone, A., Delire, C., and Noilhan, J.: Local evaluation of the interaction between soil biosphere atmosphere soil multilayer diffusion scheme using four pedotransfer functions, J. Geophys. Res.-Atmos., 116, https://doi.org/10.1029/2011JD016002, 2011.

Dee, D. P., Uppala, S. M., Simmons, A. J., Berrisford, P., Poli, P., Kobayashi, S., Andrae, U., Balmaseda, M. A., Balsamo, G., Bauer, P., and Bechtold, P.: The ERA-Interim reanalysis: configuration and performance of the data assimilation system, Q. J. Roy. Meteor. Soc., 137, 553–597, 2011.

Derin, Y., Anagnostou, E., Berne, A., Borga, M., Boudevillain, B., Buytaert, W., Chang, C. H., Delrieu, G., Hong, Y., Hsu, Y. C., and Lavado-Casimiro, W.: Multiregional satellite precipitation products evaluation over complex terrain, J. Hydrometeorol., 17, 1817–1836, https://doi.org/10.1175/JHM-D-15-0197.1, 2016.

Durand, Y., Brun, E., Merindol, L., Guyomarc'h, G., Lesaffre, B., and Martin, E.: A meteorological estimation of relevant parameters for snow models, Ann. Glaciol., 18, 65–71, 1993.

Francke, T., López-Tarazón, J. A., Vericat, D., Bronstert, A., and Batalla, R. J.: Flood-based analysis of high-magnitude sediment transport using a non-parametric method, Earth Surf. Proc. Land., 33, 2064–2077, 2008.

Gebremichael, M., Liao, G. Y., and Yan, J.: Nonparametric error model for a high resolution satellite rainfall product, Water Resour. Res., 47, W07504, https://doi.org/10.1029/2010WR009667, 2011.

Gottschalck, J., Meng, J., Rodell, M., and Houser, P.: Analysis of multiple precipitation products and preliminary assessment of their impact on global land data assimilation system land surface states, J. Hydrometeorol., 6, 573–598, https://doi.org/10.1175/JHM437.1, 2005.

Guikema, S. D., Quiring, S. M., and Han, S. R.: Prestorm estimation of hurricane damage to electric power distribution systems, Risk Anal., 30, 1744–1752, 2010.

Hamill, T. M.: Interpretation of rank histograms for verifying ensemble forecasts, Mon. Weather Rev., 129, 550–560, 2001.

Hamill, T. M. and Colucci, S. J.: Verification of Eta–RSM short-range ensemble forecasts, Mon. Weather Rev., 125, 1312–1327, 1997.

Harding, R., Best, M., Blyth, E., Hagemann, S., Kabat, P., Tallaksen, L. M., Warnaars, T., Wiberg, D., Weedon, G. P., Lanen, H. V., and Ludwig, F.: WATCH: current knowledge of the terrestrial global water cycle, J. Hydrometeorol., 12, 1149–1156, 2011.

He, J., Wanik, D., Hartman, B., Anagnostou, E., Astitha, M., and Frediani, M. E. B.: Nonparametric tree-based predictive modeling of storm damage to power distribution network, Risk Anal., 37, 441–458, https://doi.org/10.1111/risa.12652, 2017.

Hossain, F. and Anagnostou, E. N.: Assessment of current passive-microwave- and infrared-based satellite rainfall remote sensing for flood prediction, J. Geophys. Res., 109, D07102, https://doi.org/10.1029/2003JD003986, 2004.

Hossain, F. and Anagnostou, E. N.: Assessment of a multidimensional satellite rainfall error model for ensemble generation of satellite rainfall data, IEEE Trans. Geosci. Remote Sens. Lett., 3, 419–423, 2006.

Hou, A. Y., Kakar, R. K., Neeck, S., Azarbarzin, A. A., Kummerow, C. D., Kojima, M., Oki, R., Nakamura, K., and Iguchi, T.: The global precipitation measurement mission, B. Am. Meteorol. Soc., 95, 701–722, https://doi.org/10.1175/BAMS-D-13-00164.1, 2014.

Huffman, G. J., Bolvin, D. T., Nelkin, E. J., Wolff, D. B., Adler, R. F., Gu, G., Hong, Y., Bowman, K. P., and Stocker, E. F.: The TRMM Multisatellite Precipitation Analysis (TMPA): quasi-global, multiyear, combined-sensor precipitation estimates at fine scales, J. Hydrometeorol., 8, 38–55, https://doi.org/10.1175/JHM560.1, 2007.

Huffman, G. J., Adler, R. F., Bolvin, D. T., and Nelkin, E. J.: The TRMM multi-satellite precipitation analysis (TMPA), in: Satellite rainfall applications for surface hydrology, edited by: Gebremichael, M. and Hossain, F., Springer, Dordrecht, 3–22, 2010.

Joyce, R. J., Janowiak, J. E., Arkin, P. A., and Xie, P.: CMORPH: a method that produces global precipitation estimates from passive microwave and infrared data at high spatial and temporal resolution, J. Hydrometeorol., 5, 487–503, 2004.

Juban, J., Fugon, L., and Kariniotakis, G.: Probabilistic short-term wind power forecasting based on kernel density estimators, in: European Wind Energy Conference and exhibition, EWEC 2007, 7–10 May 2017, Milan, Italy, 2007.

Lakhankar, T., Ghedira, H., Temimi, M., Sengupta, M., Khanbilvardi, R., and Blake, R.: Non-parametric methods for soil moisture retrieval from satellite remote sensing data, Remote Sens.-Basel, 1, 3–21, https://doi.org/10.3390/rs1010003, 2009.

Lehner, B., Verdin, K., and Jarvis, A.: New global hydrography derived from spaceborne elevation data, Eos, 89, 93–94, 2008.

Li, L., Hong, Y., Wang, J., Adler, R. F., Policelli, F. S., Habib, S., Irwn, D., Korme, T., and Okello, L.: Evaluation of the real-time TRMM-based multi-satellite precipitation analysis for an operational flood prediction system in Nzoia basin, Lake Victoria, Africa, Nat. Hazards, 50, 109–123, https://doi.org/10.1007/s11069-008-9324-5, 2009.

Liu, Y. Y., Parinussa, R. M., Dorigo, W. A., De Jeu, R. A. M., Wagner, W., van Dijk, A. I. J. M., McCabe, M. F., and Evans, J. P.: Developing an improved soil moisture dataset by blending passive and active microwave satellite-based retrievals, Hydrol. Earth Syst. Sci., 15, 425–436, https://doi.org/10.5194/hess-15-425-2011, 2011.

Liu, Y. Y., Dorigo, W. A., Parinussa, R. M., de Jeu, R. A., Wagner, W., McCabe, M. F., Evans, J. P., and Van Dijk, A. I. J. M.: Trend-preserving blending of passive and active microwave soil moisture retrievals, Remote Sens. Environ., 123, 280–297, 2012.

Maggioni, V., Sapiano, M. R., Adler, R. F., Tian, Y., and Huffman, G. J.: An error model for uncertainty quantification in high-time-resolution precipitation products, J. Hydrometeorol., 15, 1274–1292, https://doi.org/10.1175/JHM-D-13-0112.1, 2014.

Maggioni, V., Massari, C., Brocca, L., and Ciabatta, L.: Merging bottom-up and top-down precipitation products using a stochastic error model, J. Geophys. Res., 19, 12383, 2017.

Masson, V., Le Moigne, P., Martin, E., Faroux, S., Alias, A., Alkama, R., Belamari, S., Barbu, A., Boone, A., Bouyssel, F., Brousseau, P., Brun, E., Calvet, J.-C., Carrer, D., Decharme, B., Delire, C., Donier, S., Essaouini, K., Gibelin, A.-L., Giordani, H., Habets, F., Jidane, M., Kerdraon, G., Kourzeneva, E., Lafaysse, M., Lafont, S., Lebeaupin Brossier, C., Lemonsu, A., Mahfouf, J.-F., Marguinaud, P., Mokhtari, M., Morin, S., Pigeon, G., Salgado, R., Seity, Y., Taillefer, F., Tanguy, G., Tulet, P., Vincendon, B., Vionnet, V., and Voldoire, A.: The SURFEXv7.2 land and ocean surface platform for coupled or offline simulation of earth surface variables and fluxes, Geosci. Model Dev., 6, 929–960, https://doi.org/10.5194/gmd-6-929-2013, 2013.

Mei, Y., Anagnostou, E. N., Nikolopoulos, E. I., and Borga, M.: Error analysis of satellite rainfall products in mountainous basins, J. Hydrometeorol., 15, 1778–1793, https://doi.org/10.1175/JHM-D-13-0194.1, 2014.

Mei, Y., Nikolopoulos, E. I., Anagnostou, E. N., and Borga, M.: Evaluating satellite precipitation error propagation in runoff simulations of mountainous basins, J. Hydrometeorol., 17, 1407–1423, https://doi.org/10.1175/JHM-D-15-0081.1, 2016.

Meinshausen, N.: Quantile regression forests, J. Mach. Learn. Res., 7, 983–999, 2006.

Mo, K. C., Chen, L. C., Shukla, S., Bohn, T. J., and Lettenmaier, D. P.: Uncertainties in North American land data assimilation systems over the contiguous United States, J. Hydrometeorol., 13, 996–1009, https://doi.org/10.1175/JHM-D-11-0132.1, 2012.

Mujumdar, P. P. and Ghosh, S.: Climate change impact on hydrology and water resources, J. Hydraul. Eng., 14, 1–17, 2008.

Nash, J. E. and Sutcliffe, J. V.: River flow forecasting through conceptual models part I – a discussion of principles, J. Hydrol., 10, 282–290, 1970.

Nateghi, R., Guikema, S. D., and Quiring, S. M.: Forecasting hurricane-induced power outage durations, Nat. Hazards, 74, 1795–1811, 2014.

Nikolopoulos, E. I., Anagnostou, E. N., and Borga, M.: Using high-resolution satellite rainfall products to simulate a major flash flood event in Northern Italy, J. Hydrometeorol., 14, 171–185, https://doi.org/10.1175/JHM-D-12-09.1, 2013.

Noilhan, J. and Mahfouf, J. F.: The ISBA land surface parameterisation scheme, Global Planet. Change, 13, 145–159, 1996.

Noilhan, J. and Planton, S.: A simple parameterization of land surface processes for meteorological models, Mon. Weather Rev., 117, 536–549, 1989.

Owe, M., de Jeu, R., and Holmes, T.: Multisensor historical climatology of satellite-derived global land surface moisture, J. Geophys. Res., 113, F01002, https://doi.org/10.1029/2007JF000769, 2008.

Peña-Arancibia, J. L., van Dijk, A. I., Renzullo, L. J., and Mulligan, M.: Evaluation of precipitation estimation accuracy in reanalyses, satellite products, and an ensemble method for regions in Australia and South and East Asia, J. Hydrometeorol., 14, 1323–1333, https://doi.org/10.1175/JHM-D-12-0132.1, 2013.

Quintana-Seguí, P., Le Moigne, P., Durand, Y., Martin, E., Habets, F., Baillon, M., Canellas, C., Franchisteguy, L., and Morel, S.: Analysis of near-surface atmospheric variables: validation of the SAFRAN analysis over France, J. Appl. Meteorol. Clim., 47, 92–107, https://doi.org/10.1175/2007JAMC1636.1, 2008.

Quintana-Seguí, P., Peral, M. C., Turco, M., Llasat, M.-C., and Martin, E.: Meteorological analysis systems in North-East Spain: validation of SAFRAN and SPAN, J. Environ. Inform., 27, 116–130, https://doi.org/10.3808/jei.201600335, 2016.

Quintana-Seguí, P., Turco, M., Herrera, S., and Miguez-Macho, G.: Validation of a new SAFRAN-based gridded precipitation product for Spain and comparisons to Spain02 and ERA-Interim, Hydrol. Earth Syst. Sci., 21, 2187–2201, https://doi.org/10.5194/hess-21-2187-2017, 2017.

Ruiz, J. M.: Desarrollo de un modelo hidrológico conceptual distribuido de simulación continua integrado con un sistema de información geográfica (Development of a continuous distributed conceptual hydrological model integrated in a geographic information system), PhD Thesis, ETS Ingenieros de Caminos, Canales y Puertos, Universidad Politécnica de Valencia, Spain, 1998.

Seyyedi, H., Anagnostou, E. N., Kirstetter, P. E., Maggioni, V., Hong, Y., and Gourley, J. J.: Incorporating surface soil moisture information in error modeling of TRMM passive Microwave rainfall, IEEE T. Geosci. Remote, 52, 6226–6240, 2014.

Sorooshian, S., Hsu, K. L., Gao, X., Gupta, H. V., Imam, B., and Braithwaite, D.: Evaluation of PERSIANN system satellite-based estimates of tropical rainfall, B. Am. Meteorol. Soc., 81, 2035–2046, 2000.

Stephens, G. L. and Kummerow, C. D.: The remote sensing of clouds and precipitation from space: a review, J. Atmos. Sci., 64, 3742–3765, 2007.

Taillardat, M., Mestre, O., Zamo, M., and Naveau, P.: Calibrated ensemble forecasts using quantile regression forests and ensemble model output statistics, Mon. Weather Rev., 144, 2375–2393, 2016.

Teo, C. K. and Grimes, D. I.: Stochastic modelling of rainfall from satellite data, J. Hydrol., 346, 33–50, https://doi.org/10.1016/j.jhydrol.2007.08.014, 2007.

Vidal, J. P., Martin, E., Franchistéguy, L., Baillon, M., and Soubeyroux, J. M.: A 50 year high-resolution atmospheric reanalysis over France with the Safran system, Int. J. Climatol., 30, 1627–1644, https://doi.org/10.1002/joc.2003, 2010.

Wagner, W., Dorigo, W., de Jeu, R., Fernandez, D., Benveniste, J., Haas, E., and Ertl, M.: Fusion of active and passive microwave observations to create an essential climate variable data record on soil moisture, in: ISPRS Annals of the Photogrammetry, Remote Sensing and Spatial Information Sciences, 25 August–1 September 2012, Melbourne, Australia, 315–321, 2012.

Weedon, G. P., Balsamo, G., Bellouin, N., Gomes, S., Best, M. J., and Viterbo, P.: The WFDEI meteorological forcing data set: WATCH forcing data methodology applied to ERA-Interim reanalysis data, Water Resour. Res., 50, 7505–7514, 2014.

Yenigun, K. and Ecer, R.: Overlay mapping trend analysis technique and its application in Euphrates Basin, Turkey, Meteorol. Appl., 20, 427–438, 2013.

Zimmermann, A., Francke, T., and Elsenbeer, H.: Forests and erosion: insights from a study of suspended-sediment dynamics in an overland flow-prone rainforest catchment, J. Hydrol., 428, 170–181, 2012.

10

Using measured soil water contents to estimate evapotranspiration and root water uptake profiles

M. Guderle[1,2,3] **and A. Hildebrandt**[1,2]

[1]Friedrich Schiller University, Institute for Geosciences, Burgweg 11, 07749 Jena, Germany
[2]Max Planck Institute for Biogeochemistry, Biogeochemical Processes, Hans-Knöll-Str. 10, 07745 Jena, Germany
[3]International Max Planck Research School for Global Biogeochemical Cycles, Hans-Knöll-Str. 10, 07745 Jena, Germany

Correspondence to: M. Guderle (marcus.guderle@uni-jena.de)

Abstract. Understanding the role of plants in soil water relations, and thus ecosystem functioning, requires information about root water uptake. We evaluated four different complex water balance methods to estimate sink term patterns and evapotranspiration directly from soil moisture measurements. We tested four methods. The first two take the difference between two measurement intervals as evapotranspiration, thus neglecting vertical flow. The third uses regression on the soil water content time series and differences between day and night to account for vertical flow. The fourth accounts for vertical flow using a numerical model and iteratively solves for the sink term. None of these methods requires any a priori information of root distribution parameters or evapotranspiration, which is an advantage compared to common root water uptake models. To test the methods, a synthetic experiment with numerical simulations for a grassland ecosystem was conducted. Additionally, the time series were perturbed to simulate common sensor errors, like those due to measurement precision and inaccurate sensor calibration. We tested each method for a range of measurement frequencies and applied performance criteria to evaluate the suitability of each method. In general, we show that methods accounting for vertical flow predict evapotranspiration and the sink term distribution more accurately than the simpler approaches. Under consideration of possible measurement uncertainties, the method based on regression and differentiating between day and night cycles leads to the best and most robust estimation of sink term patterns. It is thus an alternative to more complex inverse numerical methods. This study demonstrates that highly resolved (temporally and spatially) soil water content measurements may be used to estimate the sink term profiles when the appropriate approach is used.

1 Introduction

Plants play a key role in the Earth system by linking the water and the carbon cycle between soil and atmosphere (Feddes et al., 2001; Chapin et al., 2002; Feddes and Raats, 2004; Teuling et al., 2006b; Schneider et al., 2009; Seneviratne et al., 2010; Asbjornsen et al., 2011). Knowledge of evapotranspiration and especially root water uptake profiles is key to understanding plant–soil-water relations and thus ecosystem functioning, in particular efficient plant water use, storage keeping and competition in ecosystems (Davis and Mooney, 1986; Le Roux et al., 1995; Jackson et al., 1996; Hildebrandt and Eltahir, 2007; Arnold et al., 2009; Schwendenmann et al., 2014).

For estimation of root water uptake, models are prevalent in many disciplines. Most commonly, root water uptake is applied as a sink term S, incorporated in the 1-D soil water flow equation (Richards equation, Eq. 1)

$$\frac{\partial \theta}{\partial t} = \frac{\partial}{\partial z}\left[K(h)\left(\frac{\partial h}{\partial z}+1\right)\right] - S(z, t), \qquad (1)$$

where θ is the volumetric soil water content, t is time, z is the vertical coordinate, h is the soil matric potential, $K(h)$ is the unsaturated soil hydraulic conductivity and $S(z, t)$ is the sink term (water extraction by roots, evaporation, etc.). The sink term pro le $S(z, t)$ depends on root activity, which has to be known previously. Often root activity is assumed to be related to rooting profiles, represented by power laws

(Gale and Grigal, 1987; Jackson et al., 1996; Schenk, 2008; Kuhlmann et al., 2012). The parameters of those rooting profile functions are cumbersome to measure in the field, and the relevance for root water uptake distribution is also uncertain (Hamblin and Tennant, 1987; Lai and Katul, 2000; Li et al., 2002; Doussan et al., 2006; Garrigues et al., 2006; Schneider et al., 2009). Therefore, assumptions have to be made in order to determine the sink term for root water uptake in soil water flow models. The lack of an adequate description of root water uptake parameters was mentioned by Gardner (1983) and is currently still an issue (Lai and Katul, 2000; Hupet et al., 2002; Teuling et al., 2006a, b). For those reasons, methods for estimating root water uptake are a paramount requirement.

Standard measurements, for instance of soil water content profiles, are recommended to be used for estimation of evapotranspiration and root water uptake at low cost, since the evolution of soil moisture in space and time is expected to contain information on root water uptake (Musters and Bouten, 2009; Hupet et al., 2002; Zuo and Zhang, 2002; Teuling et al., 2006a). Methods using these measurements are, for instance, simple water balance approaches, which estimate evapotranspiration (Wilson et al., 2001; Schume et al., 2005; Kosugi and Katsuyama, 2007; Breña Naranjo et al., 2011) and root water uptake (Clothier and Green, 1995; Coelho and Or, 1996; Hupet et al., 2002) by calculating the difference in soil water storage between two different observation times. The advantages of these simple water balance methods are the small amount of information required and the simple methodology. However, a disadvantage is that the depletion of soil water is assumed to occur only by root water uptake and soil evaporation, and soil water fluxes are negligible. This is only the case during long dry periods with high atmospheric demand (Hupet et al., 2002).

A possible alternative which allows for the consideration of vertical soil water fluxes is the inverse use of numerical soil water flow models (Musters and Bouten, 1999; Musters et al, 2000; Vrugt et al., 2001; Hupet et al., 2002; Zuo and Zhang, 2002). Root water uptake or parameters on the root water uptake function are estimated by minimizing the differences between measured soil water contents and the corresponding model results by an objective function (Hupet et al., 2002). However, the quality of the estimation depends, on the one hand, strongly on system boundary conditions (e.g., incoming flux, drainage flux or location of the groundwater table) and soil parameters (e.g., hydraulic conductivity), which are, on the other hand, notoriously uncertain under natural conditions (Musters and Bouten, 2000; Kollet, 2009). Another problem is that the applied models for soil water flow potentially ignore biotic processes. For example, Musters et al. (2000) and Hupet et al. (2002) attempted to fit parameters for root distributions in a model determining uptake profiles from water availability, whereas empirical and modeling studies suggest that adjustment of root water uptake distribution may also be from physiological adaptations (Jackson et al., 2000; Zwieniecki et al., 2003; Bechmann et al., 2014). In order to avoid this problem, Zuo and Zhang (2002) coupled a water balance approach to a soil water model, which enabled them to estimate root water uptake without the a priori estimation of root water uptake parameters.

A second option for accounting for vertical soil water flow in a water balance approach is to analyze the soil moisture fluctuation between day and night (Li et al., 2002). In comparatively dry soil, Li et al. (2002) fitted third-order polynomials to the daytime- and nighttime-measured soil water content time series and calculated vertical soil water flow using the first derivative of the fitted polynomials during nighttime.

Up to now, little effort has been made to compare those different data-driven methods for estimating evapotranspiration and root water uptake profiles in temperate climates. In this paper, we compare those water balance methods we are aware of that do not require any a priori information of root distribution parameters. We used artificial data of soil moisture and sink term profiles to compare the quality of the estimates of the different methods. Furthermore, we investigated the influence of sensor errors on the outcomes, as these uncertainties can have a significant impact on both data-driven approaches and soil hydrological models (Spank et al., 2013). For this, we artificially introduced measurement errors to the synthetic soil moisture time series that are typical for soil water content measurements: sensor calibration error and limited precision.

Our results indicate that highly resolved soil water content measurements can provide reliable predictions of the sink term or root water uptake profile when the appropriate approach is used.

2 Material and methods

Table A1 summarizes the variable names used in this section together with their units.

2.1 Target variable and general procedure

The evapotranspiration E consists of soil evaporation E_s and the plant transpiration E_t (Eq. 2):

$$E = (E_s + E_t). \tag{2}$$

The distinction between soil evaporation and combined transpiration is not possible for any of the applied water balance methods. Therefore, the water extraction from soil by plant roots and soil evaporation is referred to as the sink term profile in the rest of the paper. The integrated sink term over

Table 1. The abbreviation and full name of the methods for further use, overviews of the four applied data-driven methods, and the required input data.

Abbreviation	Method	Method short description	Input data
sssl	Single-step, single-layer water balance	Water balance (Breña Naranjo et al., 2011)	Volumetric soil water content at a single depth, precipitation
ssml	Single-step, multi-layer water balance	Water balance over entire soil profile (Clothier and Green, 1995; Coelho and Or, 1996; Hupet et al., 2002)	Volumetric soil water content at several depths, precipitation
msml	Multi-step, multi-layer regression	Approach to use the short-term fluctuations of soil moisture (Li et al., 2002)	Volumetric soil water content at several depths, precipitation
im	Inverse model	Water balance solved iteratively with a numerical soil water flow model (Zuo and Zhang, 2002; Ross, 2003)	Soil hydraulic parameters Volumetric soil water content at several depths, precipitation

the entire soil profile results in the total evapotranspiration (Eq. 3):

$$E(t) = \int_{z=z_r}^{0} S(t,z)\mathrm{d}z \rightarrow E_j = \sum_{i=1}^{n} S_{i,j} \cdot d_{z,i}, \tag{3}$$

where z is the soil depth, $d_{z,i}$ is the thickness of the soil layer i, t is time and j is the time step. For matters of simplicity we will drop the index j when introducing the estimation methods in the following.

In this study, synthetic time series of volumetric soil water content generated by a soil water flow model coupled with a root water uptake model (Sect. 2.3) were treated as measured data and are used as the basis for all methods (Sect. 2.2) estimating the sink term $\widetilde{S}(z)$, and total evapotranspiration $\widetilde{E}$. In order to investigate the influence of sensor errors, the generated time series were systematically disturbed, as shown in Sect. 2.4. Based on these estimations, we evaluate the data-driven methods on predicting evapotranspiration $\widetilde{E}$ and sink term profiles using the quality criteria given in Sect. 2.5. As the depth at which a given fraction of root water uptake occurred is often of interest in ecohydrological studies (e.g., Clothier and Green, 1999; Plamboeck et al., 1999; Ogle et al., 2004), estimated sink term profiles were compared accordingly. Specifically, we determined up to which depths 25, 50 and 90 % ($z_{25\,\%}$, $z_{50\,\%}$ and $z_{90\,\%}$) of water extraction takes place.

2.2 Investigated data-driven methods for estimation of the sink term profile

In the following we introduce the four investigated methods. They are summarized in Table 1.

2.2.1 Single-step, single-layer (sssl) water balance

Breña Naranjo et al. (2011) derived the sink term using time series of rainfall and changes of soil water content between two observation times (single step), based on measurements at one single soil depth (single layer). The complete water balance equation for this single-layer method is

$$\widetilde{E}_{\mathrm{sssl}} = P - q - z_{\mathrm{r}}\frac{\Delta\theta}{\Delta t}, \tag{4}$$

where z_{r} is the active rooting depth, which is also the depth of the single soil layer, and is taken equal to the measurement depth of volumetric soil water content, θ. Δt indicates the length of the considered single time step. P is the rainfall and q the percolation out of the soil layer during the same time step. When rainfall occurs, infiltration as well as soil water flow takes place. It is assumed that percolation occurs only during this time and persists only up to several hours after the rainfall event (Breña Naranjo et al., 2011). Since the percolation flux is unknown, the methods cannot be applied during these wet times. During dry periods, q is set to zero and Eq. (4) simplifies to Eq. (5) (Breña Naranjo et al., 2011):

$$\widetilde{E}_{\mathrm{sssl}} = z_{\mathrm{r}}\frac{\Delta\theta}{\Delta t}. \tag{5}$$

We applied Eq. (5) to estimate evaporation (in the single-layer method equal to the sink term) from artificial soil water contents at 30 cm. Required input information is thus only time series of soil water content and active rooting depth z_{r}. Additionally, rainfall measurements are required to select dry periods, where no percolation occurs. These could start several hours up to several days after a rainfall event (Breña Naranjo et al., 2011), and the exact timing depends on the amount of rainfall and the site-location parameters like soil type and vegetation. In this study we waited until 24 h after the end of the precipitation event before applying the model.

2.2.2 Single-step, multi-layer (ssml) water balance

This method is similar to the sssl method introduced above. It calculates the sink term based on two observation times (single step), but is extended to several measurement depths (multi-layer). The water balance during dry periods of each layer is the same as in Eq. (5), and uptake in individual layers is calculated by neglecting vertical soil water fluxes and therefore assuming that the change in soil water content is only caused by root water uptake (Hupet et al., 2002):

$$\widetilde{S}_{\mathrm{ssml},\,i} = d_{z,\,i}\frac{\Delta\theta_i}{\Delta t}, \tag{6}$$

where $\widetilde{S}_{\mathrm{ssml},i}$ is the estimated sink term in soil layer i, $\Delta\theta_i$ is the change in soil water content in the soil layer i over the single time step (Δt) and $d_{z,\,i}$ is the thickness of the soil layer i. Actual evapotranspiration (E_{ssml}) is calculated by summing up $\widetilde{S}_{\mathrm{ssml},i}$ over all depths in accordance with (Eq. 3). The application of the ssml method is restricted to dry periods. It requires time series of volumetric soil water content and rainfall measurements as input to select dry periods.

2.2.3 Multi-step, multi-layer (msml) regression

The third method derives actual evapotranspiration and sink term profiles from diurnal fluctuation of soil water contents (Li et al., 2002). It uses a regression over multiple time steps (multi-step) and can be applied at several measurement depths (multi-layer).

During daytime, evapotranspiration leads to a decrease in volumetric soil water content. This extraction of soil water extends over the entire active rooting depth. Additionally, soil water flow occurs both at night and during the daytime (Khalil et al., 2003; Verhoef et al., 2006; Chanzy et al., 2012), following potential gradients in the soil profile. Thus, during dry weather conditions, the time series of soil water content shows a clear day–night signal (Fig. 1). We split up the time series by fitting a linear function to each day and night branch of the time series. The onset of transpiration is mainly defined by opening and closure of plant stomata, which is according to the supply of solar energy (Loheide, 2008; Maruyama and Kuwagata, 2008; Sánchez et al., 2013), usually 1 or 2 h after sunrise or before sunset (Lee, 2009).

Here, the basic assumption is that the soil water flow does not change significantly between day and night (Fig. S1 in the Supplement). The slope of the fitted linear functions gives the rate of root water extraction and vertical flow. This can also be shown mathematically by disassembling the Richards equation (Eq. 1) in vertical flow (subscript flow) and sink term (subscript extr) (Eq. 7), whereas the change in soil water content over time ($\partial\theta/\partial t$) integrates both fluxes:

$$\frac{\partial\theta}{\partial t} = \left.\frac{\partial\theta}{\partial t}\right|_{\mathrm{flow}} + \left.\frac{\partial\theta}{\partial t}\right|_{\mathrm{extr}} = m_{\mathrm{tot}}, \tag{7}$$

where m_{tot} corresponds to the slope of the fitted linear function for the day or night branch. Assuming that evapotranspiration during the night is negligible, the slope for the night branch is entirely due to soil water flow. During the day, uptake processes and soil water flow act in parallel:

$$\text{day}: \quad m_{\mathrm{tot}} = m_{\mathrm{flow}} + m_{\mathrm{extr}}, \tag{8a}$$

$$\text{night}: \quad m_{\mathrm{tot}} = m_{\mathrm{flow}}. \tag{8b}$$

The sink term can be calculated from Eq. (8a), assuming that m_{flow} can be estimated from Eq. (8b) and using the average of the antecedent and the preceding night. A similar procedure has previously been applied in diurnal groundwater table fluctuations (Loheide, 2008). Also the extraction will be overestimated if day and night fluxes are not separately considered. With the soil layer thickness of the respective layer i ($d_{z,i}$) taken into account, the mean daily sink term of soil layer i ($\widetilde{S}_{\mathrm{msml},i}$) is obtained:

$$\widetilde{S}_{\mathrm{msm},\,i} = (m_{\mathrm{tot},\,i} - \bar{m}_{\mathrm{flow},\,i}) \cdot d_{z,\,i}. \tag{9}$$

Since a diurnal cycle of soil moisture is only identifiable up to a time interval of 12 h, the regression method is limited to a minimum measurement frequency of 12 h. Furthermore, as rainfall causes changes of soil water content and blurs the diurnal signal, the msml regression is only applicable during dry periods. Time series of soil water content and rainfall measurements to select dry periods are required as input.

2.2.4 Inverse model (im)

The fourth approach is the most complex. The inverse model (im) estimates the average root water uptake by solving the Richards equation (Eq. 1) and iteratively searching for the sink term profile which produces the best fit between the numerical solution and measured values of soil moisture content (Zuo and Zhang, 2002). The advantage of this method is the estimation of root water uptake without the a priori estimation of rooting profile function parameters, since they are highly uncertain, as elucidated in the Introduction. We implemented the inverse water balance approach after Zuo and Zhang (2002) with the Fast Richards Solver (Ross, 2003), which is available as Fortran 90 code. We modified the original method by changing the convergence criterion. In the following section, we first introduce the iterative procedure as proposed by Zuo and Zhang (2002) and then explain the modification which we made.

The iterative procedure by Zuo and Zhang (2002) runs the numerical model over a given time step (Δt) in order to estimate the soil water content profile $\widetilde{\theta}_i^{(v=0)}$ at the end of the time step, and assuming that the sink term ($\widetilde{S}_{\text{im},i}^{(v=0)}$) is zero over the entire profile. Here $\sim$ depicts the estimated values at the respective soil layer i, and v indicates the iteration step. Next, the sink term profile $\widetilde{S}_{\text{im},i}^{(v=1)}$ is set equal to the difference between previous approximation $\widetilde{\theta}_i^{(v=0)}$ and measurements θ_i while accounting for soil layer thickness and the length of the time step for units.

In the following iterations, $\widetilde{S}_{\text{im},i}^{(v)}$ is used with the Richards equation to calculate the new soil water contents $\widetilde{\theta}_i^{(v)}$. The new average sink term $\widetilde{S}_{\text{im},i}^{(v+1)}$ is then determined with Eq. (10):

$$\widetilde{S}_{\text{im},i}^{(v+1)} = \widetilde{S}_{\text{im},i}^{(v)} + \frac{\widetilde{\theta}_i^{(v)} - \theta_i}{\Delta t} \cdot d_{z,i}. \tag{10}$$

This iteration process continues until a specified decision criterion ε_{ZZ} is reached:

$$\varepsilon_{\text{ZZ}} \geq \frac{1}{n} \sum_{i=1}^{n} \left[\frac{\widetilde{\theta}_i^{(v)} - \theta_i}{\theta_i} \right]^2, \tag{11}$$

where n is the number of soil layers in the soil column.

Since ε_{ZZ} is a normalized root-mean-square error over depth, good and poor estimations cancel between layers. This leads to termination of the iterative procedure even if the estimation of the sink term is very poor in several layers. We therefore propose a slightly adapted termination process which applies to separate soil layers as follows. The estimation of the sink term in general is applied as proposed by Zuo and Zhang (2002):

1. Calculate the difference between the estimated and measured soil water content (Eq. 12) and compare the change in this difference to the difference of the previous iteration (Eq. 13):

$$e_i^{(v)} = \left| \theta_i - \widetilde{\theta}_i^{(v)} \right|, \tag{12}$$

$$\varepsilon_{\text{GH},i}^{(v)} = e_i^{(v-1)} - e_i^{(v)}. \tag{13}$$

2. In soil layers where $\varepsilon_{\text{GH}}^{(v)} < 0$, set the root water uptake rate back to the value of the previous iteration ($\widetilde{S}_{\text{im},i}^{(v+1)} = \widetilde{S}_{\text{im},i}^{(v-1)}$), since the current iteration was no improvement. Only if $\varepsilon_{\text{GH},i}^{(v)} \geq 0$, go to step (3). This prevents acceptance of the estimated sink term $\widetilde{S}_{\text{im},i}^{(v)}$ even if it leads to a worse fit than the previous iteration.
3. If $e_i^{(v)} > 1 \times 10^{-4}$, calculate $\widetilde{S}_{\text{im},i}^{(v+1)}$ according Eq. (10); otherwise the current iteration sink term ($\widetilde{S}_{\text{im},i}^{(v+1)} = \widetilde{S}_{\text{im},i}^{(v)}$) is retained, as it results in a good fit between estimated and measured soil water contents.

The iteration process continues until the convergence criterion $\varepsilon_{\text{GH}}^{(v)}$ (Eq. 13) no longer changes between iterations (i.e., all layers have reached a satisfactory fit), or after a specified number of iterations (we chose 3000).

Besides the soil water content measurements and the rainfall, the input information required is the soil hydraulic parameters.

2.3 Generation of synthetic reference data

We used synthetic time series of volumetric soil water content with a measurement frequency of 1, 3, 6, 12 and 24 h. The time series of soil water content as well as the sink term profiles were generated with a Soil water flow model (Fast Richards Solver (Ross, 2003), same as used in Sect. 2.2 for the im). These were treated as measured data and are used as the basis for all methods. The synthetic data are based on meteorological and soil data from the Jena Biodiversity Experiment (Roscher et al., 2011). Root water uptake was calculated using a simple macroscopic root water uptake model which uses an exponential root distribution with water stress compensation (Li et al., 2001). Soil evaporation is taken as 20 % of total evapotranspiration.

The soil profile is based on the Jena Experiment, both in terms of measurement design and soil properties. The model was set up for a one-dimensional homogeneous soil profile 220 cm deep. Measurement points were set at depths of 15, 30, 60, 100, 140, 180 and 220 cm. The spatial resolution of the soil model is according to the measurement points 15, 15, 30, 40, 40, 40 and 40 cm. The advantage of the applied soil water flow model is that the water fluxes are calculated with the matrix flux potential (Kirchhoff transformation), which allows for spatial discretization with large nodal spacing (Ross, 2006). We used a maximum rooting depth of 140 cm, with 60 % of root length density located in the top 15 cm of the root zone, which corresponds to mean values measured on the field site (Ravenek et al., 2014). We used van Genuchten soil hydraulic parameters (van Genuchten, 1980) derived from the program ROSETTA (Schaap et al., 2001) based on the texture of a silty loam: $\theta_s = 0.409$ (cm^3 cm^{-3}), $\theta_r = 0.069$ (cm^3 cm^{-3}), $K_{\text{sat}} = 1.43 \times 10^{-6}$ (m s^{-1}), $\alpha = 0.6$ (m^{-1}) and $n_{\text{vG}} = 1.619$ (–).

Upper boundary conditions are derived from measured precipitation and potential evapotranspiration calculated after Penman–Monteith (Allen et al., 1998) from measurements of the climate station at the experimental site (Weather Station Saaleaue, Max Planck Institute for Biogeochemistry, http://www.bgc-jena.mpg.de/wetter/). The weather data used have a measurement resolution of 10 min. Before applying evapotranspiration and rainfall as input data to generate the

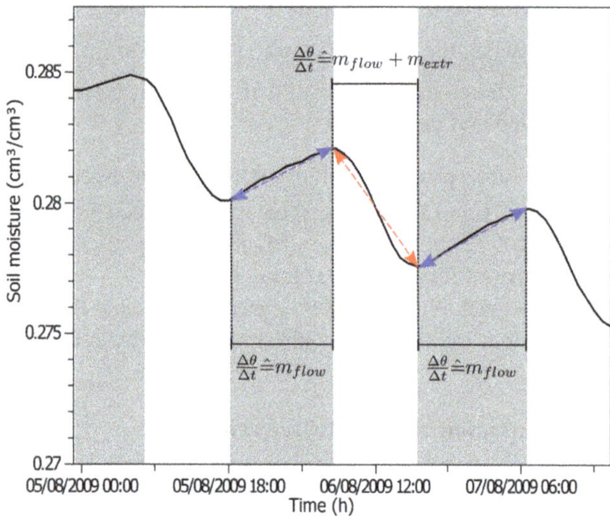

Figure 1. Short-term fluctuations in soil moisture in 15 cm depth during August 2009, showing the rewetting of soil at nighttime (blue line) and the water extraction during the day (red line); dashed lines depict the change between times with soil water extraction (white) and rewetting of soil (grey).

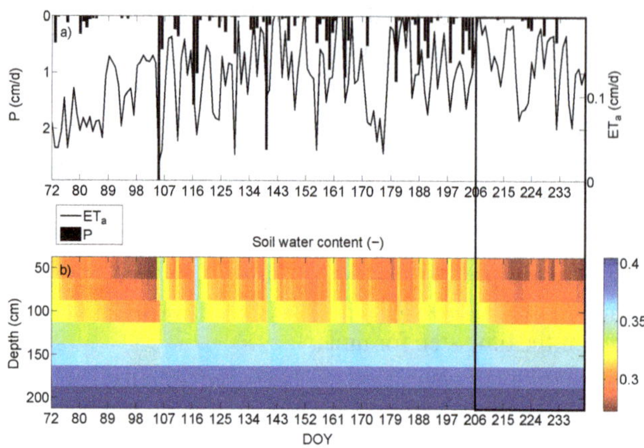

Figure 2. Actual evapotranspiration (ET_a) and precipitation (P) (cm day^{-1}) in the growing season (from March 2009 to September 2009) **(a)** and synthetic time series of soil water content **(b)** with daily resolution.

synthetic reference soil moisture and root water uptake data, both data sets were aggregated to the temporal resolutions applied for the reference run (1 h). Soil moisture and root water uptake were generated with the same temporal resolution. When translating the evapotranspiration into sink term profiles (four-digit precision), rounding errors introduce a small inaccuracy. Thus, the sum of the sink term in the reference run deviates by 0.02 % compared to the original evapotranspiration.

The lower boundary is given by the groundwater table, which fluctuates around −200 cm at the field site, but was set to constant head for simplification. Initial conditions are taken as the equilibrium (no flow) hydraulic potential profile in the soil.

We run the model with precipitation data from the field site for the year 2009, starting on 1 January to calculate time series of soil water content and the root water uptake up to September 2009. The atmospheric boundary conditions during the growing season are shown in Fig. 2a as daily values. For testing the methods, we used the period from 26 July to 28 August 2009, which covers a dry period with little rainfall (Fig. 2, black-outlined area). The times were chosen to cover a representative but dry period during the growing season and to guarantee a warm-up phase for the soil model.

The described forward simulation produces time series of soil water contents and root water uptake. Soil water content time series were used instead of measured data (synthetic measurements) as input for the investigated methods, while evapotranspiration and sink term profiles were used to evaluate them, based on the quality criteria described in Sect. 2.5.

2.4 Influence of soil moisture sensor uncertainty

Data-driven methods are as good as their input data. Therefore, we investigate and quantify the influence of common uncertainties of soil moisture sensor measurements on the estimation of sink term profiles. Sensor performance is usually characterized by three criteria, namely the accuracy, the precision and the resolution. The correctness of a measurement is described by the accuracy and for water content sensors depends greatly on the soil-specific calibration. Repeatability of many single measurements is referred to as precision, while the resolution describes the fineness of a measurement.

In this paper, we investigated the uncertainty of the applied methods stemming from calibration error (accuracy) and precision. For this we superimposed the original synthetic soil water content measurements generated in Sect. 2.3 with artificial errors. Three types of errors were implemented, as follows. (i) Precision error: the time series for each soil layer were perturbed with Gaussian noise of zero mean and standard deviation of 0.067 vol. % corresponding to a precision of 0.2 vol. %; (ii) calibration error: the perturbed time series were realigned along a new slope, which pivoted around a random point within the measurement range and a random intercept within ±1.0 vol. %; (iii) calibration and precision: perturbed series were created as a random combination of (i) and (ii), which is a common case in field studies (Spank et al., 2013). Errors were applied independently to all soil depths, and 100 new time series were created for each of the error types. We determined the quality of the estimation methods using the median of 100 ensemble simulations with the 100 perturbed input time series. The values for the applied calibration uncertainty and precision are taken from the technical manual of the IMKO TRIME©-PICO32 soil moisture sensor (http://www.imko.de/en/products/soilmoisture/soil-moisture-sensors/trimepico32).

A common procedure with environmental measurements for dealing with precision errors is smoothing of the measured time series (Li et al., 2002; Peters et al., 2013), which we also reproduced by additionally applying a moving average filter on the disturbed soil moisture time series.

2.5 Evaluation criteria

A successful model should be able to reproduce the first and second moment of the distribution of the observed values (Gupta et al., 2009), and we used a similar approach to assess the quality of the methods for estimating the total evapotranspiration and the sink term profiles. The first and the second moment refer to the mean and the standard deviation. Additionally, the correlation coefficient evaluates whether the model is able to reproduce the timing and the shape of observed time series. To compare the applicability and the quality of the four methods, we use three performance criteria suggested by Gupta et al. (2009): (i) the correlation coefficient (R), (ii) the relative variability measure (RV) and (iii) the bias (b), which are described in this section. The comparison is based on daily values.

First, we use R to estimate the strength of the linear correlation between estimated ($\tilde{}$) and synthetic values:

$$R = \frac{\mathrm{Cov}(\widetilde{x}, x)}{s_x \cdot s_{\widetilde{x}}}, \tag{14}$$

where "Cov" is the covariance of estimated and observed (synthetic) values, and s_x and $s_{\widetilde{x}}$ are the standard deviations of synthetic and estimated values, respectively. The variable x stands for any of the variables of interest, such as total evapotranspiration or $z_{25\,\%}$. R ranges between -1 and $+1$. The closer R is to 1, the better the estimate.

Second, we use the relative variability in estimated and synthetic data (RV) to determine the ability of the particular method to reproduce the observed variance (Gupta et al., 2009):

$$\mathrm{RV} = \frac{s_{\widetilde{x}}}{s_x}. \tag{15}$$

RV values around 1 indicate a good estimation procedure.

Third, we use the relative bias (b) to describe the mean systematic deviation between estimated ($\tilde{}$) and observed (synthetic) values, which is never captured by R:

$$b = \frac{\bar{\widetilde{x}} - \bar{x}}{\bar{x}} \cdot 100(\%), \tag{16}$$

where $\bar{\widetilde{x}}$ and $\bar{x}$ are the means of the estimated and synthetic data, respectively. The best model performance is reached if the bias is close to zero.

3 Results

In total, we compared synthetic evapotranspiration rates from 33 consecutive days in July/August 2009. Evapotranspiration could not be estimated for days with rainfall using either the sssl or ssml method, nor with the msml regression. Therefore, we excluded all days with rainfall from the analysis for all considered methods. In Sect. 3.1 and 3.2 we first consider the performance of the estimation methods on undisturbed synthetic time series, i.e., we ignore measurement errors or assume they do not exist. The influence of measurement errors is investigated in Sect. 3.3.

3.1 Evapotranspiration derived by soil water content measurements

The performance of the data-driven methods depends strongly on the complexity of the respective method, which substantially increases with a higher degree of complexity. However, the influence of the measurement frequency differs considerably among the four methods.

The im predicted the daily evapotranspiration for a measurement frequency of 12 h with a very small relative bias of 0.89 %, which is the best value of all investigated methods. Additionally, the im reaches the best R value ($R = 0.99$) for all measurement frequencies (Table 2), and closely follows the 1 : 1 line between synthetic and estimated evapotranspiration (Fig. 3a, b). However, the RV and the relative bias indicate better prediction with decreasing measurement frequency.

The second-best method is the msml regression, in particular when applied for high temporal resolution measurements (1 and 3 h). There, the bias is comparatively small ($\pm 20\,\%$) and the correlation between synthetic (observed) and estimated values is relatively high ($R = 0.58$ and $R = 0.71$ for 1 and 3 h resolution, respectively). Also, the msml results match the 1 : 1 line well between synthetic and estimated evapotranspiration (Fig. 3a, b).

The sssl and the ssml methods show a weaker performance compared to the more complex im and msml methods. Neither of them follows the 1 : 1 line well between synthetic and estimated evapotranspiration (Fig. 3a, b). Regardless, they could reproduce the synthetic evapotranspiration with a relatively high linear correlation (Table 2), and comparable bias to the regression method, in particular for the range of intermediate measurement frequencies. However, values for the RV are comparatively large, in particular for the ssml method. Interestingly, the model performance criteria of the simpler sssl method show only minor differences between the particular temporal resolutions, and overall the sssl method performs better than the ssml method. Note that both water balance methods (sssl and ssml) overestimate the evapotranspiration at the beginning of the study period (Fig. 3c, d), which was marked by greater vertical flow between top soil and deeper soil due to preceding rainfall events.

Our results also show that less complex data-driven methods also perform better at higher temporal resolution (1 and 3 h), except for the ssml method. In contrast, the im is better at predicting evapotranspiration when a coarse measurement

Table 2. Comparison of the model performance of the four data-driven methods for reproducing daily evapotranspiration for the particular time resolution of soil moisture measurements. The model performance is expressed as correlation coefficient R, relative variability in simulated and reference values (RV), and relative bias (b) for the period 25 July–26 August 2009. Days on which rainfall occurs were excluded for the data analysis.

	Single-step, single-layer water balance			Single-step multi-layer water balance			Multi-step, multi-layer regression			Inverse model		
Δt (h)	R	RV	b (%)	R	RV	b (%)	R	RV	b (%)	R	RV	b (%)
1	0.77	1.51	−38.6	0.64	3.32	54.2	0.58	1.54	−22.9	0.99	0.78	−41.5
3	0.75	1.54	−38.6	0.66	3.37	46.8	0.71	1.03	20.3	0.99	0.97	−18.2
6	0.75	1.69	−35.9	0.67	3.52	36.4	0.78	1.87	86.5	0.99	1.03	−7.6
12	0.75	1.44	−38.6	0.70	3.49	37.1	0.85	4.22	202.4	0.99	1.04	0.89
24	0.58	1.76	−37.3	0.53	3.72	26.4	–	–	–	0.99	1.11	3.5

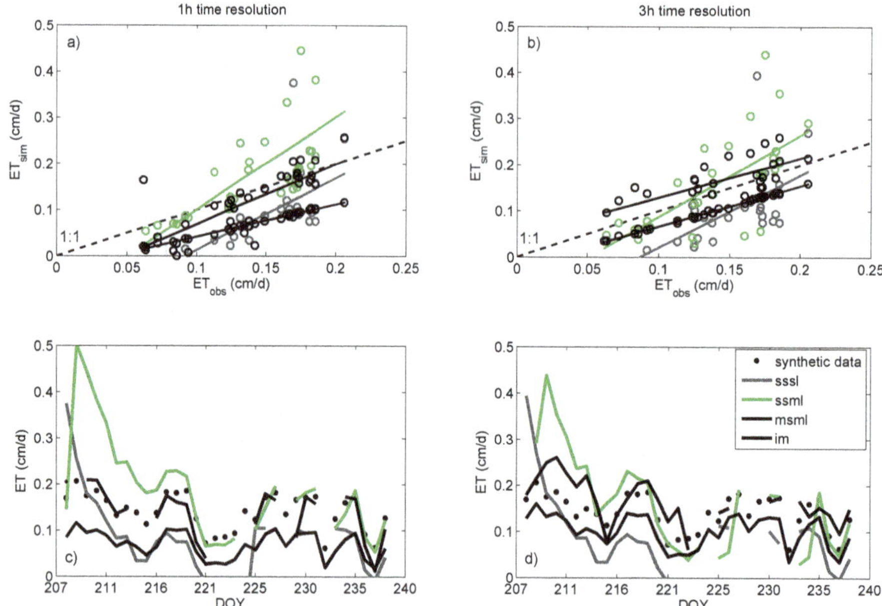

Figure 3. Top: comparison of synthetic (ET_{obs}) and estimated (ET_{sim}) values of daily evapotranspiration for hourly **(a)** and 3-hourly **(b)** observation intervals of soil water content measurements. Bottom: comparison of synthetic and estimated time series of daily evapotranspiration (ET) for hourly **(c)** and 3-hourly **(d)** observation intervals of soil water content measurements (25 July to 26 August 2009). Missing values are times when rainfall and percolation appeared. An estimation of evapotranspiration was not possible with the single-step, single-layer (sssl) water balance; the single-step, multi-layer (ssml) water balance; or the multi-step, multi-layer (msml) regression for these days.

frequency is used. Further, the results indicate that the estimated actual evapotranspiration becomes more accurate with increasing model intricacy, and with vertical flow accounted for.

3.2 Root water uptake profiles estimated with three different data-driven methods

The ssml, msml and im method appropriate for determining root water uptake profiles by inclusion of all available measurements over depth. Table 3 summarizes the model applicability to estimate the depths at which 25, 50 and 90 % of water extraction occurs (later stated as $z_{25\,\%}$, $z_{50\,\%}$ and $z_{90\,\%}$). Here, we used the standard deviation $s_{\bar{x}}$ instead of the relative variability to evaluate the observed variance. This criterion was chosen because the standard deviation of the synthetic reference values is approximately zero and thus the RV is increasing, which is not practical for the method evaluation. The criteria are shown for the respective best achieved model performance (1 h – ssml and msml; 24 h – im).

Again, the quality of predicting the sink term distribution depends on the method complexity and increases with increasing complexity. The most complex im delivers the best prediction of sink term distribution for a temporal resolution of 24 h. The depth above which 50 % of water extraction occurs ($z_{50\,\%}$) could be predicted with a bias of less than 2 % (Table 3) and for $z_{90\,\%}$; the relative bias increased only

Table 3. Comparison of model performance for reproducing the sink term profile (single-step, multi-layer water balance; multi-step, multi-layer regression; and inverse model). Depths where 25, 50 and 90 % water extraction occurs were regarded. Mean synthetic (syn.) depth and mean estimated (est.) depth describe the mean depth over 33 days where water extraction occurs. b is the relative bias and $\tilde{s}$ is the standard deviation of the estimated values. Larger width of the black arrow denotes higher accuracy of the model results.

Time resolution of measurements	Single-step, multi-layer water balance 1 h			Multi-step, multi-layer regression 1 h			Inverse model 24 h		
Criterion	$z_{25\,\%}$	$z_{50\,\%}$	$z_{90\,\%}$	$z_{25\,\%}$	$z_{50\,\%}$	$z_{90\,\%}$	$z_{25\,\%}$	$z_{50\,\%}$	$z_{90\,\%}$
Mean syn. depth (cm)	8.1	17.1	55.6	8.1	17.1	55.6	8.1	17.1	55.6
Mean est. depth (cm)	10.8	28.5	101.9	9.7	13.9	63.8	8.2	17.3	57.3
b (%)	33	74	83	−14	−21	15	0.75	1.05	2.97
$\tilde{s}$	4.07	12.31	57.89	1.69	4.01	25.83	1.81	4.08	68.26

slightly to approximately 3 %. Indeed, these comparatively accurate results are to be expected due to the two intrinsic assumptions: (1) the required soil hydraulic parameters for the implemented soil water flow model are exactly known, and (2) the measurement uncertainty of the soil sensors is zero.

The regression method (msml) also delivers good estimations of sink term profiles over the entire soil column (Table 3 and Fig. 4), although it manages without any intrinsic assumptions. Figure 4 shows that the msml regression overestimates the sink term at the intermediate depths. The maximum relative bias is about −21 % at $z_{50\,\%}$. Overall, the msml regression is applicable for determining the mean sink term distribution with an acceptable accuracy.

The ssml-estimated sink terms correspond only weakly to the synthetic ones, and the relative bias is lowest for $z_{25\,\%}$ with 33 % but increases strongly for $z_{50\,\%}$ and $z_{90\,\%}$ (Table 3). Moreover, the standard deviations of the predictions are substantial at most measurement depths (Table 3, Fig. 4). Because of these large variations in sink term distribution, the prediction of sink term profiles becomes imprecise. Thus for the chosen simulation experiment, the ssml method is not applicable for deriving the sink term from soil water content measurements.

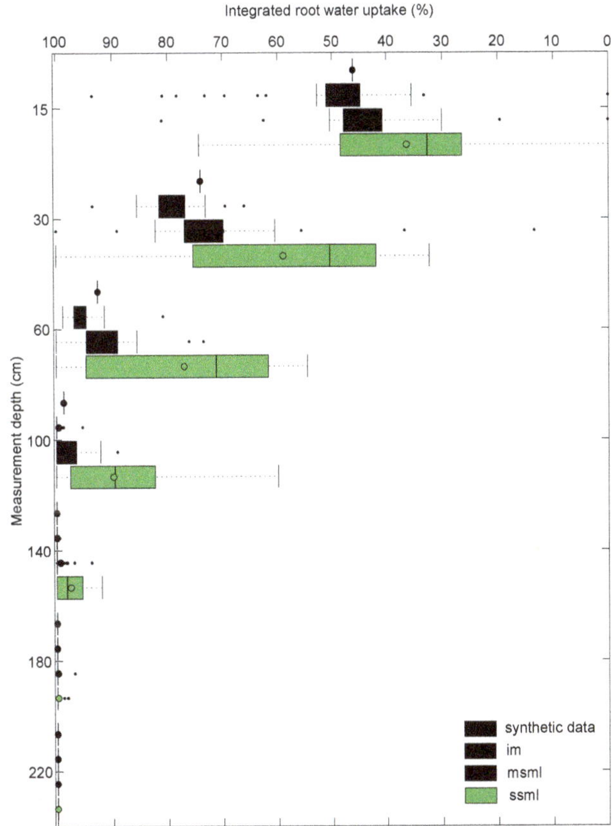

Figure 4. Box plots of the estimated daily percentage of integrated sink term. Colors are assigned as follows: synthetic values are black, the im is red, the msml regression is blue and the ssml water balance is green. The percentage of integrated sink term is shown for all measurement locations over the soil column. The circles show the mean values; the vertical line depicts the median and the 25 and 75 % percentile. Values are given for the respective underlying time resolution which achieved the best results according to Table 3 (ssml – 1 h; msml – 1 h; im – 24 h).

3.3 Influence of soil moisture sensor uncertainty on root water uptake estimation

We only evaluated the influence of measurement errors for two methods (msml and im). The single-layer approach was omitted since it does not allow for estimation of the sink term profile, and ssml was omitted since the estimation of the sink term profile was already inappropriate when ignoring measurement errors (see Sect. 3.2).

The influences of soil moisture sensor uncertainties differ considerably among the investigated methods. The msml method predicted the median daily evapotranspiration with precision uncertainty, calibration uncertainty and a combination of both reasonably well (Fig. 5). For all three types of uncertainty, the correlation between synthetic (observed)

Table 4. Comparison of the model performance with considering soil moisture measurement uncertainties for the msml regression and the im for reproducing daily evapotranspiration and the mean depths where 25, 50 and 90 % water extraction occurs. The model performance is expressed as correlation coefficient (R), relative variability in simulated and reference values (RV) and relative bias (b) for the period 25 July to 26 August 2009. The precision uncertainty is abbreviated as prec err, the calibration uncertainty as cali err, and the combined uncertainty as com err. The relative bias for reproducing evapotranspiration is abbreviated as b_{ET}, and is abbreviated as $b_{25\,\%}$, $b_{50\,\%}$ and $b_{90\,\%}$ for reproducing mean depths where 25, 50 and 90 % water extraction occurs, respectively.

Time resolution of measurements	Multi-step, multi-layer regression 1 h			Inverse model 24 h		
Criterion	prec err	cali err	com err	prec err	cali err	com err
R	0.90	0.89	0.91	−0.027	0.847	−0.054
RV	1.35	1.50	1.35	1.51	1.25	1.85
Median bias b_{ET} (%)	−6.2	−4.9	−6.1	−10.3	498.1	483.3
Median bias $b_{25\,\%}$(%)	19.6	3.6	19.5	25.2	531.1	405.1
Median bias $b_{50\,\%}$ (%)	28.0	5.4	27.7	42.0	622.4	659.1
Median bias $b_{90\,\%}$(%)	80.8	27.7	84.7	128.5	757.6	569.0

and estimated values is relatively high (around $R = 0.9$, Table 4). Also, with respect to the median relative bias (%), the three cases differ only marginally ($|b| = 7\,\%$, Table 4). Interestingly, the calibration uncertainty showed the lowest impact on the predicted evapotranspiration, with a median bias of about −5 % for the respective 100 ensemble calculations (Fig. 5).

Additionally, the bias is also used to compare the predicted relative water extraction depths ($z_{25\,\%}$, $z_{50\,\%}$ and $z_{90\,\%}$) (Fig. 6). The uncertainty caused by the calibration of the sensor shows the least differences to the observed values below 10 %. These results are similar to those from simulations with soil moisture without any introduced measurement uncertainty. Further, the uncertainties caused by the precision of the sensors have the highest impact on predicted root water uptake patterns. It turns out that the relative uncertainty increases with increasing depth (decreasing sink term or rather water extraction, Fig. 6a).

Interestingly, the im shows worse model performances than the msml regression for all three types of uncertainty. Although, the predicted evapotranspiration from soil moisture with precision uncertainty is close to the observed values (Fig. 5), it differs around days when rainfall occurs (DOY 225, 230 and 234). This results in underestimation of evapotranspiration during these times and a weak correlation (Table 4), but an acceptable relative bias of about −10 %. In contrast, for the calibration uncertainty it is the other way around. Here, the correlation is relatively high ($R = 0.85$) but evapotranspiration is greatly overestimated ($b = 498\,\%$). A combination of both uncertainty sources does not further increase the overall error, but does combine both weaknesses to an overall poor estimation (Table 4).

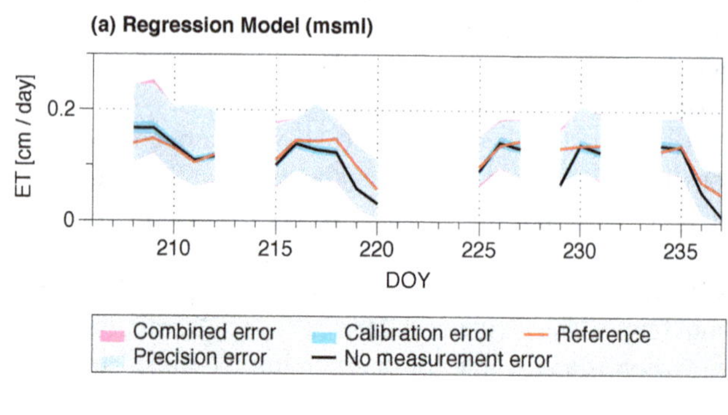

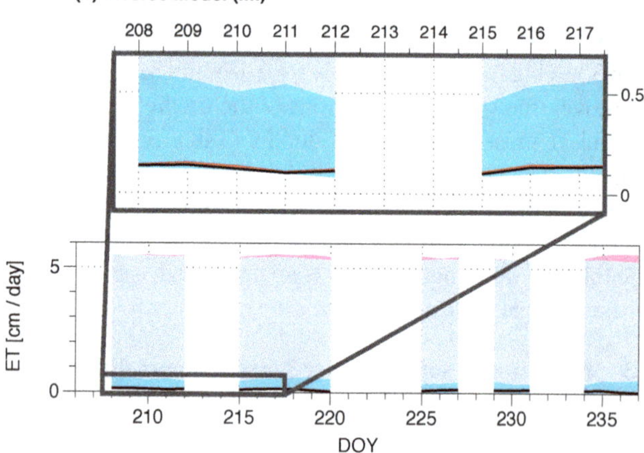

Figure 5. Influence of soil moisture uncertainty on evapotranspiration estimated with the msml regression model **(a)** and the im **(b)**. The red line is the evapotranspiration from the synthetic data (Reference). The colored bands indicate the 95 % confidence intervals.

The sensitivity to the type of uncertainty concerning prediction of sink term patterns is shown in Fig. 6b and Table 4. Similar to the msml regression, the im is able to handle uncertainties in sensor precision to predict root water uptake depths, whereas uncalibrated sensors lead to considerable increases in relative bias. Overall, the simpler msml regression method shows a higher robustness against measurement uncertainties than the more complex im.

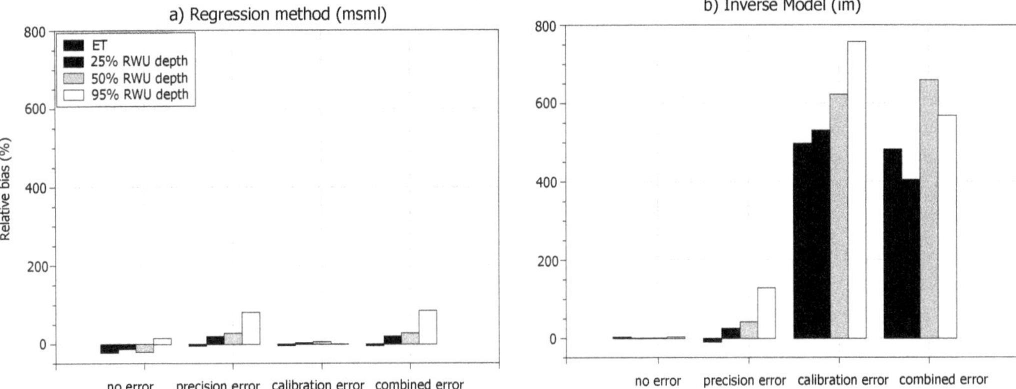

Figure 6. Comparison of the mean relative bias between synthetic and predicted values of evapotranspiration and the mean depths where 25, 50, 90 % of water extraction occurs for soil moisture time series: without uncertainty (no error), precision uncertainty (precision error), calibration uncertainty (calibration error) and precision and calibration uncertainty (combined error) for the msml regression **(a)** and the im **(b)**.

4 Discussion

We tested the application of several methods deriving, based on the soil water balance, how much water was extracted from the soil by evapotranspiration and how the extraction profile (sink term profile) changed with soil depth. The bases for all methods are time series of volumetric soil water content derived from measurements, although some methods require more information on soil properties, in particular the inverse model (im). None of the methods relies on a priori information on the shape of the sink term profile, nor do any of them make any assumptions on it being constant with time. This is the great advantage of these methods over others (Dardanelli et al., 2004; McIntyre et al., 1995; Hopmans and Bristow, 2002; Zuo et al., 2004). Since only changes in soil water content are considered, none of the investigated methods distinguish between soil evaporation and root water uptake. For the same reason, none of the water balance methods can be applied during times of fast soil water flow, for example during or after a rainfall event.

We used synthetic soil water content "observations" to validate the model results. This procedure has the great advantage that the "true" water flow and sink term profiles are perfectly known, including the nature of data uncertainty with regard to calibration error and sensor precision. However, our model only accounts for vertical matrix flow, notably neglecting horizontal heterogeneity, which may be an additional challenge for deriving evapotranspiration in real-world situations. Thus, additional tests of the methods in controlled field conditions, like with large lysimeters, and comparison with additional data, like isotope profiles, are necessary to confirm our results.

In the first part of the paper, we investigated how well all methods reproduced the sink term profile and total evapotranspiration when assuming that the measurements of soil water content were free of measurement errors, i.e., they were well calibrated and measured precisely. Even in this idealistic setting, the investigated methods performed very differently, most prominently depending on whether or not vertical flow could be accounted for by the method. The methods showing the greatest deviation between the "observed" (synthetic) evapotranspiration and sink term profiles were those not accounting for vertical flow within the soil (sssl and ssml methods). In those simpler soil water balance methods, any change in soil moisture is assigned only to root water uptake (Rasiah et al., 1992; Musters et al., 2000; Hupet et al., 2002). However, even several days after a rainfall event, the vertical matrix flow within the soil can be similar in magnitude to the root water uptake (Schwärzel et al., 2009), and this leads to considerable overestimation of the sink term when soil water flow is not accounted for. This error adds up when the sink term is integrated over depth and leads to a bias in the evapotranspiration estimate, which is the case for the ssml method.

This distinction between vertical soil water flow and water extraction is the major challenge when applying water balance methods, because these fluxes occur concurrently during daytime (Gardner, 1983; Feddes and Raats, 2004). The regression method (msml) avoids this problem by considering vertical soil water fluxes, estimated from change in soil water content during nighttime. Li et al. (2002) used a similar approach to derive transpiration and root water uptake patterns from soil moisture changes between different times of the day. This direct attribution of nighttime change in soil water content to soil water flow inherently assumes that both nighttime evapotranspiration and hydraulic redistribution are negligible. Li et al. (2002) measured nocturnal sap flow in order to ascertain that nighttime transpiration was insignificant. Also, in lysimeters, the weight changes can be used to validate the assumption. This assumption is the main drawback of this method, in contrast to the large advantage that it requires very limited input data, especially no a priori information on soil properties. In contrast, the im approach inferred evapotranspiration and sink term patterns with greater quality when soil water content measurements were free of error. However, because our analysis uses model-generated time series of soil water content in order to mimic measurements, the soil properties of the original "experiment" are completely known, which is not usually the case in natural conditions. Usually, soil hydraulic parameters have to be estimated by means of a calibration procedure. This process is non-trivial and limited by the non-uniqueness of the calibrated parameters (Hupet et al., 2003), which results in uncertainties in simulated soil water fluxes and root water uptake rates (Duan et al., 1992; Musters and Bouten, 2000; Musters et al., 2000; Hupet et al., 2002, 2003). This reliance of the im approach on precise knowledge of the soil environment is the main drawback of that approach.

Several studies on estimation of root water uptake profiles focused on uncertainties related to calibrated parameters of soil and the root water uptake models (Musters and Bouten, 2000; Musters et al., 2000; Hupet et al., 2002, 2003). When data and models are used, uncertainties arise not from soil parameter uncertainty but in fact already evolve during the measurement process of the environmental data (Spank et al., 2013). Thus, in the second part of this paper, we investigated how measurement noise (precision), wrong sensor calibration (accuracy) and their combination reflect on the derivation of evapotranspiration and sink term patterns from soil water content measurements. We only performed this analysis for the two methods which performed satisfactorily without sensor errors: the msml regression method and the im. In this more realistic setting, the simpler regression method (msml) performed much better than the im. The latter was strongly affected by inaccurate or lack of site-specific calibration. This "calibration error" renders the evolution of the vertical potential gradients and soil moisture profile inconsistent with the evolution of the vertical sink term distribution, and thus introduces grave overestimation of root water uptake and evapotranspiration for the considered time steps (Fig. S2). Generally, the prediction of the im improves when longer evaluation periods are considered (cf. Zuo and Zhang, 2002), and therefore the calibration error may become less prominent when considering time steps of several days as done in Zuo and Zhang (2002). Compared to the effect of calibration, the sensor precision had a much smaller effect. Thus, the im may be applicable and should be tested in situations where all sensors in the profile are well calibrated. A further improvement of the im could be achieved by smoothing the measured soil water content profiles via a polynomial function to get an accurate and continuous distribution of soil water contents as done in Li et al. (2002) and Zuo and Zhang (2002).

The msml regression model was overall more robust towards the investigated measurement errors. It was barely affected by calibration error but was somewhat affected by sensor precision. This is expected, since the sensor calibration only improves the absolute values of the measurements, and does not affect the course of the soil moisture desiccation. The case is different for uncertainty due to sensor precision, which results in higher deviations between observed and predicted sink term uptake patterns (Fig. 6). As this method uses linear regression on the temporal evolution of soil water contents, the quantity of root water uptake depends on the gradient of the slopes. Those slopes are strongly influenced by the random scatter of data points, which is characteristic for sensor noise. Using the smallest time step of 1 h, we could estimate the relative depth where 50 % of water extraction occurs with a bias less than 30 %. Using higher time resolution with several measurements per hour or several minutes and noise-reducing filters (Li et al., 2002; Peters et al., 2013) would likely further improve this result. This method should be further evaluated with lysimeters in order to test its application in controlled but more realistic environments.

Furthermore, our study demonstrates that measured soil moisture time series already include information on evapotranspiration and root water uptake patterns. This has already been stated by Musters and Bouten (2002) as well as Zuo and Zhang (2002). Contrary to these studies, where only temporal resolutions of 1 day or more are investigated, we additionally looked at measurement time intervals in the range of hours. Our results confirm that different methods require measurements with different temporal resolutions. The more simple msml regression model showed better applicability for measurements taken with an interval less than 6 h. These results are similar to Breña Naranjo et al. (2011) for a water balance method. The higher time resolution better reflects the temporal change in evapotranspiration, which may be considerable over the course of a day (Jackson et al., 1973). Conversely, the im works better for coarser temporal resolution for the case that soil water content measurements are error-free. If a possible measurement error is considered, coarser temporal resolutions are also better suitable to estimate evapotranspiration and root water uptake. With a higher temporal resolution (here 1 day instead of several hours) the total evapotranspiration and sink term also increases (integrated over the entire time). Therefore, the iteration of the im procedure could determine the sink term with a higher accuracy.

Another important prerequisite besides temporal resolution of the soil moisture time series is the adequate number of soil moisture measurements over the entire soil column to well capture the very nonlinear depth profile of water removal from the soil. This becomes most obvious when comparing the results from the simple single-layer water balance method (sssl) with the multi-layer (ssml) one. The prediction of the single-layer model is dominated by the specific depth at which the single sensor is located, and how much it is affected by root water uptake. In the presented case it strongly underestimated overall evapotranspiration because it observes only one part of the sink term profile and omits both the much more elevated uptake in the top soil and the deep uptake below the measurement depth. In contrast to that, the multi-layer method reproduces better the time series of evapotranspiration, because it samples the uptake profiles more holistically. Similarly, Schwärzel et al. (2009) and Clausnitzer et al. (2011) also found that high spatial resolution of water content sensors allow for a more reliable determination of evapotranspiration. Important consideration should be given to the very shallow soil depths, representative of the pure soil evaporation process ($z < 5$ cm), which are notoriously undersampled due to technical limitations. This may lead to underestimation of evaporation and therefore evapotranspiration in all investigated water balance applications.

Our results show that water balance methods have potential to be applied for derivation of water extraction profiles, but they also suggest that their application may be challenging in realistic conditions. In particular, im has great potential, in theory, but obtaining information of the soil environment with sufficient accuracy may be unrealistic. The msml regression method is particularly promising, as it requires little input and is comparably robust towards measurement errors. Further tests in controlled environments and ideally in concert with isotope studies should be conducted to further test the application of these methods in real-world conditions.

The great advantage of all considered methods is that they do not require a priori information about total evapotranspiration or the shape of the root water uptake profiles. Root water uptake moves up or down depending on soil water status (Lai and Katul, 1998; Li et al., 2002, Doussan et al., 2006; Garrigues et al., 2006), and many existing approaches are unable to account for this dynamic of root water uptake. Root water extraction profiles are central topics in ecological and ecohydrological research on resource partitioning (e.g., Ogle et al., 2004; Leimer et al., 2014; Schwendenmann et al., 2014) and drivers for ecosystem structure (Arnold et al., 2010). Water balance methods are potential tools for comparing those extraction profiles between sites and thus contributing to ecohydrological process understanding.

5 Conclusions

The aim of this study was to evaluate four water balance methods of differing complexity to estimate sink term profiles and evapotranspiration from volumetric soil water content measurements. These methods do not require any a priori information of root distribution parameters, which is the advantage compared to common root water uptake models. We used artificial data of soil moisture and sink term profiles to compare the quality of the estimates of those four methods. Our overall comparison involved the examination of the impact of measurement frequency and model intricacy, as well as the uncertainties of soil moisture sensors on predicting sink term profiles. For the selected dry period of 33 days and under consideration of possible measurement uncertainties the multi-step, multi-layer (msml) regression obtained the best estimation of sink term patterns. In general, the predictions with the four data-driven methods show that these methods have different requirements on the measurement frequency of soil moisture time series and on additional input data like precipitation and soil hydraulic parameters. Further, we were able to show that the more complex methods like the msml regression and the inverse model (im) predict evapotranspiration and the sink term distribution more accurately than the simpler single-step, single-layer (sssl) water balance and the single-step, multi-layer (ssml) water balance.

Unfortunately, the estimations of the im are strongly influenced by the uncertainty of measurements. Moreover, numerical soil water flow models like the im require a large amount of prior information (e.g., boundary conditions, soil hydraulic parameters) which is usually not available in sufficient quality. For example, the soil hydraulic parameters have to be calibrated before use, which introduces additional uncertainties in the parameter sets. It is important to keep this in mind while comparing the im with the msml regression model, especially in light of the influence of measurement uncertainties.

Our results show that highly resolved (temporal and spatial) soil water content measurements contain a great deal of information which can be used to estimate the sink term when the appropriate approach is used. However, we acknowledge that this study using numerical simulations is only a first step towards the application on real field measurements. The msml regression model has to be tested with real field data, especially with lysimeter experiments. Lysimeters allow for closing of the water balance and validation with measured evapotranspiration, while soil water content measurements can be conducted in a similar way to field experiments. With such experiments, the proposed method can be evaluated in an enhanced manner.

Appendix A

Table A1. Nomenclature.

b	relative bias (%)
d_t	length of active transpiration period over a day (h)
$d_{z,i}$	thickness of soil layer i (m)
DOY	day of year
e	difference in observed and estimated soil water content in the inverse model
E	evapotranspiration (mm h^{-1} or cm d^{-1})
E_s	bare soil evaporation (mm h^{-1})
E_t	transpiration (mm h^{-1})
$\tilde{E}$	estimated evapotranspiration (mm h^{-1})
h	soil matric potential (m)
i	soil layer index
j	time step index
$K(h)$	hydraulic conductivity (m s^{-1})
K_{sat}	saturated hydraulic conductivity (m s^{-1})
m_{tot}	slope of fitted linear function on $\theta(t)$
m_{extr}	slope of fitted linear function on $\theta(t)$ due to sink term
m_{flow}	slope of fitted linear function on $\theta(t)$ due to vertical soil water flow
n_{vG}	van Genuchten parameter (–)
NSE	Nash–Sutcliffe efficiency criterion
P	precipitation (mm h^{-1})
q	percolation (mm h^{-1})
RV	relative variability
S	sink term in Richards equation (s^{-1})
S_i	discretized sink term in the soil layer i (m s^{-1})
$\tilde{S}$	estimated sink term (m s^{-1})
s	standard deviation
t	time (s)
Δt	time step (h)
v	iteration step number (–)
$\bar{x}$	mean value
x	observed (synthetic) value
$\tilde{x}$	estimated values
z	vertical coordinate (m)
z_r	active rooting depth (cm)
$z_{25\,\%}$	depth up to which 25 % of root water uptake occurs (cm)
$z_{50\,\%}$	depth up to which 50 % of root water uptake occurs (cm)
$z_{90\,\%}$	depth up to which 90 % of root water uptake occurs (cm)
α	van Genuchten parameter (m^{-1})
θ	volumetric soil water content (m^3 m^{-3})
θ_r	residual volumetric soil water content (m^3 m^{-3})
θ_s	saturated volumetric soil water content (m^3 m^{-3})
$\tilde{\theta}$	estimated volumetric soil water content (m^3 m^{-3})
$\Delta\theta$	deviation in volumetric soil water content over time (m^3 m^{-3})
ε_{ZZ}	decision criterion for termination of the iteration process (inverse model from Zuo and Zhang, 2002)
$\varepsilon_{GH,i}$	decision criterion for termination of the iteration process in the inverse model proposed here

Acknowledgements. Financial support through the "ProExzellenz" Initiative from the German federal state of Thuringia to the Friedrich Schiller University Jena within the research project AquaDiva@Jena for conducting the research is gratefully acknowledged. This work was also financially supported by the Deutsche Forschungsgemeinschaft (DFG) within the project "The Jena Experiment". M. Guderle was also supported by the International Max Planck Research School for Global Biogeochemical Cycles (IMPRS-gBGC). We thank the editor, Nadia Ursino, for handling the manuscript and the two anonymous referees for their helpful comments. We also thank Maik Renner, Kristin Bohn, and Marcel Bechmann for fruitful discussions on an earlier version of this manuscript.

The service charges for this open access publication have been covered by the Max Planck Society.

Edited by: N. Ursino

References

Allen, R. G., Pereira, L. S., Raes, D., and Smith, M.: Crop evapotranspiration: Guidelines for computing crop requirements, FAO Irrigation and Drainage Paper No. 56, FAO, Rome, Italy, 1998.

Arnold, S., Attinger, S., Frank, K., and Hildebrandt, A.: Uncertainty in parameterisation and model structure affect simulation results in coupled ecohydrological models, Hydrol. Earth Syst. Sci., 13, 1789–1807, doi:10.5194/hess-13-1789-2009, 2009.

Asbjornsen, H., Goldsmith, G. R., Alvarado-Barrientos, M. S., Rebel, K., Van Osch, F. P., Rietkerk, M., Chen, J., Gotsch, S., Tobón, C., Geissert, D. R., Gómez-Tagle, A., Vache, K., and Dawson, T. E.: Ecohydrological advances and applications in plant-water relations research: a review, J. Plant Ecol., 4, 3–22, doi:10.1093/jpe/rtr005, 2011.

Bechmann, M., Schneider, C., Carminati, A., Vetterlein, D., Attinger, S., and Hildebrandt, A.: Effect of parameter choice in root water uptake models – the arrangement of root hydraulic properties within the root architecture affects dynamics and efficiency of root water uptake, Hydrol. Earth Syst. Sci., 18, 4189–4206, doi:10.5194/hess-18-4189-2014, 2014.

Breña Naranjo, J. A., Weiler, M., and Stahl, K.: Sensitivity of a data-driven soil water balance model to estimate summer evapotranspiration along a forest chronosequence, Hydrol. Earth Syst. Sci., 15, 3461–3473, doi:10.5194/hess-15-3461-2011, 2011.

Chanzy, A., Gaudu, J. C., and Marloie, O.: Correcting the temperature influence on soil capacitance sensors using diurnal temperature and water content cycles (Basel, Switzerland), Sensors, 12, 9773–9790, doi:10.3390/s120709773, 2012.

Chapin, F. S., Matson, P. A., and Mooney H. A.: Principles of Terrestrial Ecosystem Ecology, Springer-Verlag, New York, ISBN 0-387-95439-2, 2002.

Clausnitzer, F., Köstner, B., Schwärzel, K., and Bernhofer, C.: Relationships between canopy transpiration, atmospheric conditions and soil water availability – Analyses of long-term sapflow measurements in an old Norway spruce forest at the Ore Mountains/Germany, Agr. Forest Meteorol.,151, 1023–1034, doi:10.1016/j.agrformet.2011.04.007, 2011.

Clothier, B. E. and Green, S. R.: Rootzone processes and the efficient use of irrigation water, Agr. Water Manage., 25, 1–12, doi:10.1016/0378-3774(94)90048-5, 1994.

Coelho, F. and Or, D.: A parametric model for two-dimensional water uptake intensity by corn roots under drip irrigation, Soil Sci. Soc. Am. J., 60, 1039–1049, 1996.

Dardanelli, J. L., Ritchie, J. T., Calmon, M., Andriani, J. M., and Collino, D. J.: An empirical model for root water uptake, Field Crops Res., 87, 59–71, doi:10.1016/j.fcr.2003.09.008, 2004.

Davis, S. D. and Mooney, H. A.: Water use patterns of four co-occurring chaparral shrubs, Oecologia, 70, 172–177, doi:10.1007/BF00379236, 1986.

Doussan, C., Pierret, A., Garrigues, E., and Pagès, L.: Water uptake by plant roots: II – Modelling of water transfer in the soil root-system with explicit account of flow within the root system - Comparison with experiments, Plant Soil, 283, 99–117, doi:10.1007/s11104-004-7904-z, 2006.

Duan, Q., Sorooshian, S., and Gupta, V.: Effective and Efficient Global Optimization for Conceptual Rainfall-Runoff Models, Water Resour. Res., 28, 1015–1031, 1992.

Feddes, R. A. and Raats, P. A. C.: Parameterizing the soil-water-plant root system, in: Unsaturated-zone Modeling: Progress, Challenges and Applications, edited by: Feddes, R. A., de Rooij, G. H., and van Dam, J. C., Kluwer Academic Publishers, Dordrecht, the Netherlands, 95–141, 2004.

Feddes, R. A., Hoff, H., Bruen, M., Dawson, T., De Rosnay, P., Dirmeyer, P., Jackson, R. B., Kabat, P., Kleidon, A., Lilly, A., and Pitman, A. J.: Modeling root water uptake in hydrological and climate models, B. Am. Meteorol. Soc., 82, 2797–2809, 2001.

Gale, M. R. and Grigal, D. K.: Vertical root distributions of northern tree species in relation to successional status, Can. J. For. Res., 17, 829–834, 1987.

Gardner, W. R.: Soil properties and efficient water use: An overview, in: Limitations to efficient water use in crop production, edited by: Taylor, H. M., Jordan, W. R., and Sinclair, T. S., ASA-CSSA-SSSA, Madison, USA, 45–64, 1983.

Garrigues, E., Doussan, C., and Pierret, A.: Water Uptake by Plant Roots: I – Formation and Propagation of a Water Extraction Front in Mature Root Systems as Evidenced by 2D Light Transmission Imaging, Plant Soil, 283, 83–98, doi:10.1007/s11104-004-7903-0, 2006.

Green, S. R. and Clothier, B. E.: Root water uptake by kiwifruit vines following partial wetting of the root zone, Plant Soil, 173, 317–328, 1995.

Green, S. R. and Clothier, B. E.: The root zone dynamics of water uptake by a mature apple tree, Plant Soil, 206, 61–77, 1999.

Gupta, H. V., Kling, H., Yilmaz, K. K., and Martinez, G. F.: Decomposition of the mean squared error and NSE performance criteria: Implications for improving hydrological modelling, J. Hydrol., 377, 80–91, doi:10.1016/j.jhydrol.2009.08.003, 2009.

Hamblin, A. and Tennant, D.: Root length density and water uptake in cereals and grain legumes: how well are they correlated?, Aust. J. Agr. Res., 38, 513–527, doi:10.1071/AR9870513, 1987.

Hildebrandt, A. and Eltahir, E. A. B.: Ecohydrology of a seasonal cloud forest in Dhofar: 2. Role of clouds, soil type, and rooting depth in tree-grass competition, Water Resour. Res., 43, 1–13, doi:10.1029/2006WR005262, 2007.

Hopmans, J. W. and Bristow, K. L.: Current capabilities and future needs of root water and nutrient uptake modeling, Adv. Agron., 77, 104–175, 2002.

Hupet, F., Lambot, S., Javaux, M., and Vanclooster, M.: On the identification of macroscopic root water uptake parameters from soil water content observations, Water Resour. Res., 38, 1–14, doi:10.1029/2002WR001556, 2002.

Jackson, R. B., Candell, J., Ehleringer, J. R., Mooney, H. A., Sala, O. E., and Schulze, E. D.: A global analysis of root distributions for terrestrial biomes, Oecologia, 108, 389–411, 1996.

Jackson, R. B., Sperry, J. S., and Dawson, T. E.: Root water uptake and transport: using physiological processes in global predictions, Trends Plant Sci., 5, 482–488, 2000.

Jackson, R. D., Kimball, B. A., Reginato, R. J., and Nakayama, F. S.: Diurnal soil-water evaporation: time-depth-flux patterns, Soil Sci. Soc. Am. Pro., 37, 505–509, doi:10.2136/sssaj1973.03615995003700040014x, 1973.

Khalil, M., Sakai, M., Mizoguchi, M., and Miyazaki, T.: Current and prospective applications of Zero Flux Plane (ZFP) method, J. Jpn. Soc. Soil Phys., 95, 75–90, 2003.

Kollet, S. J.: Influence of soil heterogeneity on evapotranspiration under shallow water table conditions: transient, stochastic simulations, Environ. Res. Lett., 4, 035007, doi:10.1088/1748-9326/4/3/035007, 2009.

Kosugi, Y. and Katsuyama, M.: Evapotranspiration over a Japanese cypress forest, II. Comparison of the eddy covariance and water budget methods, J. Hydrol., 334, 305–311, 2007.

Kuhlmann, A., Neuweiler, I., van der Zee, S. E. A. T. M., and Helmig, R.: Influence of soil structure and root water uptake strategy on unsaturated flow in heterogeneous media, Water Resour. Res., 48, W02534, doi:10.1029/2011WR010651, 2012.

Lai, C. T. and Katul, G.: The dynamic role of root-water uptake in coupling potential to actual transpiration, Adv. Water Resour., 23, 427–439, doi:10.1016/S0309-1708(99)00023-8, 2000.

Lee, A.: Movement of water through plants, Pract. Hydropon. Greenhous., 50, GRODAN, http://www.grodan.com/files/Grodan/PG/Articles/2009/Movement_of_water_through_plants.pdf (last access: September 2014), 2009.

Leimer, S., Kreutziger, Y., Rosenkranz, S., Beßler, H., Engels, C., Hildebrandt, A., Oelmann, Y., Weisser, W. W., Wirth, C., Wilcke, W.: Plant diversity effects on the water balance of an experimental grassland, Ecohydrol., 7, 1378–1391, doi:10.1002/eco.1464, 2014.

Le Roux, X., Bariac, T., and Mariotti, A.: Spatial partitioning of the soil water resource between grass and shrub components in a West African humid savanna, Oecologia, 104, 147–155, 1995.

Li, K., Dejong, R., and Boisvert, J.: An exponential root-water-uptake model with water stress compensation, J. Hydrol., 252, 189–204, doi:10.1016/S0022-1694(01)00456-5, 2001.

Li, Y., Fuchs, M., Cohen, S., Cohen, Y., and Wallach, R.: Water uptake profile response of corn to soil, Plant Cell Environ., 25, 491–500, 2002.

Loheide, S. P.: A method for estimating subdaily evapotranspiration of shallow groundwater using diurnal water table fluctuations, Ecohydrology, 66, 59–66, doi:10.1002/eco.7, 2008.

Maruyama, A. and Kuwagata, T.: Diurnal and seasonal variation in bulk stomatal conductance of the rice canopy and its dependence on developmental stage, Agr. Forest Meteorol., 148, 1161–1173, doi:10.1016/j.agrformet.2008.03.001, 2008.

McIntyre, B. D., Riha, S. J., and Flower, D. J.: Water uptake by pearl millet in a semiarid environment, Field Crop. Res., 43, 67–76, 1995.

Musters, P. A. D. and Bouten, W.: Assessing rooting depths of an austrian pine stand by inverse modeling soil water content maps, Water Resour. Res., 35, 3041, doi:10.1029/1999WR900173, 1999.

Musters, P. A. D. and Bouten, W.: A method for identifying optimum strategies of measuring soil water contents for calibrating a root water uptake model, J. Hydrol., 227, 273–286, doi:10.1016/S0022-1694(99)00187-0, 2000.

Musters, P. A. D., Bouten, W., and Verstraten, J. M.: Potentials and limitations of modelling vertical distributions of root water uptake of an Austrian pine forest on a sandy soil, Hydrol. Process., 14, 103–115, 2000.

Ogle, K., Wolpert, R. L., and Reynolds, J. F.: Reconstructing plant root area and water uptake profiles, Ecology, 85, 1967–1978, 2004.

Peters, A., Nehls, T., Schonsky, H., and Wessolek, G.: Separating precipitation and evapotranspiration from noise – a new filter routine for high-resolution lysimeter data, Hydrol. Earth Syst. Sci., 18, 1189–1198, doi:10.5194/hess-18-1189-2014, 2014.

Plamboeck, A. H., Grip, H., and Nygren, U.: A hydrological tracer study of water uptake depth in a Scots pine forest under two different water regimes, Oecologia, 119, 452–460, 1999.

Rasiah, V., Carlson, G. C., and Kohl, R. A.: Assessment of functions and parameter estimation methods in root water uptake simulation, Soil Sci. Soc. Am., 56, 1267–1271, 1992.

Ravenek, J. M., Bessler, H., Engels, C., Scherer-Lorenzen, M., Gessler, A., Gockele, A., De Luca, E., Temperton, V. M., Ebeling, A., Roscher, C., Schmid, B., Weisser, W. W., Wirth, C., de Kroon, H., Weigelt, A., and Mommer, L.: Long-term study of root biomass in a biodiversity experiment reveals shifts in diversity effects over time, Oikos, 000, 1–9, doi:10.1111/oik.01502, 2014.

Roscher, C., Scherer-Lorenzen, M., Schumacher, J., Temperton, V. M., Buchmann, N., and Schulze, E. D.: Plant resource-use characteristics as predictors for species contribution to community biomass in experimental grasslands, Perspect. Plant Ecol., 13, 1–13, doi:10.1016/j.ppees.2010.11.001, 2011.

Ross, P. J.: Modeling soil water and solute transport – fast, simplified numerical solutions, Am. Soc. Agron., 95, 1352–1361, 2003.

Ross, P. J.: Fast Solution of Richards' Equation for Flexible Soil Hydraulic Property Descriptions, Land and Water Technical Report, CSIRO, 39/06, 2006.

Sánchez, C., Fischer, G., and Sanjuanelo, D. W.: Stomatal behavior in fruits and leaves of the purple passion fruit (*Passiflora edulis* Sims) and fruits and cladodes of the yellow pitaya [*Hylocereus megalanthus* (K. Schum ex Vaupel) Ralf Bauer], Agronomía Colombiana, 31, 38–47, 2013.

Schaap, M. G., Leij, F. J., and van Genuchten, M. T.: Rosetta: a computer program for estimating soil hydraulic parameters with hierarchical pedotransfer functions, J. Hydrol., 251, 163–176, doi:10.1016/S0022-1694(01)00466-8, 2001.

Schenk, H. J.: The shallowest possible water extraction profile: a null model for global root distributions, Vadose Zone J., 7, 1119–1124, doi:10.2136/vzj2007.0119, 2008.

Schneider, C. L., Attinger, S., Delfs, J.-O., and Hildebrandt, A.: Implementing small scale processes at the soil-plant interface – the role of root architectures for calculating root water uptake profiles, Hydrol. Earth Syst. Sci., 14, 279–289, doi:10.5194/hess-14-279-2010, 2010.

Schume, H., Hager, H., and Jost, G.: Water and energy exchange above a mixed European Beech – Norway Spruce forest canopy: a comparison of eddy covariance against soil water depletion measurement, Theor. Appl. Climatol., 81, 87–100, 2005.

Schwärzel, K., Menzer, A., Clausnitzer, F., Spank, U., Häntzschel, J., Grünwald, T., Köstner, B., Bernhofer, C., and Feger, K. H.: Soil water content measurements deliver reliable estimates of water fluxes: a comparative study in a beech and a spruce stand in the Tharandt forest (Saxony, Germany), Agr. Forest Meteorol., 149, 1994–2006, doi:10.1016/j.agrformet.2009.07.006, 2009.

Schwendenmann, L., Pendall, E., Sanchez-Bragado, R., Kunert, N., and Hölscher, D.: Tree water uptake in a tropical plantation varying in tree diversity: interspecific differences, seasonal shifts and complementarity, Ecohydrology, doi:10.1002/eco.1479, online first, 2014.

Seneviratne, S. I., Corti, T., Davin, E. L., Hirschi, M., Jaeger, E. B., Lehner, I., Orlowsky, B., and Teuling, A. J.: Investigating soil moisture-climate interactions in a changing Climate: a review, Earth-Sci. Rev., 99, 125–161, doi:10.1016/j.earscirev.2010.02.004, 2010.

Spank, U., Schwärzel, K., Renner, M., Moderow, U., and Bernhofer, C.: Effects of measurement uncertainties of meterological data on estimates of site water balance components, J. Hydrol., 492, 176–189, 2013.

Teuling, A. J., Uijlenhoet, R., Hupet, F., and Torch, P. A.: Impact of water uptake strategy on soil moisture and evapotranspiration dynamics during drydown, Geophys. Res. Lett., 33, L03401, doi:10.1029/2005GL025019, 2006a.

Teuling, A. J., Seneviratne, S. I., Williams, C., and Torch, P. A.: Observed timescales of evapotranspiration response to soil moisture, Geophys. Res. Lett., 33, L23403, doi:10.1029/2006GL028178, 2006b.

van Genuchten, M. T.: A closed-form equation for predicting the hydraulic conductivity of unsaturated soils, Soil Sci. Soc. Am. J., 44, 892–898, 1980.

Verhoef, A., Fernández-Gálvez, J., Diaz-Espejo, A., Main, B. E., and El-Bishti, M.: The diurnal course of soil moisture as measured by various dielectric sensors: effects of soil temperature and the implications for evaporation estimates, J. Hydrol., 321, 147–162, doi:10.1016/j.jhydrol.2005.07.039, 2006.

Vrugt, J. A., van Wijk, M. T., Hopmans, J. W., and Šimunek, J.: One-, two-, and three-dimensional root water uptake functions for transient modeling, Water Resour. Res., 37, 2457, doi:10.1029/2000WR000027, 2001.

Wilson, K. B., Hanson, P. J., Mulholland, P. J., Baldocchi, D. D., and Wullschleger, S. D.: A comparison of methods for determining forest evapotranspiration and its components: sap-flow, soil water budget, eddy covariance and catchment water balance, Agr. Forest Meteorol., 106, 153–168, 2001.

Zuo, Q. and Zhang, R.: Estimating root-water-uptake using an inverse method, Soil Sci., 167, 561–571, 2002.

Zuo, Q., Meng, L., and Zhang, R.: Simulating soil water flow with root-water-uptake applying an inverse method, Soil Sci., 169, 13–24, doi:10.1097/01.ss.0000112018.97541.85, 2004.

Zwieniecki, M. A., Thompson, M. V., and Holbrook, N. M.: Understanding the Hydraulics of Porous Pipes: Tradeoffs Between Water Uptake and Root Length Utilization, J. Plant Growth Regul., 21, 315–323, 2003.

11

Measuring and modeling water-related soil–vegetation feedbacks in a fallow plot

N. Ursino[1], G. Cassiani[2], R. Deiana[3], G. Vignoli[4], and J. Boaga[2]

[1]Department ICEA, University of Padova, Padova, Italy
[2]Department of Geosciences, University of Padova, Padova, Italy
[3]Department dBC, University of Padova, Padova, Italy
[4]Geological Survey of Denmark and Greenland, Groundwater and Quaternary Geology Mapping Department, Lyseng Alle 1, 8270 Hojbjerg, Denmark

Correspondence to: N. Ursino (nadia.ursino@unipd.it)

Abstract. Land fallowing is one possible response to shortage of water for irrigation. Leaving the soil unseeded implies a change of the soil functioning that has an impact on the water cycle. The development of a soil crust in the open spaces between the patterns of grass weed affects the soil properties and the field-scale water balance. The objectives of this study are to test the potential of integrated non-invasive geophysical methods and ground-image analysis and to quantify the effect of the soil–vegetation interaction on the water balance of fallow land at the local- and plot scale.

We measured repeatedly in space and time local soil saturation and vegetation cover over two small plots located in southern Sardinia, Italy, during a controlled irrigation experiment. One plot was left unseeded and the other was cultivated. The comparative analysis of ERT maps of soil moisture evidenced a considerably different hydrologic response to irrigation of the two plots. Local measurements of soil saturation and vegetation cover were repeated in space to evidence a positive feedback between weed growth and infiltration at the fallow plot. A simple bucket model captured the different soil moisture dynamics at the two plots during the infiltration experiment and was used to estimate the impact of the soil vegetation feedback on the yearly water balance at the fallow site.

1 Introduction

The interaction between soil, water and vegetation begins belowground where the roots grow if enough soil moisture and nutrients are available, creating preferential infiltration flow paths and providing access to water and nutrients to the plant that will grow aboveground. Vegetation type diversity, especially in arid zones, may be ascribed to differences in the way of exploiting subterranean resources and differences in root system morphology (Cody, 1986; Casper and Jackson, 1997; Schenk and Jackson, 2002). Then the interaction continues above the soil surface where the shoots and the leaves shadow the soil and limit the evaporation while transpiration begins. Vegetation reduces the soil moisture content, particularly in the hot season, but it also enhances the soil hydraulic conductivity with its root apparatus (Gish and Jury, 2004; Zimmermann et al., 2006), thus helping to replenish the subsoil water storage and creating a positive feedback system (Franz et al., 2011). When a spontaneously growing species establishes itself on bare land, water-related soil–vegetation feedbacks are often invoked to motivate field-scale soil moisture and vegetation patterns, describe patterns related to eco-hydrological processes and evaluate the associated water budget. Water-related feedbacks between the vegetation growth and the water fluxes may have a major impact on the soil moisture balance (Kefi et al., 2007), depending on climate (Baudena et al., 2009; Rietkerk et al., 2011), plant physiology and their survival strategy under water stress (Kurc and Small, 2004, 2007; Ursino, 2007, 2009).

Agricultural systems provide an opportunity to study the relevant plant–water relations and soil–vegetation feedbacks in a more controlled environment than other natural systems

(Jackson et al., 2009). The interplay between soil and vegetation locally alters the hydrologic cycle, and this is a main concern in areas where the water scarcity may become a limiting factor to agricultural production. In those countries where water is scarce, increasing root depth and local infiltration, and reducing evaporation of water from soil are key tasks (Marris, 2008). The interaction between soil and vegetation in rainfed agriculture is crucial to determining the partitioning of rainfall into runoff and infiltration, the connectivity of soil moisture patterns, the recharge of water bodies and ultimately solute transport that are hot topics in understanding the ecohydrological effects of human actions on landscape (Jackson et al., 2009). Repeated measurements of local soil moisture content in space and time can illuminate soil moisture paths, and clarify the nature of relevant spatial processes in catchment hydrology (Grayson and Bloecshl, 2000).

Image analysis is a superior choice for detecting relative change of ground cover, since it facilitates extensive data collection, and reduces human bias by limiting human judgments (Sadler et al., 2010). Image analysis has been successfully applied to ground images, with substantially different objectives, including the classification of soil structures (Gimmi and Ursino, 2004), soil texture (Graham et al., 2005, 2010) and soil cover (Laliberte et al., 2007). The whole spectrum of light may be used to analyze the vegetation responses to external stimuli (Chaerle and Van Der Straeten, 2001). Visualization techniques to monitor plant health include fluorescence (Bushman and Lichtenthaler, 1998; Oxborough, 2004), thermal (Alchanatis et al., 2010; Meron et al., 2010), magnetic resonance and reflectance (Penuelas and Filella, 1998), but among all these techniques, reflectance imaging is the most easily and cost-effectively achievable one, and a customer grade color digital camera offers a low-cost alternative to spectroscopy.

Non-invasive techniques, and particularly ground-penetrating radar (GPR) and electrical resistivity tomography (ERT) provide data at the scale and resolution necessary to understand the hydrological processes of the topsoil. The use of these techniques has been increasingly focused on their ability to measure, albeit indirectly, changes in moisture content (e.g., Binley et al., 2002; Strobbia and Cassiani, 2007; Deiana et al., 2008; Vanderborght et al., 2013 – for reviews see Huisman et al., 2003; Cassiani et al., 2006b) and solute concentration (e.g., Binley et al., 1996; Kemna et al., 2002; Cassiani et al., 2006a; Perri et al., 2012 – for a review see Kemna et al., 2006). Geophysical inspection was coupled here to local measurements of soil saturation to be obtained by Time Domain Reflectometry (TDR) (Topp et al., 1980; Roth et al., 1990), in order to quantify relevant hydrological processes related to vegetation patterns.

In our study, the response of the soil–vegetation system during an irrigation experiment was captured by plot scale maps of the soil moisture variability obtained by ERT. The interest regarding the transient behavior of the flow field following irrigation derives from the comparison of the hydrological response to irrigation of two adjacent plots of an agricultural site located in southern Sardinia, where infiltration, runoff and water storage appeared significantly different. Sardinia has a climate characterized by water deficit at most altitudes during summertime (ARPAS, 2011, 2012). One of the experimental plots was left unseeded for a year and was found barely covered by grass weed at the time of the experiment. The other one was cultivated.

Cassiani et al. (2012) ascribed the different behavior of the two plots to the different interaction between soil and vegetation, envisioning the possibility that feedbacks between water flow and vegetation growth could come into play. The fallow plot had a crusty appearance, and was dry at the surface, evidently as the result of evaporation from the top layer. The soil in the vegetated plot appeared much wetter at the surface, likely due to the shade provided by the vegetation against direct sunlight. Unlike those of the cultivated plot, the deeper soil layers of the bare plot seemed to be wet before irrigation. New experimental data and a reinterpretation of previously published data are presented here. The data presented in the following sections have been partly presented in Cassiani et al. (2012). However the analysis of these data is conducted with different methods (e.g., using time-lapse inversion of ERT data) and it is focused primarily on the behavior of the fallow plot. The re-analysis of ERT data, using ratio inversion, highlights the problem in a manner not previously appreciated using standard ERT inversion and suggests that runoff was overestimated by Cassiani et al. (2012) since, by neglecting water salinity and temperature, the soil saturation was underestimated by the ERT measurement in the fallow plot. In the following sections, we try to use for the first time a combination of VIA (Visual Image Analysis) and local TDR measurements to highlight the local infiltration to be attributed to a soil vegetation feedback and to evaluate the impact of this feedback on the yearly water balance. Crucial water-related soil–vegetation feedbacks are often conjectured and rarely quantified by dedicated experiments. The hydrological response of the fallow plot is discussed with major focus on the soil vegetation interaction and the water budget. Coupling hydrological and biological databases is a promising way to test ecohydrological modeling concepts (Garre et al., 2012). Thus, first we model the infiltration experiment and compare the model outcome with the TDR measurements, and then we extend the simulation time in order to infer the impact of the positive feedback on the annual water balance. The new bucket model presented here differs substantially from the one described by Cassiani et al. (2012), with respect to the characteristic timescale. Cassiani et al. (2012) analyzed the impact of a positive feedback between vegetation growth and infiltration on a timescale of several decades. Thus the biomass balance came into play and the authors demonstrated that if the feedback exists, there may be a characteristic root length that maximizes the biomass production in a water scarcity scenario, and the estimated optimum root length was approximatively equal to the observed one.

By simulating the yearly water balance, we suggest that the patchy vegetation that grows on the bare soil may rely on the water stored before the growing season, thanks to the preferential infiltration that occurs where the vegetation grew before, creating a discontinuity in the upper crusty soil layer. This old water, together with the new one infiltrating during the growing season when it rains, may correspond to the effective evapo-transpiration volume, provided that transpiration is restricted to the deeper soil layers below the crusty one due to typical weed root distribution (Cassiani et al., 2012). The fact that water is stored in the soil for months could support and validate our intuition of the relevant impact of salinity on the ERT measurements and advance our comprehension of the functioning and maintenance of the ecosystem under study.

2 Methods

2.1 Experimental setup

A 4-day monitoring following an irrigation test was performed at an agricultural experimental farm located in Sardinia, Italy, as part of the EU-FP7 CLIMB project (Ludwig et al., 2010), focused on the analysis of climate change impact on the hydrology of Mediterranean basins. The irrigation lasted for 1 night (approximately 8 h), with a total of 42 mm of applied artificial rainfall on both plots.

The experiment took place at the San Michele farm near Ussana, in the Rio Mannu Catchment (southern Sardinia). The basin ranges in elevation from 62 to 842 m a.s.l. (meters above the sea level) with an average of 295.5 m a.s.l. The basin is mainly covered by crop fields and grassland, while only a small percentage of its area is occupied by forests in the southeastern part of the basin. The farm area has a gentle topography and is part of the Campidano Plain. According to the World Reference Base for Soil Resources (WRB), the relevant soils present at this site are (Brown soils) Cambisols (CM), Regosols (RG) and Vertisols (VR).

The island of Sardinia has a climate characterized by a water deficit at most altitudes. In the southeastern part of Sardinia where Ussana is located, the water deficit is maximum in summer, while the soil moisture availability is at its maximum in wintertime. The hydrological regime is characterized by wet periods from October to April, where more than 90 % of the rainfall is accumulated, and very dry summers (May–September). The yearly average temperature is 16 °C. And the effective soil moisture availability ranges from 100 % of the field capacity during winter to 0–10 % in summertime (ARPAS, 2011, 2012).

The controlled irrigation experiment was undertaken in May 2010. During the period October 2009–September 2010, the cumulate rainfall at the site was 300 mm, and the temperature ranged from 10 to 30 °C.

2.2 Plot-scale soil moisture measurements

Using three ERT lines, the detailed soil moisture response of the system from the controlled irrigation was captured. Both bare and cultivated plots were irrigated with the same amount of water. Each ERT line is composed of 24 electrodes spaced 20 cm apart, for a total length of 4.6 m each, and an expected depth of investigation not exceeding 1 m. Two lines were left in place throughout the experiment until 4 days after irrigation ended in the bare plot; one was left in place in the cultivated plot. Time-lapse measurements were taken periodically, using a dipole-dipole skip 0 scheme and full acquisition of reciprocals to estimate the data error level (see e.g., Binley et al., 1995; Cassiani et al., 2006a; Monego et al., 2010). Consistently, the data inversion used an Occam inversion approach as implemented in the ProfileR/R2/R3 software package (Binley, 2011) accounting for the error level estimated from the data themselves. At each time step, between 90 and 95 % of the dipoles survived the 5 % reciprocal error threshold. In order to build a a time-consistent data set, only the dipoles surviving this error analysis for all time steps were subsequently used, reducing the number to slightly over 200 dipoles (i.e., between 80 and 90 % of the total) at all three ERT lines. The absolute inversions were run using the same 5 % error level. Time-lapse inversions were run at a lower error level equal to 2 % (consistently with the literature – e.g., Cassiani et al., 2006a).

2.3 Local soil moisture measurements

Short time monitoring of soil moisture was aimed at finding the interrelations between vegetation density, soil structure and water flow at the local scale. The local soil saturation monitoring was acquired by permanently vertically installed TDR probes (probe lengths of 32 and 50 cm) monitored with a Tektronix 1502 instrument in the two plots (Fig. 1), and by a portable TDR (Trase) Soil Moisture Measurement System (Campbell Scientific) equipped with 21 cm rod length at 15 points distributed over the bare plot that was 5 m long and 3 m wide (Fig. 2).

Repeated measurements at the 15 reference locations in the fallow plot were taken only once before irrigation (after a small rain event of about 13 mm occurred the night preceding the irrigation experiment) with a portable Trase (21 cm rod length). Unfortunately, the Trase failed after the first background measurement. During the 3-day period of time following irrigation, additional repeated measurements were acquired at the 15 reference locations of the plot over

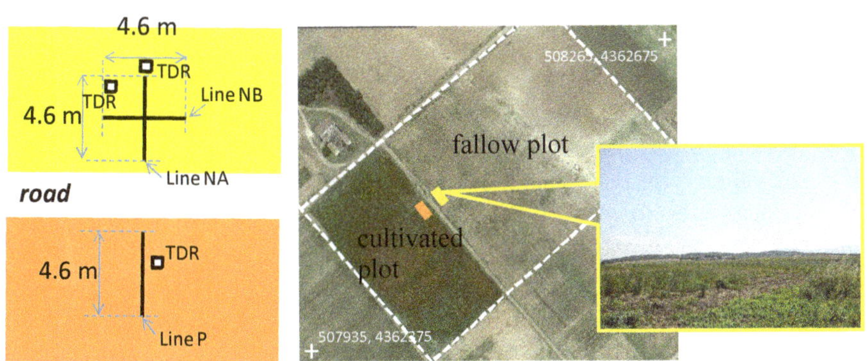

Fig. 1. The experimental site.

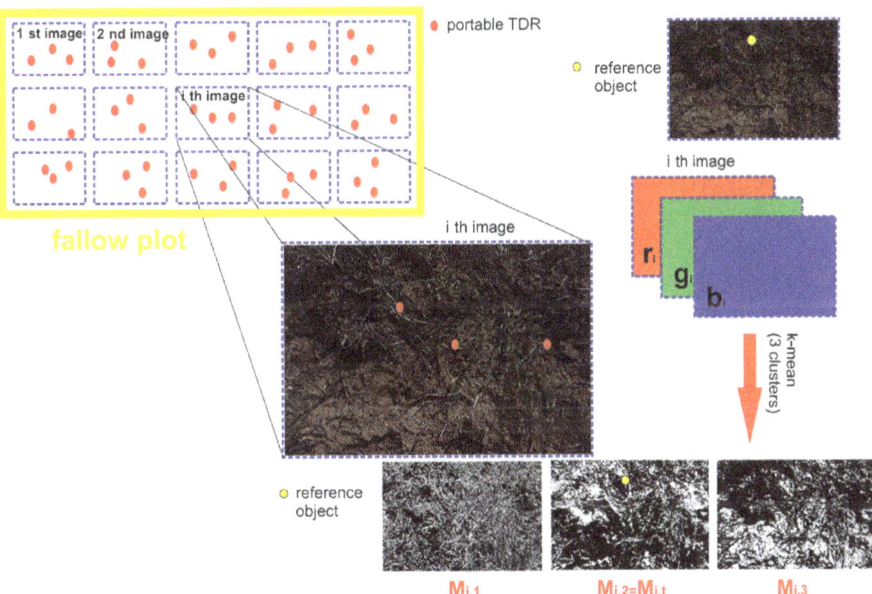

Fig. 2. Local measurements of vegetation density and soil moisture by combined VIA and repeated TDR measurements.

a different control depth (32 cm probes), always near solar noon.

2.4 Local measurements of vegetation density

The day before irrigation, a set of 15 vertical ground photographic images of a $0.9 \times 0.6\,m^2$ surface of soil were collected near solar noon in cloud-free conditions, each centered at the 15 TDR acquisition locations of the fallow plot using a Nikon D90 camera (Fig. 2). Vegetation cover was estimated by VIA and associated with the corresponding local TDR measurement of soil moisture. Note however that "there is no universal theory on color image segmentation yet", and "all of the existing color image segmentation approaches are, by nature, ad hoc." (Cheng et al., 2001)

We used the IDL7.1, a programming language developed by ITT (2009) for making automated cover detection by *k* means (MacQueen, 1967), and secondly to estimate the vegetation greenness on rangeland (that we use here to evaluate the goodness of the estimate of local vegetation density).

Each ith image was processed in order to obtain NC complementary binary images ($M_{i,j}$) that are referred to as masks, with j ranging from 1 to the number of clusters NC that was used to parameterize the k means algorithm. The optimum number of cluster (NC) is site specific. A reference object of interest (e.g., a leaf) is chosen in each picture. The targeted mask ($M_{i,t}$) that contains the reference object is used to evaluate the vegetation cover CC = $< M_{i,t} >$ as the average ($< >$) of the mask's pixel values (Fig. 2). Furthermore, the vegetation greenness is evaluated as a function of the average normalized red (r_i), green (g_i) and blue (b_i).

Borzuchowski and Schulz (2010) revise a list of vegetation spectral indices to describe plant eco-physiological parameters. None of them can be estimated in the visible spectrum, due to its restricted range of reflectance. However, we adapted the vegetation spectral index concept to the VIA data and defined the following greenness index:

$$F = \frac{< r_i \cdot M_{i,t} > - < g_i \cdot M_{i,t} >}{< r_i \cdot M_{i,t} > + < g_i \cdot M_{i,t} >}, \quad (1)$$

where $< r_i \cdot M_{i,t} >$ and $< g_i \cdot M_{i,t} >$ are the average value of the normalized red and green of the non-zero pixels in the targeted mask $M_{i,t}$ (the one that contains the reference object and is used to estimate the vegetation cover). Since the vegetation cover is discontinuous but homogeneously green at the plot scale and at the time of the experiment, we expect F to be quite homogeneous, unless the targeted mask contains intermixed soil and vegetation objects and in this sense, we used F to estimate the reliability of the segmentation procedure. Visual Image Analysis (VIA) was used here to repeatedly detect ground cover and to relate local biomass density to soil moisture dynamics in a fallow plot sparsely covered by grass weed. Repeated measurements of soil moisture and biomass density during an irrigation experiment allowed us to detect substantially different soil moisture and vegetation paths in fallow and cultivated plots.

2.5 Water balance model

We set up a simple bucket model to address the two-way interaction between plants and soil in water-controlled ecosystems according to the experimental evidence provided by ERT, TDR and VIA. Even though the focus was on the fallow plot, we used the model to also reproduce the soil moisture dynamics of the adjacent cultivated plot during the irrigation experiment for comparison.

Kurc and Small (2004, 2007), found that evapotranspiration is largely correlated with surface soil moisture, not to root zone soil moisture, and suggested that evaporation is dominant over transpiration within the top 15 cm of soil, whereas evaporation has a minor influence on soil moisture below about 15–20 cm.

We assume that after small rain events, a shallow upper soil layer (USL) acts as a temporary storage for water that is entirely returned to the atmosphere through soil evaporation.

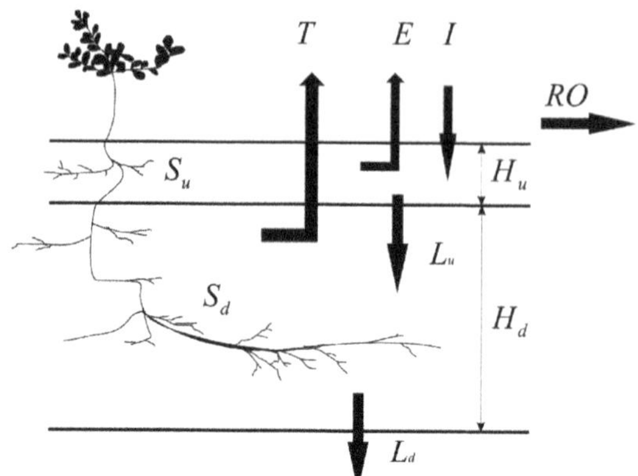

Fig. 3. Scheme of the conceptual model.

Furthermore, we assume that excess rainfall leaching from the USL supplies a deeper reservoir that we refer to as the deep soil layer (DSL) where roots have exclusive access to soil moisture leading to transpiration, even when the upper soil layer is empty. The root architecture determines which soil layer contributes more to transpiration, the choice that we made was motivated by the observation that weed roots appeared mostly concentrated in the DSL (Cassiani et al., 2012).

The growth of the vegetation is associated to the formation of macroporosity in the USL, leading to a local increase of hydraulic conductivity (soil–vegetation feedback) and leakage of excess water into the DSL, even if the USL is poorly conductive (crusty). In the absence of vegetation we do not expect any water flux to take place at the interface between the two layers due to the presence of the sealing crust. Even though cracks may develop in completely bare soils, we neglect here this possibility and compare the eco-hydrological behavior of the fallow plot with a hypothetical worse case scenario (bare soil) occurring when the DSL is disconnected from the atmosphere, in order to evaluate the impact of the soil–vegetation feedback on the annual water balance. We identify the USL's depth with $H_u = 100$ mm and the DSL's depth with $H_d = 500$ mm, according to the root depth estimated by Cassiani et al. (2012), and the effective saturation of the two layers with S_u and S_d, respectively (Fig. 3).

The daily water balance within the USL is expressed by the following differential equation:

$$\frac{\partial \theta}{\partial t} = n \cdot \frac{\partial S_u}{\partial t} = \frac{1}{H_u} \cdot (P + I - \mathrm{RO} - E - L_u), \quad (2)$$

where θ is the soil moisture content, $n = 0.4$ is the soil porosity, P is the daily precipitation, I is the irrigation, and RO is surface runoff. The evaporation E was evaluated with a modified dual crop coefficient approach (Allen et al., 1998),

$$E = \mathrm{ET}_0 \cdot [K_r \cdot (1 - K_b \cdot \mathrm{CC})], \quad (3)$$

where ET_0 is the reference evapotranspiration evaluated using the Penman–Monteith equation, K_b is the basal crop

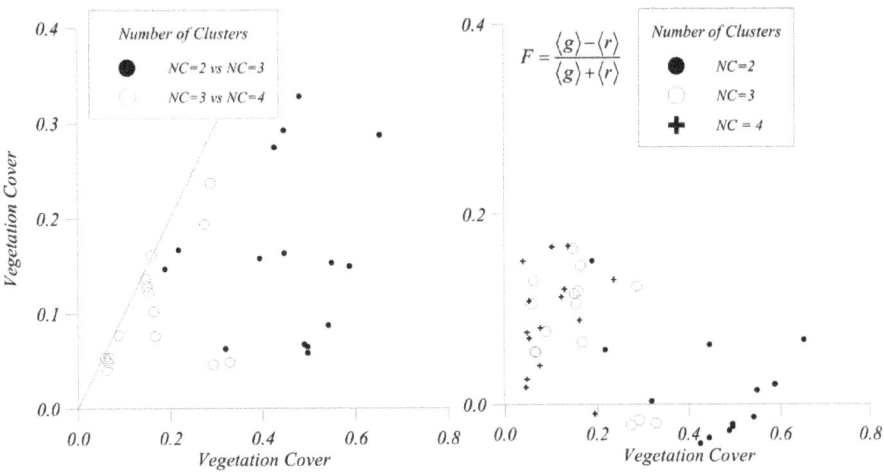

Fig. 4. Vegetation cover evaluated by image analysis with different number of clusters NC (see legend) and vegetation greenness F.

coefficient, K_r is the evaporation reduction coefficient linearly decreasing from 1 to 0 with the USL's saturation when the soil water content is $\theta_{WP} < \theta < 0.5 \cdot \theta_{FC}$; where $\theta_{FC} = 0.28$ is the soil moisture content at field capacity and $\theta_{WP} = 0.1$ is the soil moisture content at wilting point. Excess water percolates into the deeper soil layer, leading to the leakage (L_u) unless the soil is completely bare. In Eq. (3) CC is the vegetation cover that we estimated by VIA.

The daily soil moisture balance within the DSL is expressed by the following differential equation:

$$\frac{\partial \theta}{\partial t} = n \frac{\partial S_d}{\partial t} = \frac{1}{H_d} \cdot (L_u - T - L_d), \tag{4}$$

where the transpiration (T) evaluated according to the dual crop coefficient approach (Allen et al., 1998) as

$$T = ET_0 \cdot [K_b \cdot K_s \cdot CC]. \tag{5}$$

The stress coefficient K_s is a linear function of the DSL's saturation between the readily available soil water in the root zone RAW and the total available soil water in the root zone TAW. According to Allen et al. (1998) we set

$$TAW = (\theta_{FC} - \theta_{WP}) \cdot H_u \tag{6}$$

and

$$RAW = p \cdot TAW \tag{7}$$

with $p = 0.3$ in the cultivated plot, and $p = 0.7$ in the fallow plot (partially covered by grass weed) (Allen et al., 1998). Leakage out of the control volume L_d may be reasonably neglected under water scarcity conditions (Keating et al., 2002; Zhang et al., 2001).

3 Results

3.1 Estimate of vegetation cover by VIA

Repeated measurements in space of the vegetation cover CC were obtained by VIA. The optimum number of cluster NC was chosen in order to achieve positive and homogeneous estimates of the vegetation greenness F (Fig. 4, right panel). We observed that when NC = 2 (the objects are vegetation and soil) the vegetation cover may be overestimated with respect to the result obtained with a larger number of clusters (Fig. 4, left panel, bold circles) and if pixels belonging to the soil class are classified as vegetation, F varies significantly, switching from negative to positive values, meaning that the segmentation outcome is unreliable. Further increasing the NC from 3 to 4 may induce an error in the evaluation of the vegetation cover due to the fact that the pixels belonging to the vegetation class split into subclasses of slightly different color, and this results in an underestimate of the actual vegetation cover in a few cases (Fig. 4, left panel, open circles). Figure 4 (right panel) shows that when NC = 3 and 4, F varies less, indicating that objects belonging to the targeted mask could be more homogeneous and belong to the "vegetation class" as we would expect. For these reasons we set NC = 3 and estimated that the vegetation cover varies from point to point, ranging between 0 and 0.4. The estimate of CC obtained with different color representation (e.g., IHS, not shown here) was consistent with the results presented in this section.

3.2 Observed soil moisture dynamics at the plot scale

Two perpendicular ERT lines (NA and NB) were placed in the fallow plot and measurements were taken repeatedly over

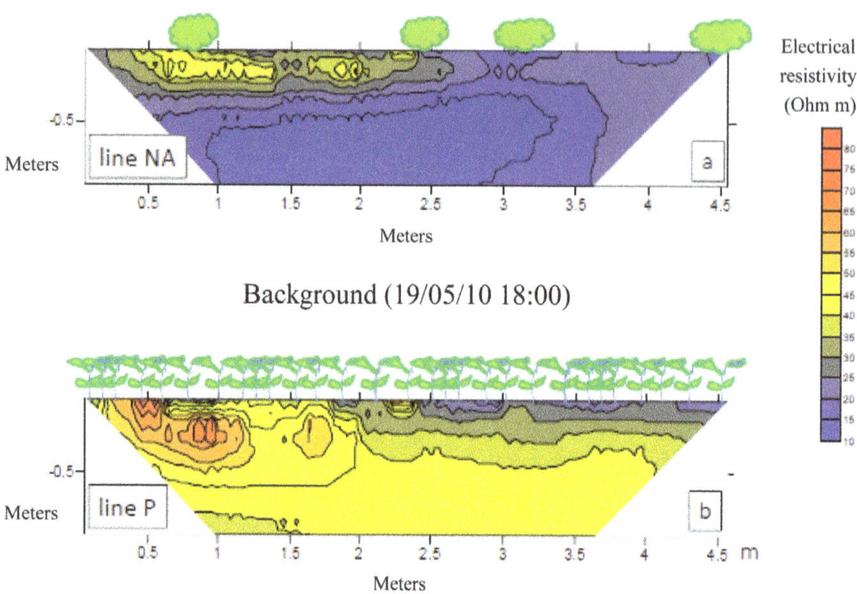

Fig. 5. Background images of the fallow **(a)** and of the adjacent cultivated **(b)**.

time before and after the irrigation experiment. The background ERT image (collected on 19 October 2010 – see e.g., line NA in Fig. 5a) shows a profile where a very resistive soil layer, about 20 cm thick and corresponding to a visually apparent crust of dry material, overlies a much more electrically conductive subsoil. This is in sharp contrast with the ERT profile acquired on the nearby cultivated plot (Fig. 5b) where the presence of vegetation cover maintains a higher moisture content (and electrical conductivity) in the top soil, whereas vegetation depletes the moisture content of the deeper layer where the roots exert their suction. This result indicates that in the fallow plot vegetation shadowing may be neglected, thus the upper soil layers are exposed to significant evaporation. Following a minor rainfall event (13 mm) the night of 19–20 May 2010 and the irrigation experiment the night of 21–22 May 2010 (42 mm), no major change in electrical resistivity was observed in this fallow plot, as opposed to the dramatic change observed in the nearby vegetated field (see Cassiani et al., 2012 for a thorough discussion). In the present paper, a new, detailed analysis of resistivity changes based on a ratio inversion approach (see e.g., Cassiani et al., 2006a) reveals the details of the subtle changes caused by irrigation to the resistivity patterns of the subsoil in the patchy plot. The results along lines NA and NB, thus coming from totally independent measurements, are consistent with each other and are shown in Fig. 6. From this figure it is apparent that (a) the natural rainfall, consisting of roughly 13 mm that occurred during the night between 19 and 20 May 2010, causes essentially no changes in the electrical resistivity profiles (see Fig. 6). We can conclude that nearly the entire precipitation must have resulted in surface runoff, with direct evaporation from local ponding in the field and along the dirt road; (b) the irrigation experiment in the night between 21 and 22 May with an amount of 42 mm of irrigated water, causes two changes in the resistivity profiles in the patchy plot: (i) a resistivity increase is apparent, albeit somehow discontinuous, in the soil layer between 10 and 50 cm depth. The increase is as high as about 25 % of the original resistivity of the same soil before irrigation; resistivity decrease, also in the 25 % range, is observed below 50 cm depth. Here too the patterns appear discontinuous. This evidence is, at first, confusing. How can the addition of water increase the resistivity of the layer in the top 10–50 cm, while decreasing at the same time the resistivity below? A first tentative explanation may attribute this result to an artefact caused by the physical size (length) of the electrodes that penetrate the ground for a depth (a few centimeters) that is non-negligible with respect to the electrode separation (20 cm). The wetting of the soil top layer, and consequently the increase of its electrical conductivity, may act as a shortcut that cannot be fully accounted for by the inversion algorithm, that assumes that the electrodes are point-like. As the current is short-circuited in the top few centimeters, the underlying soil may appear more resistive than it actually is, thus potentially causing the observed resistivity increase of the region below 10 cm. This hypothesis, however, fails to explain why the underlying decrease of resistivity below 50 cm is still perfectly detectable. Also, the changes in resistivity (±25 %) are too subtle to be attributed to an artefact caused by short-circuiting. A second, more sound, explanation is to relate these changes in soil electrical resistivity to factors other than moisture content alone. Two other factors, namely pore water salinity and temperature changes, can play a role. In fact, it is not uncommon to

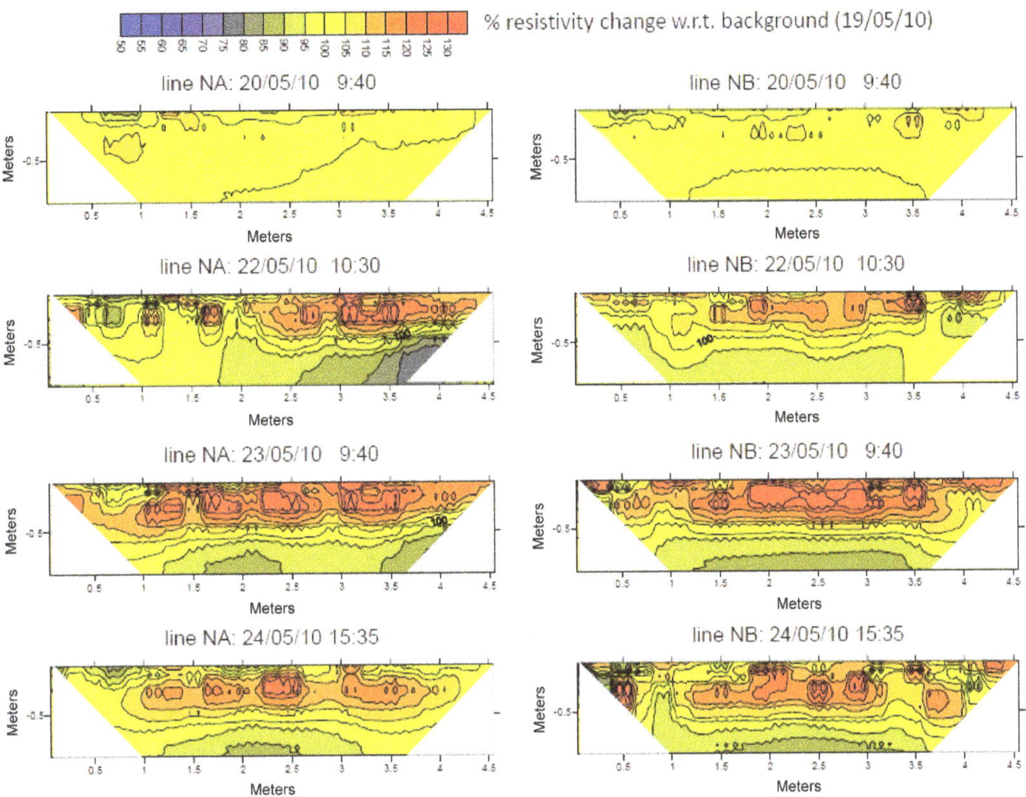

Fig. 6. ERT measurements along lines NA (left panels) and NB (right panels) in the fallow plot at different times. Top to bottom panels: background, after irrigation (22 May 2010, 10:30 LT), 1 day after irrigation (23 May 2010, 09:40 LT), more than 2 days after irrigation (24 May 2010, 15:35 LT).

observe that intense precipitation events have the effect of reducing soil bulk electrical conductivity by displacing the in situ pore water whose solutes have had the time to reach an equilibrium with the soil components, or that had different temperature (see e.g., Cassiani et al., 2006a). The incoming precipitated water pushes down the existing pore water in a sort of piston-like effect (see also Winship et al., 2006), thus causing a decrease in electrical conductivity in the upper part of the profile (Mojid and Cho, 2008) and an increase in the lower part, totally analogous of our observations here. Indeed, this hypothesis is confirmed by the evidence from the TDR probes permanently installed in the fallow plot (Fig. 7): the 50 cm-long sondes, covering the entire thickness where a resistivity increase is observed, show no average change in moisture content after rainfall or irrigation. On the contrary, the shorter sondes (32 cm) show an increase in moisture content after irrigation.

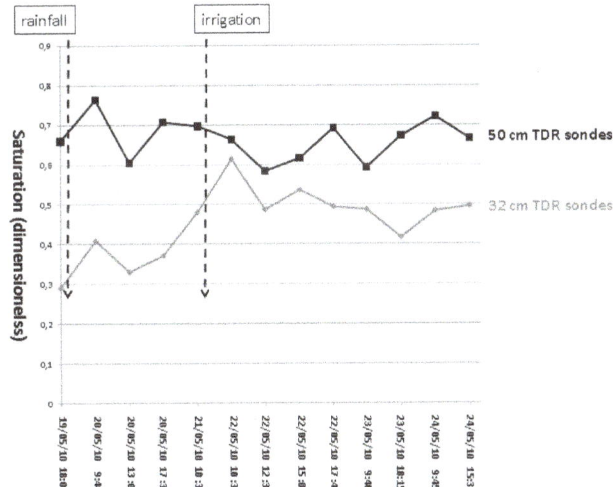

Fig. 7. Fallow plot soil saturation measured by fixed 50 and 32 cm-deep TDR probes at different times.

We observe the apparently paradoxical situation where average moisture content remains largely unchanged in the top 50 cm, while correspondingly the resistivity of the same layer is decreased by about 20 % and the resistivity of the underlying layer is increased also by roughly the same percentage. This phenomenon can find an explanation only if a second cause of resistivity change is called into play, in addition to moisture content change. The influence of this second cause is strong enough to overturn the influence of moisture content change itself. The possible causes may be substantially only two: a decrease in pore water salinity, and/or a decrease in soil temperature. We are not in a position to decide which mechanism is prevailing. In both cases, however, salinity or

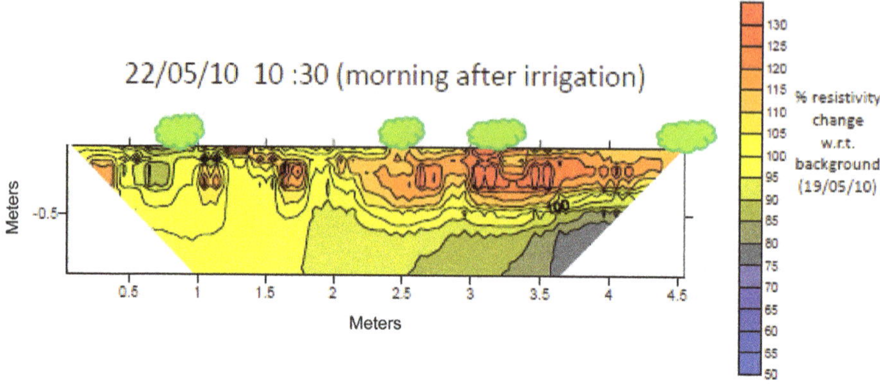

Fig. 8. Plant distribution along ERT line NA (fallow plot).

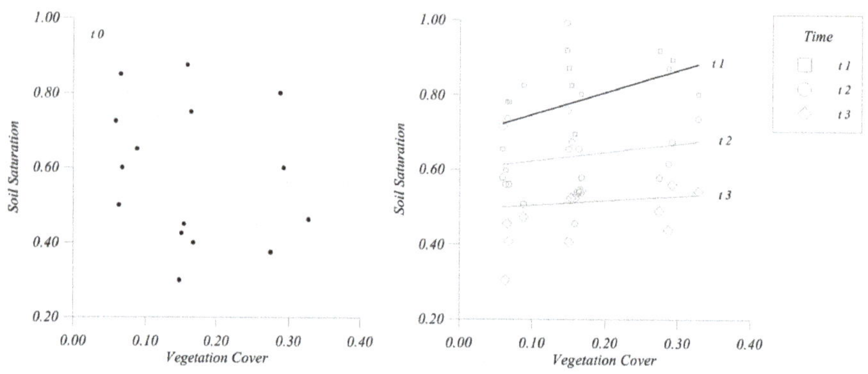

Fig. 9. Left panel: vegetation cover (CC) versus soil saturation after rainfall. Soil saturation is measured by portable TDR (rod length 21 cm). t_0: 21 May, 10:30 LT – (bold circles). Right panel: vegetation cover versus soil saturation at different times after irrigation. Soil saturation is measured by TDR (rod length = 32 cm). t_1: 22 May, 11:30 LT (squares); t_2: 23 May, 09:30 LT (open circles); t_3: 24 May, 10:30 LT (diamonds).

temperature (or both) act as tracers, marking the new water versus the old water already in the system. This "new" water pushes down the existing pore water and a mixture of old and new water reaches deeper zones, as apparent in Fig. 6.

Figure 8 shows a comparison between the location of the individual plants along the profiles and the time-lapse images along the ERT line NA, particularly the one relevant to the morning after the end of irrigation. There is some correlation between the location of major changes (especially on the right-hand side of the profile) and the plant location. Note however that this analysis neglects the 3-D effects that are possibly linked to the location of patchy vegetation off the individual ERT lines.

3.3 Feedbacks between vegetation growth and soil moisture dynamics

In order to get more evidence on the key interrelations between the spatially variable soil moisture and vegetation density, in Fig. 9 we compared the repeated measurements of vegetation cover obtained by VIA (with NC = 3) with the corresponding measurements of the soil saturation obtained by portable TDR. After the small rainfall event, the soil saturation of the upper 20 cm-thick soil layer was measured by Trase. The soil saturation appeared to be not at all correlated with the vegetation cover (Fig. 9, left panel).

In the 3 days following the irrigation, the soil saturation was measured using 32 cm-long probes. Shortly after irrigation the vegetation cover and the soil saturation were positively correlated (Fig. 9, $t = t_1$), but the correlation was very weak. Already 1 day after irrigation (Fig. 9, $t = t_2$), the results seem to indicate that redistribution took place because the soil saturation homogenized (the slope of the fit line changes) and evapotranspiration came into play (the fit lines shift downward). This result poorly supports the hypothesis that there is a positive feedback between vegetation growth and preferential infiltration, before redistribution could take place and just suggests that a positive feedback could exist. Only a larger data set could have allowed stronger conclusions on the existence of positive soil–vegetation feedbacks. The soil moisture reduction of about 0.1 in 3 days that was measured by TDR almost approaches the potential evapotranspiration that was 4.5 mm d^{-1} at the time of the experiment (the estimated evaporation and transpiration were 2 and

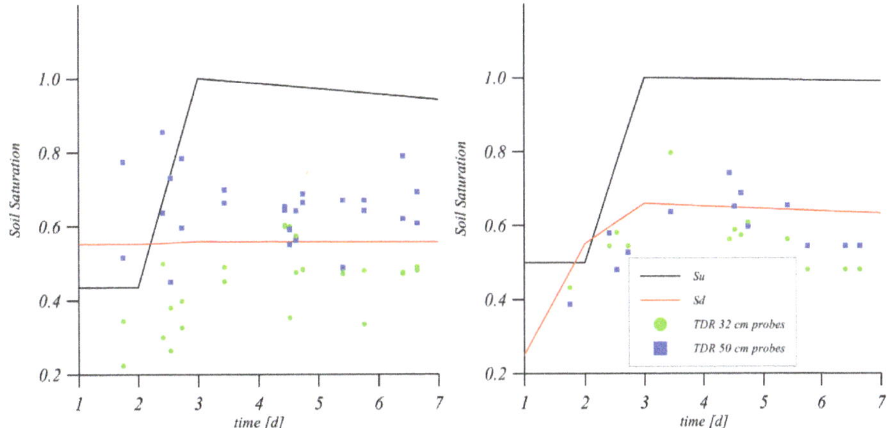

Fig. 10. Soil saturation estimated by mass balance at daily timescale, starting from 18 May 2010 (before the 13 mm rainfall event) and soil saturation measured by fixed TDR probes. Left panel: fallow plot. Right panel: cultivated plot.

2.5 mm d^{-1}, respectively) and confirms that all the irrigation water infiltrated in the fallow plot, in contrast with the previous interpretation of the infiltration experiment provided by (Cassiani et al., 2012).

3.4 Model outcome

In order to discuss our intuition on the mechanistic behavior of the plots, we modeled the dynamics of the infiltration experiment in the fallow plot and in the cultivated plot for comparison. In the fallow plot, we set the weed cover at its upper limit (evaluated by VIA), CC = 0.4. In the cultivated plot, CC = 1. The measured water content values before rainfall and irrigation were used to define the initial condition.

The model was forced by climatic data (precipitation, relative humidity, temperature, wind speed and solar radiation) recorded at a meteorological station located within the San Michele farm. The water balance of the two plots was estimated for 7 days starting 1 day before irrigation, using Eqs. (2) and (4), and the simulation results were compared with the fixed TDR measurements.

In Fig. 10, the calculated S_u and S_d are shown together with the soil moisture measurements obtained by fixed TDR. The measured soil saturation of the fallow plot was extremely variable after irrigation (Fig. 10, left panel), supporting the hypothesis that some local preferential infiltration occurred. The blue squares, corresponding to the 50 cm TDR probes, approached the calculated S_u (black line) shortly after irrigation, suggesting that water could infiltrate very quickly into the DSL. One day after irrigation the data were less scattered, and reasonably set around the calculated S_d (red line), confirming that redistribution occurred. In the cultivated soil (Fig. 10, right panel), S_d was initially low, the DSL was refilled by irrigation and slowly emptied by transpiration. The model shows how, after irrigation, the soil saturation of the two plots looks similar, according to the new interpretation of the ERT data proposed in this paper. There is at least a qualitative agreement between the TDR measurements and the simulated S_d of the two plots, suggesting that local scale processes, that are typical of the fallow plot, are missed, but the average soil moisture dynamics is reasonably captured by our simple model.

We conjectured that the presence of a crust over the bare plot could limit the water flux from the USL to the DSL, but the growth of weeds created a crust discontinuity and transformed the USL in a dual porosity layer, locally allowing the deep percolation of water in the DSL. The weed survival should be linked to this preferential local infiltration. We tried to explore the relevance of this positive feedback on the yearly water balance by running the model for the whole of 2010 (the year of the infiltration experiment). We assumed the weed to be active in between day of the year (DOY) 80 and DOY 274 and integrated Eqs. (2) and (4) between 1 January 2009 and 31 December 2010. Just the results of the calculation corresponding to the year of the experiment are shown in Fig. 11, where DOY = 1 is 1 January 2010 and DOY = 365 is 31 December 2010. The system loses memory of the initial condition after less than 200 days, thus the results shown in Fig. 11 are independent from the initial condition at 1 January 2009.

Two different scenarios are compared in Fig. 11: CC = 0.4, and the case of completely bare soil with deep percolation impeded $L_u = 0$ mm d^{-1}. In the case CC = 0 (blue and green lines) the water balance reduces to $P = \mathrm{RO} + E$ in the USL, all transpirable water was depleted at the beginning of the calculation, and since the DSL is not active, the soil saturation remains constant and represents a reference value to be compared with the calculated soil saturation in case CC = 0.4. According to our modeling assumptions, the USL is saturated by each rainfall event and slowly loses water via evaporation with $S_u = 1$ often during the winter season (blue line). During the dry summer period when the vegetation is water stressed, the saturation of the DSL in case CC = 0.4

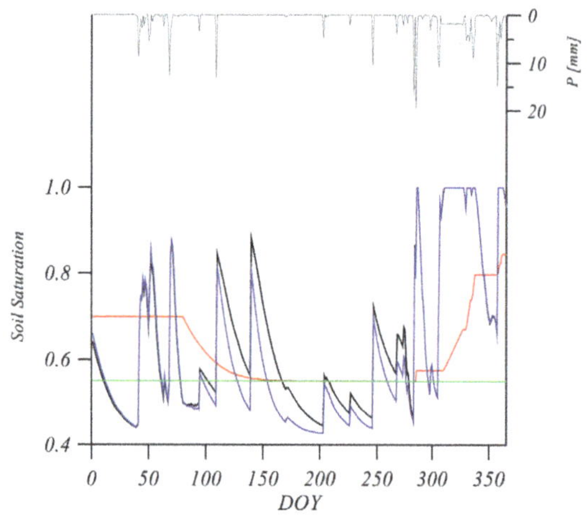

Fig. 11. Model outcome for the whole year 2010: DSL's soil saturation S_d and USL's soil saturation S_u versus time, for extreme values of the measured vegetation cover: CC = 0 and CC = 0.4.

approaches the complete depletion of the case CC = 0 that is taken here as a reference value (the red and the green lines coincide between DOY 180 and DOY 280). During the summer season, S_u is higher for CC = 0.4 (black line) than for CC = 0 (blue line) due to the vegetation shadowing, whereas during the wet season, S_u is higher when CC = 0 (blue line) due to the fact that when CC = 0.4, the USL transfers water to the DSL that acts as a reservoir and the vegetation facilitates the infiltration.

The situation that was observed at the beginning of our irrigation experiment corresponds to the calculated one at DOY 110, for CC = 0.4, with most winter rainfall stored in the DSL and $S_u < S_d$ before rainfall. According to previous studies conducted in Mediterranean catchments where transpiration and storm flow are out of phase (Brooks et al., 2003), two water compartments interact in the subsoil: a matrix with small pore with low matrix potential, and fast flow paths originating from the interaction between vegetation growth and soil structural change. The fine-grained soil matrix is filled by heavy precipitation events (possibly occurring in autumn) or irrigation (as in our experimental setup) to be dried by the vegetation during the rainless season, thus, exchanging water with the fast flow paths through absorption.

In summary, we assumed that when CC = 0.4 (black and red lines), the vegetation roots alter the structure of the USL that transfers water to the lower soil layers during the winter season where it is stored. As a consequence, the USL maintains the DSL hydrologically active, supporting the later vegetation establishment and transpiration from the DSL. By time-averaging the calculated relevant water fluxes over the whole year 2010, we found that 72 % of the mean annual rainfall evaporated and 27 % was transpired, while deep percolation was negligible, suggesting that CC = 0.4 could be the maximum achievable vegetation cover given the scarce water resources, leading to minor runoff losses. According to Cassiani et al. (2012), we suppose that CC = 0.4 could correspond to some threshold dictated by the scarce water availability, and CC could be different if more water was supplied to the environment (the fallow plot is not irrigated and the vegetation relies on rainfall only).

4 Conclusions

A combined experimental and theoretical approach was used to investigate the existence and the relevance of positive feedbacks between weed growth and infiltration on a fallow plot. The ERT data collected during an irrigation experiment (for a comprehensive description see Cassiani et al., 2012) evidenced that the infiltration flux in the fallow plot was more heterogeneous than in the cultivated plot and this fact could be dictated by the poor conductivity of the USL and by the macroporosity associated to the partial vegetation cover. Nevertheless, the fixed TDR data suggested that all the irrigation water infiltrated, and the coupled measurements of soil saturation and vegetation cover by mobile TDR and VIA did not evidence strong correlation between these two variables. Whether the infiltration is restricted by the crusty layer and enhanced by the vegetation in the fallow plot is unclear, due to the lack of a strong experimental evidence that confirm our intuition.

Relevant plant–soil–water interrelations that we tried to assess by repeated local measurements over a short timescale, were conceptualized in a modeling frame. The model captured the observed soil moisture dynamics during a 5-day irrigation experiment and was further used to investigate the impact of the positive feedback on the yearly water balance.

The results of our experimental and numerical research suggest that in the fallow plot (a) infiltration is heterogeneous and could be locally influenced by plant growth, (b) shortly after irrigation, redistribution takes place belowground where (c) roots have access to the whole active volume; (d) a positive feedback between infiltration and vegetation growth could maintain the DSL hydrologically active during the whole year; based on the model outcome, we may also state that (e) the interplay between vegetation growth and soil, which has an impact on the local hydrologic processes, affects the yearly water budget, reducing runoff and increasing the evapotranspiration, but leaving the groundwater recharge unaltered as compared to the bare soil situation.

The study of the soil–vegetation–atmosphere interaction certainly deserves special attention in arid and semiarid regions, where crop rotation, tillage and the natural transformation that the soil structure undergoes when it is left unseeded influence eco-hydrological connections that occur through lateral and overland flow, above- and belowground. Coupling hydrological and biological databases is a promising way to test modeling concepts for ecosystem dynamics and relevant processes that govern the ecosystem response to the external climate forcing, and this study may be considered a first step toward a better comprehension of the nature and relevance of possible soil–vegetation feedbacks.

Acknowledgements. We acknowledge funding from the EU FP7 Collaborative Project CLIMB "Climate Induced Changes on the Hydrology of Mediterranean Basins Reducing Uncertainty and Quantifying Risk" and from the MIUR PRIN project 2010JHF437 "Innovative methods for water resources management under hydro-climatic uncertainty scenarios".

Edited by: N. Romano

References

Alchanatis V., Cohen, Y., Cohen, S., Moller, M., Sprinstin, M., Meron, M., Tsipris, J., Saranga, Y. and Sela, E.: Evaluation of different approaches for estimating and mapping water crop status variability in cotton with thermal imaging, Prec. Agricult., 11, 27–41, doi:10.1007/s11119-009-9111-7, 2010.

Allen, R. G., Pereira, L. S., Raes, D.,and M. Smith, M.: Crop evapotranspiration: Guidelines for computing crop water requirements, Irr. and Drain. Paper 56, UN-FAO, Rome, Italy, 1998.

ARPA Sardegna: Analisi Agro meteorologica e climatologica (Ottobre 2009–settembre 2010), 2001.

ARPA Sardegna: Analisi Agro meteorologica e climatologica (Ottobre 2010–settembre 2011), 2012.

Baudena, M., Andrea, F. D., and Provenzale, A.: A model for soil-vegetation-atmosphere interactions in water-limited ecosystems, Water Resour. Res., 45, W02701, doi:10.1029/2008WR007172, 2009.

Binley, A.: http://www.es.lancs.ac.uk/people/amb/Freeware/freeware.htm (last access: 4 December 2011), 2011.

Binley, A., Ramirez, A., and Daily, W.: Regularised image reconstruction of noisy electrical resistance tomography data, in: Process Tomography – 1995, edited by: Beck, M. S., Hoyle, B. S., Morris, M. A., Waterfall, R. C., and Williams, R. A., Proceedings of the 4th Workshop of the European Concerted Action on Process Tomography, 6–8 April 1995, Bergen, 401–410, 1995.

Binley, A., Henry-Poulter, S., and Shaw, B.: Examination of solute transport in an undisturbed soil column using electrical resistance tomography, Water Resour. Res., 32, 267, 125–146, doi:10.1029/95WR02995, 1996.

Binley, A. M., Cassiani, G., Middleton, R., and Winship, P.: Vadose zone flow model parameterisation using cross-borehole radar and resistivity imaging, J. Hydrol., 267, 147–159, 2002.

Borzuchowski, J. and Schulz, K. Retrieval of Leaf Area Index (LAI) and Soil Water Content (WC) Using Hyperspectral Remote Sensing under Controlled Glass House Conditions for Spring Barley and Sugar Beet, Remote Sens., 2, 1702–1721, doi:10.3390/rs2071702, 2010.

Brooks, J. R., Barnard, H., Coulombe, R., and McDonnell, J.: Ecohydrologic separation of water between trees and streams in a Mediterranean climate, Nat. Geosci., 3, 100–104, doi:10.1038/NGEO722, 2010.

Bushman, C. and Lichtenthaler, H. K.: Principles and characteristics of multi-color fluorescence imaging of plants, J. Plant Physiol., 152, 297–314, 1998.

Casper, B. B. and Jackson, R. B.: Plant competition underground, Annu. Rev. Ecol. System., 28, 545–570, 1997

Cassiani, G., Ursino, N., Deiana, R., Vignoli, G., Boaga, J., Rossi, M., Perri, M. T., Blaschek, M., Duttmann, R., Meyer, S., Ludwig, R., Soddu, A., Dietrich, P., and Werban, U.: Non-invasive monitoring of soil static characteristics and dynamic states: a case study highlighting vegetation effects, Vadose Zone J., doi:10.2136/vzj2011.0195, in press, 2012.

Cassiani, G., Bruno, V., Villa, A., Fusi, N., and Binley, A. M.: A saline trace test monitored via time-lapse surface electrical resistivity tomography, J. Appl. Geophys., 59, 244–259, 2006a.

Cassiani, G., Binley, A. M., and Ferré, T. P. A.: Unsaturated zone processes, in: Applied Hydrogeophysics, NATO Sci, Ser. 51, edited by: Vereecken, H., Binley, A., Cassiani, G., Kharkhordin, I., Revil, A., and Titov, K., Springer-Verlag, Berlin, 75–116, 2006b.

Chaerle, L. and Van Der Straeten, D.: Seeing is believing: imaging techniques to monitor planet health, Biochim. Biophys. Acta, 1519, 153–166, 2001.

Cheng, H. D., Jiang, X. H., Sun, Y., and Wang, J.: Color image segmentation: advances and prospects, Pattern Recognit., 34, 2259–2281, 2001.

Cody, M. L.: Roots in plant ecology, Tree, 1, 76–78, 1986.

Deiana, R., Cassiani, G., Villa, A., Bagliani, A., and Bruno, V.: Calibration of a Vadose Zone Model Using Water Injection Monitored by GPR and Electrical Resistance Tomography, Vadose Zone J., 7, 215–226, 2008.

Franz, T. E., King, E. G., Caylor, K. K., and Robinson, D. A.: Coupling vegetation organization patterns to soil resource heterogeneity in a central Kenyan dryland using geophysical imagery, Water Resour. Res., 47, W07531, doi:10.1029/2010WR010127, 2011.

Garre, S. I., Coteur, C., Wongleecharoen, T., Kongkaew, J., Diels, J., and Vanderborght, J.: Noninvasive Monitoring of Soil Water Dynamics in Mixed Cropping Systems: A Case Study in Ratchaburi Province, Thailand, Vadose Zone J., 37, 1–28, doi:10.2136/vzj2012.0129, 2012.

Gimmi, T. and Ursino, N.: Mapping Material Distribution in a Heterogeneous Sand Tank by Image Analysis, Soil Sci. Soc. Am. J., 68, 1508–1514, 2004.

Gish, T. J. and Jury, W. A.: Effect of Plant Roots and Root Channels on Solute Transport, T. ASABE, 26, 440–444, 1983.

Graham, D. J., Reid, I., and Rice, S. P.: Automated Sizing of coarse grained sediments: Image Processing Procedures, Math. Geol., 37, 1–28, doi:10.1007/s11004-005-8745-x, 2005.

Graham, D. J., Rollet, A. J., Piegay, H., and Rice, S. P.: Maximizin the accuracy of image-based surface sediment sampling tecniques, Water Resour. Res., 46, W02508, doi:10.1029/2008WR006940, 2010.

Grayson, R. and Bloecshl, G.: Spatial patterns in catchment hydrology, Observations and modelling, Cambridge University Press, 2000.

Huisman, J. A., Hubbard, S. S., Redman, J. D., and Annan, A. P.: Measuring soil water content with ground penetrating radar: a review, Vadose Zone J., 2, 477–491, 2003.

ITT Visual Information Solutions: IDL 7.1, www.ittvis.com (last access: December 2011), 2009.

Jackson, R. B., Jobbagy, E. G., and Nosetto, M. D.: Ecohydrology Bearings – Invited Commentary Ecohydrology in a human-dominated landscape, Ecohydrology, 2, 383–389, doi:10.1002/eco.81, 2009.

Keating, B. A., Gaydon, D., Huth, N. I., Probert, M. E., Verburg, K., Smith, C. J., and Bond, W.: Use of modelling to explore the water balance of dryland farming systems in the Murray-Darling Basin, Australia, Eur. J. Agron., 18, 159–169, 2012.

Kefi, S., Rietkerk, M., Alados, C. L., Pueyo, Y., Papanastasis, V. P., ElAich, A., and de Ruiter, P. C.: Spatial vegetation patterns and imminent desertification in Mediterranean arid ecosystems, Nature, 449, 213–215, doi:10.1038/nature06111, 2007.

Kemna, A., Vanderborght, J., Kulessa, B., and Vereecken, H.: Imaging and characterisation of subsurface solute transport using electrical resistivity tomography (ERT) and equivalent transport models, J. Hydro., 267, 125–146, 2002.

Kemna, A., Binley, A., Day-Lewis, F., Englert, A., Tezkan, B., Vanderborght, J., Vereecken, H., and Winship, P.: Solute transport processes, in: Applied Hydrogeophsics, edited by: Vereecken, H., Springer-Verlag, Berlin, 117–159, 2006.

Kurc, S. A. and Small, E. E.: Dynamics of evapotranspiration in semiarid grassland and shrubland during the summer monsoon season, central New Mexico, Water Resour. Res., 40, W09305, doi:10.1029/2004WR003068, 2004.

Kurc, S. A. and Small, E. E.: Soil moisture variations and ecosystem-scale fluxes of water and carbon in semiarid grassland and shrubland, Water Resour. Res., 43, W06416, doi:10.1029/2006WR005011, 2007.

Laliberte A.S., A. Rango, J.E. Herrick, Ed L. Fredrockson and L. Burkett (2007), An object-based image annalysis approach for determining fractional cover of senescent and green vegetation with digital plot photography, *J. of Arid Environment*, *69*, 1–14, DOI:10.1016/j.jaridenv.2006.08.016.

Ludwig, R., Soddu, A., Duttmann, R., Baghdadi, N., Benabdallah, S., Deidda, R., Marrocu, M., Strunz, G., Wendland, F., Engin, G., Paniconi, C., Prettenthaler, F., Lajeunesse, I., Afifi, S., Cassiani, G., Bellin, A., Mabrouk, B., Bach, H., and Ammerl, T.: Climate-induced changes on the hydrology of mediterranean basins – a research concept to reduce uncertainty and quantify risk, Fresenius Environ. Bull., 19, 2379–2384, 2010.

MacQueen, J. B.: Some Methods for classification and Analysis of Multivariate Observations, Proceedings of 5-th Berkeley Symposium on Mathematical Statistics and Probability, Berkeley, University of California Press, Berkeley, CA, 281–297, 1967.

Marris, E.: More crop per drop, Nature, 452, 273–277, doi:10.1038/452273a, 2008.

Meron, M., Tsipris, J., Orlov, V., Alchanatis, V., and Cohen, Y.: Crop water stress mapping for site-specific irrigation by thermal imagery and artificial reference surfaces, Precis. Agricult., 11, 148–162, doi:10.1007/s11119-009-9153-x, 2010.

Mojid, M. A. and Cho, H.: Wetting Solution and Electrical Double Layer Contributions to Bulk Electrical Conductivity of Sand-Clay Mixtures, Vadose Zone J., 7, 972–980, doi:10.2136/vzj2007.0141, 2008.

Monego, M., Cassiani, G., Deiana, R., Putti, M., Passadore, G., and Altissimo, L.: Tracer test in a shallow heterogeneous aquifer monitored via time-lapse surface ERT, Geophysics, 75, WA61–WA73, doi:10.1190/1.3474601, 2010.

Oxborough, K.: Imaging of chlorophyll a fluorescence: theoretical and practical aspects of an emerging technique for the monitoring of photosynthetic performance, J. Exp. Bot., 55, 1195–1205, doi:10.1093/jxb/erh145, 2004.

Penuelas, J. and Filella, I.: Visible and near-infrared reflectance techniques for diagnosing plant physiological status, Trends Plant Sci., 3, 151–156, 1998.

Perri, M. T., Cassiani, G., Gervasio, I., Deiana, R., and Binley, A. M.: A saline tracer test monitored via both surface and cross-borehole electrical resistivity tomography: comparison of time-lapse results, J. Appl. Geophys., 79, 6–16, doi:10.1016/j.jappgeo.2011.12.011, 2012.

Rietkerk, M., Brovkin, V., van Bodegom, P. M., Claussen, M., Dekker, S. C., Dikstra, H. A., Goryachkin, S. V., Kabat, P., van Nes, E. H., Neutel, A. M., Nicholson, S. E., Nobre, C., Petoukhov, V., Provenzale, A., Scheffer, M., and Seneviratne, S. I.: Local ecosystem feedbacks and critical transitions in the climate, Ecol. Complex., 8, 223–228, 2011.

Roth, K., Schulin, R., Flühler, H., and Attinger, W.: Calibration of time domain reflectometry for water content measurement using a composite dielectric approach, Water Resour. Res., 26, 2267–2273, doi:10.1029/WR026i010p02267, 1990.

Sadler, R. J., Hazelton, M., Boer, M. B., and Grierson, P.: Deriving state-and-transition models of semi-arid grassland dynamics using imagery, Ecol. Modell., 221, 433–444, 2010.

Schenk, H. J. and Jackson, R. B.: Rooting depths, lateral root spreads, and belowground/aboveground allometries of plants in water limited ecosystems, J. Ecol., 90, 480–494, doi:10.1046/j.1365-2745.2002.00682.x, 2002.

Strobbia, C. and Cassiani, G.: Multilayer ground-penetrating radar guided waves in shallow soil layers for estimating soil water content, Geophysics, 72, J17–J29, 2007.

Topp, G. C., Davis, J. L., and Annan, A. P.: Electromagnetic determination of soil water content: measurements in coaxial transmission lines, Water Resour. Res., 16, 574–582, 1980.

Ursino, N.: Modeling banded vegetation patterns in semiarid regions: Interdependence between biomass growth rate and relevant hydrological processes, Water Resour. Res., 43, W04412, doi:10.1029/2006WR005292, 2007.

Ursino, N.: Above and below ground biomass patterns in arid lands, Ecol. Modell., 220, 1411–1418, 2009.

Vanderborght, J., Huisman, J. A., van der Kruk, J., and Vereecken, H.: Geophysical Methods for Field-Scale Imaging of Root Zone Properties and Processes, in: Soil-Water-Root Processes: Advances in Tomography and Imaging, SSSSA Special Publication 61, edited by: Anderson, S. H. and Hopmans, J. W., SSSA, Madison, USA, doi:10.2136/sssaspecpub61.c12, 2013.

Winship, P., Binley, A., and Gomez, D.: Flow and transport in the unsaturated Sherwood Sandstone: characterization using cross-borehole geophysical methods, in: Fluid Flow and Solute Movement in Sandstones: The Onshore UK Permo-Triassic Red Bed Sequence, edited by: Barker, R. D. and Tellam, J. H., Special Publications, Geological Society, London, 263, 219–231, 2006.

Zhang, L., Dawes, W. R., and Walker, G. R.: Response of mean annual evapotranspiration changes at catchment scale, Water Resour. Res., 57, 701–708, 2001.

Zimmermann, B., Elsenbeer, H., and Moraes, J. M.: The influence of land-use changes on soil hydraulic properties: implications for runoff generation, Forest Ecol. Manage., 222, 29–38, 2006.

12

Obtaining sub-daily new snow density from automated measurements in high mountain regions

Kay Helfricht[1], **Lea Hartl**[1], **Roland Koch**[2], **Christoph Marty**[3], **and Marc Olefs**[2]

[1]IGF – Institute for Interdisciplinary Mountain Research, Austrian Academy of Sciences, Innsbruck, 6020, Austria
[2]ZAMG – Zentralanstalt für Meteorologie und Geodynamik, Climate research department, 1190 Vienna, Austria
[3]WSL Institute for Snow and Avalanche Research SLF, 7260 Davos, Switzerland

Correspondence: Kay Helfricht (kay.helfricht@oeaw.ac.at)

Abstract. The density of new snow is operationally monitored by meteorological or hydrological services at daily time intervals, or occasionally measured in local field studies. However, meteorological conditions and thus settling of the freshly deposited snow rapidly alter the new snow density until measurement. Physically based snow models and nowcasting applications make use of hourly weather data to determine the water equivalent of the snowfall and snow depth. In previous studies, a number of empirical parameterizations were developed to approximate the new snow density by meteorological parameters. These parameterizations are largely based on new snow measurements derived from local in situ measurements. In this study a data set of automated snow measurements at four stations located in the European Alps is analysed for several winter seasons. Hourly new snow densities are calculated from the height of new snow and the water equivalent of snowfall. Considering the settling of the new snow and the old snowpack, the average hourly new snow density is $68\,\mathrm{kg\,m^{-3}}$, with a standard deviation of $9\,\mathrm{kg\,m^{-3}}$. Seven existing parameterizations for estimating new snow densities were tested against these data, and most calculations overestimate the hourly automated measurements. Two of the tested parameterizations were capable of simulating low new snow densities observed at sheltered inner-alpine stations. The observed variability in new snow density from the automated measurements could not be described with satisfactory statistical significance by any of the investigated parameterizations. Applying simple linear regressions between new snow density and wet bulb temperature based on the measurements' data resulted in significant relationships ($r^2 > 0.5$ and $p \leq 0.05$) for single periods at individual stations only. Higher new snow density was calculated for the highest elevated and most wind-exposed station location. Whereas snow measurements using ultrasonic devices and snow pillows are appropriate for calculating station mean new snow densities, we recommend instruments with higher accuracy e.g. optical devices for more reliable investigations of the variability of new snow densities at sub-daily intervals.

1 Introduction

In mountain regions there is an increasing demand for high-quality analysis, nowcasting and short-range forecasts of the spatial distribution of snowfall. Operational services, concerning avalanche warning, road maintenance and hydrology, as well as hydropower companies and ski resorts, need reliable information on the depth of new snow (HN) and the water equivalent (HNW) of snowfall. Therefore the new snow density (ρ_{HN}) is needed to convert HN into HNW and vice versa. Information on HN is especially relevant for cold and windy conditions, when measuring HNW is a difficult task because conventional rain gauge measurements are prone to large errors (e.g. Goodison et al., 1998). Recent results of the Solid Precipitation Intercomparison Experiment (SPICE; Nitu et al., 2012) reveal that these errors still exist in standard meteorological measurements (e.g. Buisán et al., 2016; Pan et al., 2016). Many snow cover models calculate HN from HNW at sub-daily time intervals, although reliable HNW input data are difficult to obtain (Egli et al., 2009), and thus the new snow density is needed in equal temporal resolution to convert between HNW and HN (e.g. Lehning et al., 2002; Roebber et al., 2003; Olefs et al., 2013). Additionally, ρ_{HN} has a considerable effect on the snow bulk density of the total snowpack (e.g. Schöber et al., 2016).

Since the 1960s ultrasonic rangers have become more common for observing snow depth changes automatically even at sub-hourly time intervals (e.g. Gubler, 1981; Goodison et al., 1984; Lundberg et al., 2010). They have the advantage of a more objective method compared to subjective manual measurements of snow depth (Ryan et al., 2008). Although high-accuracy optical snow depth sensors have been more frequently used in practice over recent years (e.g. Mair and Baumgartner, 2010; Helfricht et al., 2016), longer time series of snow depths exist from ultrasonic measurements. Beside snow depth (HS), the water equivalent of the snowpack (SWE) is observed operationally using weighing devices such as lysimetric snow pillows (e.g. Serreze et al., 1999; Egli et al., 2009; Lundberg et al., 2010; Krajči et al., 2017) and snow scales (e.g. http://www.sommer.at/en/products/snow-ice/snow-scales-ssg-2, last access: 3 May 2018). Upward-looking GPR (e.g. Heilig et al., 2009) and GPS techniques (e.g. Koch et al., 2014; McCreight et al., 2014) and the combination of both (Schmid et al., 2015) have been applied in scientific studies to monitor the depth, SWE and liquid water content of the snowpack. However, these techniques are rather expensive or not yet in use for long-term observations by operational services. In general, automatic measurements of SWE are prone to a high relative uncertainty and require a certain degree of maintenance, which makes them complex and labour-intensive (Smith et al., 2017). Due to such constraints, SWE measurement instrumentation is installed at considerably fewer stations compared to HS instruments, and only at sites with easy access for appropriate maintenance. Recent studies present the performance of cosmic ray neutron sensors (e.g. Schattan et al., 2017), which are partly used for long-term observations such as e.g. from Col de Porte (Morin et al., 2012).

The density of new snow is influenced by the shape and size of the snow crystals (e.g. Nakaya, 1951). Relationships between predominant snow crystal type, riming properties and snowfall density were already reported by Power et al. (1964) from snowstorm observations in Canada. Once the snow crystals have accumulated at the snow surface, the density of the fresh snow starts to increase depending on prevailing weather conditions and compaction caused by overlaying of snow. A common mean ρ_{HN} used to convert between HN and HNW is $100\,\mathrm{kg\,m^{-3}}$. Many studies analysed ρ_{HN} values on a daily basis and confirmed this 10 : 1 rule as applicable for a first estimate (e.g. Roebber et al., 2003; Egli et al., 2009; Teutsch, 2009). However, ρ_{HN} values span a wide range, and values from 10 to $350\,\mathrm{kg\,m^{-3}}$ have been reported from American and European mountain ranges, with mean values between 70 and $110\,\mathrm{kg\,m^{-3}}$ (e.g. Diamond and Lowry, 1954; LaChapelle, 1962; Power et al., 1964; Judson, 1965; McKay et al., 1981; Meister, 1985; Judson and Doesken, 2000; Valt et al., 2014). Most of the ρ_{HN} data analysed in these studies were observed using readings on a snow board. The density is calculated from HN measured with a ruler and HNW is derived from an external precipitation device or from weighing the new snow either in solid or melted form (Fierz et al., 2009).

Several studies have shown that measured ρ_{HN} can be related to meteorological parameters, although with different time intervals and different degrees of determination. Gold and Power (1952) showed that the crystal type is related to its estimated formation temperature. Diamond and Lowry (1954) and Simeral et al. (2005) built an empirical calculation that ascertained relationships between ρ_{HN} and air temperature at the 700 mb level. Teutsch (2009) also concluded that ρ_{HN} of 12 h intervals at valley stations is best correlated to the wet bulb temperature at mountain stations in close vicinity ($r^2 = 0.86$). Judson and Doesken (2000) found that near-surface air temperature and new snow density at mountain stations could explain 52 % of the variance in snow density. Wetzel et al. (2004) presented a similar degree of correlation of ρ_{HN} to temperature at three high-elevation sites. Alcott and Steenburg (2010) showed that ρ_{HN} is correlated with near-crest-level temperature and wind speed particularly for high-SWE events. Wright et al. (2016) presented a statistical analysis of data from 42 seasons of manual daily snow density measurements along with air temperature and wind speed to derive parameterizations to estimate new snow density. However, they end up with a low coefficient of determination.

On the basis of data from seven stations in Switzerland located between 1250 and 1800 m a.s.l., Meister (1985) concluded that ρ_{HN} does not correlate with the amount of new snow (HN), that it does not depend on altitude and that air temperature does not accurately determine ρ_{HN}. Nevertheless, binning the data into temperature classes results in a statistical equation with a correlation coefficient of 0.85. Further, he recommended considering wind speed in addition to air temperature, at least for stations higher than 1800 m a.s.l. On the basis of data sets from Schmidt and Gluns (1991) and the US Army Corps of Engineers (1956), Hedstrom and Pomeroy (1998) developed a power function using the air temperature, for which they found a coefficient of determination of 0.84 and a standard error of estimate of $9.3\,\mathrm{kg\,m^{-3}}$. Jordan et al. (1999) introduced an algorithm for assigning ρ_{HN} within the SNTHERM snow cover model. They added wind dependence to the temperature parameterization of Meister (1985). This achieved a reduction of the error, but a significant scatter remained between observed and parameterized ρ_{HN} values. Lehning et al. (2002) built an empirical calculation for ρ_{HN} valid for a time interval of 30 to 60 min in the framework of the snow model SNOWPACK. They used air temperature, surface temperature, relative humidity and wind speed for the regression analysis and achieved an approximate multiple coefficient of determination of 0.83. Schmucki et al. (2014) used another empirical power relation, including air temperature, wind speed and relative humidity, to calculate the ρ_{HN} using SNOWPACK simulations for three contrasting sites in Switzerland. ρ_{HN} was analysed in short time intervals of 1–2 h by Ishizaka et al. (2016). They measured even lower densities in comparison to ρ_{HN} estimates obtained using the SNOWPACK density model, especially for aggregated snow crystal types. On the basis of data from Col de Porte (1325 m altitude, French Alps), Pahaut et al. (1976) developed a statistical relationship including the melting point of water, air temperature and wind speed. This parameterization is used to calculate the density of new snow in the snow cover model CROCUS (Vionnet et al., 2012).

Settling of the new snow by its weight and destructive metamorphism may reduce HN and hence increase ρ_{HN} between snowfall and the HN reading and has to be considered when computing new snow density (e.g. Anderson, 1976; Lehning et al., 2002; Steinkogler, 2009; Vionnet et al., 2012). The contribution of settling to snow depth changes is highest in the first hours after snowfall. Wind drift and radiation input to the snow surface after the snowfall may increase ρ_{HN} in comparison to ρ_{HN} at the time of snowfall. However, direct measurements of ρ_{HN} at the time of snowfall are laborious and difficult to align with the hours of peak snowfall rates.

Whereas most of the studies have analysed daily and sub-daily, manual ρ_{HN} measurements, to the best of our knowledge no extensive analysis of automated ρ_{HN} measurements in hourly intervals over several winter seasons exists. The aim of this study is to assess the value of automated measurements of hourly HN and HNW for the calculation of ρ_{HN} at different stations and at hourly time intervals. Therefore we examine the following questions.

1. Are automated measurements of HN and HNW suitable for the calculation of ρ_{HN} at hourly intervals?
2. How do the mean and the variability of observed ρ_{HN} differ between distinct study sites?
3. How well do established density parameterizations represent observed hourly ρ_{HN} values?

To this end, we calculated ρ_{HN} from hourly snow depth changes (HN) and hourly SWE changes (HNW). The mean values and the variability of hourly ρ_{HN} are discussed for observations at four different meteorological stations and compared to calculations using established ρ_{HN} parameterizations. A critical assessment with an outlook for next-generation measurement techniques is given in the discussion.

2 Data and methods

Data from four automatic weather stations (AWSs) were used in this study (Fig. 1, Table 1). A prerequisite for the station selection was the combined measurement of HS and SWE at each station in addition to the standard meteorological measurements of air temperature, relative humidity, precipitation, wind speed and global radiation. HS data are measured using ultrasonic rangers. SWE data are recorded using snow pillows. Details regarding the instruments at each AMS and the exact location of each AWS, as well as the start and end dates of the available data coverage, are presented in Table 1.

The Kühroint station (Germany) is operated by the Bavarian Avalanche Warning Service. It is a well-equipped and maintained station for snow climate at the northern fringe of the eastern Alps. It is located in a meadow below the tree line.

The Kühtai station (Austria) is operated by the Tiroler Wasserkraft AG (TIWAG). It is located south of the Inntal valley, but north of the Alpine main ridge, and it is situated in a wind-sheltered location.

The station at Wattener Lizum (Austria) is operated by the Austrian Research Centre for Forests (BFW) of the Federal Ministry of Agriculture, Forestry, Environment and Water Management. This station is situated in a south–north-oriented high alpine valley above the tree line near to the Alpine main ridge. This station has an exceptionally long time series of snow-hydrological measurements (Krajči et al., 2017; Parajka, 2017).

The station at Weissfluhjoch (Switzerland) is operated by the Institute for Snow and Avalanche Research (SLF), which is part of the Swiss Federal Institute for Forest, Snow and Landscape Research (WSL). The station is presented in more detail by Marty and Meister (2012). Weissfluhjoch is the highest elevated station considered in this study.

On the basis of coinciding data availability we consider four time periods as presented in Table 1. Data outputs of the AWSs are logged at time intervals ranging from 2 to 30 min. Hourly values were computed for global radiation, relative humidity, air temperature and wind speed. The hourly value is the mean of the previous hour. For precipitation it is the sum of the previous hour. To account for noise in the ultrasonic signal, HS and SWE were smoothed using a centred moving average over three values in the original data resolution. The hourly values for HS and SWE are the values from the smoothed time series.

The thermodynamic wet bulb temperature (T_w) was computed applying the psychrometric equation (Sonntag, 1990) and an exact iterative approach presented by Olefs et al. (2010). A standard barometric equation was used to determine air pressure based on the station elevation. Air pressure dependency of T_w is generally minor and only relevant for air temperatures larger than $+2\,°C$ (Olefs et al., 2010).

A necessary condition for all further analysis of the time series was the presence of a precipitation signal at the heated precipitation gauges in combination with positive snow depth changes. Then, the hourly height of new snow (HN) and the water equivalent of snowfall (HNW) were computed as the change in HS and SWE. Within the next filtering step, only

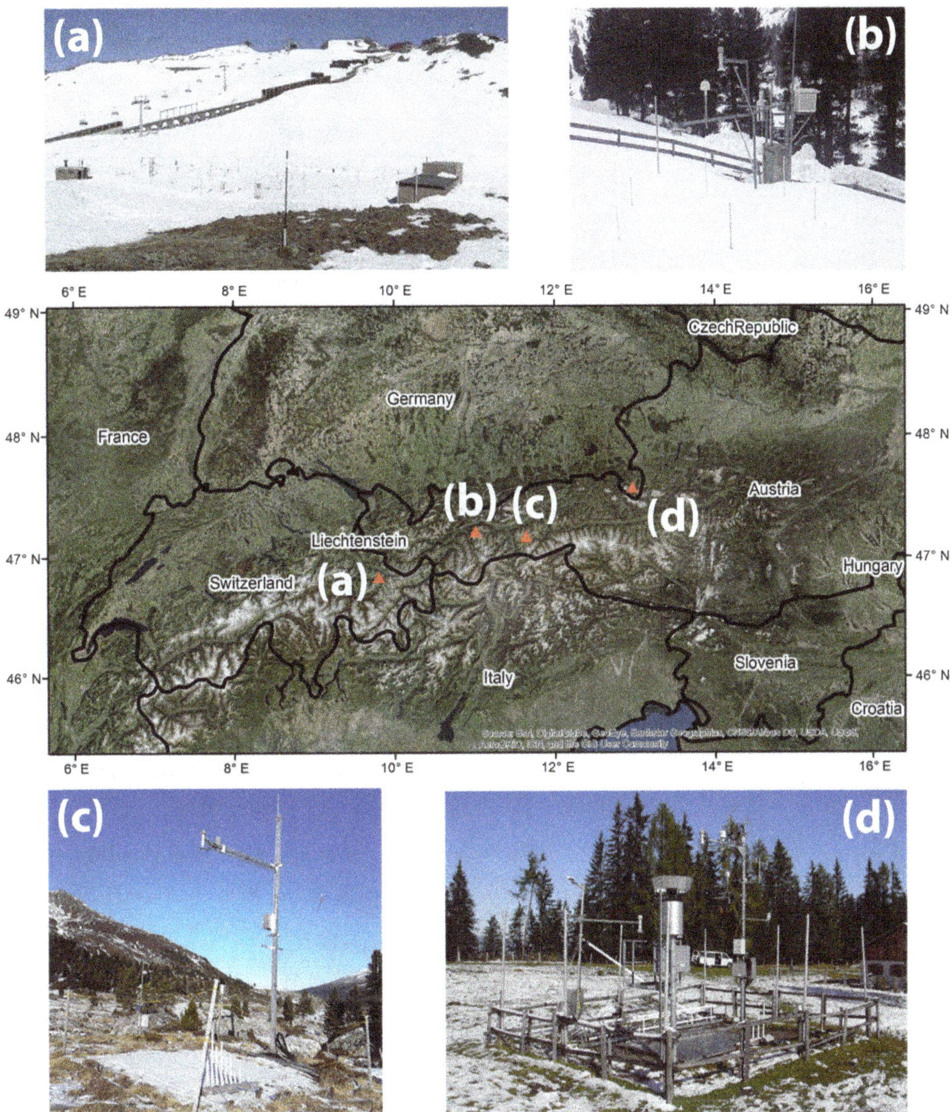

Figure 1. Map of the station locations. Pictures are given for **(a)** Weissfluhjoch station, **(b)** Kühtai station, **(c)** Wattener Lizum station and **(d)** Kühroint station.

HN and HNW values with T_w less than 0 °C and a wind speed (u) of less than 5 m s^{-1} were considered.

Constraints have to be set in order to avoid low values of HNW and HN, which are prone to large relative errors due to random and systemic measurement uncertainties in HN and SWE, but a minimum of approx. 100 remaining samples for statistical analysis must be ensured.

To investigate the influence of different minimum HNW and HN limits, a distribution matrix was calculated by varying the minimum HNW and HN limits in steps of 0.5 mm for HNW and 0.5 cm for HN, respectively. To account for settling during ongoing snowfall, the compaction correction described in Anderson (1976) was applied. The approach was simplified with respect to HS, SWE and snow density by considering only two layers of the snowpack: the new snow and the total snowpack of the previous time step. Destructive settling (S) of HN is considered for each time step in which the snow depth increases (Eq. 1). The destructive settling of the new snow (S_{HN}) for each time step is calculated by

$$S_{\mathrm{HN}} = -0.000002777 \cdot e^{(0.04 \cdot T)} \left\{ \rho_{\mathrm{HN}} \leq 150\,\mathrm{kg\,m^{-3}} \right\} \quad \text{(1a)}$$

$$S_{\mathrm{HN}} = S_{\mathrm{HN}} \cdot e^{(0.046 \cdot T \cdot (\rho_{\mathrm{HN}} - 150))} \left\{ \rho_{\mathrm{HN}} \geq 150\,\mathrm{kg\,m^{-3}} \right\}, \quad \text{(1b)}$$

where T is the air temperature. Settling of the new snow layer caused by the weight of the ongoing snow accumulation is not taken into account.

Settling within the old snowpack is computed considering the total snow depth (HS). The destructive settling within the old snow layer (S_{HS}) is calculated using Eq. (1), substituting HS for HN and using the bulk density of the old snowpack

Table 1. Coordinates and data availability are given for the four snow stations. The instrumentation for measuring snow depth (HS), snow water equivalent (SWE), temperature (T), relative humidity (RH), precipitation (P), wind speed (u) and global radiation (r) is listed.

Station abbreviation		Kühroint	Kühtai	Wattener Lizum	Weissfluhjoch
		KRO	KTA	WAL	WFJ
Location	East	12°57′35.5″	11°00′21.6″	11°38′18.6″	9°48′35.7″
	North	47°34′12.4″	47°12′25.6″	47°10′05.5″	46°49′46.4″
	z (m a.s.l.)	1420	1970	1994	2540
Data		1 Jan 2011–2 Dec 2015	27 Feb 1987–20 May 2015	1 Oct 2010–30 Dec 2016	1 Oct 2013–29 Sep 2015
Instruments	HS	Sommer USH 8	Sommer USH 8	Sommer USH 8	Campbell Scientific SR50A
	SWE	Sommer Snow Scale SSG	OTT Thalimedes Shaft Encoder, Endress + Hauser Deltapilot M	Sommer Snowpillow	Sommer Snowpillow
	T	Rotronic MP408	Kroneis NTC	Vaisala HMP45C	Rotronic Hydroclip S3
	RH	Rotronic MP408	Pernix hair hygrometer	Vaisala HMP45C	Rotronic Hydroclip S3
	P	Sommer NIWA/Med-K505	Ott Pluvio since 2001, custom built tipping bucket before	Sommer NIWA/Med-K505	Lambrecht Pluvio 1518 H3
	u	Young 05103	Kroneis cup anemometer + vane	YOUNG Wind Monitor	Young 05103
	r	Schenk 8101	Schenk 8101	Kipp&Zonen CM21	Kipp&Zonen CM21
Comments			Data gap winter 2012/13, wind regionalized from 1999	Meteorological measurements at 2041 m a.s.l.	

(ρ_{HS}) calculated from HS and total SWE of the previous time step. Settling within the old snowpack caused by the weight of the snowpack (S_{wHS}) is given as

$$S_{\mathrm{wHS}} = -248.976 \cdot \frac{\mathrm{HN}}{3\,600\,000} \cdot e^{0.8 \cdot T} \cdot e^{-0.021 \cdot \rho_{\mathrm{HS}}}. \quad (2)$$

The resulting settling factors of S_{HN}, S_{HS} and S_{wHS} are multiplied by HS and HN to adjust HN accordingly.

New snow density (ρ_{HN}) was obtained from the ratio of HN to HNW. Outliers below the 5 % percentile and higher than the 95 % percentile were excluded. The ρ_{HN} data were grouped by wet bulb temperature and wind speed, using bins of 1 °C and 0.5 ms^{-1} respectively. A least squares regression was carried out using both the ungrouped data and the median of the grouped data to quantify possible correlations of ρ_{HN} with T_{w} and u.

The ρ_{HN} values were compared to the following parameterizations developed in previous studies. In these parameterizations, ρ_{HN} is a function of meteorological parameters such as air temperature (T), wind speed (u) and relative humidity (RH). The time interval for ρ_{HN} readings of the respective study is given in brackets.

$$\rho_{\mathrm{HP}} = 67.92 + 51.52 \cdot e^{\frac{T}{2.59}} \text{ (Hedstrom and Pomeroy 1998, event/daily)} \quad (3)$$

$$\rho_{\mathrm{D}} = 119 + 6.48\,T \text{ (Diamond and Lowry 1954, frequent interval during event)} \quad (4)$$

$$\rho_{\mathrm{LC}} = 50 + 1.7 \cdot (T + 15)^{1.5} \text{ (LaChapelle 1962, event)} \quad (5)$$

$$\rho_J = 500 \cdot \left(1 - 0.951 \cdot \mathrm{e}^{-1.4 \cdot (5-T)^{-1.15} - 0.008u^{1.7}}\right) \left\{-13\,°\mathrm{C} < T \le 2.5\,°\mathrm{C}\right\} \quad (6a)$$

$$\rho_J = 500 \cdot \left(1 - 0.904 \cdot e^{-0.008u^{1.7}}\right) \left\{T \le 13\,°\mathrm{C}\right\} \text{ (Jordan et al., 1999, event/daily)} \quad (6b)$$

$$\rho_V = 109 + 6 \cdot \left(T - T_f\right) + 26u^{0.5} \text{ (Vionnet et al., 2012, event/daily)} \quad (7)$$

$$\rho_S = 10^{3.28 + 0.03T - 0.36 - 0.75 \cdot \arcsin(\sqrt{0.01\,\mathrm{RH}} + 0.03 \cdot \log_{10} u)} \left\{T \ge -14\,°\mathrm{C}\right\} \quad (8a)$$

$$\rho_S = 10^{3.28 + 0.03T - 0.75 \cdot \arcsin(\sqrt{0.01\,\mathrm{RH}} + 0.03 \cdot \log_{10} u)} \left\{T < -14\,°\mathrm{C}\right\} \text{ (Schmucki et al., 2014, event/hourly)} \quad (8b)$$

$$\rho_{\mathrm{L}} = 70 + 6.5\,T + 7.5\,T_{\mathrm{s}} + 0.26\,\mathrm{RH} + 13\,u - 4.5\,T\,T_{\mathrm{s}} - 0.65\,T\,u - 0.17\,\mathrm{RH}u + 0.06\,T\,T_{\mathrm{s}}\mathrm{RH} \text{ (Lehning et al., 2002, event/hourly)} \quad (9)$$

The melting point of snow (T_f) in Eq. (7) was approximated as 0 °C (Vionnet et al., 2012). Following Schmucki et al. (2014), we limited the parameter range and set RH to a constant value of 0.8 (80 %) during snowfall and the lower boundary for the wind speed to 2 ms^{-1}.

The temperature of the snow surface (T_{s}) is required in Eq. (9). As this was not available for each station, we used the approximation $T_{\mathrm{s}} = T$. We argue that T_{s} could not considerably exceed 0° because of the maximum T_{w} of 0 °C. Since only precipitation events are considered, RH can be expected to be high, and thus the difference between T_{w} and T is small.

The uncertainty of ultrasonic measurements on snow can be assumed to be in the range of ±1 cm, which is partly a consequence of changes in signal velocity due to meteorological conditions. However, we used the original HS data logged in millimetre resolution to avoid the effects caused by rounding to full centimetre when calculating HN. Likewise, we used the tenths of millimetre SWE data logged at the pillows. Another documented error source of the HS measurement is signal blocking by e.g. dense snowfall or drifting snow, which causes peaks of the HS. However, with the filtering procedure applied in this study, no such spikes were left in the analysis.

A source of uncertainty is the spatial offset between the HS measurements and the SWE measurements. HS is measured directly above the SWE measurement at Kühtai station, Kühroint station and Wattener Lizum station (Fig. 1). However, the footprint of the snow depth sensor may be smaller than the surface area of the pillow, and it decreases with increasing HS. A spatial variability of HS on the pillow can be caused by snow drift and differing snow settling or snowmelt.

For the calculations within this study we used the changes in HS and SWE over the time period of snowfall only. Errors due to spatial variability in HS and SWE caused by spatial differences in energy consumption and snow drift between precipitation events are reduced. This is especially valid for the HS and SWE measurements at the stations with matching HS and SWE measurements. The snow depth sensor and the snow pillow of Weissfluhjoch station are separated by 9 m. Schmid et al. (2014) suggest a small-scale variability in HS of ±4.3 % at the Weissfluhjoch station. Again, the error may be smaller due to using temporally limited changes of HS, but an additional uncertainty of ±5 % can be assumed here.

A well-known issue with snow pillows is bridging effects (e.g. Serreze et al., 1999; Johnson and Schaefer, 2002). Dense snow layers and crusts within the snowpack sustain the weight of the new snow so that HNW, and thus ρ_{HN}, are underestimated. We cannot exclude such data explicitly. However, all filtering conditions have to be fulfilled to include values in the analysis, so that data without or with lagged HN increase were not considered. Additionally, the chosen snow stations are well maintained in case of implausible data due to their overall good accessibility; e.g. trenches are dug out around the base area of the snow pillow at Kühtai station to cut off the measured part of the snowpack to avoid bridging effects.

Nevertheless, the measurement uncertainty is ±1 cm for HN and 0.1 cm for HNW. Considering mean HN (Table 2) and HNW values, the uncertainty is ±25 kg m^{-3} or 37 % of the mean density. This value is lower considering higher HN, but increases to 80 % for the combination of minimum HN and minimum HNW of 1.6 and 0.2 mm respectively.

3 Results and discussion

3.1 Data filtering, correction of settling and evaluation

Figure 2 presents the median new snow density (ρ_{HN}) data calculated from all filtered HN and HNW values exceeding the respective minimum HN and HNW limits. This presentation highlights the variability of ρ_{HN} by using different minimum limits with respect to the high relative uncertainty of low HN and HNW values. Changing the minimum limits for HN and HNW affects the resulting ρ_{HN} considerably. However, increasing the minimum limits for HN and HNW results in a distinct lowering of the number of data remaining for the subsequent analysis (Fig. 2). There are certain differences between the stations for high minimum HNW limits. Calculated ρ_{HN} decrease when low minimum HN and high minimum HNW limits are applied at Kühtai and Wattener Lizum station. In contrast, ρ_{HN} values increase for equal minimum limits at Kühroint and Weissfluhjoch stations. At Kühtai and Wattener Lizum stations, high HNW values of more than 3 mm HNW are accompanied by a rather high HN (Fig. 2). In contrast, a low HN occurring with a high HNW at Kühroint and Weissfluhjoch causes a high ρ_{HN}. However, these results are based on a small number of values only. In general, the calculated median ρ_{HN} values are rather constant following the 1 : 1 line of minimum HNW and HN limits (Fig. 2).

In order to avoid low values of HNW and HN, but ensuring an appropriate number of approx. 100 samples and with respect to the results of the Fig. 2, we decided to use a minimum limit of 1.5 mm in HNW and 2.0 cm in HN. This leads to the exclusion of on average 94 % of all data points that have a precipitation signal and positive snow depth changes (Table 2). Frequency distributions for HN, HNW, T_w and u of the unfiltered and filtered data are presented for each sta-

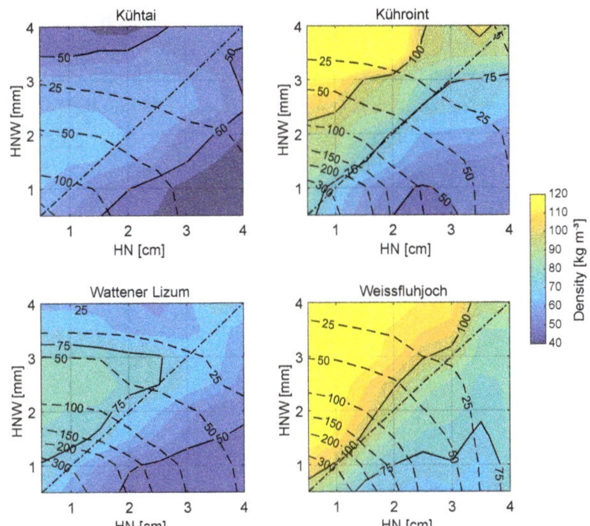

Figure 2. Median new snow densities (colour scale) calculated using all data exceeding different minimum limits of the height of new snow (HN) and the water equivalent of snowfall (HNW) for period 1 (1 October 2013–20 May 2015). Note that multiples of 25 kg m^{-3} are highlighted with red contour lines. The labelled black dashed lines give the count of the hourly data remaining after filtering. The straight dotted and dashed lines show results for equal minimum limits of HN (cm) and HNW (mm).

Table 2. Time periods analysed in this study with the mean and the median of hourly values for the height of new snow (HN), wet bulb temperature (T_W), wind speed (u), calculated densities from observed values (ρ) and calculated densities corrected for settling of the snowpack (ρ_{HN}). The results are valid for the filtered data values (n_{th}) with HN > 2 cm, HNW > 1.5 mm, T_W < 0 °C and u < 5 m s^{-1} as a subset of all data that have a precipitation signal and positive HS change (n_P).

Station	Period		Count data		HN (cm)		T_W (°C)		u (m s^{-1})		ρ (kg m^{-3})		ρ_{HN} (kg m^{-3})	
	no.		n_P	n_{th}	mean	median	mean	median	mean	median	mean	median	mean	median
KRO	1	1 Oct 2013–20 May 2015	1139	91	3.2	3.1	−3.9	−3.0	1.1	0.9	82	73	73	67
	2	1 Oct 2011–30 Sep 2013	1576	118	3.4	3.1	−4.2	−4.2	1.0	0.9	87	77	74	69
KTA	1	1 Oct 2013–20 May 2015	579	53	3.8	3.3	−3.4	−3.4	0.8	0.8	70	69	61	61
	2	1 Oct 2011–30 Sep 2013	506	36	3.3	2.8	−4.8	−4.0	0.8	0.7	75	66	60	54
	3	1 Oct 1999–30 Sep 2011	5293	252	3.5	3.2	−3.5	−3.2	0.8	0.8	74	74	64	64
	4	27 Feb 1987–30 Sep 1999	7958	387	3.7	3.3	−3.6	−3.4	0.8	0.7	74	75	61	59
WAL	1	1 Oct 2013–20 May 2015	1248	111	3.6	3.4	−4.3	−4.8	1.3	1.3	76	72	68	66
	2	1 Oct 2011–30 Sep 2013	1588	126	3.9	3.5	−4.3	−3.6	1.7	1.7	71	69	62	58
WFJ	1	1 Oct 2013–20 May 2015	1619	100	3.0	2.7	−4.9	−4.0	2.2	2.0	95	86	91	83

tion and for each time period in the Supplement Figs. S01 to S09.

The exclusion of high wind speeds only has a small effect at the lower stations and is more noticeable at the more wind-exposed stations of Wattener Lizum and Weissfluhjoch. Considering period 1 comprising all stations, the filtering process causes the highest filtering rate for Weissfluhjoch station, with 6 % of data remaining after applying the filtering. The overall highest amount of data reduction is found at Kühtai station, with 5 % of the data remaining after filtering of the longer periods 3 and 4 (Table 2). There was a considerable fraction of data with positive HS changes, a precipitation signal and positive T_W. Most of these data seem to be paired with very small HS changes and are eliminated for the final data set when the HN minimum limit is applied.

The correction of the HN underestimation caused by settling of the snowpack during snowfall leads to an average reduction of mean ρ_{HN} of 10.2 kg m^{-3}, with a standard deviation (σ) of 2.6 kg m^{-3} (Table 2). This corresponds to 13.5 % with a σ of 3.7 %. The compaction correction causes noticeably less change in ρ_{HN} at Weissfluhjoch in period 1 (5 % reduction of mean ρ_{HN}) than in the other time periods and other stations. The next closest is Kühroint, also in period 1, with a reduction in ρ_{HN} of 7 %. Unless otherwise stated in the text, ρ_{HN} always refers to the corrected densities hereafter. Based on a 15-year data set of Weissfluhjoch (WSL Institute for Snow and Avalanche Research SLF, 2015, https://doi.org/10.16904/1) from 1 September 1999 to 31 December 2015, the contribution of settling relative to HN was calculated using the multi-layer SNOWPACK model (e.g. Lehning et al., 2002) and the approach from Anderson (1976) to compare the results of this study to a more physically based estimate. Results are presented in Fig. 3. While a median relative contribution of settling to HN by 19 % was calculated with SNOWPACK, the approach of Anderson (1976) resulted in lower values of 5 % in median and 9 % in mean. Thus, the settling considered for the presented data can be assumed to be a lower estimation. However, higher contributions of settling would result in lower ρ_{HN} values, with an increased HN assuming a fixed HNW.

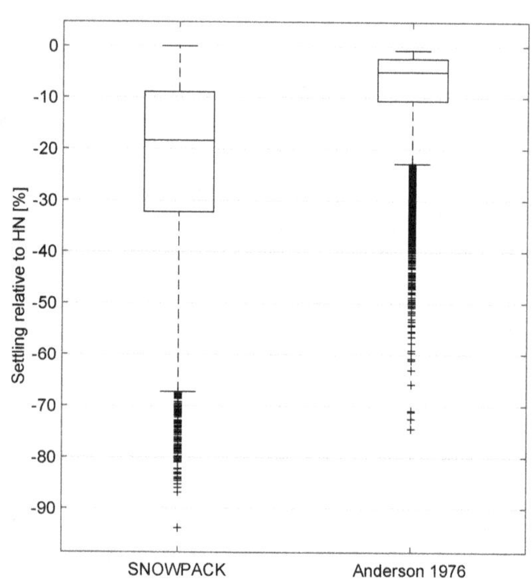

Figure 3. Box plots (median, 25 and 75 % percentiles, 1.5 × interquartile ranges, outliers) of settling relative to hourly new snow heights (HN) modelled with SNOWPACK and using the approach presented by Anderson (1976).

Figure 4 shows the distribution of ρ_{HN} values obtained from the filtered data at Kühroint station as representative of all stations and periods (Figs. S10 to S17). In general, the ρ_{HN} values show high variability at all stations. Nevertheless, ρ_{HN} values are within a reasonable range of less than 200 kg m^{-3}. The histograms of ρ_{HN} show one-tailed distribution towards higher ρ_{HN}. Median ρ_{HN} values of the different stations and for different periods range between 66 and 86 kg m^{-3} for uncorrected values and between 54 and 83 kg m^{-3} for ρ_{HN} corrected for settling (Table 2).

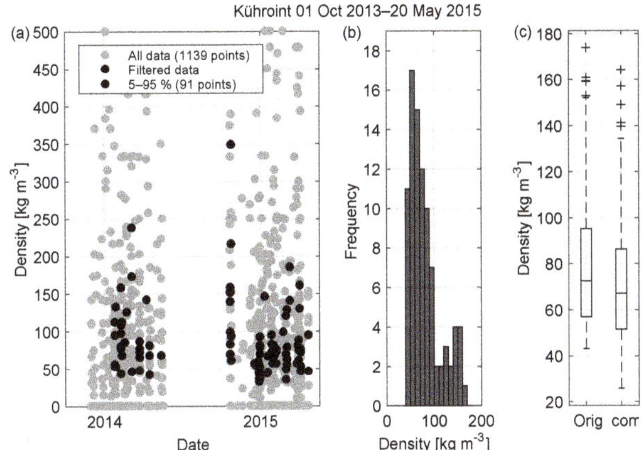

Figure 4. Distribution of calculated new snow densities at Kühroint station for period 1 (1 October 2013–20 May 2015). **(a)** All data have a precipitation signal and positive HS change, all data are filtered with HN > 2 cm, HNW > 1.5 mm, $T_w < 0\,°C$ and $u < 5\,ms^{-1}$) and filtered data are reduced by cutting off at 5 and 95 % percentiles. **(b)** Histogram of all filtered densities. **(c)** The box plot showing the median and 25 and 75 % interquartile range of uncorrected densities and densities corrected for settling of the snowpack. Note that similar figures are available in the Supplement (Figs. S10–S17) for all stations and all time periods considered in this study.

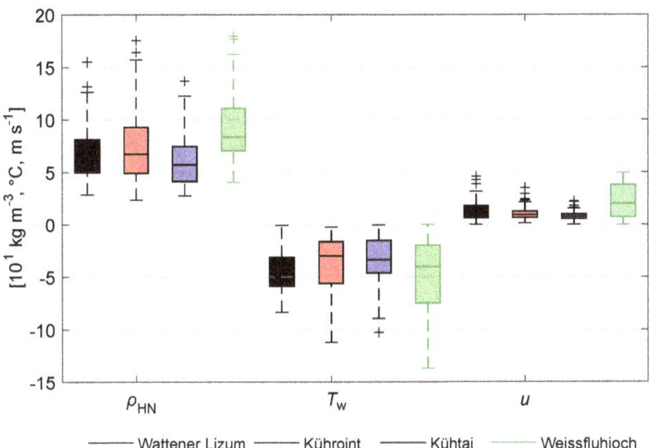

Figure 5. Box plot (median, 25 and 75 % percentiles, 1.5 × interquartile ranges, outliers) of calculated new snow densities (ρ_{HN}) based on observations, wet bulb temperature (T_w) and wind speed (u) for filtered snowfall events (Table 2) at all four stations within period 1 (1 October 2013–20 May 2015).

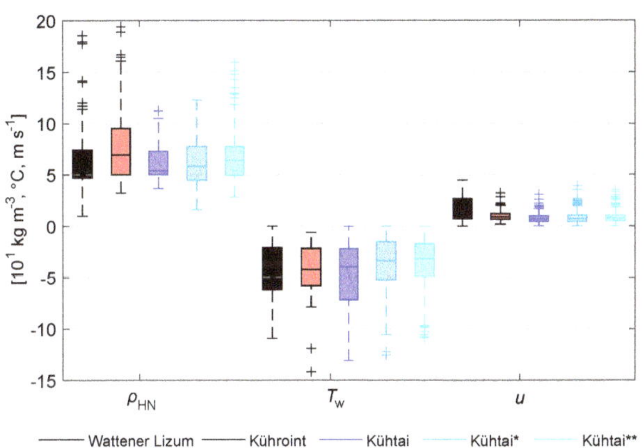

Figure 6. Box plots (median, 25 and 75 % percentiles, 1.5 × interquartile ranges, outliers) of calculated new snow densities (ρ_{HN}) based on observations, wet bulb temperature (T_w) and wind speed (u) for filtered snowfall events (Table 2) at three stations within period 2 (1 October 2011–1 October 2013) and at Kühtai station within period 3 (index*, 1 October 1999–30 September 2011) and period 4 (index**, 27 February 1987–30 September 1999).

3.2 Station-dependent differences

The distributions of ρ_{HN}, T_w and u during all filtered snowfall data are presented in Figs. 5 and 6 and in Table 2. The lowest T_w and highest wind speeds were observed during snowfall at Weissfluhjoch station. However, the range and distribution of T_w at Weissfluhjoch station result in a higher median T_w during snowfall compared to T_w at Wattener Lizum station. With respect to wind speeds, Wattener Lizum is second. The lowest wind speeds at Kühtai station occur together with the lowest ρ_{HN}. Weissfluhjoch station has the highest median ρ_{HN} by a large margin with 83 kg m^{-3} in period 1 compared to, respectively, 67, 61 and 66 kg m^{-3} at Kühroint, Kühtai and Wattener Lizum stations.

Wind influence may be the reason for higher ρ_{HN} at Weissfluhjoch station. Snow grains are fragmented by snow drift (e.g. Sato et al., 2008), and thus more packed into the layer of new snow during windy conditions even over the course of only 1 h. The Kühtai station shows the lowest ρ_{HN}, and the difference of the mean ρ_{HN} is 17 kg m^{-3} between Weissfluhjoch and Kühtai stations for period 1. Median ρ_{HN} and median T_w of the different periods show a relationship between the periods at Kühtai station, with a higher ρ_{HN} for a higher T_w (Fig. 6, Table 2).

The overall mean hourly ρ_{HN} of all stations and time periods is 68 kg m^{-3}, with a standard deviation of 9 kg m^{-3}. In general, this is considerably lower than new snow densities from daily measurements (e.g. Roebber et al., 2003; Egli et al., 2009; Teutsch, 2009). Meister (1985) measured ρ_{HN} lower than 100 kg m^{-3} on a daily basis, analysing data with a HN of more than 0.1 m. In contrast, the presented ρ_{HN} values are closer to the time of the snowfall event, and density changes over several hours due to e.g. energy exchanges and wind drift at the uppermost snow layer can be excluded. On the basis of ρ_{HN} in situ measurements in hourly resolution Lehning et al. (2002) emphasized that at sub-daily time intervals, lower densities in comparison to daily new snow densities have to be applied. Comparatively low ρ_{HN} values close to 50 kg m^{-3} were also presented by Ishizaka et al. (2016), with an average ρ_{HN} of 52 kg m^{-3} for aggregated snowflakes and 55 kg m^{-3} for small hydrometeors. They further found a

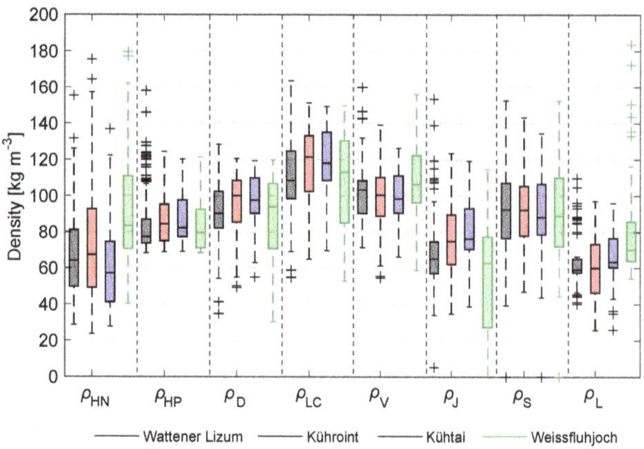

Figure 7. Box plots (median, 25 and 75 % percentiles, 1.5 × interquartile ranges, outliers) of calculated new snow densities (ρ_{HN}) based on observations and densities calculated using parameterizations developed in previous studies (Eqs. 3–9) at all four stations within period 1 (1 October 2013–20 May 2015).

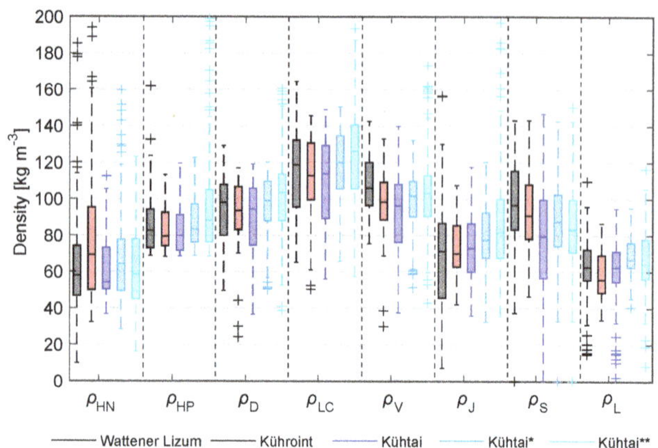

Figure 8. Box plots (median, 25 and 75 % percentiles, 1.5 × interquartile ranges, outliers) of calculated new snow densities (ρ_{HN}) based on observations and densities calculated using parameterizations developed in previous studies (Eqs. 3–9) at three stations within period 2 (1 October 2011–1 October 2013) and at Kühtai station within period 3 (index*, 1 October 1999–30 September 2011) and period 4 (index**, 27 February 1987–30 September 1999).

mean ρ_{HN} of 72 kg m^{-3} for a second group of smaller crystals and 99.4 kg m^{-3} for graupel-type hydrometeors.

The observed inter-station variability shows the importance of differing ρ_{HN} between more windy mountain stations and less windy stations in the valleys.

3.3 Density parameterizations

A simple linear regression analysis showed that the short-term variability of ρ_{HN} cannot be explained with corresponding changes in T_w or u (Table 3, Figs. 7 and S18 to S26). An increase of ρ_{HN} with increasing T_w can be identified in the figures, and the slopes of the least squares regressions show an increase of ρ_{HN} with an increase of wet bulb temperature for all stations (Table 3). However, no consistent relationship between ρ_{HN} and u could be found, either for single stations or for different periods at one station. The binned analysis based on T_w showed a considerable r^2 of more than 0.5 on a 0.01 significance level at Kühroint and Kühtai station, with intercepts of 70 to 80 kg m^{-3} and gradients of about 3 to 4 kg m^{-3} per 1 °C.

Although the regressions generally show the expected trends, it must be noted that the variability of ρ_{HN} remains unexplained. This could partly be attributed to the measurement uncertainties. However, the variability caused by measurement uncertainties is assumed to be equalized, only considering the mean and median of ρ_{HN} values for total time periods. Relationships between ρ_{HN} and T_w were recognized for distinct periods and stations only, but with similar coefficients of determination in comparison to the results of e.g. Judson and Doesken (2000), Wetzel et al. (2004) or Wright et al. (2016).

Testing multiple regressions using additional meteorological parameters did not increase the statistical significance. Instead, a comparison to existing parameterizations of ρ_{HN} was performed for all stations and periods.

Considering the various parameterizations, which use meteorological parameters to approximate new snow density (Eqs. 3 to 9), it is evident that the observed variability of ρ_{HN} is not correlated to the variability of parameterized new snow densities (Table 4). Most of the seven parameterizations overestimate the median of the observed ρ_{HN} values (Figs. 3, 7 and 8 and Table 4). However, some parameterizations produce considerably better results than others for median ρ_{HN} values. The parameterizations of LaChapelle (1962), Diamond and Lowry (1954) and Vionnet et al. (2012) consistently overestimate ρ_{HN}. The parameterization of Hedstrom and Pomeroy (1998) overestimates ρ_{HN} at Kühroint, Kühtai and Wattener Lizum stations (Figs. 7 and 8), but converges with the median ρ_{HN} at Weissfluhjoch station for period 1 (Fig. 7, Table 4). In general, the ρ_{HN} values simulated using the parameterization of Jordan et al. (1999) are closer to calculated ρ_{HN}, but median ρ_{HN} values are underestimated for Weissfluhjoch station. Median ρ_{HN} values and the range of ρ_{HN} at Weissfluhjoch are well simulated using the parameterization of Schmucki et al. (2014), but it overestimates median ρ_{HN} of Kühroint, Kühtai and Wattener Lizum stations (Figs. 3 and 7 and Table 4). However, this parameterization was fitted to original density data from Weissfluhjoch.

The lowest root mean squared error (RMSE) was achieved for Weissfluhjoch station with the parameterization of Diamond and Lowry (1954). The parameterizations of Lehning et al. (2002) and Jordan et al. (1999) result in the lowest RMSE (Table 4) compared to ρ_{HN} at Kühroint, Kühtai and

Table 3. Results of a single linear regression between the corrected densities (ρ_{HN}) as a dependent variable and wet bulb temperature (T_w) and wind speed (u) as explanatory variables for the class median values based on all filtered data points binned into 0.5° K classes and classes of 0.5 ms^{-1}, respectively. The corresponding coefficient of determination (r^2) and the p value are presented.

Station	Period no.	T_w				u			
		Intercept	$\delta\rho/\delta T_w$	r^2	p	Intercept	$\delta\rho/\delta u$	r^2	p
KRO	1	82.07	4.00	0.65	0.00	45.12	19.10	0.35	0.16
	2	76.54	0.99	0.11	0.35	64.84	1.29	0.00	0.90
KTA	1	66.37	1.84	0.12	0.44	72.59	−14.44	0.41	0.36
	2	55.15	−0.37	0.02	0.75	54.25	3.37	0.53	0.17
	3	68.18	1.51	0.56	0.01	64.81	−3.82	0.39	0.10
	4	72.41	3.75	0.82	0.00	49.31	9.41	0.30	0.26
WAL	1	78.84	2.88	0.47	0.06	65.28	1.32	0.02	0.71
	2	64.58	0.97	0.17	0.21	59.43	1.50	0.05	0.57
WFJ	1	92.68	0.71	0.04	0.53	92.88	−2.91	0.18	0.23

Wattener Lizum stations, with slightly lower density values using the parameterization of Lehning et al. (2002) fitting best to the low median ρ_{HN} values of the Kühtai station.

Thus, the parameterization of Lehning et al. (2002) appears to be the first choice regarding the calculation of hourly new snow densities for high elevations and inner-alpine regions. This parameterization requires multiple input parameters. Where such data are not available, the parameterization of Jordan et al. (1999), requiring temperature and wind data only, might be a good alternative. Even though correlations are low in general, some of the highest Pearson correlation values (r, Table 4) were achieved by applying the simpler, linear equations by Diamond and Lowry (1954), LaChapelle (1962) and Vionnet et al. (2012). In addition to the regressions presented in Table 3, this shows again the identifiable relation between snow density and temperature.

Mair et al. (2016) evaluated some of the parameterizations also considered in this study. Using a distinctly larger time window for smoothing their HS data (5 h average), they calculated median ρ_{HN} between 75 and 100 $kg\,m^{-3}$ using the parameterizations of Jordan et al. (1999) and Hedstrom and Pomeroy (1998), which is close to the results presented in this study. They also found that using the parameterization of LaChapelle (1962) results in a mean ρ_{HN} higher than 100 $kg\,m^{-3}$. In general they concluded that using a constant ρ_{HN} of 100 $kg\,m^{-3}$ caused an overestimation of seasonal precipitation by up to 30 %. Conversely, a mean ρ_{HN} of 70 $kg\,m^{-3}$ will result in better SWE estimations. This is in accordance with the resulting average ρ_{HN} of 68 $kg\,m^{-3}$ calculated from automated measurements within our study.

4 Conclusion

The aim of this study was to assess the value of automated measurements of snow depth (HS) and snow water equivalent (SWE) to compute new snow density (ρ_{HN}) on an hourly time interval. Complementary data sets of HS and SWE measurements using ultrasonic devices and snow pillows from four mountain stations were used to calculate the height of new snow (HN) and the water equivalent of snowfall (HNW). Subsequently, ρ_{HN} was calculated from HN and HNW, considering potential underestimation of HN by settling of the snowpack.

The snow measurements using ultrasonic devices and snow pillows were found to be appropriate for the calculation of station average hourly ρ_{HN} values. However, the observed variability in ρ_{HN} from the automated measurements could not be described with appropriate statistical significance by any of the investigated algorithms. An average ρ_{HN} of 68 $kg\,m^{-3}$ with a standard deviation of 9 $kg\,m^{-3}$ was calculated considering all stations and time periods. The average ρ_{HN} for individual stations in a common period ranged from 61 to 83 $kg\,m^{-3}$, with a higher ρ_{HN} at more windy locations. Thus, wind speed is a crucial parameter for the inter-station variability of ρ_{HN}.

Seven existing parameterizations for estimating new snow densities were tested, and most calculations overestimate ρ_{HN} in comparison to the results from the hourly automated measurements. Two of the tested parameterizations were capable of simulating low ρ_{HN} at sheltered inner-alpine stations. This reveals that it has to be carefully considered which parameterization should be used for which application and environment.

Nevertheless, the natural variability of ρ_{HN} is masked using the combination of ultrasonic ranging and snow pillow data for ρ_{HN} calculation because of the limited accuracy of the sensors and snow depth changes due to settling of the snowpack and wind drift. We conclude that the value of the analysed data is given by the mean and median ρ_{HN} and its variation between different stations and time periods, and the

Table 4. Comparison of corrected density values (ρ_{HN}, (kg m^{-3})) and parameterizations, applying Eqs. (3) to (9) presented in Sect. 2. Median values (m, (kg m^{-3})) are shown together with the Pearson correlation coefficient (r) and the root mean squared error (RMSE, (kg m^{-3})) between the respective calculations and ρ_{HN}. Best values of the performance measures are highlighted for each station and time period using bold and italic numbers.

Station	Period	ρ_{HN}	ρ_{HP}			ρ_D			ρ_{LC}			ρ_V			ρ_J			ρ_S			ρ_L		
	no.	m	m	r	RMSE	m	r	RMSE	m	r	RMSE	m	r	RMSE	m	r	RMSE	m	r	RMSE	m	r	RMSE
KRO	1	67	85	0.28	14.4	100	0.45	23.3	121	0.44	44.4	101	***0.47***	25.3	75	0.29	***0.5***	92	0.36	18.9	60	0.40	15.8
	2	69	79	0.18	8.7	94	0.18	18.8	113	0.19	38.8	99	0.13	25.7	70	***0.20***	***1.0***	91	0.10	19.8	56	***0.20***	14.5
KTA	1	61	82	0.35	28.7	98	***0.38***	40.1	118	***0.38***	62.3	98	0.33	41.7	76	0.37	23.0	88	0.26	32.2	63	0.35	***8.9***
	2	54	80	0.14	22.4	94	0.21	33.2	114	0.21	53.2	96	0.27	35.6	73	0.12	14.3	79	***0.36***	22.1	62	0.05	***4.9***
	3	64	83	0.21	22.0	99	***0.25***	32.1	120	0.24	53.2	102	0.19	34.5	77	0.24	14.5	88	0.09	20.3	67	0.06	***4.8***
	4	59	88	0.25	26.7	103	***0.32***	35.7	126	0.31	57.1	103	***0.32***	37.6	82	0.25	19.5	83	0.10	24.2	64	0.26	***5.4***
WAL	1	66	77	0.26	16.1	90	***0.33***	23.9	108	0.32	43.9	103	0.25	32.8	65	0.24	***0.7***	92	0.04	17.9	59	0.10	5.9
	2	58	83	0.08	24.0	98	0.14	31.8	119	0.13	52.5	106	***0.15***	45.5	71	0.06	6.9	97	−0.09	28.9	63	0.08	***1.7***
WFJ	1	83	79	0.08	8.0	94	***0.10***	***2.1***	113	***0.10***	17.6	106	0.00	19.1	63	0.09	34.6	89	0.01	2.7	70	−0.03	14.6

considerably lower ρ_{HN} values in contrast to ρ_{HN} calculated on daily or event-based measurements.

The study shows the potential of collocated measurements of HS and SWE for determining ρ_{HN} automatically. However, recent developments in optical distance sensors and weighing devices increase the accuracy of such snow measurements and hence decrease the uncertainty of subsequent calculations. We therefore recommend the use of high-accuracy sensors for the determination of ρ_{HN} at sub-daily intervals.

Data availability. A processed set of SNOWPACK input data from Weissfluhjoch station is available in WSL Institute for Snow and Avalanche Research SLF (2015) (WFJ_MOD: Meteorological and snowpack measurements from Weissfluhjoch, Davos, Switzerland) at https://doi.org/10.16904/1.

Detailed information about the Weissfluhjoch data set can be found in WSL Institute for Snow and Avalanche Research SLF (2015) and in Marty and Meister (2012). Data of Kühtai station are published by Krajči et al. (2017) and are available from the Zenodo repository at https://doi.org/10.5281/zenodo.556110 (Parajka, 2017).

Data of Kühroint station are available on request from the Bavarian Avalanche Warning Service (https://www.lawinenwarndienst-bayern.de/organisation/lawinenwarnzentrale/kontakt.php, last access: 2 May 2018).

Data of Wattener Lizum station are available on request from the Austrian Federal Research and Training Centre for Forests, Natural Hazards and Landscape (BFW; https://bfw.ac.at/rz/bfwcms.web?dok=6057, last access: 2 May 2018).

Author contributions. KH is the main investigator of this study. LH performed snow density analysis within the pluSnow project. RK performed initial quality control, provision and set-up of the project database for all station and meta-data. CM prepared the data of Weissfluhjoch station, contributed fruitful discussions and helped to hone the focus of the analysis and the manuscript. MO contributed significantly to analysis and discussions as the main project partner within the framework of the pluSnow project.

Competing interests. The authors declare that they have no conflict of interest.

Acknowledgements. The pluSnow project is financed by the Gottfried and Vera Weiss Science Foundation (WWW). The project funding is managed in trust by the Austrian Science Fund (FWF): P 28099-N34. The project duration was October 2015–September 2018. The authors want to thank the colleagues of the Tiroler Wasserkraft AG (TIWAG), of the Federal Research and

Training Centre for Forests (BFW) and of the Bavarian Avalanche Warning Service for data provision. In particular we are grateful for the close collaboration with Johannes Schöber (TIWAG) and Reinhard Fromm (BFW). We also want to thank Michael Lehning, Charles Fierz and the two reviewers for their helpful comments and fruitful discussion of the results.

Edited by: Thom Bogaard

References

Alcott, T. I. and Steenburgh, W. J. : Snow-to-Liquid Ratio Variability and Prediction at a High-Elevation Site in Utah's Wasatch Mountains, Weather Forecast., 25, 323–337, https://doi.org/10.1175/2009WAF2222311.1, 2010.

Anderson, E. A.: A point energy and mass balance model of a snow cover, NOAA Tech. Rep. NWS-19, 150 pp., 1976.

Buisán, S. T., Earle, M. E., Collado, J. L., Kochendorfer, J., Alastrué, J., Wolff, M., Smith, C. D., and López-Moreno, J. I.: Assessment of snowfall accumulation underestimation by tipping bucket gauges in the Spanish operational network, Atmos. Meas. Tech., 10, 1079–1091, https://doi.org/10.5194/amt-10-1079-2017, 2017.

Diamond, M. and Lowry, W. P.: Correlation of density of new snow with 700-millibar temperature, J. Meteorol., 11, 512–513, 1954.

Egli, L., Jonas, T., and Meister, R.: Comparison of different automatic methods for estimating snow water equivalent, Cold Reg. Sci. Technol., 57, 107–115, https://doi.org/10.1016/j.coldregions.2009.02.008, 2009.

Fierz, C., Armstrong, R. L., Durand, Y., Etchevers, P., Greene, E., McClung, D. M., Nishimura, K., Satyawali, P. K., and Sokratov S. A.: The International Classification for Seasonal Snow on the Ground, IHP-VII Technical Documents in Hydrology No. 83, IACS Contribution No. 1, UNESCO-IHP, Paris, 2009.

Gold, L. W. and Power, B. A.: Correlation of snow crystal type with estimated temperature of formation, J. Meteorol., 9, 447–447, https://doi.org/10.1175/1520-0469(1952)009<0448:COSCTW>2.0.CO;2, 1952.

Goodison, B. E., Wilson, B., We, K., and Metcalfe, J. R.: An inexpensive remote snow depth gauge: an assessment, The 52th Western Snow Conference, Sun Valley, Idaho, 17–19 April 1984.

Goodison, B. E., Louie, P. Y. T., and Yang, D.: WMO solid precipitation measurement intercomparison. Final Report, World Meteorological Organization, No. 872, 212 pp., 1998.

Gubler, H.: An Inexpensive Remote Snow-Depth Gauge based on Ultrasonic Wave Reflection from the Snow Surface, J. Glaciol., 27, 157–163, https://doi.org/10.3189/S002214300001131X, 1981.

Hedstrom, N. R. and Pomeroy, J. W.: Measurements and modeling of snow interception in the boreal forest, Hydrol. Process., 12, 1611–1625, https://doi.org/10.1002/(SICI)1099-1085(199808/09)12:10/11<1611::AID-HYP684>3.0.CO;2-4, 1998.

Heilig, A., Schneebeli, M., and Eisen, O.: Upward-looking ground-penetrating radar for monitoring snowpack stratigraphy, Cold Reg. Sci. Technol., 59, 152–162, https://doi.org/10.1016/j.coldregions.2009.07.008, 2009.

Helfricht, K., Koch, R., Hartl, L., and Olefs, M.: Potential and Challenges of an extensive operational use of high accuracy optical snow depth sensors to minimize solid precipitation undercatch, Proceedings of the 16th International Snow Science Workshop ISSW, Breckenridge, Colorado, 3–7 October 2016, 631–635, 2016.

Ishizaka, M., Motoyoshi, H., Yamaguchi, S., Nakai, S., Shiina, T., and Muramoto, K.-I.: Relationships between snowfall density and solid hydrometeors, based on measured size and fall speed, for snowpack modeling applications, The Cryosphere, 10, 2831–2845, https://doi.org/10.5194/tc-10-2831-2016, 2016.

Johnson, J. B. and Schaefer, G. L.: The influence of thermal, hydrologic, and snow deformation mechanisms on snow water equivalent pressure sensor accuracy, Hydrol. Process., 16, 3529–3542, https://doi.org/10.1002/hyp.1236, 2002.

Jordan, R. E., Andreasand, E. L., and Makshtas, A. P.: Heat budget of snow-covered sea ice at North Pole, J. Geophys. Res., 104, 7785–7806, https://doi.org/10.1029/1999JC900011, 1999.

Judson, A.: The weather and climate of a high mountain pass in the Colorado Rockies, Research Paper RM-16, USDA Forest Service, Fort Collins, CO, 28 pp., 1965.

Judson, A. and Doesken, N.: Density of Freshly Fallen Snow in the Central Rocky Mountains, B. Am. Meteorol. Soc., 81, 1577–1587, https://doi.org/10.1175/1520-0477(2000)081<1577:DOFFSI>2.3.CO;2, 2000.

Koch, F., Prasch, M., Schmid, L., Schweizer, J., and Mauser, W.: Measuring Snow Liquid Water Content with Low-Cost GPS Receivers, Sensors, 14, 20975–20999, https://doi.org/10.3390/s141120975, 2014.

Krajči, P., Kirnbauer, R., Parajka, J., Schöber, J., and Blöschl G.: The Kühtai data set: 25 years of lysimetric, snow pillow, and meteorological measurements, Water Resour. Res., 53, 5158–5165, https://doi.org/10.1002/2017WR020445, 2017.

LaChapelle, E. R.: The density distribution of new snow, USDA Forest Service, Alta Avalanche Study Center, Project F, Progress Rep. 2, Salt Lake City, UT, 13 pp., 1962.

Lehning, M., Bartelt, P., Brown, B., and Fierz, C.: A physical SNOWPACK model for the Swiss avalanche warning. Part III: meteorological forcing, thin layer formation and evaluation, Cold Reg. Sci. Technol., 35, 169–184, https://doi.org/10.1016/S0165-232X(02)00072-1, 2002.

Lundberg, A., Granlund, N., and Gustafsson, D.: Towards automated "Ground truth" snow measurements: a review of operational and new measurement methods for Sweden, Norway, and Finland, Hydrol. Process., 24, 1955–1970, https://doi.org/10.1002/hyp.7658, 2010.

Mair, E., Leitinger, G., Della Chiesa, S., Niedrist, G., Tappeiner, U., and Bertoldi, G.: A simple method to combine snow height and meteorological observations to estimate winter precipitation at sub-daily resolution, Hydrolog. Sci. J., 61, 2050–2060, https://doi.org/10.1080/02626667.2015.1081203, 2016.

Mair, M. and Baumgartner, D. J.: Operational experience with automatic snow depth sensors – ultrasonic and laser principle, TECO, WMO, Helsinki, Finland, WMO, 2010.

Marty, C. and Meister, R.: Long-term snow and weather observations at Weissfluhjoch and its relation to other high-altitude observatories in the Alps, Theor. Appl. Climatol., 110, 573–583, https://doi.org/10.1007/s00704-012-0584-3, 2012.

McCreight, J. L., Small, E. E., and Larson, K. M.: Snow depth, density, and SWE estimates derived from GPS reflection data: Validation in the western U. S., Water Resour. Res., 50, 6892–6909, https://doi.org/10.1002/2014WR015561, 2014.

McKay, G. A. and Gray, D. M.: The distribution of snow cover. In Handbook of Snow. Principles, Processes, Management and Use, edited by: Gray, D. M. and Male, D. H., Pergamon Press, Toronto, 1981.

Meister, R.: Density of New Snow and its Dependence of Air Temperature and Wind, Workshop on the Correction of Precipitation Measurements, 1–3 April 1985, Zurich, Volume: B, edited by: Sevruk, B., Correction of Precipitation Measurements, Zürcher Geographische Schriften, No. 23, 1985.

Morin, S., Lejeune, Y., Lesaffre, B., Panel, J.-M., Poncet, D., David, P., and Sudul, M.: An 18-yr long (1993–2011) snow and meteorological dataset from a mid-altitude mountain site (Col de Porte, France, 1325 m alt.) for driving and evaluating snowpack models, Earth Syst. Sci. Data, 4, 13–21, https://doi.org/10.5194/essd-4-13-2012, 2012.

Nakaya, U.: The Formation of Ice Crystals, in: Compendium of Meteorology, edited by: Malone, T. F., American Meteorological Society, Boston, MA, 207–220, https://doi.org/10.1007/978-1-940033-70-9_18, 1951.

Nitu, R., Aulamo, O., Baker, B., Earle, M., Goodison, B., Hoover, J., Hendrikx, J., Joe, P., Kochendorfer, J., Laine, T., Lanza, L., Landolt, S., Rasmussen, R., Roulet, Y. A., Smith, C., Samanter, A., Sabatini, F., Vuerich, E., Vuglinsky, V., Wolff, M., and Yang, D.: WMO Intercomparison of instruments and methods for the measurement of precipitiation and snow on the ground, in: 9th International Workshop on Precipitation in Urban Areas, 6. St. Moritz, Switzerland, 2012.

Olefs, M., Fischer, A., and Lang, J.: Boundary conditions for artificial snow production in the Austrian Alps, J. Appl. Meteorol. Clim., 49, 1096–113, https://doi.org/10.1175/2010JAMC2251.1, 2010.

Olefs, M., Schöner, W., Suklitsch, M., Wittmann, C., Niedermoser, B., Neururer, A., and Wurzer. A.: SNOWGRID – A New Operational Snow Cover Model in Austria, in: International Snow Science Workshop, Grenoble – Chamonix Mont-Blanc, 2013.

Pahaut, E.: La métamorphose des cristaux de neige (Snow crystal metamorphosis), 96, Monographies de la Météorologie Nationale, Météo France, 1976.

Pan, X., Yang, D., Li, Y., Barr, A., Helgason, W., Hayashi, M., Marsh, P., Pomeroy, J., and Janowicz, R. J.: Bias corrections of precipitation measurements across experimental sites in different ecoclimatic regions of western Canada, The Cryosphere, 10, 2347–2360, https://doi.org/10.5194/tc-10-2347-2016, 2016.

Parajka, J.: The Kühtai dataset: 25 years of lysimetric, snow pillow and meteorological measurements https://doi.org/10.5281/zenodo.556110, 2017.

Power, B. A., Summers, P. W., and D'Avignon, J.: Snow crystal forms and riming effect as related to snowfall density and general storm conditions, J. Atmos. Sci., 21, 300–305, https://doi.org/10.1175/1520-0469(1964)021<0300:SCFARE>2.0.CO;2, 1964.

Roebber, P. J., Bruening, S. L., Schultz, D. M., and Cortinas, J.V.: Improving Snowfall Forecasting by Diagnosing Snow Density, Weather Forecast., 18, 264–287, https://doi.org/10.1175/1520-0434(2003)018<0264:ISFBDS>2.0.CO;2, 2003.

Ryan, W. A., Doesken, N. J., and Fassnacht, S. R.: Evaluation of Ultrasonic Snow Depth Sensors for U.S. Snow Measurements, J. Atmos. Ocean. Tech., 25, 667–684, https://doi.org/10.1175/2007JTECHA947.1, 2008.

Sato, T., Kosugi, K., Mochizuki, S., and Nemoto, M.: Wind speed dependences of fracture and accumulation of snowflakes on snow surface, Cold Reg. Sci. Technol., 51, 229–239, https://doi.org/10.1016/j.coldregions.2007.05.004, 2008.

Schattan, P., Baroni, G., Oswald, S. E., Schöber, J., Fey, C., Kormann, C., Huttenlau, M., and Achleitner, S.: Continuous Monitoring of Snowpack Dynamics in Alpine Terrain by Above-Ground Neutron Sensing, Water Resour. Res., 53, 1–20, 2017.

Schmid, L., Koch, F., Heilig, A., Prasch, M., Eisen, O., Mauser, W., and Schweizer, J.: A novel sensor combination (upGPR-GPS) to continuously and nondestructively derive snow cover properties, Geophys. Res. Lett., 42, 3397–3405, https://doi.org/10.1002/2015GL063732, 2015.

Schmid, L., Heilig, A., Mitterer, C., Schweizer, J., Maurer, H., Okorn, R., and Eisen, O.: Continuous snowpack monitoring using upward-looking ground-penetrating radar technology, J. Glaciol., 60, 509–525, https://doi.org/10.3189/2014JoG13J084, 2014.

Schmidt, R. A. and Gluns, D. R.: Snowfall interception on branches of three conifer species, Can. J. Forest Res., 21, 1262–1269, https://doi.org/10.1139/x91-176, 1991.

Schmucki, E., Marty, C., Fierz, C., and Lehning, M.: Evaluation of modelled snow depth and snow water equivalent at three contrasting sites in Switzerland using SNOWPACK simulations driven by different meteorological data input, Cold Reg. Sci. Technol., 99, 27–37, https://doi.org/10.1016/j.coldregions.2013.12.004, 2014.

Schöber, J., Achleitner, S., Bellinger, J., Kirnbauer, R., and Schöberl, F.: Analysis and modelling of snow bulk density in the Tyrolean Alps., Hydrol. Res., 47, 419–441, https://doi.org/10.2166/nh.2015.132, 2016.

Serreze, M. C., Clark, M. P., Armstrong, R. L., McGinnis, D. A., and Pulwarty, R. S.: Characteristics of the western United States snowpack from snowpack telemetry (SNOTEL) data, Water Resour. Res., 35, 2145–2160, https://doi.org/10.1029/1999WR900090, 1999.

Simeral, D. B.: New snow density across an elevation gradient in the Park Range of northwestern Colorado, M.A. thesis, Department of Geography, Planning and Recreation, Northern Arizona University, 101 pp., 2005.

Smith, C. D., Kontu, A., Laffin, R., and Pomeroy, J. W.: An assessment of two automated snow water equivalent instruments during the WMO Solid Precipitation Intercomparison Experiment, The Cryosphere, 11, 101–116, https://doi.org/10.5194/tc-11-101-2017, 2017.

Sonntag, D.: Important new values of the physical constants of 1986, vapour pressure formulations based on the ITS-90, and psychrometer formulae, Z. Meteorol., 70, 340–344, 1990.

Steinkogler, W.: Systematic Assessment of New Snow Settlement in SNOWPACK, MA thesis, University of Innsbruck, Institute of Meteorology and Geophysics, 96 pp., 2009.

Teutsch, C.: Neuschneedichtenanalyse in den Ostalpen, MA thesis, Institute of Meteorology and Geophysics, Innsbruck, Austria, University of Innsbruck, 2009.

US Army Corps of Engineers: Snow Hydrology: Summary Report of the Snow Investigations, North Pacific Division, Portland, Oregon, 437 pp., 1956.

Valt, M., Chiambretti, I., and Dellavedova, P.: Fresh snow density on the Italian Alps, Geophysical Research Abstracts, 16, EGU2014-9715, 2014.

Vionnet, V., Brun, E., Morin, S., Boone, A., Faroux, S., Le Moigne, P., Martin, E., and Willemet, J.-M.: The detailed snowpack scheme Crocus and its implementation in SURFEX v7.2, Geosci. Model Dev., 5, 773–791, https://doi.org/10.5194/gmd-5-773-2012, 2012.

Wetzel, M., Meyers, M., Borys, R., McAnelly, R., Cotton, W., Rossi, A., Frisbie, P., Nadler, D., Lowenthal, D., Cohn, S., and Brown, W.: Mesoscale Snowfall Prediction and Verification in Mountainous Terrain, Weather Forecast., 19, 806–828, https://doi.org/10.1175/1520-0434(2004)019<0806:MSPAVI>2.0.CO;2, 2004.

Wright, P. J., Comey, B., McCollister, C., and Rheam, M.: Estimation of the new snow density using 42 seasons of meteorological data from Jackson Hole Mountain Resort, Wyoming, Proceedings, International Snow Science Workshop, Breckenridge, Colorado, 1180–1185, 2016.

WSL Institute for Snow and Avalanche Research SLF: WFJ_MOD: Meteorological and snowpack measurements from Weissfluhjoch, Davos, Switzerland, WSL Institute for Snow and Avalanche Research SLF, https://doi.org/10.16904/1, 2015.

13

Closing the water balance with cosmic-ray soil moisture measurements and assessing their relation to evapotranspiration in two semiarid watersheds

A. P. Schreiner-McGraw[1], E. R. Vivoni[1,2], G. Mascaro[3], and T. E. Franz[4]

[1]School of Earth and Space Exploration, Arizona State University, Tempe, AZ 85287, USA
[2]School of Sustainable Engineering and the Built Environment, Arizona State University, Tempe, AZ 85287, USA
[3]Julie Ann Wrigley Global Institute of Sustainability, Arizona State University, Tempe, AZ 85287, USA
[4]School of Natural Resources, University of Nebraska-Lincoln, Lincoln, NE 68583, USA

Correspondence to: E. R. Vivoni (vivoni@asu.edu)

Abstract. Soil moisture dynamics reflect the complex interactions of meteorological conditions with soil, vegetation and terrain properties. In this study, intermediate-scale soil moisture estimates from the cosmic-ray neutron sensing (CRNS) method are evaluated for two semiarid ecosystems in the southwestern United States: a mesquite savanna at the Santa Rita Experimental Range (SRER) and a mixed shrubland at the Jornada Experimental Range (JER). Evaluations of the CRNS method are performed for small watersheds instrumented with a distributed sensor network consisting of soil moisture sensor profiles, an eddy covariance tower, and runoff flumes used to close the water balance. We found a very good agreement between the CRNS method and the distributed sensor network (root mean square error (RMSE) of 0.009 and 0.013 $m^3 m^{-3}$ at SRER and JER, respectively) at the hourly timescale over the 19-month study period, primarily due to the inclusion of 5 cm observations of shallow soil moisture. Good agreement was also obtained in soil moisture changes estimated from the CRNS and watershed water balance methods (RMSE of 0.001 and 0.082 $m^3 m^{-3}$ at SRER and JER, respectively), with deviations due to bypassing of the CRNS measurement depth during large rainfall events. Once validated, the CRNS soil moisture estimates were used to investigate hydrological processes at the footprint scale at each site. Through the computation of the water balance, we showed that drier-than-average conditions at SRER promoted plant water uptake from deeper soil layers, while the wetter-than-average period at JER resulted in percolation towards deeper soils. The CRNS measurements were then used to quantify the link between evapotranspiration and soil moisture at a commensurate scale, finding similar predictive relations at both sites that are applicable to other semiarid ecosystems in the southwestern US.

1 Introduction

Soil moisture is a key land surface variable that governs important processes such as the rainfall–runoff transformation, the partitioning of latent and sensible heat fluxes and the spatial distribution of vegetation in semiarid regions (e.g., Entekhabi, 1995; Eltahir, 1998; Vivoni, 2012). Semiarid watersheds with heterogeneous vegetation in the southwestern United States (Gibbens and Beck, 1987; Browning et al., 2014) exhibit variations in soil moisture that challenge our ability to quantify land–atmosphere interactions and their role in hydrological processes (Dugas et al., 1996; Small and Kurc, 2003; Scott et al., 2006; Gutiérrez-Jurado et al., 2013; Pierini et al., 2014). Moreover, accurate measurements of soil moisture over scales relevant to land–atmosphere interactions in watersheds are difficult to obtain. Traditionally, soil moisture is measured continuously at single locations using techniques such as time domain reflectometry and then aggregated in space using a number of methods (Topp et al., 1980; Western et al., 2002; Vivoni et al., 2008b). Soil moisture is also estimated using satellite-based techniques, such as passive or active microwave sensors (e.g., Kustas et al.,

1998; Moran et al., 2000; Kerr et al., 2001; Bartalis et al., 2007; Narayan and Lakshmi, 2008; Entekhabi et al., 2010), but spatial resolutions are typically coarse and overpass times infrequent as compared to the spatiotemporal variability of soil moisture occurring within semiarid watersheds.

One approach to address the scale gap in soil moisture estimation is through the use of cosmic-ray neutron sensing (CRNS) measurements (Zreda et al., 2008, 2012) that provide soil moisture with a measurement footprint of several hectares (Desilets et al., 2010). Developments of the CRNS method have focused on understanding the processes affecting the measurement technique, for example, the effects of vegetation growth (Franz et al., 2013a; Coopersmith et al., 2014), atmospheric water vapor (Rosolem et al., 2013), soil wetting and drying (Franz et al., 2012a), and horizontal heterogeneity (Franz et al., 2013b). To date, the validation of the CRNS technique has been performed using single site measurements, spatial aggregations of different measurement locations, and particle transport models (Desilets et al., 2010; Franz et al., 2013b; Zhu et al., 2015). Distributed sensor networks measuring the water balance components of small watersheds and the spatial variability of soil moisture within a watershed offer the opportunity to test the accuracy of the CRNS method through multiple, independent approaches. For instance, the CRNS technique can be validated based upon the application of the watershed water balance, as performed for the eddy covariance (EC) technique, which is often used to measure surface turbulent fluxes (Scott, 2010; Templeton et al., 2014). Once validated, CRNS soil moisture estimates can be used to apply the water balance equation in a continuous fashion with the aim of quantifying hydrological fluxes during storm and interstorm periods, including the occurrence of percolation to deep soils or the transfer of water from the deeper vadose zone to the atmosphere.

An important advantage of the CRNS technique is that its measurement scale is comparable to the footprint of evapotranspiration (ET) measurements based on the EC technique, whose extent depends on wind speed and direction, atmospheric stability, and instrument and surface roughness heights (e.g., Hsieh et al., 2000; Kormann and Meixner, 2001; Falge et al., 2002). Furthermore, the relation between ET and soil moisture is an important parameterization in land surface models (e.g., Laio et al., 2001; Rodríguez-Iturbe and Porporato, 2004; Vivoni et al., 2008a) and, in most cases, has been investigated using EC measurements of ET and soil moisture observations at single sites. A number of studies, however, have shown that accounting for the spatial variability of land surface states is important to properly identify the linkage with EC measurements (e.g., Detto et al., 2006; Vivoni et al., 2010; Alfieri and Blanken, 2012). In other words, aggregated turbulent fluxes should be compared to spatially averaged surface states obtained at commensurate measurement scales. As a result, CRNS soil moisture estimates could be useful to improve the characterization of the relation between evapotranspiration flux and soil moisture. To our knowledge, soil moisture estimates from the CRNS technique have only been recently used to study the hydrological processes occurring in small watersheds that overlap with the CRNS measurement footprint or for improving the parameterization of land surface models (Shuttleworth et al., 2013; Rosolem et al., 2014).

In this contribution, we study the soil moisture dynamics of small semiarid watersheds in Arizona and New Mexico each instrumented with a cosmic-ray neutron sensor, eddy covariance tower, runoff flume, and a network of soil moisture sensor profiles. The watersheds represent the heterogeneous vegetation and soil conditions observed in the Sonoran and Chihuahuan deserts of the southwestern US (Templeton et al., 2014; Pierini et al., 2014). We first compare the CRNS method with the distributed sensor network and estimates from a novel method based on closing the water balance at each site. Given the simultaneous observations during the study period (March 2013 to September 2014, 19 months), we quantify the variations in hydrological processes (e.g., infiltration, evapotranspiration, percolation) that differentially occur at each site in response to varying precipitation. Combining these measurement techniques also affords the capacity to construct and compare relationships between the spatially averaged CRNS estimates and the spatially averaged ET obtained from the EC method. To our knowledge, this is the first study where CRNS measurements are validated via two independent methods at the small watershed scale and used to make new inferences about watershed hydrological processes.

2 Study areas and data sets

2.1 Study sites and their general characteristics

The two study sites are long-term experimental watersheds in semiarid ecosystems of the southwestern United States. Watershed monitoring began in 1975 at the Santa Rita Experimental Range (SRER), located 45 km south of Tucson, Arizona, in the Sonoran Desert (Fig. 1), as described by Polyakov et al. (2010) and Scott (2010). Precipitation at the site varies considerably during the year, with 54 % of the long-term mean amount (364 mm yr^{-1}) occurring during the summer months of July–September due to the North American monsoon (Vivoni et al., 2008a; Pierini et al., 2014). Soils at the SRER site are a coarse-textured sandy loam (Anderson, 2013) derived from Holocene-aged alluvium from the nearby Santa Rita Mountains. The savanna ecosystem at the site consists of the velvet mesquite tree (*Prosopis velutina* Woot.), interspersed with grasses (*Eragrostis lehmanniana*, *Bouteloua rothrockii*, *Muhlenbergia porteri*, and *Aristida glabrata*), and various cacti species (*Opuntia spinosior*, *Opuntia engelmannii*, and *Ferocactus wislizeni*). Similarly, watershed monitoring began in 1977 at the Jornada Experimental Range (JER), located 30 km north of Las Cruces,

Table 1. Watershed and precipitation characteristics at the SRER and JER sites. Precipitation values are long-term averages (1923–2014 at SRER and 1915–2006 at JER) for annual and seasonal quantities, defined as fall (October–December), winter (January–March), spring (April–June), and summer (July–September). Note that individual vegetation species have been generalized into three functional types.

Characteristic (unit)	Value	SRER	JER
Watershed area (m^2)		12 535	46 734
Elevation (m)	mean	1166.6	1458.3
	max	1171.1	1467.5
	min	1160.9	1450.5
Slope (degree)	mean	3.2	3.9
	max	19.2	45
	min	2.1	0
Drainage density ($1\,m^{-1}$)		0.04	0.03
Major vegetation type (%)	shrubs	32 %	27 %
	cacti	6 %	1 %
	grasses	37 %	6 %
	bare soil	25 %	66 %
	annual	364	251
	fall	72	54
Precipitation (mm)	winter	69	31
	spring	26	32
	summer	197	134

New Mexico, in the Chihuahuan Desert (Fig. 1), as described by Turnbull et al. (2013). Mean annual precipitation at the JER is considerably lower than SRER (251 mm yr^{-1}), with a similar proportion (53 %) occurring during the summer monsoon (Templeton et al., 2014). Soils at the JER site are primarily sandy loam with high gravel contents (Anderson, 2013) transported from the San Andres Mountains. The mixed shrubland ecosystem at the site consists of creosote bush (*Larrea tridentata*), honey mesquite (*Prosopis glandulosa* Torr.), several grass species (*Muhlenbergia porteri*, *Pleuraphis mutica*, and *Sporobolus cryptandrus*), and other shrubs (*Parthenium incanum*, *Flourensia cernua*, and *Gutierrezia sarothrae*). Figure 2 presents a vegetation classification at each site grouped into major categories: (1) SRER has velvet mesquite (labeled mesquite), grasses, cacti (*Opuntia engelmannii* or prickly pear), and bare soil, while (2) JER has honey mesquite (labeled mesquite), creosote bush, other shrubs, grasses, and bare soil. Table 1 presents the vegetation and terrain properties for the site watersheds obtained from 1 m digital elevation models (DEMs) and 1 m vegetation maps (Fig. 2). Pierini et al. (2014) and Templeton et al. (2014) described the image acquisition and processing methods employed to derive these products at SRER and JER, respectively.

Table 2. Energy balance closure at SRER and JER using 30 min net radiation (R_n), ground (G), latent (λE), and sensible (H) heat fluxes. The parameters m and b are the slope and intercept in the relation $\lambda E + H = m(R_n - G) + b$, while the ratio of the sum of $(\lambda E + H)$ to the sum of $(R_n - G)$ is a measure of how much available energy is accounted for in the turbulent fluxes.

Site	$\lambda E + H = m\,(R_n - G) + b$		$\frac{\sum \lambda E + H}{\sum R_n - G}$
	m	b	
SRER	0.72	17	0.85
JER	0.72	9.9	0.82

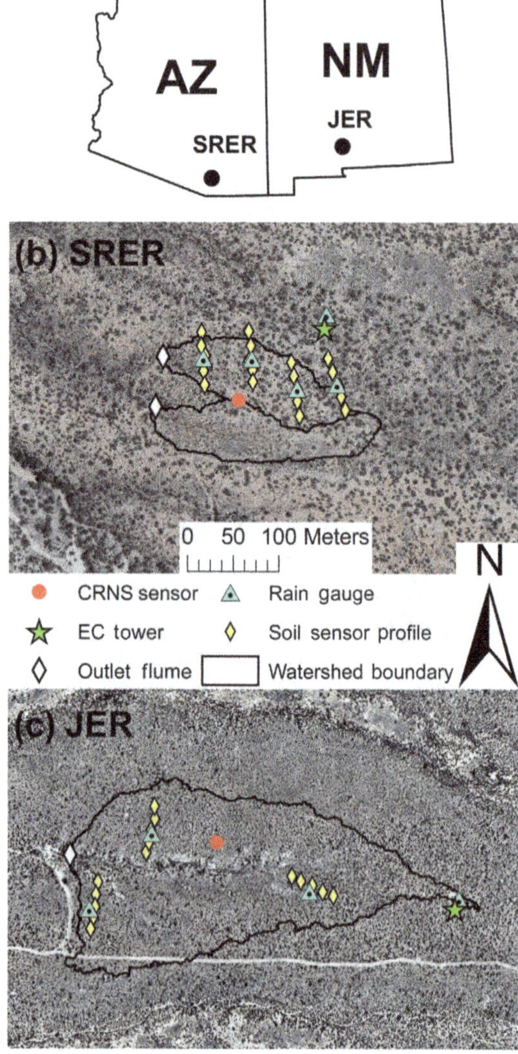

Figure 1. **(a)** Location of the study sites in Arizona and New Mexico. Watershed representations and sensor locations at **(b)** SRER and **(c)** JER, shown at the same scale.

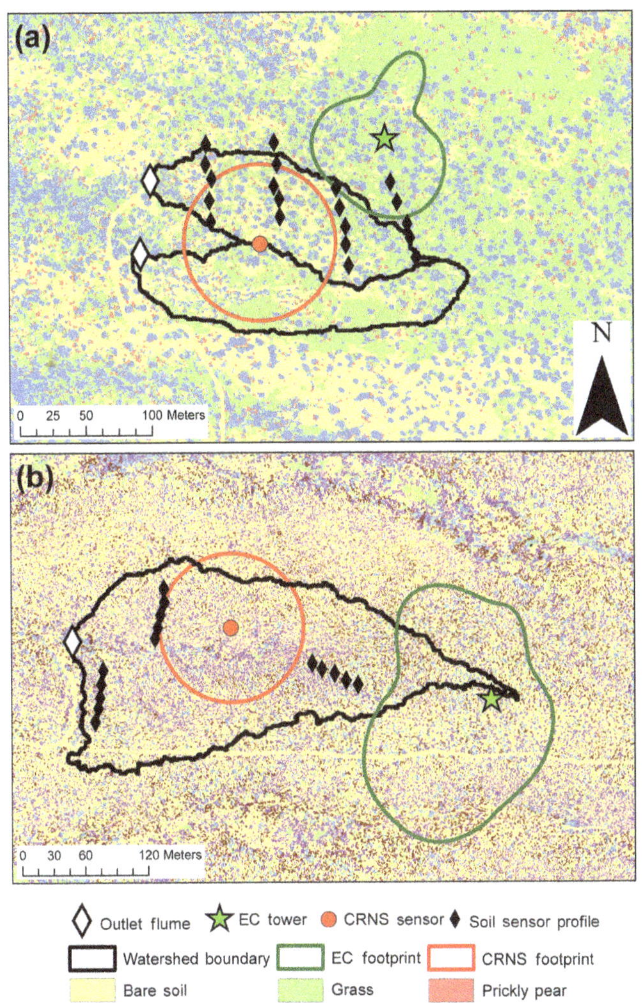

Figure 2. Vegetation classification for **(a)** SRER and **(b)** JER derived from aerial image analyses along with sensor locations and the 50 % contributing areas of the CRNS and EC footprints.

2.2 Distributed sensor networks at the small watershed scale

Long-term watershed monitoring at the SRER and JER sites consisted of rainfall and runoff observations at Watersheds 7 and 8 (SRER, 1.25 ha) and the Tromble Weir (JER, 4.67 ha). Pierini et al. (2014) and Templeton et al. (2014) describe recent monitoring efforts using a network of rainfall, runoff, soil moisture, and temperature observations, as well as radiation and energy balance measurements at EC towers, commencing in 2011 and 2010 at SRER and JER, respectively. This brief description of the distributed sensor networks is focused on the spatially averaged measurements used for comparisons to the CRNS method. Precipitation (P) was measured using up to four tipping-bucket rain gauges (TE525MM, Texas Electronics) to construct a 30 min resolution spatial average based on Thiessen polygons within the watershed boundaries. At the watershed outlets, streamflow (Q) was estimated at Santa Rita supercritical runoff flumes (Smith et al., 1981) using a pressure transducer (CS450, Campbell Scientific Inc.) and an in situ linear calibration to obtain 30 min resolution observations. ET was obtained at 30 min resolution using the EC technique that employs a three-dimensional sonic anemometer (CSAT3, Campbell Scientific Inc.) and an open-path infrared gas analyzer (LI-7500, LI-COR Inc.) installed at 7 m height on each tower. Flux corrections for the EC measurements followed Scott et al. (2004) and were verified using an energy balance closure approach reported in Table 2 for the study period. Energy balance closure at both sites is within the reported values across a range of other locations where the ratio of $\Sigma(\lambda E+H)\,/\,\Sigma(R_n-G)$ has an average value of 0.8 (Wilson et al., 2002; Scott, 2010). To summarize these observations, Fig. 3 shows the spatially averaged P, Q, and ET (mm h^{-1}), each aggregated to hourly resolution, at each study site during 1 March 2013 to 30 September 2014, along with seasonal precipitation amounts. While the results compare favorably to previous measurements (Turnbull et al., 2013; Pierini et al., 2014; Templeton et al., 2014), it should be noted that ET and Q data are assumed to represent the spatially averaged watershed conditions, despite the small mismatch between the watershed boundaries and EC footprints (Fig. 2) and the summation of Q in the two watersheds at SRER.

Distributed soil moisture measurements were obtained using soil dielectric probes (Hydra Probe, Stevens Water) organized as profiles (sensors placed at 5, 15 and 30 cm depths) in each study site. Profiles were originally installed at multiple locations along transects to investigate the different primary controls on soil moisture at each site: (1) at SRER we installed four transects of five profiles each located under different vegetation classes (mesquite, grass, prickly pear and bare soil), and (2) at JER we established three transects of five profiles each installed along different hillslopes (north-, south- and west-facing), as shown in Fig. 1. Individual sensors measure the impedance of an electric signal, as described in Campbell (1990), through a 40.3 cm^3 soil volume (5.7 cm in length and 3.0 cm in diameter; see Stevens Water Monitoring System, 1998) to determine the volumetric soil moisture (θ) in $m^3\,m^{-3}$ and soil temperature in °C as 30 min averaged values. A *loam* calibration equation was used in the conversion to θ (Seyfried et al., 2005) and corrected using relations established through gravimetric soil sampling at each study site (a power-law relation at SRER with $R^2=0.99$ and a linear relation at JER with $R^2=0.97$), following Pierini (2013). Given that sensors were originally installed to conduct watershed studies, spatial averaging was performed using site-specific weighting schemes accounting for the main controls on the soil moisture distribution. Thus, (1) at SRER we utilized the percentage area of each vegetation class (Table 1) and the associated sensor locations within each type (Pierini et al., 2014), and (2) at JER we accounted for the aspect and elevation at the sensor locations and used

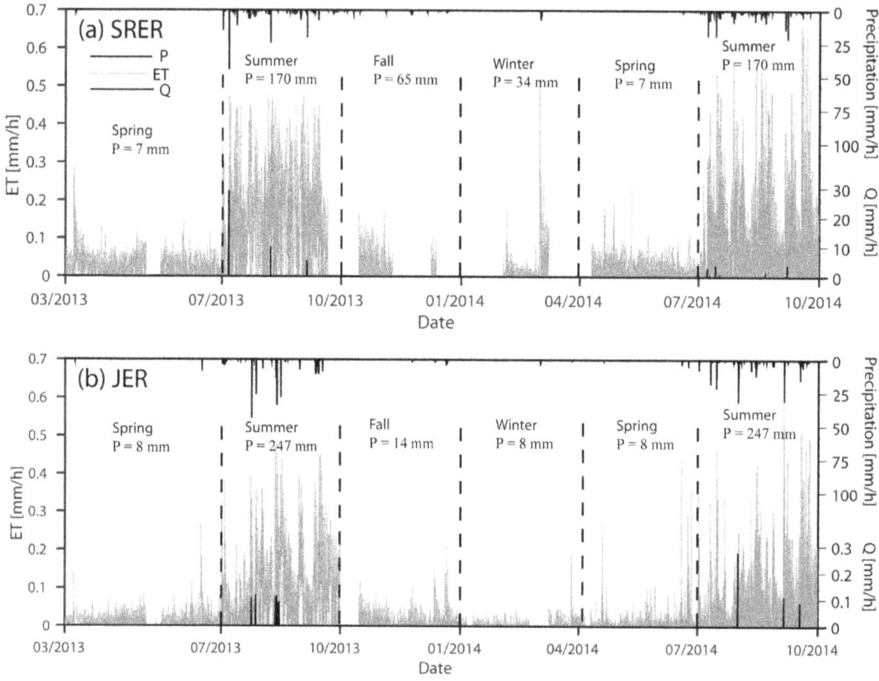

Figure 3. Hourly precipitation, streamflow, and evapotranspiration at the **(a)** SRER and **(b)** JER sites during the study period (March 2013 to September 2014). Gaps in ET data indicate periods of EC tower malfunction due to equipment failures, data collection problems, or vandalism. Vertical dashed lines indicate the seasonal definitions and their corresponding total precipitation.

these to extrapolate to other locations with similar characteristics based on the 1 m DEM (Templeton et al., 2014).

2.3 Cosmic-ray neutron sensing method for soil moisture estimation

The CRNS method relates soil moisture to the density of fast or moderated neutrons (Zreda et al., 2008) measured above the soil surface. A cosmic-ray neutron sensor (CRS-1000/B, Hydroinnova LLC) was installed in each watershed in January 2013 to record neutron counts at hourly intervals. We selected the study period (1 March 2013 to 30 September 2014) to coincide with the availability of data from the distributed sensor networks. While the theory of using neutrons for soil moisture measurements has a long history (e.g., Gardner and Kirkham, 1952), recent developments in the measurement of neutrons generated from cosmic rays has increased the horizontal scale, reduced the need for manual sampling, and led to a non-invasive approach. Zreda et al. (2008) and Desilets and Zreda (2013) described the horizontal scale as having a radius of ~ 300 m at sea level and a vertical aggregation scale ranging from 12 to 76 cm depending on soil wetness, while the work of Köhli et al. (2015) found a smaller horizontal scale with a radius of ~ 230 m at sea level. Since the travel speed of fast neutrons is $> 10\,\mathrm{km\,s^{-1}}$, neutron mixing occurs almost instantaneously in the air above the soil surface (Glasstone and Edlund, 1952), providing a well-mixed region that can be sampled with a single detector.

Using a particle transport model, Desilets et al. (2010) found a theoretical relationship between the neutron count rate at a detector and soil moisture for homogeneous SiO_2 sand:

$$\theta(N) = \frac{0.0808}{\left(\frac{N}{N_o}\right) - 0.372} - 0.115, \tag{1}$$

where θ ($\mathrm{m^3\,m^{-3}}$) is volumetric soil moisture (adjusted from gravimetric content to account for the soil bulk density), N is the neutron count rate (counts $\mathrm{h^{-1}}$) normalized to the atmospheric pressure and solar activity level, and N_o (counts $\mathrm{h^{-1}}$) is the count rate over a dry soil under the same reference conditions. The corrections applied to the neutron count rate are detailed in Desilets and Zreda (2003) and Zreda et al. (2012) and are applied automatically in the COSMOS website (http://cosmos.hwr.arizona.edu/). Additionally, since neutron counts are affected by all sources of hydrogen in the support volume, we apply a correction (C_{WV}) for atmospheric water vapor that was derived by Rosolem et al. (2013) as

$$C_{WV} = 1 + 0.0054\left(\rho_v^o - \rho_v^{ref}\right), \tag{2}$$

where ρ_v^o ($\mathrm{g\,m^{-3}}$) and ρ_v^{ref} ($\mathrm{g\,m^{-3}}$) are absolute water vapors at current and reference conditions. To estimate N_o, we performed a manual soil sampling at 18 locations within the CRNS footprint (sampled every 60° at radial distances of

25, 75, and 200 m from the detector) at six depths (0–5, 5–10, 10–15, 15–20, 20–25, 25–30 cm) for a total of 108 samples per site. Gravimetric soil moisture measurements were made following oven drying at 105 °C for 48 h (Dane and Topp, 2002) and converted to volumetric soil moisture using the soil bulk density (1.54 ± 0.18 g cm^{-3} at SRER and 1.3 ± 0.15 g cm^{-3} at JER). The spatially averaged volumetric soil moisture was related to the average neutron count obtained for the same time period (6 h average) resulting in $N_o = 3973$ at SRER and $N_o = 3944$ at JER, considered to be in line with the expected amounts given the elevations of both sites. Table 3 compares the gravimetric measurements and the CRNS soil moisture estimates during the calibration dates and provides further details on the soil properties at the two sites. We applied a 12 h boxcar filter to the measured count rates to remove the statistical noise associated with the measurement method (Zreda et al., 2012). On days where soil moisture changed by more than 0.06 m^3 m^{-3} due to rainfall, the boxcar filter was not applied. We note that additional terms to the calibration accounting for variations in lattice water, soil organic carbon, and vegetation have been proposed (Zreda et al., 2012; Bogena et al., 2013; McJannet et al., 2014; Coopersmith et al., 2014). However, given the relatively small amount of biomass ($\sim$2.5 kg m^{-2} at SRER; Huang et al., 2007; $\sim$0.5 kg m^{-2} at JER, Huenneke et al., 2001), low soil organic carbon (4.2 mg C g^{-1} soil at SRER; 2.7 mg C g^{-1} soil at JER; Throop et al., 2011), and low clay percent (5.2 % at SRER; 4.9 % at JER; Anderson, 2013), and thus low lattice water amounts (Greacen, 1981), we have neglected these terms in the analysis.

Figure 2 presents the horizontal aggregation scale of the CRNS method in comparison to the watershed boundaries and to the EC footprints obtained for summer 2013 (Anderson, 2013). Since both the CRNS and EC footprints have horizontally decaying contributions, we limited the size of the analysis region to the 50 % contribution or source area to enhance the overlap with the watershed boundaries and sensor networks. The footprints for both the CRNS method and the EC method vary considerably (Anderson, 2013; Köhli et al., 2015), with temporal changes occurring in the amount of overlap with the watersheds and between each other. Nevertheless, the vegetation distributions sampled in the CRNS, EC, and watershed areas (Fig. 2) are nearly the same (Vivoni et al., 2014), and the soils have low spatial variability (Anderson, 2013; Table 3), such that CRNS and EC measurements are considered representative of the watershed conditions. In addition to the changing horizontal scale, the CRNS method measures a time-varying vertical scale that depends on the soil water content. Franz et al. (2012b) used a particle transport model to determine that the CRNS measurement depth, z^*, varied with soil moisture as

$$z^*(\theta) = \frac{5.8}{\rho_b \tau + \theta + 0.0829}, \tag{3}$$

Table 3. Soil properties at SRER and JER. Soil moisture values correspond to conditions during the CRNS calibration dates (February 13, 2013 at SRER and February 10, 2013 at JER) for the gravimetric sampling at 18 locations with six depths (θ_G), CRNS (θ_{CRNS}), and the sensor network (θ_{SN}), each expressed as volumetric soil moisture using the soil bulk density (ρ_b) and soil porosity (φ) of the samples. Mean values of θ_G, ρ_b, and φ are shown along with the ± 1 standard deviations. Particle size distributions were obtained from soil auger sampling of the top 45 cm at 20 locations at each site (Anderson, 2013). Mean values of percent clay, silt, sand, and gravel are shown along with the ± 1 standard deviations.

Property (unit)	SRER	JER
Soil moisture calibration		
θ_G (m^3 m^{-3})	0.114 ± 0.023	0.056 ± 0.013
θ_{CRNS} (m^3 m^{-3})	0.114	0.056
θ_{SN} (m^3 m^{-3})	0.105	0.016
ρ_b (g cm^{-3})	1.54 ± 0.18	1.30 ± 0.15
φ (m^3 m^{-3})	0.42 ± 0.07	0.51 ± 0.06
Particle size distribution		
Clay (%)	5.2 ± 1.3 %	4.9 ± 1.1 %
Silt (%)	13.0 ± 2.2 %	28.5 ± 5.0 %
Sand (%)	72.5 ± 5.7 %	34.9 ± 8.3 %
Gravel (%)	9.3 ± 5.1 %	34.7 ± 11.5 %

where ρ_b is bulk density of the soil (Table 3) and τ is the weight fraction of lattice water in the mineral grains and bound water. Lattice water must be considered here since a local calibration of Eq. (3) is not possible. As a result, lattice water content was established at 0.02 g g^{-1} at each site given the weathered soils and the measurements from Franz et al. (2012b). To account for the temporal variation of z^*, the sensor profiles representing different soil layers (0–10, 10–20, and 20–40 cm in depth) were weighted based on z^* at each hourly time step according to

$$wt(z) = a\left(1 - \left(\frac{z}{z^*}\right)^{b}\right) \text{ for } 0 \le wt \le z^*, \tag{4}$$

where $wt(z)$ is the weight at depth z, a is a constant defined to integrate the profile to unity $\left(a = 1/\left(z^* - \left\{z^{*b+1} / [z^{*b}(b+1)]\right\}\right)\right)$, and b controls the shape of the weighting function. For simplicity, we assumed a value of $b = 1$ leading to a linear relationship (Franz et al., 2012b).

3 Methods

3.1 Comparison of CRNS to distributed network of soil moisture sensors

The CRNS method was first validated against the distributed network of soil moisture sensors. As done in previous studies, we compared hourly soil moisture observations obtained from the CRNS method (θ_{CRNS}) to estimates from the distributed sensor network (θ_{SN}) that have been averaged in space (i.e., based on vegetation type at SRER and elevation/aspect location at JER) and depth-weighted according to the time-varying CRNS measurement depth (z^*). We used several metrics to quantitatively assess the comparisons, including root mean square error (RMSE), correlation coefficient (CC), bias (B) and standard error of estimates (SEE). We performed an additional test of the CRNS technique by comparing relations between the mean soil moisture ($<\theta>$), obtained from either θ_{CRNS} or θ_{SN}, and the spatial standard deviation (σ) of soil moisture measured in the distributed sensor network. This relation has been studied previously with the goal of evaluating the role of heterogeneities related to vegetation, terrain position, and soil properties (Famiglietti et al., 1999; Lawrence and Hornberger, 2007; Fernández and Ceballos, 2003; Vivoni et al., 2008b; Mascaro et al., 2011; Qu et al., 2015). Based on Famiglietti et al. (2008), we fitted an empirical function to the observations at each site:

$$\sigma = k_1 \langle\theta\rangle e^{-k_2\langle\theta\rangle}, \tag{5}$$

where k_1 and k_2 are regression parameters, and compared these to prior studies in the region (e.g., Vivoni et al., 2008b; Mascaro and Vivoni, 2012; Stillman et al., 2014).

3.2 CRNS water balance analyses methods

In small watersheds of comparable size to the CRNS measurement footprint, the water balance can be expressed as

$$z^* \frac{\Delta\theta}{\Delta t} = P - \mathrm{ET} - Q - L, \tag{6}$$

where $\Delta\theta$ is the change in volumetric soil moisture over the time interval Δt, P is precipitation, ET is evapotranspiration, Q is streamflow, and L is leakage or deep percolation, with all of the terms expressed as spatially averaged quantities and valid over the effective soil measurement depth (z^*). The water balance was applied to validate the accuracy of the CRNS observations using measurements of the spatially averaged fluxes (P, ET, and Q) for a set of storm events. For each event, we computed the change in soil moisture measured by the CRNS, $\Delta\theta_{\mathrm{CRNS}}$, and the change calculated from the water balance, $\Delta\theta_{\mathrm{WB}}$. In both cases, changes were computed as the difference between the pre-storm soil moisture and the peak amount due to a rainfall event. For the application of Eq. (6), the soil measurement depth z^* was calculated as the average value over the duration of the soil moisture response to each individual storm. Note that, during a storm, ET is very low and the use of z^* in Eq. (6) instead of the plant rooting depth is justified. In addition, since this comparison is performed over a short time interval during the rising limb of the soil moisture response, we assumed no leakage (i.e., $L = 0$). To test the validity of this hypothesis, we analyzed the soil moisture records measured at the EC towers, where sensors were installed to measure the profile up to 1 m (i.e., a depth larger than z^*). We found that the percolation beyond a depth of ~40 cm is infrequent at both sites during summer monsoon storms, thus sustaining our assumption. However, percolation can occur on a timescale of several days during winter precipitation (e.g., Franz et al., 2012b; Templeton et al., 2014; Pierini et al., 2014). Although there are large amounts of bare soil in the watersheds, shrub and tree roots have been shown to extend laterally for 10 m or more (Heitschmidt et al., 1988), such that most of contributing area will be under the influence of both bare soil evaporation and plant transpiration.

Once validated against the distributed sensors and the application of the water balance, the CRNS estimates were subsequently used to determine the daily spatially averaged fluxes into and out from the measurement depth (z^*) as proposed by Franz et al. (2012b):

$$f_{\mathrm{CRNS}}(t) = \left(\theta_{\mathrm{CRNS},t} - \theta_{\mathrm{CRNS},t-1}\right) \min(z_t^*, z_{t-1}^*) / \Delta t. \tag{7}$$

In Eq. (7), f_{CRNS} is the daily flux (mm day^{-1}), Δt is the time step (1 day), and $\min(z_t^*, z_{t-1}^*)$ represents the minimum daily-averaged measurement depth between the 2 days being compared. Positive values of f_{CRNS} indicate an increase in soil moisture and, thus, represent net infiltration ($f_{\mathrm{CRNS}} = I$) into the measurement depth, usually occurring after a rainfall event. As a result, assuming negligible plant interception, daily P data can be used to estimate Q as $P - I$, which in turn can be compared to the runoff measurements in the watersheds. On the other hand, negative values of f_{CRNS} are equal to the net outflow ($f_{\mathrm{CRNS}} = O$), which can occur either as evapotranspiration or leakage. Using the EC method to obtain daily ET, $L = O - \mathrm{ET}$ can be determined as a measure of exchanges between the soil layers above and below z^*: L is positive when there is drainage to deeper soil layers and negative when deeper water is being drawn to support plant transpiration.

3.3 Relation between evapotranspiration and soil moisture at commensurate scale

Soil moisture at single locations is typically linked to ET in hydrologic models (e.g., Chen et al., 1996; Ivanov et al., 2004) and empirical studies (e.g., Small and Kurc, 2003; Vivoni et al., 2008a) using relations such as $\mathrm{ET} = f(\theta)$. For example, a commonly used approach is based on a piecewise linear relation between daily ET and θ (Rodríguez-Iturbe and

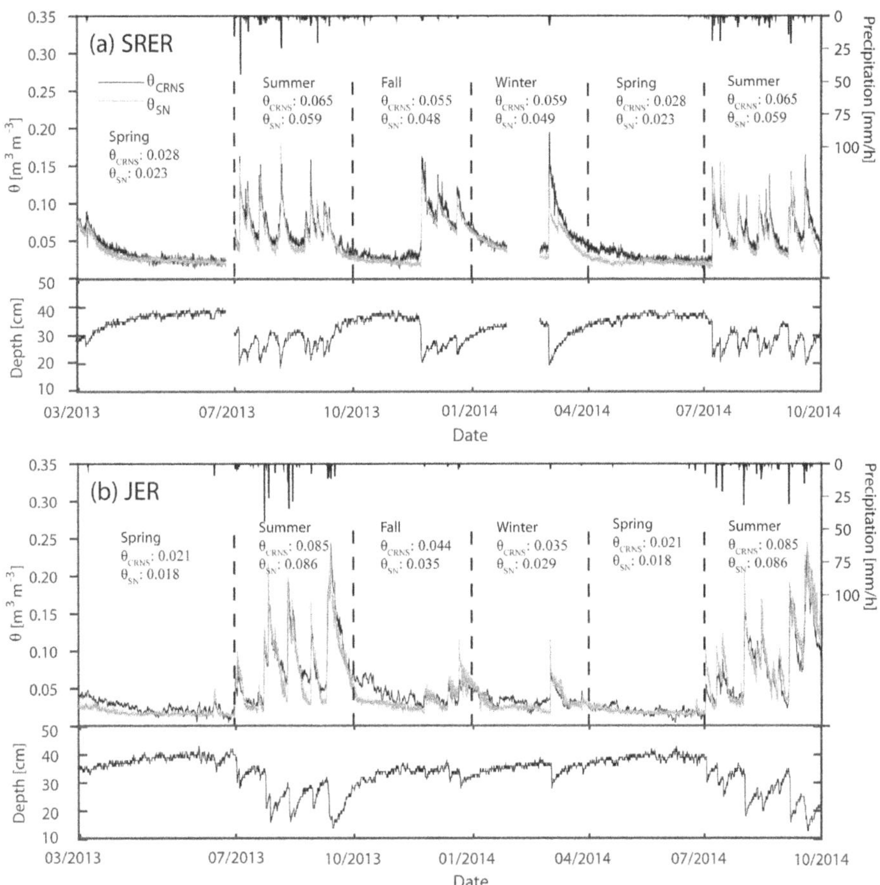

Figure 4. Comparison of the spatially averaged, hourly soil moisture ($m^3\,m^{-3}$) from CRNS method (θ_{CRNS}, black lines) and distributed sensor network (θ_{SN}, gray lines) at **(a)** SRER and **(b)** JER, along with spatially averaged, hourly precipitation during 1 March 2013 to 30 September 2014. Vertical dashed lines indicate the seasonal definitions and their corresponding seasonally averaged θ_{CRNS} and θ_{SN} in $m^3\,m^{-3}$. Also shown are the time-varying measurement depths (z^*).

Porporato, 2004):

$$\mathrm{ET}(\theta)=\begin{cases} 0 & 0<\theta\le\theta_h \\ E_w\dfrac{\theta-\theta_h}{\theta_w-\theta_h} & \theta_h<\theta\le\theta_w \\ E_w+(\mathrm{ET}_{max}-E_w)\dfrac{\theta-\theta_h}{\theta^*-\theta_h} & \theta_w<\theta\le\theta^* \\ \mathrm{ET}_{max} & \theta^*<\theta\le\varphi \end{cases}, \tag{8}$$

where E_w is soil evaporation, ET_{max} is maximum evapotranspiration, θ_h, θ_w, and θ^* are the hygroscopic, wilting, and plant stress soil moisture thresholds, and φ is the soil porosity. Vivoni et al. (2008a) applied Eq. (8) to observations of ET from the EC method and θ at single locations to derive the relation parameters using a nonlinear optimization algorithm (Gill et al., 1981). We evaluate this approach using the spatially averaged soil moisture estimates (θ_{CRNS}, and θ_{SN}) whose spatial scale is more commensurate with the ET measurements than single measurement sites.

4 Results and discussion

4.1 Comparison of CRNS method to distributed sensor network

Figure 4 presents a comparison of the spatially averaged, hourly soil moisture obtained from the CRNS method (θ_{CRNS}) and the distributed sensor network (θ_{SN}), as well as the time-varying measurement depth (z^*) of CRNS. Relative to the long-term summer precipitation (Table 1), the study period had below average (188 and 153 mm in 2013 and 2014) and significantly above average (246 and 247 mm) rainfall at SRER and JER, respectively. The fall–winter period in the record had below average precipitation (99 mm) at SRER and significantly below average amounts (21 mm) at JER. Overall, the spring periods were dry, consistent with the long-term averages. In response, the temporal variability of soil moisture clearly shows the seasonal conditions at the two sites, with relatively wetter conditions during the summer monsoons. Seasonally averaged θ_{CRNS} compares favorably with seasonally averaged θ_{SN} (Fig. 4), with both estimates

Table 4. Statistical comparisons of CRNS method with distributed sensor network and water balance estimates based on the standard error of estimates (SEE), root mean square error (RMSE), bias (B), and correlation coefficient (CC), described in Vivoni et al. (2008b). Values in parentheses for JER indicate metrics when large rainfall events are excluded.

Metric (unit)	SRER	JER
θ_{CRNS} versus θ_{SN}		
RMSE ($m^3\ m^{-3}$)	0.009	0.013
CC	0.949	0.946
B	1.117	1.019
SEE ($m^3\ m^{-3}$)	0.012	0.013
$\Delta\theta_{CRNS}$ versus $\Delta\theta_{WB}$		
RMSE ($m^3\ m^{-3}$)	0.001	0.082 (0.019)
CC	0.949	0.940 (0.945)
B	0.936	0.543 (0.903)
SEE ($m^3\ m^{-3}$)	0.024	0.095 (0.020)

showing relatively large differences between wetter summer conditions (0.065 and 0.085 $m^3\ m^{-3}$ at SRER and JER) and drier spring values (0.028 and 0.021 $m^3\ m^{-3}$ at SRER and JER, respectively). As shown in prior studies (e.g., Zreda et al., 2008; Franz et al., 2012b), the CRNS method tracks the sensor observations very well. Nevertheless, there is an indication that θ_{CRNS} has a tendency to dry less quickly during some rainfall events (i.e., overestimate soil moisture during recession limbs). This might be due to landscape features such as nearby channels (Fig. 1) and their associated zones of soil water convergence that remain wetter than areas measured by the distributed sensor network. Overall, however, there is an excellent match between θ_{CRNS} and θ_{SN} in terms of capturing the occurrence and magnitude of soil moisture peaks across the different seasons, thus reducing some issues noted by Franz et al. (2012b) with respect to a purported oversensitivity of θ_{CRNS} for small rainfall events (<5 mm). We attribute this improvement to the use of a 5 cm sensor in each profile that tracks important soil moisture dynamics occurring in the shallow surface layer within semiarid ecosystems.

To complement this, Fig. 5 compares θ_{CRNS} and θ_{SN} as a scatter plot along with the sample size (N) and the SEE, which quantify the deviations from the 1 : 1 line. Table 4 provides the full set of statistical metrics for the comparison of θ_{CRNS} versus θ_{SN} at the two study sites. The correspondence between both methods is very good, with low RMSE and SEE, a high CC, and a bias close to 1. These values are comparable to previous validation efforts where the RMSE was found to be 0.011 $m^3\ m^{-3}$ (Franz et al., 2012b) and less than 0.03 $m^3\ m^{-3}$ (Bogena et al., 2013; Coopersmith et al., 2014;

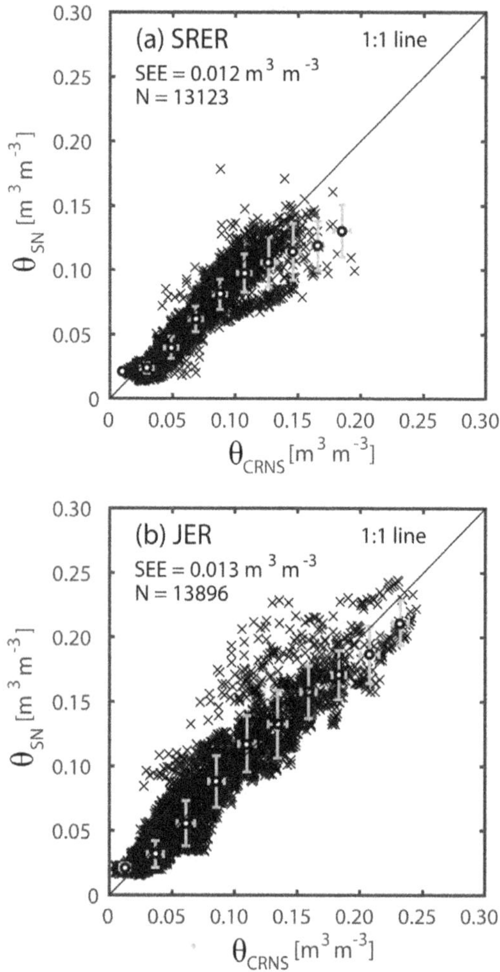

Figure 5. Scatter plots of the spatially averaged, hourly soil moisture ($m^3\ m^{-3}$) from CRNS method (θ_{CRNS}) and distributed sensor network (θ_{SN}) at **(a)** SRER and **(b)** JER. The SEE and the number of hourly samples (N) are shown for each site. Bin averages and ± 1 standard deviation are shown (circles and error bars) for bin widths of 0.025 $m^3\ m^{-3}$.

Zhu et al., 2015). The comparison of the semiarid sites is also illustrative of the ability of the CRNS method to estimate soil moisture over a range of conditions. Despite the more arid climate at JER (Table 1), the study period consisted of higher precipitation (247 mm) and higher soil moisture values during the summer (0.085 $m^3\ m^{-3}$), as compared to SRER (170 mm, 0.065 $m^3\ m^{-3}$), indicating a more active monsoon in the Chihuahuan Desert. In contrast, the fall–winter period is generally drier at JER (21 mm, 0.039 $m^3\ m^{-3}$), as compared to SRER (99 mm, 0.057 $m^3\ m^{-3}$), where high P and low ET in the winter promoted infiltration below the CRNS measurement depth, as observed at a 1 m sensor profile at SRER (not shown). These two effects lead to a larger range of soil moisture at JER as compared to SRER in Fig. 5. As a result, the CRNS method is found to be a reliable method

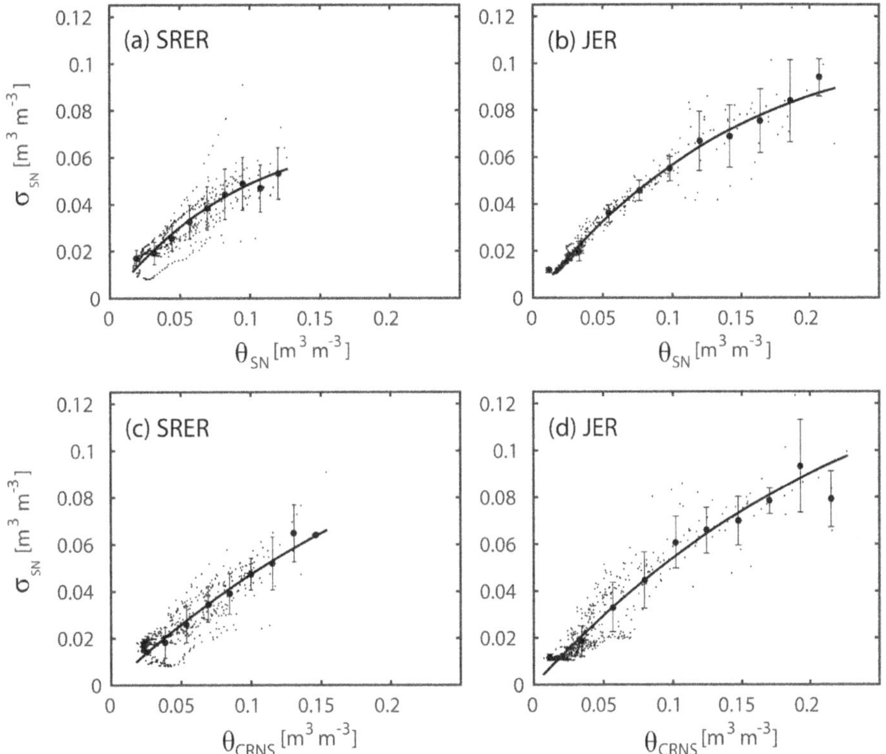

Figure 6. Soil moisture spatial variability as a function of the spatially averaged distributed sensor network (θ_{SN}, top) and the CRNS method (θ_{CRNS}, bottom) for **(a, c)** SRER and **(b, d)** JER. Bin averages and $\pm$ 1 standard deviation are shown (circles and error bars) for bin widths of 0.015 m^3 m^{-3} at SRER and 0.025 m^3 m^{-3} at JER. Regressions for the relations of σ with $<\theta>$ are valid for the entire data set.

for measuring soil moisture in the observed range of values at SRER and JER during the study period.

To further test the CRNS method against the distributed sensor network, Fig. 6 depicts the relations between the spatial variability of soil moisture (σ) and the spatially averaged conditions ($<\theta>$). For illustration purposes, bin averages and standard deviations are also presented for each relation. Least-squares regressions of Eq. (5) based on hourly observations were applied to estimate k_1 and k_2 for the relations σ vs. θ_{SN} ($k_1 = 0.75$ and $k_2 = 4.23$ at SRER; $k_1 = 0.74$ and $k_2 = 2.75$ at JER) and these parameters were adopted to interpret the relations of σ vs. θ_{CRNS}. The RMSE are very low and similar in both cases (RMSE $= 0.007$ and 0.008 m^3 m^{-3} at SRER and 0.005 and 0.008 m^3 m^{-3} at JER for the relation with θ_{SN} and θ_{CRNS}, respectively), thus confirming the good correspondence between the two methods. As shown in prior efforts in semiarid ecosystems using sensor networks or aircraft observations (e.g., Fernández and Ceballos, 2003; Vivoni et al., 2008b; Mascaro et al., 2011; Stillman et al., 2014), there is a general increase in σ with $<\theta>$, explained by the role played by local heterogeneities (e.g., vegetation types, surface soil variations, topography) as well as the bounded nature of the soil moisture process at the driest state. The similar relations derived in these different sites might be broadly applicable to other semiarid ecosystems in the southwestern US.

4.2 Validation of CRNS method with water balance estimates

Figure 7 presents the comparison of the spatially averaged $\Delta\theta_{CRNS}$ and $\Delta\theta_{WB}$ as a scatter plot for approximately 40 rainfall events with a total depth larger than 10 mm and durations ranging from 0.5 to 31 h (mean of 6 h). The statistical metrics are presented in Table 4. The correspondence between the methods is very good, with low RMSE and SEE, a high CC, and a bias close to 1, with a closer match at SRER. For example, the SEE at SRER (0.024 m^3 m^{-3}) is significantly less than the value at JER (0.095 m^3 m^{-3}) and close to the SEE of the comparison of θ_{CRNS} and θ_{SN}. This suggests that the three approaches (i.e., CRNS, sensor network, water balance) are in agreement at the SRER. For the JER, the lower correspondence between $\Delta\theta_{CRNS}$ and $\Delta\theta_{WB}$ is attributed to five large events where $\Delta\theta_{WB}$ is above 0.2 m^3 m^{-3}. Removing these events lowers the SEE at JER to 0.020 m^3 m^{-3}, in line with SRER and the comparison of θ_{CRNS} and θ_{SN} at JER. A closer inspection of the soil moisture response at JER allows for investigating the physical reasons causing the different behavior of these five events. Figure 8 shows the soil moisture change ($\Delta\theta_{SN}$) at different sensor depths averaged for the selected large events and for the remaining events, as well as the mean of CRNS measure-

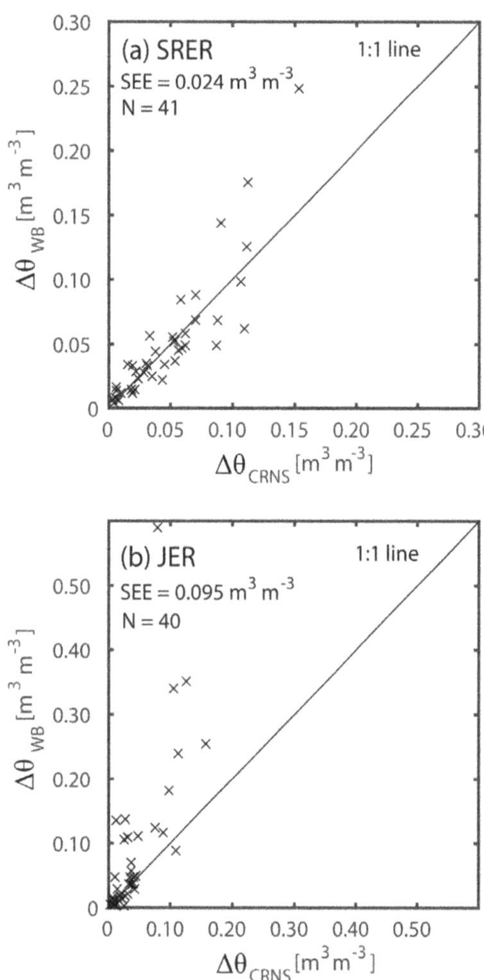

Figure 7. Scatter plots of the spatially averaged change in soil moisture ($m^3\,m^{-3}$) derived from CRNS method ($\Delta\theta_{CRNS}$) and the application of the water balance ($\Delta\theta_{WB}$) at **(a)** SRER and **(b)** JER. The SEE and the number of event samples (N) are shown for each site.

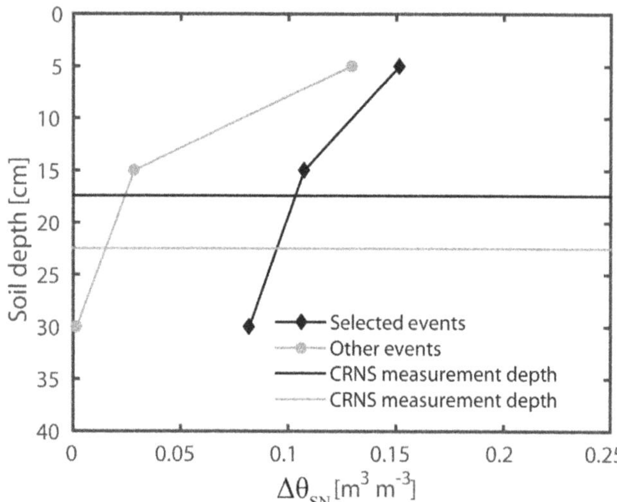

Figure 8. Change in soil moisture ($\Delta\theta_{SN}$) at depths of 5, 15, and 30 cm at the JER for the five large events (selected events) and the remaining cases (other events). Horizontal lines are the time-averaged CRNS measurement depths averaged over selected events (black; standard deviation of 3.8 cm) and other events (gray; standard deviation of 6.5 cm).

ment depths (z^*) for each case. The five large events exhibit high soil moisture changes at 30 cm depth (i.e., 0.08 $m^3\,m^{-3}$) below z^* (i.e., 17 cm), while other events have soil moisture changes near zero at 30 cm and are captured well within z^*. This indicates that infiltration fronts during the larger events penetrated beyond z^* and were not entirely captured by the CRNS method, leading to an underestimate of $\Delta\theta_{WB}$. For these events, the assumption $L = 0$ in Eq. (6) is not fully supported. In contrast, the better correspondence at SRER suggests that infiltration fronts were contained within z^*. This is plausible given the less rocky soil and flatter terrain at SRER as compared to JER (Anderson, 2013). At JER, soil water movement to deeper layers can be promoted by higher gravel contents and the presence of calcium carbonate and undulated terrain, which facilitate lateral water transfer to sandy channel beds (Templeton et al., 2014).

4.3 Utility of CRNS for investigating hydrological processes

Given the confidence gained with respect to the CRNS estimates, we utilized these observations to quantify the water balance fluxes during storm and interstorm periods at the two sites. Figure 9 shows the cumulative f_{CRNS} and the cumulative, spatially averaged P and ET measured by the distributed sensor network. An overall drying trend is present at SRER during the study period (i.e., cumulative f_{CRNS} becomes more negative), while JER exhibits a relatively small change in cumulative f_{CRNS}, both in response to the below average (SRER) and above average (JER) precipitation. An important contrast at the sites is the overall water balance (Table 5), where higher P, lower ET, and lower Q at JER (measured ET / $P = 0.54$, $Q / P = 0.01$) implies that more soil water is available for leakage to deeper soil layers. This is reflected in a large positive difference between cumulative outflow (O = ET $+L$) and ET at JER (i.e., $L > 0$ from z^*, soil water movement to lower layers, as depicted in the soil water balance diagram). In contrast, SRER exhibits a higher ET / $P = 0.96$ and $Q / P = 0.14$, such that negative differences occur between O and ET (i.e., $L < 0$ into z^*, movement from lower layers, as depicted in the soil water balance diagram). This is particularly important during the summers when vegetation is active and produces more ET than the outflow from the CRNS measurement depth, indicating that soil water is obtained from deeper soil layers that are readily accessed by velvet mesquite roots (e.g., Snyder and Williams, 2003; Scott et al., 2008; Potts et al., 2010). This is consistent

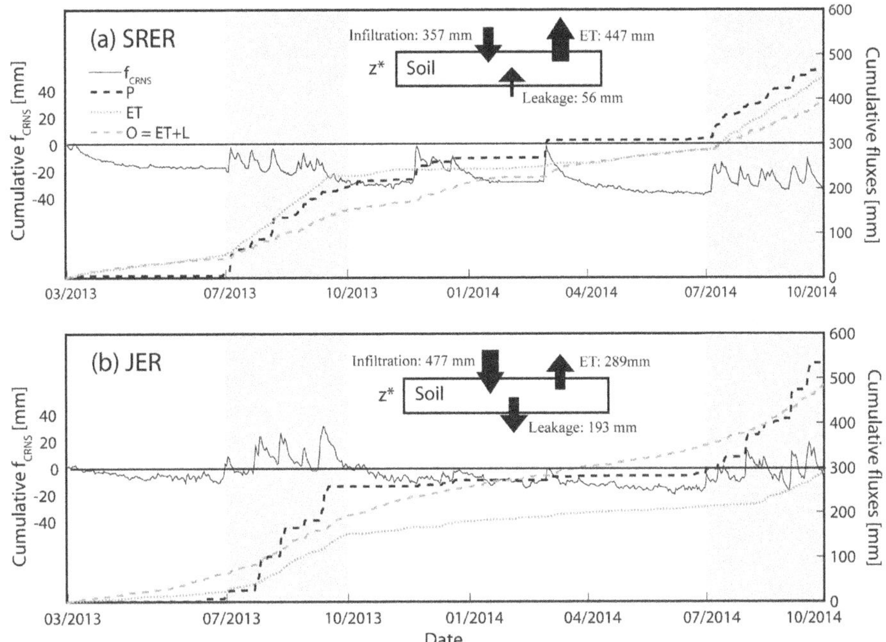

Figure 9. Comparison of cumulative f_{CRNS} and measured water balance fluxes (P and ET) during study period. CRNS estimates of infiltration (I), outflow (O), and leakage (L) are either depicted as cumulative fluxes ($O = \mathrm{ET} + L$) or as total amounts during the study period (I and L) as arrows in the soil water balance box of depth z^*. Shaded regions indicate the summer seasons (July–September). The horizontal line represents $f_{\mathrm{CRNS}} = 0$.

with the sustained ET during interstorm periods in the summer season at SRER despite the low θ_{CRNS}, while JER exhibits sharp declines in ET when θ_{CRNS} is reduced between storms.

Overall, the soil water balance from the CRNS method shows stark ecosystem differences at the two sites during the study period. The mesquite savanna at SRER extracted substantial amounts of water from deeper soil layers during the summer season such that losses to runoff and the atmosphere are in excess of seasonal precipitation. Deeper soil water is recharged beyond the CRNS measurement depth during winter periods, as observed by Scott et al. (2000), and subsequently accessed by deep-rooted trees during the summer (Scott et al., 2008). In contrast, the mixed shrubland at JER lost a substantial amount of precipitation to deeper soil layers throughout the year, due to the low values of runoff and evapotranspiration, and the soil, terrain, and channel conditions promoting recharge (Templeton et al., 2014). Winter recharge is fostered by the lack of ET from drought-deciduous plants that lose their leaves in the wintertime. We hypothesize that deep percolation is likely occurring in the channels, since (i) soil moisture observations in the hillslopes (i.e., far from the channel) show a lack of deep percolation; (ii) the runoff ratio decreases with the basin contributing area, indicating transmission losses along the channel (Templeton et al., 2014); and (iii) one sensor profile installed in a channel at SRER shows that the wetting front frequently reaches at least 30 cm depth. Furthermore, the f_{CRNS} ap-

Table 5. Total water flux estimates from daily CRNS soil water balance method (f_{CRNS}) and daily sensor measurements during study period at the SRER and JER sites. P is from rain gauge measurements in both cases. L in CRNS is computed as O – ET where ET is from EC method, while L in sensor estimates is calculated from solving the water balance.

Water flux	SRER	JER
CRNS estimates		
Precipitation (P, mm)	464	533
Infiltration (I, mm)	357	477
Outflow (O, mm)	391	482
Leakage (L, mm)	−56	193
Outflow ratio (O / P)	0.84	0.90
Runoff ratio (Q / P)	0.23	0.11
Sensor measurements		
Precipitation (P, mm)	464	533
Storage change ($\Delta\theta$, mm)	−13	26
Outflow (O, mm)	437	506
Leakage (L, mm)	−10	217
Evapotranspiration (ET, mm)	447	289
Evaporation ratio (ET / P)	0.96	0.54
Streamflow (Q, mm)	64	5
Runoff ratio (Q / P)	0.14	0.01

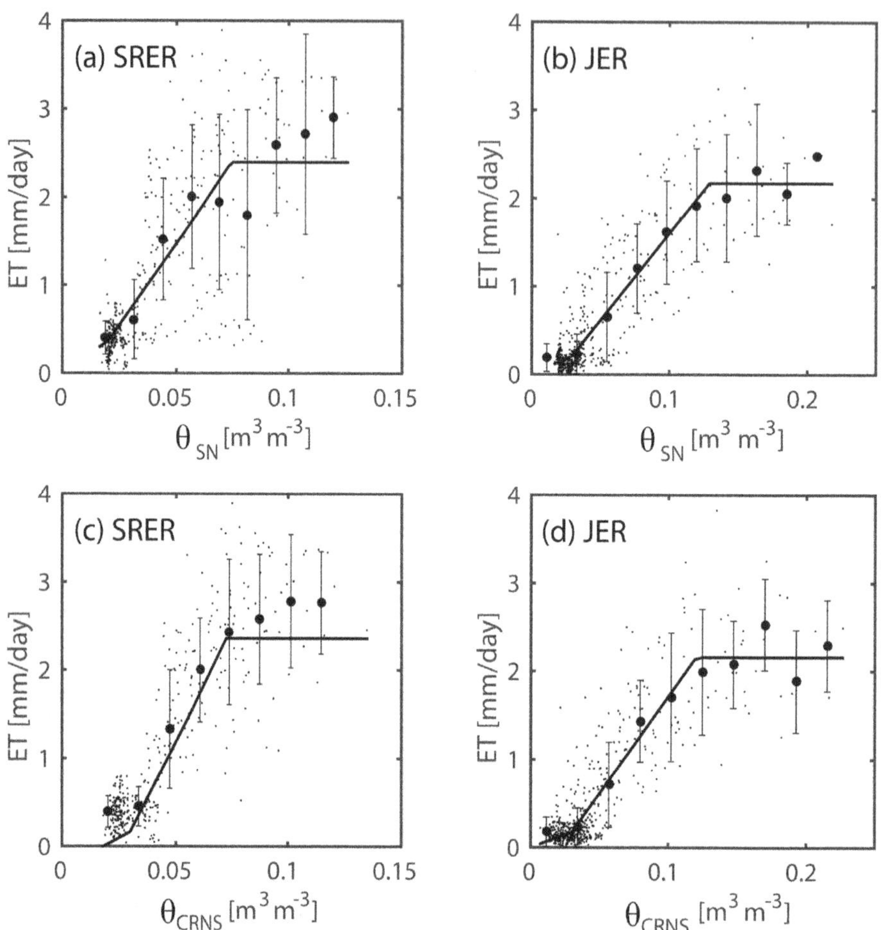

Figure 10. Evapotranspiration relation with the spatially averaged distributed sensor network (θ_{SN}, top) and the CRNS method (θ_{CRNS}, bottom) for **(a, c)** SRER and **(b, d)** JER. Bin averages and ± 1 standard deviation are shown (circles and error bars) for bin widths of 0.015 m^3 m^{-3} at SRER and 0.025 m^3 m^{-3} at JER. Regressions for the relations of ET with $<\theta>$ are valid for the entire data set.

Table 6. Regression parameters for the relations of evapotranspiration and soil moisture (θ_{SN} and θ_{CRNS}) at the SRER and JER sites along with the RMSE of the regressions. $\theta_h = 0$ in all cases.

Site	Relation	ET_{max} (mm day^{-1})	E_w (mm day^{-1})	θ_w (m^3 m^{-3})	θ* (m^3 m^{-3})	RMSE (mm day^{-1})
SRER	ET–θ_{SN}	2.61	0.41	0.03	0.07	1.15
	ET–θ_{CRNS}	2.40	0.36	0.02	0.08	0.55
JER	ET–θ_{SN}	2.16	0.18	0.03	0.12	0.34
	ET–θ_{CRNS}	2.17	0.21	0.03	0.13	0.34

proach provided estimates that can be compared to the watershed water balance since these are at a similar spatial scale (Table 5). Estimates of outflow (O) from the measurement depth and leakage (L) are higher when calculated with θ_{SN}, consistent with more rapid drying as compared to the CRNS method. On the other hand, the CRNS method results in higher values of the runoff ratio (Q / P) than observed in the distributed sensor network, in particular for JER. This is likely due to the daily scale of the CRNS analysis, which limits the suitability of the runoff estimate for semiarid watersheds characterized by runoff responses lasting minutes to hours.

4.4 Utility of CRNS for improving ET estimates

Figure 10 compares the relationships between the measured daily ET using the EC method and the spatially averaged soil moisture values (θ_{SN} and θ_{CRNS}) at the SRER and JER sites along with the piecewise linear regressions estimated using Eq. (8) and a nonlinear optimization approach. Following Vivoni et al. (2008a), regression parameters related to soil and vegetation conditions are presented in Table 6. For illustration purposes, bin averages and standard deviations are also shown. Clearly, the piecewise linear relation is a suitable approach for capturing the ET–θ observations, yielding a relatively low RMSE at the two sites. A lower RMSE for the relation using θ_{CRNS} as compared to θ_{SN} at SRER is attributed to its ability to detect a wider range of dry conditions and the improved match in the spatial scales of ET and θ_{CRNS}, in an analogous fashion to the comparison between a single sensor and the distributed sensor network (Templeton et al., 2014). In addition, the CRNS method represents soil evaporation (E_w) in a more realistic way as it discriminates differences in drier states, illustrated by the realistic gradual increase of bare soil evaporation with increasing soil water (Fig. 10). For ET and θ_{SN}, the dry portions of the relations have too steep of a slope and do not represent well how bare soil evaporation changes with soil moisture. When comparing both sites through the ET–θ relation, the SRER has a larger E_w and ET_{max} and lower θ^*, as compared to JER, tested to be significantly different at the 95 % confidence level using a bootstrap approach. Together, these parameters indicate that SRER has a higher overall ET, consistent with higher extractions from the CRNS measurement depth due to the mesquite trees, extensive grass cover, and higher soil evaporation.

5 Summary and conclusions

In this study, we utilized distributed sensor networks to examine the CRNS soil moisture method at the small watershed scale in two semiarid ecosystems of the southwestern US. To our knowledge, this is the first study to compare CRNS measurements to two complementary approaches for obtaining spatially averaged soil moisture at a commensurate scale: (1) a distributed set of sensor profiles weighted in the horizontal and vertical scales within each watershed, and (2) a watershed-averaged quantity obtained from closing the water balance. We highlighted a few novel advantages of the CRNS method revealed through the comparisons, including the ability to resolve the shallow soil moisture dynamics and to match the estimates obtained from closing the water balance for most rainfall events. In the distributed sensor comparisons, we found that the CRNS method overestimated soil moisture during the recession limbs of rainfall events, possibly due to landscape features such as nearby channels remaining wet. In the water balance comparisons, we identified that our assumption of no leakage beneath z^* was not met during large rainfall events and the CRNS method was not able to capture all of the soil water present. We attribute this to rapid bypassing of the measurement depth due to soil and terrain characteristics. Due to this observed bypass flow, we suggest that future studies using the CRNS method include a few soil moisture sensor profiles below z^* to detect leakage events.

The CRNS soil moisture estimates were used in combination with the various measurement methods to explore the relative magnitudes of the water balance components at each site given the different precipitation amounts during the study period. The drier than average conditions in the mesquite savanna ecosystem at SRER lead to drier surface soils incapable of supporting the measured evapotranspiration unless supplemented by plant water uptake from deeper soil layers. In contrast, wetter than average summer periods in the mixed shrubland at JER had wet surface soils that promoted leakage into the deeper vadose zone, which was subsequently unavailable for runoff and evapotranspiration losses. Comparisons across different seasons also suggested that carryover of soil water from winter leakage toward deeper soil layers is consumed during the summer season by active plants. These novel inferences within the two ecosystems relied heavily on the application of the CRNS method and its limited measurement depth to discriminate between shallow and deeper vadose zone processes as well as on the direct measurement of the water balance components, in particular evapotranspiration. It is important to keep in mind, however, that the ability to resolve watershed-scale hydrological processes, such as the interaction between shallow and deep soil layers attributed to plant water uptake and leakage, depends to a large degree on the accuracy and representativeness of the distributed sensor network measurements and how their horizontal and vertical scales overlap with the CRNS measurement footprint. We expect these limitations to be especially critical in semiarid ecosystems with high spatial heterogeneity induced by vegetation and bare soil patches.

The collocation of a distributed sensor network within the CRNS measurement footprint also allowed us to examine important process-based relations that are often incorporated into hydrologic models or remote sensing analyses (e.g., Famiglietti and Wood, 1994; Famiglietti et al., 2008). The spatial variability of soil moisture is linked to the spatially averaged conditions through predictable relations that do not vary significantly across the study sites. For higher mean soil moisture, we observed a nearly linear increase in spatial variability followed by an asymptotic behavior attributed to the seasonally wet conditions during the North American monsoon. Based on these relations ($k_1$and k_2), the spatial variability within a CRNS measurement footprint can be approximated for other semiarid ecosystems in the region. In addition, combining fixed and mobile CRNS methods can establish landscape-scale (10^2 to 10^3 km^2) soil moisture monitoring networks at grid sizes (~ 1 km^2) comparable to land surface modeling (Franz et al., 2015). Similarly, intermediate-scale soil moisture sensing can be linked effectively to daily evapotranspiration and used to obtain soil and vegetation parameters (E_w, ET_{max}, θ_h, θ_w, and θ^*) tailored to each ecosystem. In terms of the ET–θ relation, the CRNS method has the potential to significantly improve land–atmosphere interaction studies since it possesses a measurement scale that is commensurate to the sampling area of the EC technique.

Acknowledgements. We thank Heye Bogena and three anonymous reviewers for their useful comments that helped to improve the manuscript. We also thank Mitch P. McClaran and Mark Heitlinger from the University of Arizona for help at the Santa Rita Experimental Range and John Anderson, Al Rango and other staff members at the USDA-ARS Jornada Experimental Range for their assistance. We thank funding from the US Army Research Office (grant 56059-EV-PCS) and the Jornada Long-Term Ecological Research project (National Science Foundation grant DEB-1235828). We are also grateful to Nicole A. Pierini and Cody A. Anderson for help with field activities.

Edited by: N. Romano

References

Alfieri, J. G. and Blanken, P. D.: How representative is a point? The spatial variability of surface energy fluxes across short distances in a sand-sagebrush ecosystem, J. Arid Environ., 87, 42–49, 2012.

Anderson, C. A.: Assessing land-atmosphere interactions through distributed footprint sampling at two eddy covariance towers in semiarid ecosystems of the southwestern U.S. Masters of Science in Civil, Environmental and Sustainable Engineering, Arizona State University, 243 pp., 2013.

Bartalis, Z., Wagner, W., Naeimi, V., Hasenauer, S., Scipal, K., Bonekamp, H., Figa, J., and Anderson, C.: Initial soil moisture retrievals from the METOP-A Advanced Scatterometer (ASCAT), Geophys. Res. Lett., 34, L20401, doi:10.1029/2007GL031088, 2007.

Bogena, H. R., Huisman, J. A., Baatz, R., Franssen, H. J. H., and Vereecken, H.: Accuracy of the cosmic-ray soil water content probe in humid forest ecosystems: The worst case scenario, Water Resour. Res., 49, 5778–5791, 2013.

Browning, D. M., Franklin, J., Archer, S. R., Gillan, J. K., and Guertin, D. P.: Spatial patterns of grassland-shrubland state transitions: a 74-year record on grazed and protected areas, Ecol. Appl., 24, 1421–1433, 2014.

Campbell, J. E.: Dielectric properties and influence of conductivity in soils at one to fifty Megahertz, Soil Sci. Soc. Am. J., 54, 332–341, 1990.

Chen, F., Mitchell, K., Schaake, J., Xue, Y., Pan, H.-L., Koren, V., Duan, Q. Y., Ek, M., and Betts, A.: Modeling of land surface evaporation by four schemes and comparisons with FIFE observations, J. Geophys. Res., 101, 7251–7268, 1996.

Coopersmith, E. J., Cosh, M. H., and Daughtry, C. S. T.: Field-scale moisture estimates using COSMOS sensors: A validation study with temporary networks and Leaf-Area-Indices, J. Hydrol., 519, 637–643, 2014.

Dane, J. H. and Topp, C. G.: Methods of soil analysis. Part 4. Physical methods, SSSA Book Ser. 5, SSSA, Madison, WI, 2002.

Desilets, D. and Zreda, M.: Spatial and temporal distribution of secondary cosmic-ray nucleon intensities and applications to in-situ cosmogenic dating, Earth Planet. Sci. Lett., 206, 21–42, 2003.

Desilets, D. and Zreda, M.: Footprint diameter for a cosmic-ray soil moisture probe: Theory and Monte Carlo simulations, Water Resour. Res., 49, 3566–3575, 2013.

Desilets, D., Zreda, M., and Ferré, T. P. A.: Nature's neutron probe: Land surface hydrology at an elusive scale with cosmic rays, Water Resour. Res., 46, W11505, doi:10.1029/2009WR008726, 2010.

Detto, M., Montaldo, N., Albertson, J. D., Mancini, M., and Katul, G.: Soil moisture and vegetation controls on evapotranspiration in a heterogeneous Mediterranean ecosystem on Sardinia, Italy, Water Resour. Res., 42, W08419, doi:10.1029/2005WR004693, 2006.

Dugas, W. A., Hicks, R. A., and Gibbens, R. P.: Structure and function of C3 and C4 Chihuahuan Desert plant communities: Energy balance components, J. Arid Environ., 34, 63–79, 1996.

Eltahir, E. A. B.: A soil moisture rainfall feedback mechanism 1. Theory and observations, Water Resour. Res., 34, 765–776, 1998.

Entekhabi, D.: Recent advances in land-atmosphere interaction research, Rev. Geophys., 33, 995–1004, 1995.

Entekhabi, D., Njoku, E. G., O'Neill, P. E., Kellogg, K. H., Crow, W. T., Edelstein, W. N., Entin, J. K., Goodman, S. D., Jackson, T. J., Johnson, J., Kimball, J., Piepmeier, J. R., Koster, R. D., Martin, N., McDonald, K. C., Moghaddam, M., Moran, S., Reichle, R., Shi, J. C., Spencer, M. W., Thurman, S. W., Tsang, L., and Van Zyl, J.: The soil moisture active passive (SMAP) mission, Proc. IEEE, 98, 704–716, 2010,

Falge, E., Baldocchi, D., Tenhunen, J., Aubinet, M., Bakwin, P., Berbigier, P., Bernhofer, C., Burba, G., Clement, R., Davis, K. J., Elbers, J. A., Goldstein, A. H., Grelle, A., Granier, A., Gudmundsson, J., Hollinger, D., Kowalski, A. S., Katul, G., Law, B. E., Malhi, Y., Meyers, T., Monson, R. K., Munger, J. W., Oechel, W., Paw, K. T., Pilegaard, K., Rannik, U., Rebmann, C., Suyker, A., Valentini, R., Wilson, K., and Wofsy, S.: Seasonality of ecosystem respiration and gross primary production as derived from FLUXNET measurements, Agr. Forest Meteorol., 113, 53–74, 2002.

Fernández, J. M. and Ceballos, A.: Temporal stability of soil moisture in a large-field experiment in Spain, Soil Sci. Soc. Am. J., 67, 1647–1656, 2003.

Famiglietti, J. S. and Wood, E. F.: Multiscale modeling of spatially variable water and energy balance processes, Water Resour. Res., 30, 3061–3078, 1994.

Famiglietti, J. S., Devereaux, J. A., Laymon, C. A., Tsegaye, T., Houser, P. R., Jackson, T. J., Graham, S. T., Rodell, M., and van Oevelen, P. J.: Ground-based investigation of soil moisture variability within remote sensing footprints during the Southern Great Plains 1997(SGP97) Hydrology Experiment, Water Resour. Res., 35, 1839–1851, 1999.

Famiglietti, J. S., Ryu, D., Berg, A. A., Rodell, M., and Jackson, T. J.: Field observations of soil moisture variability across scales, Water Resour. Res., 44, W01423, doi:10.1029/2006WR005804, 2008.

Franz, T. E., Zreda, M., Ferré, T. P. A., Rosolem, R., Zweck, C., Stillman, S., Zeng, X., and Shuttleworth, W. J.: Measurement depth of the cosmic-ray soil moisture probe affected by hydrogen from various sources, Water Resour. Res., 48, W08515, doi:10.1029/2012WR011871, 2012a.

Franz, T. E., Zreda, M., Rosolem, R., and Ferré, T. P. A.: Field validation of a cosmic-ray neutron sensor using a distributed sensor network, Vadose Zone J., 11, doi:10.2136/vzj2012.0046, 2012b.

Franz, T. E., Zreda, M., Rosolem, R., Hornbuckle, B. K., Irvin, S. L., Adams, H., Kolb, T. E., Zweck, C., and Shuttleworth, W. J.: Ecosystem-scale measurements of biomass water using cosmic ray neutrons, Geophys. Res. Lett., 40, 3929–3933, 2013a.

Franz, T. E., Zreda, M., Ferré, T. P. A., and Rosolem, R.: An assessment of the effect of horizontal soil moisture heterogeneity on the area-average measurement of cosmic-ray neutrons, Water Resour. Res., 49, 6450–6458, 2013b.

Franz, T. E., Wang, T., Avery, W., Finkenbiner, C., and Brocca, L.: Combined analysis of soil moisture measurements from roving and fixed cosmic ray neutron probes for multiscale real-time monitoring, Geophys. Res. Lett., 42, 3389–3396, doi:10.1002/2015GL063963, 2015.

Gardner, W. H. and Kirkham, D.: Determination of soil moisture by neutron scattering, Soil Sci., 73, 391–401, 1952.

Gibbens, R. P. and Beck, R. F.: Increase in number of dominant plants and dominance-classes on a grassland in the northern Chihuahuan Desert, J. Range Manage., 40, 136–139, 1987.

Gill, P. E., Murray, W., and Wright, M. H.: Practical Optimization. Academic Press, London, UK, 402 pp., 1981.

Glasstone, S. and Edlund, M. C.: Elements of Nuclear Reactor Theory, Van Nostrand, New York, 416 pp., 1952.

Greacen, E. L.: Soil Water Assessment by the Neutron Method, CSIRO, Melbourne, Australia, 148 pp., 1981.

Gutiérrez-Jurado, H. A., Vivoni, E. R., Cikoski, C., Harrison, J. B. J., Bras, R. L., and Istanbulluoglu, E. I.: On the observed ecohydrologic dynamics of a semiarid basin with aspect-delimited ecosystems, Water Resour. Res., 49, 8263–8284, 2013.

Heitschmidt, R. K., Ansley, R. J., Dowhower, S. L., Jacoby, P. W., and Price, D. L.: Some observations from the excavation of honey mesquite root systems, J. Range Manage., 41, 227-231, 1988.

Hsieh C.-I., Katul, G., and Chi, T.: An approximate analytical model for footprint estimation of scalar fluxes in thermally stratified atmospheric flows, Adv. Water Resour., 23, 765–772, 2000.

Huang, C., March, S. E., McClaran, M. P., and Archer, S. R.: Post-fire stand structure in a semiarid savanna: cross-scale challenges estimating biomass, Ecol. Appl., 17, 1899–1910, 2007.

Huenneke, L. F., Clason, D., and Muldavin, E.: Spatial heterogeneity in Chihuahuan Desert vegetation: implications for sampling methods in semi-arid ecosystems, J. Arid Environ., 47, 257–270, 2001.

Ivanov, V. Y., Vivoni, E. R., Bras, R. L., and Entekhabi, D.: Catchment hydrologic response with a fully-distributed triangulated irregular network model, Water Resour. Res., 40, W11102, doi:10.1029/2004WR003218, 2004.

Kerr, Y. H., Waldteufel, P., Wigneron, J. P., Martinuzzi, J. M., Font, J., and Berger, M.: Soil moisture retrieval from space: The Soil Moisture and Ocean Salinity (SMOS) mission, IEEE T. Geosci. Remote Sens., 39, 1729–1735, 2001.

Köhli, M., Schrön, M., Zreda, M., Schmidt, U., Dietrich, P., and Zacharias, S.: Footprint characteristics revised for field-scale soil moisture monitoring with cosmic-ray neutrons, Water Resour. Res., 51, 5772–5790, 2015.

Kormann, R. and Meixner, F. X.: An analytical footprint model for non-neutral stratification, Bound. Layer Meteorol., 99, 207-224, 2001.

Kustas, W. P., Zhan, X., and Schmugge, T. J.: Combining optical and microwave remote sensing for mapping energy fluxes in a semiarid watershed, Remote Sens. Environ., 64, 116–131, 1998.

Laio, F., Porporato, A., Ridolfi, L., and Rodríguez-Iturbe, I.: Plants in water-controlled ecosystems: active role in hydrologic processes and response to water stress II. Probabilistic soil moisture dynamics, Adv. Water Resour., 24, 707-723, 2001.

Lawrence, J. E. and Hornberger, G. M.: Soil moisture variability across climate zones, Geophys. Res. Lett., 34, L20402, doi:10.1029/2007GL031382, 2007.

Mascaro, G. and Vivoni, E. R.: Utility of coarse and downscaled soil moisture products at L-band for hydrologic modeling at the catchment scale, Geophys. Res. Lett., 39, L10403, doi:10.1029/2012GL051809, 2012.

Mascaro, G., Vivoni, E. R., and Deidda, R.: Soil moisture downscaling across climate regions and its emergent properties, J. Geophys. Res., 116, D22114, doi:10.1029/2011JD016231, 2011.

McJannet, D., Franz, T. E., Hawdon, A., Boadle, D., Baker, B., Almeida, A., Silberstein, R., Lambert, T., and Desilets, D.: Field testing of the universal calibration function for determination of soil moisture with cosmic-ray neutrons, Water Resour. Res., 50, 5235–5248, 2014.

Moran, M. S., Hymer, D. C., Qi, J. G., and Sano, E. E.: Soil moisture evaluation using multi-temporal synthetic aperture radar (SAR) in semiarid rangeland, Agr. Forest Meteorol., 105, 69–80, 2000.

Narayan, U. and Lakshmi, V.: Characterizing subpixel variability of low resolution radiometer derived soil moisture using high resolution radar data, Water Resour. Res., 44, W06425, doi:10.1029/2006WR005817, 2008.

Pierini, N. A.: Exploring the ecohydrological impacts of woody plant encroachment in paired watersheds of the Sonoran Desert, Arizona. Master of Science Thesis in Civil, Environmental and Sustainable Engineering, Arizona State University, Tempe, AZ, 160 pp., 2013.

Pierini, N. P., Vivoni, E. R., Robles-Morua, A., Scott, R. L., and Nearing, M. A.: Using observations and a distributed hydrologic model to explore runoff thresholds linked with mesquite encroachment in the Sonoran Desert, Water Resour. Res., 50, 8191–8215, doi:10.1002/2014WR015781, 2014.

Polyakov, V. O., Nearing, M. A., Nichols, M. H., Scott, R. L., Stone, J. J., and McClaran, M. P.: Long-term runoff and sediment yields from small semiarid watersheds in southern Arizona, Water Resour. Res., 46, W09512, doi:10.1029/2009WR009001, 2010.

Potts, D. L., Scott, R. S., Bayram, S., and Carbonara, J.: Woody plants modulate the temporal dynamics of soil moisture in a semi-arid mesquite savanna, Ecohydrology, 3, 20–27, 2010.

Qu, W., Bogena, H. R., Huisman, J. A., Vanderborght, J., Schuh, M., Priesack, E., and Vereecken, H.: Predicting sub-grid variability of soil water content from basic soil information, Geophys. Res. Lett., 42, 789–796, 2015.

Rodríguez-Iturbe, I., and Porporato, A.: Ecohydrology of Water-Controlled Ecosystems, 442 pp., Cambridge Univ. Press, Cambridge, UK, 2004.

Rosolem, R., Shuttleworth, W. J., Zreda, M., Franz, T., Zeng, X., and Kurc, S. A.: The effect of atmospheric water vapor on neutron count in the cosmic-ray soil moisture observing system, J. Hydrometeorol., 14, 1659–1671, 2013.

Rosolem, R., Hoar, T., Arellano, A., Anderson, J. L., Shuttleworth, W. J., Zeng, X., and Franz, T. E.: Translating aboveground cosmic-ray neutron intensity to high-frequency soil moisture profiles at sub-kilometer scale, Hydrol. Earth Syst. Sci., 18, 4363–4379, doi:10.5194/hess-18-4363-2014, 2014.

Scott, R. L., Shuttleworth, W. J., Keefer, T. O., and Warrick, A. W.: Modeling multi-year observations of soil moisture recharge in the semiarid American Southwest, Water Resour. Res., 36, 2233–2247, 2000.

Scott, R. L.: Using watershed water balance to evaluate the accuracy of eddy covariance evaporation measurements for three semiarid ecosystems, Agr. Forest Meteorol., 150, 219–225, 2010.

Scott, R. L., Edwards, E. A., Shuttleworth, W. J., Huxman, T. E., Watts, C., and Goodrich, D. C.: Interannual and seasonal variation in fluxes of water and carbon dioxide from a riparian woodland ecosystem, Agr. Forest Meteorol., 122, 65–84, 2004.

Scott, R. L., Huxman, T. E., Williams, D. G., and Goodrich, D. C.: Ecohydrological impacts of woody-plant encroachment: seasonal patterns of water and carbon dioxide exchange within a semiarid riparian environment, Global Change Biol., 12, 311–324, 2006.

Scott, R. L., Cable, W. L., and Hultine, K. R.: The ecohydrologic significance of hydraulic redistribution in a semiarid savanna, Water Resour. Res., 44, W02440, doi:10.1029/2007WR006149, 2008.

Seyfried, M. S., Grant, L. E., Du, E., and Humes, K.: Dielectric loss and calibration of the Hydra probe soil water sensor, Vadose Zone J., 4, 1070–1079, 2005.

Shuttleworth, J., Rosolem, R., Zreda, M., and Franz, T.: The COsmic-ray Soil Moisture Interaction Code (COSMIC) for use in data assimilation, Hydrol. Earth Syst. Sci., 17, 3205–3217, doi:10.5194/hess-17-3205-2013, 2013.

Small, E. E. and Kurc, S. A.: Tight coupling between soil moisture and the surface radiation budget in semiarid environments: Implications for land-atmosphere interactions, Water Resour. Res., 39, 1278, doi:10.1029/2002WR00129, 2003.

Smith, R. E., Chery, D. L., Renard, K. G., and Gwinn, W. R.: Supercritical flow flumes for measuring sediment-laden flow, Tech. Bull. 1655, 70 pp., US Gov. Print. Off., Washington, D. C., 1981.

Snyder, K. A. and Williams, D. G.: Defoliation alters water uptake by deep and shallow roots of *Prosopis velutina* (velvet mesquite), Funct. Ecology, 17, 363–374, 2003.

Stevens Water Monitoring System: Comprehensive Stevens Hydra Probe User Manual, 62 pp., 1998.

Stillman, S., Ninneman, J., Zeng, X., Franz, T., Scott, R. L., Shuttleworth, W. J., and Cummins, K.: Summer soil moisture spatiotemporal variability in southeastern Arizona, J. Hydrometeorol., 15, 1473–1485, 2014.

Templeton, R. C., Vivoni, E. R., Méndez-Barroso, L. A., Pierini, N. A., Anderson, C. A., Rango, A., Laliberte, A. S., and Scott, R. L.: High-resolution characterization of a semiarid watershed: Implications on evapotranspiration estimates, J. Hydrol., 509, 306–319, 2014.

Throop, H. L., Archer, S. R., Monger, H. C., and Waltman, S.: When bulk density methods matter: Implications for estimating soil organic carbon pools in rocky soils, J. Arid Environ., 77, 66–71, 2011.

Topp, G. C., Davis, J. L., and Annan, A. P.: Electromagnetic determination of soil water content: Measurements in coaxial transmission lines, Water Resour. Res., 16, 574–582, 1980.

Turnbull, L., Parsons, A. J., and Wainwright, J.: Runoff responses to long-term rainfall variability in creosotebush-dominated shrubland, J. Arid Environ., 91, 88–94, 2013.

Vivoni, E. R., Moreno, H. A., Mascaro, G., Rodríguez, J. C., Watts, C. J., Garatuza-Payán, J., and Scott, R. L.: Observed relation between evapotranspiration and soil moisture in the North American monsoon region, Geophys. Res. Lett., 35, L22403, doi:10.1029/2008GL036001, 2008a.

Vivoni, E. R., Gebremichael, M., Watts, C. J., Bindlish, R., and Jackson, T. J.: Comparison of ground-based and remotely-sensed surface soil moisture estimates over complex terrain during SMEX04, Remote Sens. Environ., 112, 314–325, 2008b.

Vivoni, E. R.: Spatial patterns, processes and predictions in ecohydrology: Integrating technologies to meet the challenge, Ecohydrology, 5, 235–241, 2012.

Vivoni, E. R., Watts, C. J., Rodriguez, J. C., Garatuza-Payan, J., Mendez-Barroso, L. A., and Saiz-Hernandez, J. A.: Improved land-atmosphere relations through distributed footprint sampling in a subtropical scrubland during the North American monsoon, J. Arid Environ., 74, 579–584, 2010.

Vivoni, E. R., Rango, A., Anderson, C. A., Pierini, N. A., Schreiner-McGraw, A. P., Saripalli, S., and Laliberte, A. S.: Ecohydrology with unmanned aerial vehicles, Ecosphere 5, 130, doi:10.1890/ES14-00217.1, 2014.

Western, A. W., Grayson, R. B., and Blöshl, G.: Scaling of soil moisture: A hydrologic perspective, Ann. Rev. Earth Planet. Sci., 30, 149–180, 2002.

Wilson, K., Goldstein, A., Falge, E., Aubinet, M., Baldocchi, D., Berbigier, P., Bernhofer, C., Ceulemans, R., Dolman, H., Field, C., Grelle, A., Ibrom, A., Law, B. E., Kowalski, A., Meyers, T., Moncrieff, J., Monson, R., Oechel, W., Tenhunen, J., Valentini, R., and Verma, S.: Energy balance closure at FLUXNET sites, Agr. Forest Meteorol., 113, 223–243, 2002.

Zhu, Z., Tan, L., Gao, S., and Jiao, Q.: Observation on soil moisture of irrigation cropland by cosmic-ray probe. IEEE Geosci. Remote Sens. Lett., 12, 472–476, doi:10.1109/LGRS.2014.2346784, 2015.

Zreda, M., Desilets, D., Ferre, T. P. A., and Scott, R. L.: Measuring soil moisture content non-invasively at intermediate spatial scale using cosmic-ray neutrons, Geophys. Res. Lett., 35, L21402, doi:10.1029/2008GL035655, 2008.

Zreda, M., Shuttleworth, W. J., Zeng, X., Zweck, C., Desilets, D., Franz, T., and Rosolem, R.: COSMOS: the COsmic-ray Soil Moisture Observing System, Hydrol. Earth Syst. Sci., 16, 4079–4099, doi:10.5194/hess-16-4079-2012, 2012.

Permissions

All chapters in this book were first published in HESS, by Copernicus Publications; hereby published with permission under the Creative Commons Attribution License or equivalent. Every chapter published in this book has been scrutinized by our experts. Their significance has been extensively debated. The topics covered herein carry significant findings which will fuel the growth of the discipline. They may even be implemented as practical applications or may be referred to as a beginning point for another development.

The contributors of this book come from diverse backgrounds, making this book a truly international effort. This book will bring forth new frontiers with its revolutionizing research information and detailed analysis of the nascent developments around the world.

We would like to thank all the contributing authors for lending their expertise to make the book truly unique. They have played a crucial role in the development of this book. Without their invaluable contributions this book wouldn't have been possible. They have made vital efforts to compile up to date information on the varied aspects of this subject to make this book a valuable addition to the collection of many professionals and students.

This book was conceptualized with the vision of imparting up-to-date information and advanced data in this field. To ensure the same, a matchless editorial board was set up. Every individual on the board went through rigorous rounds of assessment to prove their worth. After which they invested a large part of their time researching and compiling the most relevant data for our readers.

The editorial board has been involved in producing this book since its inception. They have spent rigorous hours researching and exploring the diverse topics which have resulted in the successful publishing of this book. They have passed on their knowledge of decades through this book. To expedite this challenging task, the publisher supported the team at every step. A small team of assistant editors was also appointed to further simplify the editing procedure and attain best results for the readers.

Apart from the editorial board, the designing team has also invested a significant amount of their time in understanding the subject and creating the most relevant covers. They scrutinized every image to scout for the most suitable representation of the subject and create an appropriate cover for the book.

The publishing team has been an ardent support to the editorial, designing and production team. Their endless efforts to recruit the best for this project, has resulted in the accomplishment of this book. They are a veteran in the field of academics and their pool of knowledge is as vast as their experience in printing. Their expertise and guidance has proved useful at every step. Their uncompromising quality standards have made this book an exceptional effort. Their encouragement from time to time has been an inspiration for everyone.

The publisher and the editorial board hope that this book will prove to be a valuable piece of knowledge for researchers, students, practitioners and scholars across the globe.

List of Contributors

Z. Zhang, H. Hu and P. Yang
State Key Laboratory of Hydroscience and Engineering, Department of Hydraulic Engineering, Tsinghua University, Beijing, 100084, China

F. Tian
State Key Laboratory of Hydroscience and Engineering, Department of Hydraulic Engineering, Tsinghua University, Beijing, 100084, China
State Key Laboratory of Hydroscience and Engineering, Department of Hydraulic Engineering, Tsinghua University, Beijing, China

Stanislaus J. Schymanski, Daniel Breitenstein and Dani Or
Department of Environmental Systems Science, ETH Zurich, 8092 Zurich, Switzerland

P. B. Kirchner
Sierra Nevada Research Institute, UC Merced, Merced, CA, USA
Joint Institute for Regional Earth System Science and Engineering, UCLA, Los Angeles, CA, USA

R. C. Bales, J. Flanagan and Q. Guo
Sierra Nevada Research Institute, UC Merced, Merced, CA, USA

N. P. Molotch
Department of Geography and the Institute of Arctic and Alpine Research, University of Colorado at Boulder, Boulder, CO, USA
Jet Propulsion Laboratory, California Institute of Technology, Pasadena, CA, USA

Erik Gregow, Antti Mäkelä and Elena Saltikoff
Meteorological Research, Finnish Meteorological Institute, 00101 Helsinki, Finland

Antti Pessi
Applied Meteorology, Vaisala, 3 Lan Dr., Westford, MA 01886, USA

Marie-Claire ten Veldhuis
Delft University of Technology, Water Management Department, Delft, the Netherlands
Princeton University, Hydrometeorology Group, Princeton, USA

Marc Schleiss
Delft University of Technology, Geosciences and Remote Sensing Department, Delft, the Netherlands
Princeton University, Hydrometeorology Group, Princeton, USA

V. Couvreur
Earth and Life Institute, Université catholique de Louvain, Croix du Sud, 2, bte L7.05.02, 1348 Louvain-la-Neuve, Belgium
Department of Land, Air and Water Resources, University of California, 1 Shields Ave., Davis, CA 95616, USA

J. Vanderborght
Institute of Bio- und Geosciences, IBG-3: Agrosphere, Forschungszentrum Juelich GmbH, 52425 Juelich, Germany

L. Beff
Earth and Life Institute, Université catholique de Louvain, Croix du Sud, 2, bte L7.05.02, 1348 Louvain-la-Neuve, Belgium

M. Javaux
Earth and Life Institute, Université catholique de Louvain, Croix du Sud, 2, bte L7.05.02, 1348 Louvain-la-Neuve, Belgium
Institute of Bio- und Geosciences, IBG-3: Agrosphere, Forschungszentrum Juelich GmbH, 52425 Juelich, Germany

Qian Zhang
University of Maryland Center for Environmental Science, US Environmental Protection Agency Chesapeake Bay Program Office, 410 Severn Avenue, Suite 112, Annapolis, Maryland 21403, USA

Ciaran J. Harman
Department of Environmental Health and Engineering, Johns Hopkins University, 3400 North Charles Street, Baltimore, Maryland 21218, USA

James W. Kirchner
Department of Environmental System Sciences, ETH Zurich, Universitätstrasse 16, 8092 Zurich, Switzerland
Swiss Federal Research Institute WSL, Zürcherstrasse 111, 8903 Birmensdorf, Switzerland
Department of Earth and Planetary Science, University of California, Berkeley, Berkeley, California 94720, USA

Tingting Gong, Huimin Lei, Dawen Yang, Yang Jiao and Hanbo Yang
State Key Laboratory of Hydroscience and Engineering, Department of Hydraulic Engineering, Tsinghua University, Beijing, 100084, China

Md Abul Ehsan Bhuiyan and Emmanouil N. Anagnostou
Department of Civil and Environmental Engineering, University of Connecticut, Storrs, CT, USA

Efthymios I. Nikolopoulos
Department of Civil and Environmental Engineering, University of Connecticut, Storrs, CT, USA
Innovative Technologies Center S.A., Athens, Greece

Pere Quintana-Seguí
Ebro Observatory, Ramon Llull University - CSIC, Roquetes (Tarragona), Spain

Anaïs Barella-Ortiz
Ebro Observatory, Ramon Llull University - CSIC, Roquetes (Tarragona), Spain
Castilla-La Mancha University, Toledo, Spain

A. Hildebrandt
Friedrich Schiller University, Institute for Geosciences, Burgweg 11, 07749 Jena, Germany
Max Planck Institute for Biogeochemistry, Biogeochemical Processes, Hans-Knöll-Str. 10, 07745 Jena, Germany

M. Guderle
Friedrich Schiller University, Institute for Geosciences, Burgweg 11, 07749 Jena, Germany
Max Planck Institute for Biogeochemistry, Biogeochemical Processes, Hans-Knöll-Str. 10, 07745 Jena, Germany
International Max Planck Research School for Global Biogeochemical Cycles, Hans-Knöll-Str. 10, 07745 Jena, Germany

N. Ursino
Department ICEA, University of Padova, Padova, Italy

G. Cassiani and J. Boaga
Department of Geosciences, University of Padova, Padova, Italy

R. Deiana
Department dBC, University of Padova, Padova, Italy

G. Vignoli
Geological Survey of Denmark and Greenland, Groundwater and Quaternary Geology Mapping Department, Lyseng Alle 1, 8270 Hojbjerg, Denmark

Kay Helfricht and Lea Hartl
IGF - Institute for Interdisciplinary Mountain Research, Austrian Academy of Sciences, Innsbruck, 6020, Austria

Roland Koch and Marc Olefs
ZAMG - Zentralanstalt für Meteorologie und Geodynamik, Climate research department, 1190 Vienna, Austria

Christoph Marty
WSL Institute for Snow and Avalanche Research SLF, 7260 Davos, Switzerland

A. P. Schreiner-McGraw
School of Earth and Space Exploration, Arizona State University, Tempe, AZ 85287, USA

E. R. Vivoni
School of Earth and Space Exploration, Arizona State University, Tempe, AZ 85287, USA
School of Sustainable Engineering and the Built Environment, Arizona State University, Tempe, AZ 85287, USA

G. Mascaro
Julie Ann Wrigley Global Institute of Sustainability, Arizona State University, Tempe, AZ 85287, USA

T. E. Franz
School of Natural Resources, University of Nebraska-Lincoln, Lincoln, NE 68583, USA

Index

www.ingramcontent.com/pod-product-compliance
Lightning Source LLC
Chambersburg PA
CBHW080523200326
41458CB00012B/4315
* 9 7 8 1 6 4 1 1 6 2 9 9 9 *